T0265904

Life and Language Beyond Earth

Have you ever wondered whether we are alone in the universe, or if life forms on other planets might exist? If they do exist, how might their languages have evolved? Could we ever understand them, and indeed learn to communicate with them? This highly original, thought-provoking book takes us on a fascinating journey over billions of years, from the formation of galaxies and solar systems, to the appearance of planets in the habitable zones of their parent stars, and then to how biology and, ultimately, human life arose on our own planet. It delves into how our brains and our language developed, in order to explore the likelihood of communication beyond Earth and whether it would evolve along similar lines. In the process, fascinating insights from the fields of astronomy, evolutionary biology, palaeoanthropology, neuroscience and linguistics are uncovered, shedding new light on life as we know it on Earth, and beyond.

Raymond Hickey is Adjunct Professor at the University of Limerick, Ireland and former Professor at the University of Duisburg and Essen, Germany. His main research interests are varieties of English, language contact, variation and change and issues in phonology. Some of his recent publications include *Listening to the Past* (2017), *The Cambridge Handbook of Areal Linguistics* (2017), *English in Multilingual South Africa* (2020) and *The Handbook of Language Contact* (2020).

'This book is a great read that grips you from the start - it's an absolute Wunderkammer of enthralling facts and informed speculations that build the case around possible life and language on planets beyond our Solar System.'
Kate Burridge, Professor of Linguistics at Monash University, Australia

'Raymond Hickey offers fascinating surveys of two very different fields: astronomy and linguistics. But his book is specially valuable and farsighted because these two topics may some day be linked: rapid advances in exobiology lead optimists to conjecture that extraterrestrials could be discovered this century.'
Lord Martin Rees, Astronomer Royal

'This fascinating book discusses the question of the nature of the languages spoken by possible intelligent extraterrestrials in an engaging and interesting way. The book is impressively wide-ranging, covering cosmology, evolution, biology, linguistics and many other fields. This is the most thorough treatment of these issues I am aware of and it will certainly be of interest to anyone curious to find out more about these fascinating questions.'
Ian Roberts, Professor of Linguistics at the University of Cambridge and IUSS Pavia

'Ray Hickey's work has always surprised me in the ever-widening range of complex topics he's tackled, but this book is a quantum leap beyond his earlier work. The story he tells here is amazing and along the way offers a crisp introduction to how linguists understand human language and our cognitive capacity for it.'
Joe Salmons, Professor of Linguistics at the University of Wisconsin–Madison, USA

'This book takes us on a wondrous tour of the universe, as we explore exotic worlds and the alien civilizations that might flourish throughout our galaxy. By pondering the evolution of intelligence and language in a cosmic context, we are forced to rethink what it means to be human.'
Douglas Vakoch, President of METI International, editor of *Xenolinguistics: Towards a Science of Extraterrestrial Language*

Life and Language Beyond Earth

Raymond Hickey

Shaftesbury Road, Cambridge CB2 8EA, United Kingdom

One Liberty Plaza, 20th Floor, New York, NY 10006, USA

477 Williamstown Road, Port Melbourne, VIC 3207, Australia

314–321, 3rd Floor, Plot 3, Splendor Forum, Jasola District Centre, New Delhi – 110025, India

103 Penang Road, #05–06/07, Visioncrest Commercial, Singapore 238467

Cambridge University Press is part of Cambridge University Press & Assessment, a department of the University of Cambridge.

We share the University's mission to contribute to society through the pursuit of education, learning and research at the highest international levels of excellence.

www.cambridge.org
Information on this title: www.cambridge.org/9781009226417

DOI: 10.1017/9781009229272

First published 2023

Printed in the United Kingdom by TJ Books Limited, Padstow Cornwall

A catalogue record for this publication is available from the British Library.

A Cataloging-in-Publication data record for this book is available from the Library of Congress.

ISBN 978-1-009-22641-7 Hardback

Contents

Tables and Figures

Preface

The working title for this book, which as a linguist I had for a number of years, was 'Language beyond Earth'. But it became clear to me after a while that it would not be possible to discuss language without also considering life, the framework in which it is embedded. If one is looking beyond Earth then any questions about language can only be sensibly addressed after one has ascertained if life on planets elsewhere might be possible, what this life might be like and at what degree of development. Hence the present title, 'Life and Language Beyond Earth'.

This book contains information from five large branches of science: astronomy/astrophysics, evolutionary biology, palaeoanthropology, neuroscience and linguistics. My area of academic expertise, my professional comfort zone, if you like, is the last of these but the goal of the interdisciplinary approach I have adopted here has meant that these other areas have had to be afforded roughly equal weight. So, this book can be seen to be about life beyond Earth and language beyond Earth, with a clear link between the two. The rise of human language, and hence speculation about similar systems of communication on other planets, is closely linked to the story of evolution on our planet and the appearance of our species, *Homo sapiens*. Our knowledge of this story is constantly changing as new fossils are found and techniques for genetic analysis are developed. No doubt, this will render many things said here out of date before too long, meaning that what the book offers is a snapshot of what we know now (mid-2022) and how our knowledge might develop in future.

Not written primarily for scientists, this book avoids technical discussions of scientific matters as much as possible while providing enough information to understand the issues at hand. However, a certain amount of terminology is unavoidable. The scientific terms used in the book are explained throughout (with pointers to the relevant pages in the index) and many are also given short definitions in the glossary at the end. That way it is hoped that readers can understand all the issues and concepts discussed here even if they have had no contact with them before.

Some readers may wish to pursue matters dealt with in this book at a later stage. To this end, a list of general and of linguistic references is given at the back to help them on their way. The notes for each chapter contain additional information which might be of interest to some readers but which would interrupt the flow of the text if included in the latter.

A book like this, and the literature on exoplanets and possible life on them in general, is about scenarios which are at present completely hypothetical. There is a lot of speculation here, but it is worth asking questions, even though one does not yet have answers – the questions themselves can set us thinking in directions which could well be insightful, indeed beneficial, at a future date. Whether the hypothetical situations described here will change in the near to mid-term future it is not possible to say. Whether intelligent life elsewhere will be discovered is currently unknown and this book is not about making predictions concerning that issue. Rather it is a realistic consideration of how life and language arose on Earth and how this might happen or have happened on exoplanets. It is something which cannot be ruled out and hence it is, in my opinion, a rewarding enterprise to think about what life forms beyond our Earth might be like and whether we could in principle communicate with them. What I am submitting here is input from a linguist, providing a perspective on the issue of exo-life which to date has not been offered in this form and which colleagues in other sciences will hopefully find useful as a complement to their own research.

When writing this book I got much valuable feedback from various people, especially from my colleague of long-standing, Prof. Laura Wright, who was interested in the project and supportive from the beginning. The publishers, Cambridge University Press, also commissioned several reviews, given the broad nature of the themes discussed. The anonymous reviewers were very helpful, especially 'Reviewer D', who read and commented on

two versions and whose detailed feedback was invaluable while moving towards the final version of the book. During the production of the book, Isabel Collins, Stephanie Taylor, Ruth Boyes and Zoë Lewin provided much assistance and expertise. Last but not least, my thanks go to Rebecca Taylor, Commissioning Editor in Linguistics at Cambridge University Press, for her unflagging patience and continuous encouragement.

How to Use This Book

The current book has been written for anyone interested in the question of whether intelligent life forms beyond Earth exist and how such beings, if found on planets outside our Solar System, might engage in communication, namely what kind of language they might have.

If one were to ask what the narrative in this book is, what the common thread running through the various sections is, the answer would be assessing the likelihood that life forms beyond Earth exist and what they might be like in principle, going on what we know about how life and language evolved on our planet. To this end the book has been organised into six large sections, which strive to cover the main areas of overall relevance (see Table 0.1).

Table 0.1 Structure of the book

Part number	Part title	Intended readership
I	Introduction	All readers
II	The Universe We Live In	General readers, not necessarily for astronomers/astrophysicists
III	Our Story on Earth	General readers, not necessarily for evolutionary biologists and palaeoanthropologists
IV	The Runaway Brain	General readers, not necessarily for neuroscientists
V	Language, Our Greatest Gift	General readers, not necessarily for linguists
VI	Life and Language, Here and Beyond	All readers

Part I is an introduction in which I try to outline the basic questions about exolife, what we can justifiably assume about possible exobeings and what conclusions could be drawn from the detection of a signal from beyond our Solar System. The section closes with remarks on how science operates and how it comes to conclusions along with some thoughts on possible 'weird life'.

Part II is dedicated to recent astronomy and its technology, especially in relation to the discovery of exoplanets. This section is in a way the least speculative as the insights of astronomy are derived from concrete observations of extrasolar star systems and their planets. We know that there are thousands of exoplanets in our immediate galactic surroundings. The question is which of these, if any, could harbour life comparable to human life on Earth.

Part III is concerned with the evolution of life on Earth. This is important because obviously any life forms on other planets will have evolved from earlier, simpler forms, which will have evolved from simple, single-celled organisms. This holds no matter how far ahead of us exolife forms might be in any possible development away from their biological origins.

Part IV explores the assumption that, if there are beings on exoplanets, who are comparably intelligent to us and who could in theory contact us via some means of interstellar communication, the intelligence of such beings would have a physical substrate which is functionally and structurally comparable to our brains. But before we begin speculating about the brains of beings on exoplanets we should consider how ours developed, what consciousness is and how it may have arisen.

Part V is dedicated to the nature of human language. Here, general information is offered for readers who have not had any prior contact with linguistics. One of the essential issues addressed here is how the human language faculty – the ability to acquire and speak language – arose in the *Homo* species. How this compares with communication systems found in animals is also examined.

Part VI reviews all the major issues dealt with throughout the book. Just how similar might exobeings be to us – in their physique and physiology, in their sciences, in their societies, in their cultures – are questions which are examined critically. How likely is it that the pathways which we have taken in our evolution would be similar for beings on exoplanets, assuming that in some cases they will have developed languages, the precondition for any societies beyond Earth? And what are the chances that we might successfully communicate with such beings in the foreseeable future? This section is indeed quite speculative compared to the preceding ones. Many questions remain open, but it is important to ask them and start thinking about possible answers.

The discussions of language beyond Earth revolve around the issue of whether what we know about our world and the structure and origins of human language is likely to apply to beings on exoplanets as well. Depending on the answers presented to these questions readers can judge for themselves how likely beings with language are to exist beyond our Solar System. An essential distinction, which will appear repeatedly in this book, is that between the *language faculty*, the biological endowment all of us have which allows us to acquire language natively in early childhood, and *languages*, which are the outcomes of our using this language faculty.

For the discussions in this book I frequenly talk about an exolanguage on an exoplanet, using the singular, as I am referring to a typical instance. But it is fair to assume that if there is an exoplanet with beings who have a language faculty then there would be many exolanguages on such a planet, assuming a large enough population with an attendant geographical distribution. Why can one assume this? The exoplanets, where we are likely to find intelligent exolife, would be so-called 'rocky worlds' with a mixture of land and sea. This would imply a complex geography on a planet, with continents, mountain ranges and large islands. In turn, this would suggest the existence of several societies in different parts of such a planet. Furthermore, the analogous history of societies on Earth would imply a similar diversification of languages arising on an exoplanet over time, comparable to the way languages split up and went their separate ways on Earth.

Many comments on life and language beyond Earth are embedded in the treatment of other topics throughout this book. These comments are contained in text boxes like the following whereas longer discussions have dedicated text sections.

WOULD LINGUISTICS EXIST AS A DISCIPLINE ON AN EXOPLANET?

If exoplanets had highly structured societies in which branches of knowledge were dealt with in various disciplines, linguistics might well be one of these, though just what the discipline would be called and how it would be organised is completely unknown.

At the back of the book there is a glossary of the main terms used in discussions of life and language along with a comprehensive bibliography, which can be consulted when looking for further information about the issues discussed. The notes for each chapter offer more detailed information on individual points.

The topic of life beyond Earth has a long history to it. For anyone interested in this background I suggest consulting some literature on the matter; see the sections entitled Life beyond Earth and History of the Topic in the bibliography at the back of the book.

Part I

Introduction

Seeking life in the universe involves a realistic appraisal of what is possible, the distances involved, just across the Milky Way galaxy, and the time we are living in, our present, which may be in the distant past or the remote future of life forms on other planets. We also need to consider, in as much detail as possible, how similar to or different from us exobeings might be.

1

Approaching the Topic

How do you write a book about things, life and language on planets outside our own Solar System, when you do not even know if they exist or not? The only sensible answer is by looking at how life developed on Earth[1] and then considering how this could manifest itself on exoplanets, those beyond the planets which orbit our star, the Sun. This is because when considering possible Earth-like planets (see Section 8.8 for 10 criteria) we have, at our present state of knowledge, a set consisting of only one member, our Earth; this is the 'set of one' issue (Figure 1.1).

All you can do in this situation is assess likelihoods on the basis of what we know about how life evolved here on Earth. For that reason, this book is concerned with key aspects of the evolution of terrestrial life and how parallel aspects might manifest themselves on exoplanets (see Section 2.7 for a classification of different types). Equally, any consideration of language beyond Earth must start by considering how we came to have such a successful system for thought and communication – our language

1 In this book 'Earth' with a capital first letter is the planet on which we live. The Solar System is the system of planets with the Sun, our parent star, at its centre. The names of our planets and their moons are also written in capitals. By analogy, the word Moon, with a capital M, refers to the single moon orbiting our Earth. All lowercase spellings refer to objects and systems beyond our Solar System.

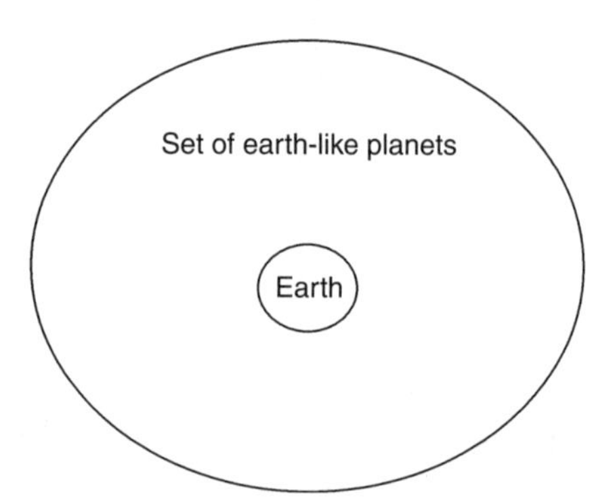

Figure 1.1 Our Earth as a set of one

faculty and its many realisations as languages in our world (see Chapter 32). It is this language faculty which enables us to acquire and speak language, so its evolution on Earth should be considered in detail when speculating about how a similar ability could arise with comparably intelligent beings on other planets.

This book is not about estimating the probability of life and language beyond Earth – we simply know too little at present to do that – but about the conditions under which life and language might arise, and the manner in which they might manifest themselves, based on careful consideration of all the factors which we can identify as relevant. It is up to readers themselves to reflect on how likely life and language may be beyond our Solar System. At the moment this question is only theoretical, and is likely to remain so for some time to come unless, and this is a big proviso, the efforts of scientists suddenly yield results, either by discovering an artificial signal from somewhere in our section of the galaxy (a technosignature) or by telescope observations showing beyond reasonable doubt that life forms exist on an exoplanet (a biosignature). Again, we do not know whether such a discovery is just around the corner or in the distant future or, indeed, will ever be made. But there is a huge body of investigative scientific work on the structure of our universe, the laws of physics, the nature of chemistry, possible biology beyond our Earth (the concern of astrobiology) and, above all, on the nature of planets outside our Solar

System. This body of research is growing daily. And it is within the framework of this research that the present book is to be understood.[2]

1.1 Four Basic Questions

Whether the present book is justified depends on how one views the consideration of four basic questions, each more specific in an ascending order.

Four basic questions about life and language beyond Earth

1. Is there any life beyond Earth?
2. Is there intelligent life beyond Earth?
3. Is this life technologically advanced enough to communicate with us?
4. Does such life have a communication system which we would recognise as language?

The first question is the most basic and may well be answered in the near future by scientists examining planets and moons within our Solar System, assuming that such life does not have the same source as that on Earth. For instance, studying Mars may provide evidence of previous life on Mars in its distant past when conditions were similar to those on Earth, with liquid water and a thicker oxygen-bearing atmosphere. The question of just what constitutes life is discussed in detail in Part III.

The second question above concerns intelligent life, which here refers to sentient beings on an exoplanet with cognitive abilities comparable to those

2 It should be said that this book is not connected in any way with science fiction or Hollywood space movies. These are genres of writing and film, valid in their respective realms, but which have precious little to do with the subject matter of this book. For that reason, the term 'alien' is not used anywhere in this book.

of humans on Earth. The physical substrate for their intelligence would have to be something functionally similar to our brains, similar in the amount of internal structure, and capable of complex computations. Whether the physical substrate of their intelligence would consist of tens of billions of neurons and trillions of connections linked to a nervous system controlling the body, as with us humans, is something we simply do not know.

The third question above has more facets to it than might appear at first sight. To be able to communicate across the vast distances of interstellar space, exobeings would have to have developed a technology which allowed them to manipulate radio or light waves for this purpose. Such technology would imply a whole raft of further facts about exobeings (see Section 1.2) and their lives on exoplanets.

The fourth question above can only be considered if at least the first two are answered positively. Allowing for extraterrestrial intelligence one can then ask whether exobeings would also have language like humans. For most of the discussion in this book I will refer to exobeings as if there were only one group of them from one exoplanet speaking one exolanguage. Reality may prove to be very different. If we continue to discover exoplanets at the current rate (over 5,000 confirmed, mid-2022), we may well find more than one planet which might harbour intelligent life (though this has not happened yet). And, indeed, there may be exoplanets with exolanguages spoken by beings with differing degrees and types of intelligence, which we on Earth are not aware of.

1.2 Working Backwards for a Moment

There are many ways by which we could learn of exobeings on an exoplanet (see the detailed discussion of possible scenarios in the final chapter of the book). Consider the following situation, which is probably the most likely of the diverse options. We discover an artificial signal, apparently emanating from a planet in another solar system, what conclusions would we be justified in making then?

WHAT ARE 'EXOBEINGS'?

A number of terms are essential regardless of how one reads this book. Each of these begins with the prefix *exo-*, which in the present context means 'beyond our Solar System'. Apart from *exoplanet*, a planet outside our Solar System, the most important is *exobeings*, used throughout this book with a very specific meaning. It is a shorthand for intelligent forms of life on an Earth-like planet beyond our Solar System, which are technologically advanced enough to engage, in principle, in interstellar communication. This implies that exobeings would use a system of communication in their social interactions, an *exolanguage*, which is functionally comparable to human language. This also means that they would be, again in principle, in a position to communicate with the inhabitants of different planets, either by sending and possibly receiving meaningful signals to and/or from others.

Possible conclusions to be drawn about exobeings on discovering an artificial signal from a different solar system

1. They would have to be able to construct objects and so have considerable cognitive abilities and dexterous limbs, functionally comparable to our hands.
2. They would have to produce metals and alloys out of which to make their artefacts.
3. They would have to know electricity and be able to generate it at will as a source of energy for their artefacts.
4. They would need to understand the science of wave transmission in order to undertake interstellar communication.
5. They would need language to interact with each other when constructing and operating their complex instruments; and, of course, in order to pass on accumulated knowledge down through the generations.

It should be remembered that there may well be exoplanets with societies which are culturally and artistically advanced but without science and technology to match. Here one can recognise that the term 'advanced' is open to a multitude of interpretations and its use has to do with the assumptions and expectations of those who employ the term. In this book, I am using 'advanced' specifically to refer to a society of exobeings who would possess technology allowing them to manipulate waves of the electromagnetic spectrum for the purposes of potential communication with beings on other planets. So, for this book, 'advanced' implies complex technology. Any exobeings on an exoplanet without such technology would remain hidden from us, at least given our current level of science and engineering. Needless to say, the term 'advanced' is not intended in any evaluative sense.

1.3 Questions, Questions, Questions

Would exobeings have a carbon-based biology like our own, would their sensory organs be similar to ours, would they have eyes, ears, noses, mouths and a sense of touch like ours? What would their physical form be like, what intake of energy would they have? Would they breathe and eat like we do? Would exobeings come in two flavours, comparable with our male and female, using sexual reproduction in which half of the genetic material in offspring comes from the mother and half from the father?

What about other characteristics of intelligent forms of life elsewhere. Would they be curious and continually seek answers to things they do not understand? Would they show great mobility, a high degree of self-awareness and reflection, the ability to think logically, solve problems and design artefacts? Would they have mastered digital[3] technology, indeed nanotechnology, in a manner similar to the way we have and are striving to do right now?

3 The term 'digital' is used here somewhat in the popular sense, that is to say, more as a shorthand for 'scientifically advanced and using computer technology'. This could also include sophisticated analogue equipment, but that is not the point in this general use of 'digital'.

Would they behave in ways which we could understand and feel an affinity with? Would their societies be comparable to ours? Would there be a similar stratification and division of labour on their planet? Would the transmission of accumulated knowledge across the generations be comparable to our own? Intelligent forms of life are likely to be organised into entities comparable to societies, and the complexity of the latter would be a consequence of the intricate communication system they use. Without a complex communication system, a complex society cannot exist. The latter presupposes the former.

When speculating about exobeings and their language, other questions arise. Would they have a language faculty similar to ours enabling them to acquire and process language? Would they use sounds within our hearing range for communication? Would they breathe and have similar vocal organs to us with vocal folds to generate voice when exhaling air from their lungs? Would their languages be similarly structured to ours, making them – in principle – comprehensible to us? Could we learn them, or could exobeings learn any of our languages?

Other broader issues would depend on the precise nature of an exoplanet and how exobeings evolved there. For instance, would exobeings have a day–night biological rhythm like humans do? Not if their planet were tidally locked to their star, like our Moon is to Earth, always showing the same side to us.[4] A further question concerns rest and sleep: humans use sleep for essential memory filtering and consolidation, a process which could be just as essential for exobeings as it is for us.

These questions are all worthy of consideration but there are no definitive answers, rather probabilities and likelihoods. So, this book is not to be interpreted as a claim for the existence of exobeings – there is no evidence for this as yet. The book is about considering the possibility that they

4 In this case, exobeings would probably not have a waking–sleeping cycle like we do. But for a planet to be in the habitable zone of its solar system it has to be some distance from its star (assuming that the star is fairly hot) and at this distance the star's gravity would probably not be strong enough to prevent the planet from rotating around its own axis, which means the star would not keep it tidally locked.

might exist and the possible nature of the system(s) of communication they would use amongst themselves. Of course, if there is no life beyond Earth then there are no languages beyond it either, and this book would be pointless. But we simply do not know if there are exobeings comparable to ourselves on other planets. Unless we have evidence to the contrary, it is legitimate to think about their possible existence and the kinds of language they might have.

From the above, it is immediately apparent that even the simplest question about life on exoplanets involves a whole series of further questions which need to be considered. These are addressed in more detail throughout this book as part of a realistic consideration of possible life on exoplanets. In addition, the conceivable evolutionary paths which might have been taken by exobeings are looked at in detail.

The Issue of Bias

For some readers, the approach I have adopted might appear too anthropocentric, showing a bias towards life forms similar to us Earthlings. But we need to bear in mind that life forms on an exoplanet will have arisen through Darwinian evolution over hundreds of millions of years and, in this long process, organisms will have developed which are functionally comparable to those on Earth. And if there are planets with exobeings then the latter will also have evolved structures which are functionally comparable to our brains. And for any planet with societies complex enough to develop advanced technology, language is an absolute precondition.

Just how life and language might be manifested on an exoplanet is unknown to us but, in terms of structural and functional organisation, exolife on exoplanets will most probably share basic similarities with life on Earth. To assume otherwise, to assume that exolife would be completely different from terrestrial life, both in principle and in realisation, would shift the burden of proof. Then it would be necessary to show what these differences might be like and how they would engender intelligent life capable of mastering complex technology.

1.4 An Unlikely Story

Imagine the following scenario. It is completely unrealistic but bear with me for a moment. A fleet of 10 interstellar spacecraft are sent out from Earth for one year to investigate various planets in star systems around our galaxy, the Milky Way. The teams have decided in advance which planets they are going to see: only those which have a mixture of land and sea, an atmosphere with about a fifth to a quarter of oxygen and surface temperatures comparable to those on Earth, say on average around 20 °C in the more temperate zones. Now each spacecraft can manage to visit two planets per week, gathering essential data, taking samples along with some videos into the bargain. This means that in their year zipping across the galaxy to their stellar destinations they collect data from 1,000 Earth-like exoplanets (those beyond our Solar System).

When the teams return home there is an international press conference where they present summaries of their findings. A select group of journalists are invited to the exclusive presentation and the astronauts reveal the results of their missions:

Finding 1: On all planets there were microorganisms in the sea and on the land there were tiny tough plants hanging out in crevices.

Finding 2: On most planets there were fish-like creatures in the sea, really strange shapes and sizes, but they were sea animals able to move around independently and react to stimuli from the environment.

Finding 3: On some planets there was complex vegetation with atmospheres of water clouds and rain. There were creatures moving around on land, some big, some small. Again, a great variety. Some of the animals were herbivores and others were carnivores.

Finding 4: On two planets there were creatures who seemed fairly intelligent and dexterous with hand-like limbs. They had built simple huts and used raft-like structures for moving on water.

During question time, one of the journalists asks the scientists, “So, you didn’t find any planet with life like on Earth, right?” The spokesperson concedes this, however adds: “But we only looked at 1,000 planets and there are at least 50 billion of the right type in our galaxy alone”. Then one of the younger astronauts asks for the microphone: “Sure, we still have to keep looking. However, when on our mission we did some time-travel experiments as well, and one of these yielded interesting results”.

Here’s what happened: when their spacecraft was passing one of the stellar systems, they saw it contained a planet about the size of the Earth. It was not on their list of planets to investigate, but they thought it was worth taking a look at anyway. However, it turned out to be a hot ball of molten rock, constantly bombarded by meteorites and asteroids. One of the seniors on board the spacecraft saw that a young astronaut wasn’t occupied at that moment and said that they should do a time-travel test on the planet. Sitting at a screen and scrolling on a dial, the scientists could cycle through millions, even hundreds of millions of years into the planet’s future, in a matter of seconds. They could see that the planet eventually cooled down and the bombardment stopped. The atmosphere contained water and, on the surface, there was also a lot of water, apparently delivered by comet strikes in the planet’s early years. Scrolling through nearly two billion years, the scientists just saw green slime all over the place, but nothing else. “That planet ain’t in a hurry to go anywhere”, the younger scientist said, “will we keeping scrolling on?”. “Okay, give it another few billion years, just to be sure”, the other replied. They saw that there was a change in atmosphere: oxygen turned up and, zooming in on the planet to microscopic levels, they saw that cells had moved from simple structures to more complex things. A bit later creatures appeared in the sea and, after about four billion years, life forms suddenly got more complex with predation becoming a means of gaining food. Later again, some creatures moved onto land and they began to proliferate. The younger scientist was getting impatient, “Still nothing to write home about on this planet, let’s go for a coffee break”. The senior scientist replied, “Just a few more minutes, to see if these animals are getting up to anything interesting”. Nothing turned up, but the moment before the two scientists were going to turn off the time-travel computer, the planet suddenly came alive with cracks and pops from all sorts of radio signals, and small objects could be seen orbiting the planet. "Wow, what's happening down there? Zoom in on that place again". Zooming down to the surface, a vista of large artefacts, apparently clustered at various points across the surface of the planet, could be seen. Objects were moving across the land and the sea along certain routes. In the air, craft could be seen going from one place in the planet to another. But the time-travel program was having difficulties moving further forward or displaying more detail. The computer was beginning to falter: it was already 4.5 billion years into the future of the nascent planet and that seemed to be the limit of what the time-travel software could manage.

1.5 Back to Reality

The above is pure fantasy. There is no such thing as time travel into the future and you cannot zip around the galaxy from one solar system to the next in a matter of days. To give you a flavour for distances in the Milky Way galaxy, consider this: the nearest star to us is Alpha Centauri (actually a three-star system) and, travelling at the rate of our present fastest spacecraft (*c.* 60,000 kilometres per hour), it would take about 75,000 years to get there. You cannot travel faster than light (according to Einstein's theory of relativity), not even at a considerable fraction of this. There are some theoretical options, such as using wormholes or Alcubierre drives, but they are just that: theoretical ideas far removed from present reality.

However, the above anecdote illustrates two important points: the first is that the aim of much exoplanet research – investigating planets beyond our Solar System – is to find planets similar to Earth. Following on that is the further aim of many scientists, which at present is beyond our investigative potential, namely, to find out whether they might harbour intelligent life forms. The second point, made by the fictitious time-travel experiment, is that finding life on other planets depends crucially on biological evolution there being roughly in sync with our own human evolution. The planet being described in the little story is, of course, our Earth: 4.55 billion years ago it was a ball of molten rock constantly hit by objects around it. From the perspective of human beings (and only from that vantage point) not much happened during most of Earth's history. The rise of humans occurred only at the very end, in the last few hundred thousand years. When searching for exoplanets, astronomers will find many barren, inhospitable worlds but, given how tenacious life forms are, some of these worlds may well come to harbour intelligent beings in their distant futures, times completely outside our reach. And, of course, there may have been similar situations in the opposite direction: maybe planets exist on which complex civilisations flourished but which are long since gone, either because of self-destruction or through some untoward external event such as an asteroid strike, a nearby supernova explosion or a direct hit by a gamma-ray burst.

2

Looking beyond Earth

2.1 Are We Alone in the Universe?

A key question that is often posed is, 'Are we alone in the universe?' If the answer to this question were simply 'Yes' or 'No', this book would probably not have been written. But what looks like a simple question is actually a complex and multifaceted set of issues which can hopefully be elucidated by discussing the many aspects involved.

First, by 'the universe' we can only mean the small corner of the galaxy which we inhabit, say within about a 100-light-year or, at maximum, 1,000-light-year radius. This is a tiny fraction of our Milky Way galaxy (at least 100,000 light years in diameter) and we cannot even see half of the galaxy which is beyond the central bulge of the disk on the other side. The Milky Way is an infinitesimally small part of the entire universe. There are hundreds of billions of galaxies in the observable universe, consider all the galaxies to be seen in the Hubble Deep Field and the similar fields pictured by the James Webb Space Telescope (see Chapter 4), each with hundreds of billions of stars, most of which have planets. If there was one Earth analogue with intelligent life forms per galaxy, there would be billions of exoplanets with exobeings, but they would all live alone in so-called 'island universes', galaxies out of reach for each other.

Second, what do we mean by 'alone'? This word would seem to be used, in general discussions, for example on the internet, as a shorthand for the question, 'Are there beings comparable to ourselves in their level of general intelligence on other planets?' This is a huge question and most of this book is concerned with deconstructing the range of issues and implications involved here and offering tentative answers or at least directions in which answers might be found.

Third, asking the question, 'Are we alone in the universe?', implies that we would know the answer if there were beings on other planets capable of communicating with us across the vast distances between solar systems in our galaxy. For this to ever happen, such exobeings would have had to have mastered digital technology and so be in a position – in theory – to communicate with other beings, such as us on planet Earth (who would be exobeings for them). So, there may be intelligent life forms on exoplanets but without the technology to communicate this to beings on other planets.

Fourth, the above question refers to time, to our 'now'. This is a critical consideration as shown by the following basic possibilities.

The time factor when searching for exoplanets with life

1. There may have been exocivilisations with advanced technology which are long since gone (for whatever reason, see Section 6.4 for further discussion).
2. There may be planets with forms of life which, in their future, may evolve into intelligent beings with science and technology capable of interstellar communication.
3. There may be planets with civilisations which are already well into their digital age, even by millions of years.

What a civilisation in scenario (3) might look like is often the subject of throw-away comments, even by scientists, such as 'they would be so different from us, we cannot even begin to imagine what they would be like'.

That may well be true (see Section 3.8 on 'weird life' below), but adopting that stance will stifle all attempts at rational speculation about exobeings who all will, no matter how advanced, have stemmed from simple-celled life forms through a process of Darwinian evolution. It is true that this is slow, but the last bit for us on Earth, our development from our common ancestor with the chimpanzees, only took about seven million years (estimates vary somewhat here). I say 'only' because that is just 0.15 per cent of the age of the Earth. This means that, on any planet with animals similar to primates on Earth, it would probably require less than 1 per cent of the age of that planet for intelligent life to arise, assuming that the conditions were right for such a development. That would mean that exobeings could in principle evolve many times on an exoplanet which has a stable environment, thus increasing the chances that we might contact them at a time when they were at approximately the same level of development as ourselves.

ARE WE ALONE IN THE UNIVERSE?

Imagine we find a planet on which there are beings which have a brain size approximately two-thirds of ours, say around 800–900 cubic centimetres. They live in caves and hunt for food using spears with flint heads and maybe can make fire to stay warm, keep predators at bay and cook meat. They might have a simple language consisting of, say, about a hundred words with literal meanings, which they string together to form simple sentences.

How would we then answer the question, 'Are we alone in the universe?' Now consider that this scenario is actually what obtained on Earth over a million years ago while *Homo erectus*, the main predecessor of *Homo sapiens*, was evolving (though positing language here is contestable). What this would mean is that if we visited our hypothetical planet in over a million years' time we might find a flourishing digital civilisation. It also means that if intelligent exobeings happened to have taken a look at Earth over a million years ago, they might not have been impressed and would have just moved on to consider other planets.

2.2 What We Know about Exoplanets

Let's turn to the question, 'Are there beings comparable to ourselves on other planets?' We can get a handle on this by considering what evidence of the various stages necessary for life on other planets we would have to find in order to answer the question. We have some information already, some we could find within our own Solar System, some has been found by astronomers researching exoplanets and more will doubtlessly become available during the present, twenty-first century. So what do we already know about the preconditions for life on other planets?[1]

What we know about exoplanets, general

1. Any elements found on exoplanets would be the same as those we know from Earth (about 90 occurring naturally). Their properties would lead to the same classification, such as gases, metals, etc.
2. The fundamental laws of physics and the constants of nature would apply on exoplanets exactly as on Earth.
3. The aggregation of atoms to molecules to macromolecules would apply just as on Earth.
4. The principles and organisation of chemistry, for example, the types of molecules, such as acids, bases, salts, sugars, fats as well as classes of processes, such as redox reactions (involving the gain or loss of electrons), would obtain on an exoplanet just as on Earth.
5. Intermolecular (electrostatic) forces, holding huge numbers of molecules together, such as objects we can observe on our human size scale, would apply just as on Earth.

1 There is an interactive website at NASA called *Eyes on Exoplanets*, which can be accessed at https://exoplanets.nasa.gov/eyes-on-exoplanets. It is an excellent example of scientific outreach, allowing the general public to participate in the insights gained from exoplanet research in the past decade or so.

6. Each exoplanet would likely avail of light and heat from its parent star[2] as sources of energy, though internal sources could also be exploited.
7. Life on an exoplanet would be subject to the gravity of the planet and would adapt accordingly.
8. Life on an exoplanet would require a continuous source of energy to fuel its metabolism. Oxygen is a prime candidate for such a source, though not the only option.
9. Life on an exoplanet would, in its advanced forms, most likely be heterotrophic. This means that they would gain energy from nutrients, which they would take in through a process analogous to eating as for animals on Earth.[3]

Exobeings would have developed means of dealing with situations like those listed in (6) and (7). They would be able to utilise heat from their star and will most likely have biologies which can cope with gravity, for instance by developing capillary action[4] by means of which water can move against gravity in very thin tubes, similar to the mechanism used by trees on Earth. On (9) one can note that organisms which are autotrophic, like trees, are generally rooted in the ground, others, like algae/bacteria, are in constant physical contact with their source of nutrients. Our level of mobility means that this is not the case, hence our heterotrophy.

The above are very general statements. There are also some more detailed ones we can make about chemistry on an exoplanet. For instance, given

2 Here, I am assuming that an exoplanet would be orbiting a single star, although binary star systems are very common in our galaxy.
3 It is true that terrestrial plants, along with some bacteria and algae, are autotrophs, making glucose with light as energy for this process. But plants are bound to one location and the other autotrophic organisms are simple forms of life. Exobeings, like all animals on Earth, would move around and probably show activity at times when there is little light, so it is most unlikely that they would be autotrophic like terrestrial plants.
4 There is a phenomenon called adhesion by which particles from dissimilar surfaces tend to cling together. This, along with surface tension in a liquid such as water, can lead to a movement in the opposition direction to the pull of gravity. Capillary action is essential for biological systems, moving liquids around organisms against gravity.

that atoms have the same internal structures across the universe, procedures for aggregation such as covalent and ionic bonding to form molecules and the 'octet rule', whereby atoms prefer sets of eight electrons in their outer shells and combine in order to achieve this (as with carbon dioxide (CO_2) for instance), would also apply beyond Earth. In addition, we can make general statements about the physical and chemical nature of exoplanets, such as those listed below.

What we know about exoplanets, details

1. Rocky exoplanets in the habitable zones (see Section 8.2) of their stars exist in abundance.
2. Such exoplanets can contain (i) molecular oxygen and (ii) liquid water.
3. Complex organic (carbon-based)[5] molecules have been found outside Earth and are known to exist in the great clouds of dust (proto-planetary disks) from which stars and their planets form.

Some phenomena known on Earth will also apply on exoplanets because the foundations for these are the same everywhere. Consider that virtually all biology is driven by energy gradients, realised by differences in charge between ions (atoms/molecules with either a positive or negative[6] charge), and the flow of electrons (negative) and protons (positive) across these gradients. This would apply everywhere because all atoms and molecules have a value for electric charge and this value can change by donating or receiving electrons (via redox reactions).

For exobeings to have a mature understanding of science they would need to have grasped these processes. Of course, the words and labels they

5 Not all scientists share the assumption that carbon-based biology would be the default across the universe and some criticise this attitude as 'carbon-chauvinism'.

6 These are traditional terms, first used by Benjamin Franklin (1706–1790). Like the labels 'masculine' and 'feminine' for grammatical gender, they are arbitrary, but fixed by convention.

would use in their science would not be those we use, such as positive for protons and negative for electrons, but their understanding would, in principle, be the same.

2.3 Exobeings and Their Planetary Environment

Recall that exobeings, if they exist, will have evolved from the simplest of organisms over hundreds of millions of years. Darwinian evolution is the path they would take, allowing us to make certain statements about the environment on their planet.

Evolution on an exoplanet

1. An exoplanet with exobeings will have a biosphere of which these beings will be an integral part. If such exobeings have transitioned to a post-biological stage then the relationship to the biosphere and its environmental conditions may well have been redefined, leading to greater independence from it. At this stage of their evolution, how exobeings would need to interact with the plant and animal life forms on their planet, not least as a source of food, would be key. If post-biological exobeings and their original natural environment were completely dissociated, they could in principle occupy planets whose environment would have been dangerous for them in their biological stage, for example, in having high levels of radiation, no source of water, etc.
2. Exobeings would go through stages of growth, from being born to their time of death. If there was no cycle of birth, life and death, there would be no evolution and exobeings could not arise. Assuming the latter would be necessary for a turnover of the generations, a precondition for evolution. Whether biotechnology and medication could overcome senescence is very much an open question. Death for exobeings could come from other sources: predation, accidents and maybe from disease, though how that would manifest itself is quite uncertain.

3. Exobeings would likely have childhoods comparable, in principle, to those of humans on Earth, given the longer gestation periods of animals with large brains, biological maturation and maternal investment in the rearing of young.[7] It is uncertain how childhood would manifest itself for exobeings. However, from Earth we know that more intelligent forms of life (dolphins, elephants, humans) generally bear fewer young and these then have longer childhoods. One reason for the extended childhoods of humans is to give us time to develop the cognitive skills, above all language, which we are capable of, given the size and organisation of our brains.

It would be interesting to consider the possible relationship of exobeings to the environment of their planet. Would exobeings have discovered fossil organic compounds, like petroleum (crude oil) on Earth, as a source of energy on their planet (assuming an oxygen-rich atmosphere in which to burn such fuel)? If they burned these compounds as fuels, they would be injecting greenhouse gases into their atmosphere, creating disequilibrium which could be observed by human-made space telescopes. An exoplanet could suffer from global warming and climate change just as Earth is doing today and might end up like Venus with lethal atmospheric warming. If they had advanced chemistry, they might have mastered the production of synthetic polymers, which are the basis for different kinds of durable and flexible plastic in today's world. Such materials would be advantageous in their technical artefacts but also be a source of pollution, just witness how our rivers, seas and oceans are choking in plastic made by humans. By and large, when left to their own devices, animals do not pollute their environment to any comparable extent (intensive livestock breeding, which is responsible for the release of large quantities of methane, is organised by humans).

7 There are basically two competing views on the life histories of primates: (i) the *cognitive buffer hypothesis* suggests that the long juvenile period is adaptive, in that larger brains enable greater flexibility in behaviour, and hence provides 'buffering' against challenges from the environment thereby improving chances of a longer lifespan; (ii) the *developmental costs hypothesis* suggests that longer maturational periods are necessary for the development of large brains. See Powell, Barton and Street (2019).

2.4 Exobeings and Humans on Earth

Because scientific speculation starts with what we know and proceeds in small steps to consider what might be the case, one should by default assume that exobeings would be comparable to ourselves and move on from there. Of course, one could say that exobeings might be completely different from us (see Section 3.8). True, it is possible, but that is a dead end: you cannot continue the discussion beyond that point. Hence the view taken in this book is that the exobeings we can sensibly speculate about would probably be functionally fairly similar to ourselves, in principle. Even so, there are sufficient respects in which they could be different and these are the subject of various discussions throughout this book.

For example, silicon-based biologies are conceivable though they would not have the flexibility of carbon-based ones because the bonds which silicon atoms can undergo are not as stable as those of carbon (with silicon, the first bond is much stronger than the other possible bonds and this creates imbalances in silicon compounds; silicon also reacts too readily with oxygen to form silicon dioxide (SiO_2). Carbon is much more common across the universe than silicon, as is it formed through stellar fusion in even quite small stars, whereas silicon is only created in the cores of large stars. But on Earth this is not true, as silicon is more abundant (it is the basic element in rock minerals such as quartz, which consists of silicon dioxide, a silicate). Nonetheless, life is opportunistic and so one can expect that, again in principle, life beyond Earth would not be radically different in its structural organisation, though its manifestation would be dissimilar to that on Earth.

The word 'manifestation' is important here. Exobeings would most likely be very different from us in their phenotype, a term from genetics to refer to the actual appearance of an organism. For example, cetaceans (dolphins, porpoises and whales) look very different from land-based mammals like bears, horses or dogs. But their internal anatomy and physiology is essentially similar: all these animals breathe air into lungs, they bear live young and suckle them. Furthermore, cetaceans have flippers which evolved from arms and have tiny internal limbs which are, in terms of evolution, the same as the hind legs of a cat, a cow and other quadrupeds. Conversely, the phenotypical similarity of, say, whales and sharks, belies the fact that the latter are fish, which extract oxygen from water via gills and whose young mature in eggs.[8]

Now to exobeings: they will have some organ which, in its ability to perform highly complex computations, will be functionally similar to our brains. Assuming that they have mastered science and technology and have built artefacts, they would have limbs similar to our hands, which can grasp in two ways, via a power grip, with our fists, and with a precision grip, via the tip of the thumb and index finger.

2.5 From Knowns to Unknowns

Informed speculation is the process by which one proceeds step by step from what is known to what is unknown but thought to perhaps exist. Thinking outside the box is certainly useful but you should stay anchored in what you know, otherwise you just drift aimlessly.

We know what exoplanetary geology would be like, we know what volcanoes are, how different types of rocks form. We know what steam, liquid water and ice are. We know what glaciers are and how they can form landscapes. We know what gravity is; we know what atmospheres are, what oxygen, hydrogen, carbon, nitrogen, etc. are; we know what carbon compounds like methane (CH_4) or nitrogen compounds like ammonia (NH_3) are. In addition, we know the periodic table of elements so we know which of the latter can in principle occur on an exoplanet and how these elements can interact, to form molecules and to produce substances on a macro level. We know what rotation, heat, wind, water vapour, condensation, etc. are and hence what weather could be like on other planets.

8 With most sharks the young hatch from the egg within the mother shark resulting in live birth. This system is termed ovoviviparity.

Importantly, we know what evolution is, how it works and what it can achieve given time and a stable planetary environment. We know what natural selection and random mutation are. We know how organisms arrange their cells during reproduction to achieve diversity and hence further evolution.

Actually, this is quite a lot: these facts will influence the manner in which exobeings arise in fairly precise ways. So while thinking about life beyond Earth is speculation, we nonetheless have a pretty good idea of the framework within which such life might evolve. And again, see Section 3.8 on 'weird life' below.

2.6 Sources of Energy and Biological Evolution

Oxygen is a good source of energy for biological processes and liquid water is an ideal biosolvent. This would mean that exobeings could have a continuous intake of oxygen (with exhalation of waste products like carbon dioxide). Given that a range of substances are necessary for metabolism, they could well have a dual system with an intake of food at regular intervals alongside the constant breathing in of oxygen.

Exoplanets, fulfilling the preconditions just listed above, would be well placed to have complex biologies. However, there are things we do not know, matters of principle, which would be very relevant when considering the question of life beyond Earth.

Basic questions for biological evolution

1. Do cells form spontaneously by organic membranes creating enclosures to contain relevant functional parts?
2. Do DNA molecules (or their functional equivalents), the carriers of genetic information, form spontaneously under favourable conditions (perhaps via simpler molecules as with RNA on Earth)?
3. Do cells maintain themselves and replicate spontaneously, leading over time to the rise of a large biosphere?

4. Do cells naturally increase in internal complexity (possibly by incorporating other cells and co-opting them into cooperation as happened on Earth), leading over time to the evolution of increasingly more complex life forms?
5. Do the principles of natural selection and random mutation apply to biological systems on exoplanets?

Questions (1) to (5) above are interrelated and the key word for nearly all of them is 'spontaneously'. On Earth it seems that, when the conditions for life were met, it began; at least we cannot identify a specific trigger for life, such as the maintenance, division and reduplication of cells, and this cannot be simulated under laboratory conditions with the building blocks of cells. The beginning of life, technically known as abiogenesis, seems to be a spontaneous process which happened, under the relevant conditions on Earth between 3.5 and 4 billion years ago. Does this also apply beyond Earth? We may in fact be able to answer this question through discoveries within our Solar System. With the probes, which are now investigating Mars in detail, signs of previous life may be identified. With probes planned for the near future, we may discover signs of life forms existing under the ice sheets of Europa or Enceladus, moons of Jupiter and Saturn, respectively (see Section 8.12 below). If either of these scenarios returns positive results, we could see life on Earth, in its simplest unicellular form, not as a unique freak in our universe but a common occurrence across solar systems and galaxies, on the assumption that life forms in such locations are genuine instances of 'second genesis', which means they are not related to life forms on Earth.

Of course, there are details which we cannot be sure of, for example, whether cell-based biologies would use adenosine triphosphate (ATP) as a universal source of energy, as is done on Earth, and whether this would be produced in cells by the mechanisms we recognise in biologies on our planet. As always in this book we are considering likelihoods, so recall that ATP is produced during the process of cellular respiration, during which glucose and oxygen combine to produce water and carbon dioxide, yielding energy in the process. This energy is then used to bind a phosphate atom to an adenosine diphosphate (ADP) molecule giving ATP, which can later be used to yield energy by releasing the third phosphate atom. Given that oxygen is a highly reactive substance and that glucose consists of carbon, hydrogen and oxygen, all very common elements in the universe, it is not unlikely that cellular respiration involving these elements would develop elsewhere. This line of thinking can be used in other areas of (bio-)chemistry, such as atomic bonding between molecules involving hydrogen, the simplest and most common element in the universe. This is speculation, yes, but it is informed by established knowledge concerning biology on Earth.

2.7 The Brains of Exobeings

The above considerations will help us for much of the way in answering initial questions concerning possible life beyond Earth. However, we are still confronted with an important question for this book, indeed for the entire search for extraterrestrial intelligence: 'Would forms of life beyond Earth have brains, or similar organs, comparable in complexity and functional sophistication to ours?' Of course, cognitive power is not one dimensional and non-human animals on Earth have problem-solving strategies which often bear comparison with ours. But for the purpose of this book, the question of cognitive power is related to technological achievements,

EVOLUTION AND THE POST-BIOLOGICAL

On an exoplanet, as on Earth, evolution would take place in the biosphere in which exobeings would be embedded. If these moved to a post-biological phase, evolution as we know it would probably stop because the variations and mutations which are the basis for biological evolution would no longer occur, or at least not apply to post-biological beings. Changes would likely be manufactured, that is, they would stem from conscious decisions to alter aspects of such post-biological forms of life.

as it is these which would make interstellar communication possible, in principle.

So, to approach the question of cognitive ability, recall that animals generally have a degree of intelligence, based on a corresponding physical brain, which is appropriate for survival in their environment. Snakes, cattle, horses, salmon, flies all have the cognitive power that allows them to lead successful lives in their environment and to continue as species across thousands of generations. True, there are substantial differences in intelligence across the animal world; just consider that between dolphins and sharks or tuna fish. In addition, it should be said that we have no way of determining how consciousness is experienced by animals, after all they have no direct way of communicating this to us. If sparrows or goats or eels had intelligence comparable to humans, would they not, with time, end up doing many other things apart from searching for food, eating and reproducing?

It would appear that we humans have developed intelligence far in excess of what is necessary to survive in our original environment on the savannah plains and uplands of eastern and southern Africa.[9] Just why we should have developed such intelligence is the subject of a large part of this book. Any exobeings with cognitive abilities comparable to ours will have gone through similar stages of evolution in their long progression from less complex forms of life. Importantly, they will also have developed intelligence probably far in excess of what would be necessary merely to survive in their natural environment on an exoplanet. This contention can be challenged,[10] but the opinion being represented here is that exobeings capable of building radio telescopes and spaceships will possess intelligence on a level far beyond that which is required for survival in their biological environment. Why humans have done that on Earth is one of the great mysteries of evolution and an issue which needs to be discussed in some depth before speculating about the situation on any exoplanets.

9 This is also true of other animals, such as cetaceans, elephants and other primates, but none of these has advanced to develop a communication system comparable to human language.

10 And was by one of the reviewers of this book for the publishers.

2.8 Emergence and Consciousness

Any exobeings attempting communication outside their solar system would be conscious. Exobeings building technological artefacts would have minds which were the product of whatever structures their bodies possess and which would be analogous to our brains. This would mean that the hard problem of consciousness (see Section 17.1) would apply, or at least would have applied, to them (they might have solved the problem, after all). Furthermore, there would probably be a gradient for degrees of consciousness across their animal kingdom, much as there is on Earth.

Consciousness with exobeings would likely manifest itself as a subjective awareness of their surroundings, an ability to reflect on themselves, remember their past and plan their future. With consciousness similar to ours, they would have the capacity for abstract thought with which they could solve scientific problems and attain high levels of technological achievement.

In the universe as we have so far observed it – our planet Earth and what we know of other celestial objects around us – it is the human brain, perhaps along with the brains of cetaceans (whales, dolphins and porpoises) and elephants, which represents a pinnacle of complexity, both in organisation and functionality. And it is the physical human brain which engenders consciousness, a phenomenon which emerges from the electrochemical activity of billions of neurons and their trillions of connections. Again, there is a key word in this last sentence, namely 'emerges': the phenomenon of emergence is central to the ever-increasing complexity of biological systems. It refers to the fact that systems show features on higher levels of organisation which are not observable on lower levels. If the high-level features can be causally connected to those on the lower-level, one is dealing with weak emergence. If this is not the case, one has a case of strong emergence: the appearance of higher-level features which are not predictable from lower levels.

Probably the most striking example of strong emergence is human consciousness. Would consciousness arise with complex forms of life on exoplanets? We may be in a position to answer this question, in principle, if we discover some life forms on moons like Europa or Enceladus, as hinted at above (see also Section 8.12 below). If the underwater worlds of these moons contain marine life around hydrothermal vents on their sea floors, similar to the life forms found at such locations on Earth, we could examine such life forms to see if they exhibited very basic levels of consciousness, such as being self-aware, showing mobility, processing input from their surroundings, reacting to such input and making simple decisions when searching for food/prey and when avoiding danger. If this were the case then we would be justified, again in principle, in assuming greater degrees of consciousness for higher forms of life on exoplanets.

If strong emergence and consciousness were to be found with life forms beyond Earth, we could then ask the intriguing question, 'Is there an upper limit to cognitive complexity in the universe and can we determine what this limit is or might be?' There are scholars who believe that exobeings could be cognitively more advanced than we are. We have no reason to assume that we represent the zenith of cognitive power across the universe. We will just have to wait and see what we might discover.

3

Striving to Understand

Information, knowledge and understanding are closely related concepts but with clear differences between them. First of all, information refers to single facts and is independent of any human agent. It is a fact that the Sun is just under 150 million kilometres from the Earth; that is a piece of information. An individual may know that. Furthermore, this individual might know many other facts about our Solar System, and so have a coherent and structured amount of astronomical information, in which case one speaks of that person possessing knowledge about astronomy.

Understanding is a step further and has more to do with grasping the implications and consequences of information and the interaction and interdependence of elements within a system of which an individual has knowledge. For instance, we know why the Sun shines: it releases photons (visible light) which stem from nuclear fusion in its core. But to understand this process one needs to know a lot more about how systems of fusion work, for instance, how much pressure is required inside a star so that the nuclei of atoms can overcome electrostatic interaction[1] and get sufficiently close so as to fuse together, releasing the slight loss of mass

1 This is called the Coulomb barrier after the French physicist Charles-Augustin de Coulomb (1736–1806).

as a large amount of energy according to Einstein's theory of relativity. Furthermore, we must know what elements are involved in fusion, what new ones arise, how long the process can last given an initial volume for a star as well as what will happen when the star begins to run out of nuclear fuel. When we coherently structure this information in our memory and grasp its interconnectedness then we can begin to say that we understand the system. But just how much knowledge of individual phenomena is necessary to understand a system cannot be said for certain at the outset. When can one say one understands something? Perhaps ultimately when one reaches freedom from perplexity regarding some phenomenon and its ramifications, when one is no longer baffled by something and ultimately if one can make predictions about how the system might develop or behave in the future. Whether and when this point is reached will vary from individual to individual and depend on the subject matter involved.

And maybe in the end the goal might not be to reach definitive answers but to better understand the types of questions one should ask and the paths to take towards attempting answers. This is the principle which has guided me in writing this book.

3.1 What Is Scientific Speculation?

Speculation is the consideration of a subject matter for which there is no data or virtually none. Scientific speculation is a kind which starts from an established base of knowledge and seeks parallels to what has already been empirically shown in this outset base, moving forward, step by small step, in the hope of reaching reasonable conclusions about the subject of speculation. Why should one do this? The main reason is that the very process of speculation can help to heighten our awareness for the subject and related issues and so hone our perception of this matter without there being any evidence for its actual existence. One might ask, 'Is this at all possible?' The answer is a cautious 'Yes.' Take the phenomenon of human language. This exists for us and we know a lot about its structure. Hence, if we are considering the nature of forms of intelligent life beyond our planet, it is reasonable to assume that they have complex systems of communication functionally comparable to our own. Otherwise they could not engage in intricate interaction and advanced societies and civilisations, in any way comparable to what we humans have and have had, could not arise.

Models in branches of science are often built on what at the time is speculation, based on unproven assumptions, but where the vindication of a model often comes later, for example, when unexpected or previously unavailable observations confirm the model. A good example is the phenomenon of gravitational waves, ripples in the fabric of space–time, predicted by Einstein's theory of general relativity but only conclusively shown to exist fairly recently, when such waves, which had resulted from the merging of two distant black holes, were registered in February 2016 by appropriate instruments at the LIGO observatory in the USA.

Without wishing to directly compare the great achievements of modern physics with the consideration of exolanguage(s), the principle is nonetheless the same: by considering the language faculty and languages among humans on Earth we can consider possibilities about what languages on exoplanets might be like, if these exist. In this context it is important not to be overly speculative, above all to keep in mind what scientists call *Occam's Razor*, named after the medieval English scholar William of Occam (*c.* 1285–1349), who formulated this principle explicitly. This is a keystone of much scientific theory building, which states that one should not assume more than is necessary for explanation. It is debatable whether it should apply in every single case, but as a methodological approach to theorising and problem solving it has a certain a priori value.

As the British Astronomer Royal Martin Rees has said, we can share in the subjective sense of awe and wonder at the natural world, which many individuals experience and express. However, when we wish to contribute to our understanding of unanswered questions, only objective, data-driven science, based on observation, experiments and evidence, can hope to progress our understanding of unexplained phenomena. This is the only approach which can hold out a prospect of increasing our knowledge of the universe of which we are a part, however small and insignificant.

3.2 What Counts as Proof?

On a most basic level the scientific method requires that there be proof for any assumptions we make about the physical world. Now just what constitutes proof is a much-discussed question, among both scientists and philosophers, and many stress that scientific knowledge is, by its very nature, provisional and continuously under revision. Furthermore, certain things cannot be proved: a negative, for example, there is no multiverse; or something in the future, such as predicting that the Sun will explode on 1 January 2050: however, if that date passes and nothing happens, the previous assumption will be manifestly wrong.

And if one does not have proof for something, one should at least have a maximum amount of observational and/or experimental evidence, preferably independent items of evidence. Take the following case. Astronomers assume nowadays that the universe began about 13.8 billion years ago with a sudden expansion termed the Big Bang. Two main items of evidence are offered in support of this view. The first is the fact that the distance between galaxies is increasing because the very fabric of space is being stretched. If the universe occupies more space now than it did, say, 10,000 years ago, it is clearly expanding. This means that there was a time when it was, say, only 80 per cent of its present size; and when it was only 50, 20, 5 per cent, etc. If one follows the path of time further and further back, one comes to a stage where the universe was small, very, very small, indeed back to a point of infinite density and temperature. The second item of evidence is provided by the cosmic microwave background, heat radiation remaining since what is called the recombination period (the time at which electrons and protons combined to form hydrogen atoms) about 380,000 years after the Big Bang. This radiation was discovered in 1964 by two scientists, Arno Penzias and Robert Wilson working at Bell Telephone Laboratories in New Jersey. It now has a temperature of a few kelvins[2] (just

2 Kelvin is a temperature scale, named after its inventor, the Irish-Scottish Lord Kelvin (William Thomson, 1824–1907), which takes as zero the lowest possible temperature, negative 273 degrees Celsius. This temperature is that at which all molecular motion has ceased and cannot be reached in practice but scientists have conducted experiments in supercooling in which the temperature is just a few kelvins.

above the lowest possible temperature) and is present in the universe in all directions. Furthermore, the cosmic microwave background is remarkably smooth, but contains small irregularities, so-called anisotropies, which reflect the density fluctuations in the universe at a time when it was very small compared to today. These two items, the expansion of the universe and the cosmic microwave background radiation, are now generally taken as sufficient scientific evidence of the Big Bang, which has come to be termed simply the 'standard cosmology'. The European Space Agency's Planck satellite, which was operative between 2009 and 2013, collected an enormous amount of data, mapping the entire sky at microwave wavelengths, thus providing a picture of the universe a mere 380,000 years after the Big Bang, thus confirming the standard cosmology interpretation of the universe's origin.

But what would happen if the scientific method was abandoned, allowing people to make claims without sufficient evidence? The answer is, 'Anything goes'. One might have people claiming that the Earth is not more than 10,000 years old, that it is flat, that the Sun and the stars revolve around the Earth, etc. People could claim that complex structures, like biological organisms, arise in an instant and have arisen suddenly in the past. The ultimate difficulty in abandoning scientific principles is that there are then no constraints on what someone can claim. For instance, someone could maintain that the Moon is a huge piece of cheese hanging in the sky. This is patently ridiculous but no more so than claiming that the Earth is not older than 10,000 years. There is no evidence for either assumption, in fact there is a massive amount of evidence that neither claim is true.

Science and the Zeitgeist

The scientific method is a great achievement of humans and has benefited us enormously in technology, engineering and medicine. But it would be wrong to imagine that it is entirely independent of society and of assumptions and beliefs which hold at a given time. Standards of evidence have varied over the millennia with a notable increase in rigour in the last two centuries or so, with the great advances in the physical sciences. There is also a basic, though not watertight distinction between empirical and non-empirical academic disciplines. Philosophy is essentially a non-empirical science and the quality of research is not measured by demonstrations of proof but rather by the standards of argument and deduction used.

Models are used in sciences and their value, for instance in linguistics, is often judged by the ease with which they explain a series of observed phenomena. Indeed, very often it is precisely the manner in which a new model accounts for seemingly intractable data which convinces researchers in the field of the validity of a model. This was the case with Einstein's theory of special relativity, which was able to account for what is called Brownian motion; this plays a role in the manner in which Mercury orbits the Sun.

3.3 What Do Scientists Know and Not Know?

This is a tricky question. If you ask a linguist how language is stored in the brain or ask an astrophysicist what dark matter is, they may very well say, 'We don't know'. This is strictly speaking true, but it does not mean that scientists have absolutely no idea of the answer. The term 'dark matter' refers to matter postulated to account for the following observations: stars in galaxies move around the core as the galaxies rotate, but the speed at which stars move is approximately the same close to the centre and further out towards the edge, about 200 kilometres per second for the Milky Way. So the Sun is orbiting the centre of our galaxy at about the same speed as stars much closer in and much further out towards the edge. This observation was confirmed in the 1970s for many galaxies by the American astronomer Vera Rubin (1928–2016) and is known as the galaxy rotation problem.

To account for the similarity in rotational speed of the stars, scientists have posited[3] a huge amount of additional matter smoothly distributed

3 There are alternative theories for the galactic rotation problem, for instance, modified Newtonian dynamics (MOND), which involves a slight alternation in general relativity to reach a solution.

beyond the visible edge of galaxies.[4] This 'dark matter' would then interact with visible matter gravitationally (accounting for rotational speeds) but would not be detectable anywhere in the electromagnetic spectrum (as visible light, infrared, ultraviolet radiation, etc.). It is a particularly speculative assumption but one which accounts for the observed astronomical anomalies.

Trying to Find Answers

The pursuit of an answer usually involves ruling out certain possibilities. Hence the linguist can say how language is probably stored in the brain, going on careful analysis of the structure of languages, the manner in which they are acquired in childhood and used by adults. Astrophysicists have a good idea of where dark matter is (in the vast haloes surrounding galaxies) and of its properties, such as the lack of interaction with itself (it does not clump and form observable astronomical objects) and its susceptibility to gravity.

3.4 How Accurate Are Facts?

What are facts and how accurate are they? Consider the estimate for the number of stars in the Milky Way. We don't know the exact figure and could never count them, quite apart from the issue of defining what objects are to be considered stars. For instance, most stars in our galaxy seem to be

4 This means that most of the matter of galaxies is not observable. So where is all the missing mass? Some suggestions have been: (i) cold gas between the stars (nowhere near enough to account for the observations made); (ii) weakly interacting massive particles (WIMPs), which do not interact electromagnetically with the known particles of the standard model of elementary particles; (iii) axions, hypothetical particles with low mass but high occurrence, which can help solve other problems in particle physics; or (iv) the loosely defined massive compact halo objects, which include black holes, neutron stars, brown dwarfs along with giant planets. However, the latter would produce clumpiness in the gravitational field around galaxies, but the matter is smooth, so observations do not support (iv) either. There are other features of dark matter which require explanation by any theory, such as the fact that it is cold, stable in its distribution and slow-moving compared to the speed of light.

faintly glowing red stars. If these are included, the figure increases dramatically. So the figures you will find for the number of stars in the Milky Way centre around 100 billion to 400 billion depending on what objects are counted as stars. Whatever figure an astronomer uses will just be a working approximation and for many tasks that is sufficient. In linguistics there are similar issues, such as how many languages are spoken on Earth. Again, it would be well nigh impossible to count these, not least because of the demarcation problem, namely when is one dealing with a language and when with a dialect of a language, that is, not a separate language. The figure usually given is 6,000–7,000 languages and that suffices when discussing issues like language endangerment, planning and maintenance.

3.5 What We Still Cannot Explain

Although the sciences have made great strides in the past two centuries or so, there are still a number of intractable phenomena which we cannot explain. These are phenomena which came to the fore with the advances science made in the twentieth century. The following is a short list of six phenomena/questions which have proved especially difficult to account for/supply answers to.

Some unexplained facts of the universe

1. Why did the Big Bang occur and is there is a multiverse in which it was embedded as assumed by models of eternal inflation?
2. Why can physicists – as yet – not unify general relativity with quantum mechanics and why is there still not a generally accepted theory of quantum gravity?
3. How is space created between galaxies and galaxy clusters, pushing the expansion of the universe at an ever-increasing rate (dark energy problem)?
4. What is dark matter is and how does it relate to the observable matter of galaxies?

5. What triggered the start of life, technically known as abiogenesis?
6. How does the biochemistry of our brains give rise to consciousness?

Very often when presented with such questions, especially ones like the last two, scientists pretend not to be responsible for a possible answer and claim that 'this is really a philosophical question.' In a way, this is just shifting the onus for explanation to colleagues in a different field. However, this does in fact lead to a distinction made by some scientists when classifying the nature of issues they are confronted with.

3.6 Problems and Mysteries

Some scientists, such as the American linguist Noam Chomsky (1928–), make a distinction in science between problems and mysteries (Chomsky 1991). Problems allow one to ask valid questions and reach possible answers, but mysteries are intractable. Of course, things which are regarded as mysteries can become 'mere' problems with the advancement of science. For instance, lightning was regarded before the scientific era as a mystery because people had no idea what it was or what caused it. But, with increasing understanding of natural phenomena, it became a problem: a weather phenomenon caused by storm clouds colliding. Then, with the discovery of electricity and electromagnetic radiation, it was understood as a build-up of a huge charge which is suddenly neutralised by conduction down to earth or between clouds, visible as a flash of lightning. Thus, it largely ceased to be a problem (except perhaps for ball lightning).

The question really is, are there phenomena which will always remain mysteries? The best candidate for that at the moment is consciousness (item (6) in 'Some unexplained facts of the universe' above) because we cannot even imagine what the actual connection between the physical substrate of the brain and our subjective experience of consciousness might look like. However, we cannot be sure if mysteries will always remain such and we do not know whether our mysteries would also be mysteries for exobeings.

3.7 The Nature of Exceptions

When considering answers offered to scientific questions it is important to understand the value of generalisations which scientists make. Finding one exception or a few does not invalidate a generalisation.[5] For instance, it is a valid generalisation that animals in the sea extract oxygen from water via their gills. But there are animals, like whales and dolphins, which breathe air through a hole on their head when they surface at regular intervals. In fact, one can see here, as so often with exceptions, that there is a principled explanation for them. In this case it is that air-breathing animals in the sea are mammals from land whose remote ancestors migrated back into the sea. Other times, exceptions actually fit into the general picture, and are due to some special development. For instance, the vast majority of birds fly (a valid generalisation), but there are exceptions: ostriches, penguins, etc., which have lost the ability to fly in the course of their evolution; ostriches still have wings and use them for balance and courtship displays. During their evolution penguins refunctionalised their wings as flippers for swimming. And pheasants illustrate the transition to flightlessness: they can fly short distances with their reduced wings, but prefer to run on the ground.

3.8 What About 'Weird Life'?

When considering possible scenarios for life beyond Earth much has been made of the possibility of 'weird life', forms which would exist under conditions radically different from mainstream life on Earth. There are examples of 'weird life' on Earth: thermophiles (= 'heat lovers') who thrive in boiling pools, such as Grand Prismatic Spring in Yellowstone National

5 The opposite is also true: correlations do not have to be causally linked. Consider the fact that all the following composers were under 5′ 5″ (1.65 m): Mozart, Beethoven, Schubert, Wagner, Grieg, Mahler, Ravel, Stravinsky. The exception among early composers is Johann Sebastian Bach, whose height has been estimated at 5′ 11″ (1.8 m), impressive for his time. So, are height and power of composition causally linked? Of course not, these instances are just coincidences.

Park, or tardigrades (eight-legged microanimals, less than 1 mm long), which are virtually indestructible: they can survive extremes of heat and pressure, food deprivation and even the vacuum of space.[6] Such life forms are undoubtedly interesting to biologists and astrobiologists. However, as the Nobel-laureate Jack Szostak has pointed out, extremophiles are adaptations from earlier forms of life which arose in less extreme environments through Darwinian evolution (though there are contradictory views on this issue). This means that such extremophiles are unlikely to represent the beginning of life (at least on Earth or comparable planets), but are rather a later development under very particular conditions.

What such extremophiles tell us is that life forms can eke out an existence in very unfavourable circumstances, they may testify to the tenacity of life, once established. Extreme conditions of heat or cold would not be conducive to the rise of complex life. This also means that much of possible 'weird life' can be excluded from the concerns of the present book. Consider that Saturn's moon Titan probably has a subsurface liquid-water ocean, which is being investigated by astrobiologists as a possible abode for life. On its surface there are lakes of liquid methane

HOW WEIRD MIGHT EXOBEINGS BE?

To achieve a certain level of technological achievement, at least comparable to what has characterised recent developments on Earth, exobeings would have to exhibit both powerful cognition to invent complex artefacts and dexterity of their limbs to construct such devices with which they could engage in interstellar communication. True, astronomical devices can be cooled to very low operational temperatures, such as on the James Webb Space Telescope, but they would require much higher temperatures, closer to those found on Earth, for their construction.

6 These organisms belong to the group of extremophiles which can survive in environments usually detrimental to life, such as extreme salinity, acidity, alkalinity or exposure to radiation. There is a dedicated journal, *Extremophiles. Microbial Life Under Extreme Conditions*, publishing relevant research from this area.

and the atmosphere consists largely of nitrogen. Liquid water under the surface could perhaps provide a source of oxygen thus providing the four basic elements for organic compounds, carbon, hydrogen, nitrogen and oxygen. But the surface temperature on Titan is about –180 °C. That is slightly above the boiling point of oxygen (–183 °C) so it would just about be present in gaseous form; water, on the other hand, would be frozen as hard as granite and so could not be used as a biosolvent. While it may be that some kind of microbial life might exist in the methane lakes of Titan, it is hard to imagine that a technologically advanced civilisation could arise in such an environment. Even assuming for a moment that exobeings could arise and survive at this or any comparable temperature, it is difficult to imagine how they could build artefacts like spacecraft or radio telescopes for interplanetary travel or interstellar communication at –180 °C. The conclusion from this and similar environments, such as the subglacial ocean on the Jovian moon Europa or the upper atmosphere of Venus, is that while 'weird life' is interesting to many astrobiologists for providing very different scenarios from the common life forms on Earth, it would very challenging for a technologically advanced civilisation to arise in environments like these.

3.9 How Different Could They Be from Us?

Considerations of 'weird life' do not just involve extreme environments for possible life forms but also concern the appearance and functionality of exobeings. There are two ways in which they could, in principle, be different from us, outlined below.

Differences between possible exobeings and humans

1. Darwinian evolution on their exoplanet may have resulted in beings which are physically organised very differently from us, with brains which we would not immediately recognise as such, maybe with a more distributed nervous system, like an octopus.

2. The continuing evolution of exobeings, including technological advances, might have resulted in changes to their life forms into realms which we cannot imagine, such as digital extensions to biology, which we have no inkling of.

In any particular case we might be dealing with combinations of both (1) and (2) above, which would render any exobeings more different from us than anything we might be expecting. However, we are still left with at least the following aspects which exobeings might be assumed to share with us.

Shared features between possible exobeings and humans

1. Bodies which would allow them free and flexible movement in the three-dimensional space of their planets.
2. Limbs with analogues to our hands showing a power and a precision grip. Otherwise, they could not construct anything.
3. The ability to receive and process sensory input from their surroundings, at least visual and auditory data, though probably tactile data and possibly olfactory and gustatory data as well. These abilities would have arisen during their evolution.
4. Complex physical structures capable of the sophisticated computation which human brains achieve.
5. Conscious awareness of themselves and their surroundings, with memory of their past and the ability to plan for their future. Exobeings would have the potential for learning and at least some of them would be curious to discover more about their environment and advance science on their planet.
6. A system for organising their thoughts and communicating with others, which would be functionally equivalent to human language. This would be based on an evolved internal faculty which would be expressed as perhaps one of many exolanguages. There are cogent reasons for assuming that these would avail of sound, but other modalities are also possible.

These considerations clearly suggest that while forms of life on exoplanets may be very different from those on Earth at the lowest level, further developed forms of life could be more like us in their structure and function though not in terms of their phenotype, their actual appearance.

3.10 Two Other Questions

It is a common theme in science-fiction film and literature: aliens are watching us and/or aliens have already landed on Earth and are among us. But there is a serious side to such considerations.

Are We Being Observed by Them?

Just as scientists on Earth continually scan the heavens with their telescopes, investigating all kinds of astronomical questions, there could just as well be exobeings from other planets doing the same thing but in reverse.[7] The question of interest then would be, 'Have they realised that our Earth is a life-bearing planet?' If so, what consequence would that have for them? Here we have no answers as we have no way of knowing what is the case. Of course, we might also ask whether, if they knew about us, they would be interested in contacting us. Again, we have no answer to such a question.

Did They Ever Visit Us?

Whether exobeings ever visited Earth is something which is extremely unlikely to have ever happened. For one thing, the vast distances between solar systems in our galaxy would present the same challenge to possible exobeings with spacecraft as it would to us on Earth. And for another, there is absolutely no evidence that exobeings have ever been on Earth.

7 See the article 'Aliens Might Already Be Watching Us' (*Scientific American*, June 2021), www.scientificamerican.com/article/aliens-might-already-be-watching-us/.

Part II

The Universe We Live In

Any search for exobeings should begin with considering the nature and structure of the universe, what it is composed of, how stars and planets form, the different kinds we can recognise and their assumed properties. Exobeings would live in the same universe as we do and hence would be constrained and moulded by the same forces that we have been subject to.

4

Trying to Grasp Size

Learning about the universe can be life changing. When you realise that our galaxy is at least 100,000 light years in diameter, that it contains several hundred billion stars, most with planets, and when you learn that the observable universe may contain anything up to one to two trillion[1] such galaxies you cannot pretend that you do not know this and retreat to some earlier period of knowledge to recover a state of innocence when you thought Earth was all that mattered. True, it is where our lives are based, but there are no words to describe how utterly insignificant the Earth is to the universe which contains it.

Our Solar System is located in an arm of a not uncommon type of galaxy in a universe with innumerable such galaxies. To get an idea of the sizes involved here one can give figures as above. But our brains are not made for dealing with such numbers. However, here is a comparison you could

1 The American-British astronomer Christopher Conselice and his team estimated the number of galaxies in the observable universe to be around two trillion going on a re-examination of data from the Hubble space telescope. Note that in this book three common terms for very large numbers will be used: (i) a million, which is a thousand thousand: 1,000,000 or 10^6; (ii) a billion, a thousand million: 1,000,000,000 or 10^9 and (iii) a trillion, a million million:1,000,000,000,000 or 10^{12}. Anything beyond a trillion cannot be realistically imagined by humans, for example 10^{80}, an approximate figure for the number of atoms in the observable universe.

perhaps relate to. Imagine you are on a very long sandy beach; you have dunes behind and the sea in front of you and to your left and right several miles of strand stretching to the horizon in both directions. Now bend down and try to pick up a single grain of sand. The proportion of that grain to all the sand on that huge beach would not even come close to the proportion of Earth to all the celestial objects in the observable universe.

Now imagine you remove that single grain of sand from the huge beach. Would that be of any significance to the beach? Of course not. And that comparison might help to convey how insignificant our planet Earth is to the universe. If Earth simply disappeared tomorrow along with the nearly eight billion people living on it, not only would that be of little relevance to our Solar System but also of none to our neighbouring stars, the rest of our galaxy and certainly not to any of the other galaxies.[2]

4.1 Astronomy and History

Throughout human history there has been a long and detailed involvement with astronomy[3] by observers, in different parts of the world, both using the unaided eye and assisted by landscapes and/or architectural features, such as megaliths, earthworks, etc. These kinds of observations are the historical precursors to instrumental astronomy, which led the way to where the field is today. Mapping the stars and planets which can be seen from Earth is an activity reaching far back into antiquity. The Nebra sky disk, found in Germany in 1999, shows stars in gold on a bronze disk and would appear to have been used about 2000 BCE and buried in 1600 BCE. In China, India, Mesopotamia, Egypt, Greece and the area of Mayan culture in present-day Mexico and Central America, astronomy was an active area of investigation and one which often had regal status. In both the

2 As the Canadian anthropologist Kathryn Denning has remarked, the discoveries of astronomy in the past few centuries have essentially been about decentring us and our planet.

3 Not to be confused, of course, with astrology, a pseudoscientific practice which maintains that there are links, on the one hand, between personalities and events here on Earth and the position of celestial bodies, such as stars and planets, on the other.

Islamic and the medieval Christian worlds, astronomy occupied a central place and was closely linked to religion, for example, astronomical phenomena, such as eclipses, were often associated with the wrath of a god or gods, both by the officials and the practitioners of religions. Throughout all these periods and in all these places the geocentric view of astronomy generally prevailed, albeit with different details of how this was conceived.

Heliocentric and Geocentric Views

That the Earth orbits around the Sun is a view which we find with the Greek astronomer Aristarchus of Samos (*c.* 310–*c.* 230 BCE), but his ideas were ignored by astronomers like Ptolemy (*c.* 100–*c.* 170 CE) and the philosopher and scientist Aristotle (384–322 BCE), who maintained that the heavens revolved around the Earth, and so this fallacy became scientific orthodoxy for well over 1,500 years.[4]

But the Polish astronomer Nicolaus Copernicus (1473–1543) put the heliocentric view of the Solar System on a firm footing with his *De Revolutionibus* (Of the Revolutions), although without empirical evidence, and he also acknowledged Aristarchus' heliocentric view. Soon afterwards, other scientists followed his work and expanded it. Two of these are especially well known: the German Johannes Kepler (1571–1630),[5] who developed a theory of planetary motion, and the Italian Galileo Galilei (1564–1642), who built a telescope with which he could observe the phases of Venus, the four larger moons of Jupiter (now called Galilean moons in his honour) and even sunspots. With his telescope Galileo could clearly observe the appearance of Venus as a full disc or a sickle with all the intermediary positions. The conclusion he drew from this was that Venus, Jupiter and the other planets rotated around the Sun in orbits of their own, independent of Earth. On this followed the obvious conclusion that the Earth also orbited the Sun. With that, the centuries-old view of a geocentric universe

4 There was, however, a view called 'cosmic pluralism,' a stance from classical antiquity to the modern era, that there may well be other earths with other instances of humankind.

5 Kepler predicted the motions of the planets more accurately than Copernicus by assuming that they moved in slightly elliptical orbits.

was abandoned and the path was open for the development of modern astronomy, although Galileo was not to enjoy the fruits of his discoveries. Instead, his heliocentric view of the Solar System, along with other opinions, led to conflict with the Catholic church and to his being accused of heresy by the Roman inquisition and put under house arrest in the latter years of his life.

4.2 How Has the Universe Developed?

Nowadays, it is common knowledge that the universe started as a Big Bang, a popular term, coined by the British astronomer Fred Hoyle (1915–2001), who in fact did not believe in this story of its origin.[6] But what is meant by this moniker and why should we assume such an origin? The universe is between 13.7 and 13.8 billion years old. This age has been calculated by astronomers based on the rate of expansion, the age of galaxies and theories of star formation from clouds of gas consisting largely of hydrogen, the simplest atom formed in huge quantities in the early phase of the universe.

The details of the all-important first second (sic!) of the universe are not entirely clear. But in the now standard model of how the universe came to be as it is, that is, structurally very uniform in every direction, it has been postulated that just after the beginning, between 10^{-37} and 10^{-35} seconds, the universe went through about 100 iterations of exponential expansion, called inflation,[7] within this fraction of a second. Furthermore, in the very

6 During the middle of the twentieth century there was much disagreement on this issue. For example, there was considerable rivalry between Fred Hoyle and the Ukrainian-American physicist George Gamov (1904–1968), who favoured the steady state theory of the universe versus the Big Bang theory.

7 The American astrophysicist Alan Guth (1947–) maintains that this expansion happened during the epoch in which all the fundamental forces of the universe (bar gravity) were still united. This happened within a tiny fraction of a single second, to be precise, between 10^{-37} and 10^{-35} seconds after the Big Bang. The Russian-American physicist Andrei Linde (1948–), at Stanford University in California, says that when cosmologists were trying to explain the uniformity of our universe in all directions they were led to a theory of inflation to explain this. However, there are cosmologists, such as the British physicist Roger Penrose, who do not accept inflation theory, or at least demand that its supporters explain how inflation could have come about given the myriad other means by which the universe could have proceeded in the first second of its existence.

early universe there were continuous annihilations of antimatter and matter. All of the latter, which forms the stuff of the universe, including ourselves, resulted from the tiny imbalance in favour of matter, less than one part in a billion.

What Is Our Cosmic Neighbourhood?

Our galaxy – the Milky Way – began to form about 600 million years after the Big Bang, that is, about 13.2 billion years ago, and contains several hundred billion stars. As an estimate that figure has been increasing, especially since we now know that M red dwarfs are the most common types of stars (they probably comprise 75 per cent of the stars in our galaxy). These are far less luminous that our Sun, a G dwarf star, a type which accounts for about 3 per cent, and so the red dwarfs are more difficult to observe than brighter stars but can be detected, of course, by using various telescopes.

Our Solar System is located on the inside of the Orion Arm towards the (visible) edge of the Milky Way between the outer Perseus Arm and the inner Carina-Sagittarius Arm (Figure 4.1). It is over 26,000 light years from

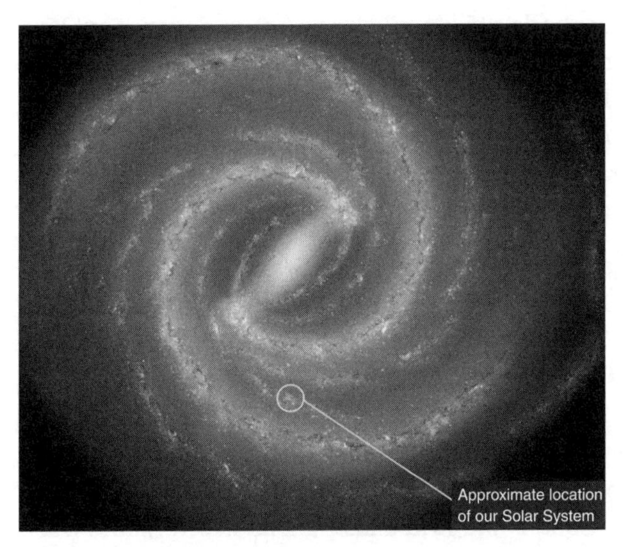

Figure 4.1 The Milky Way with approximate location of our Solar System (the small circle is several thousand light years in diameter)

the centre of the galaxy. The whole system rotates around this point which contains a supermassive black hole, Sagittarius A*, which has a mass several million times that of our Sun.

What Is Now?

The notion of 'now' derives from the way we manage time by dividing it into regular units (seconds, minutes, hours, days, weeks, months, years), which are taken to apply everywhere on Earth, albeit in different time zones. So we talk about a certain date, say 12 March 2022, and this is taken to apply everywhere on this planet, apart from the parts of the world beyond the international dateline where the time tips over 24 hours and we claim that the date is one day further on.

But in the universe there is no privileged notion of 'now' at any one point. 'Now' really only applies to us on Earth. Notions of 'now' on other planets would be out of sync with us. Take, for instance, the star Betelgeuse, the orange upper-left point in the Orion constellation. It is about 500 light years away and a prime candidate for a Type II supernova explosion (one caused by the collapse of the star's core) in the not-too-distant future. But say it had exploded on the day William Shakespeare was born in April 1564 we still would not know about this because the light of such an explosion, travelling at 300,000 kilometres per second, would still not have reached us. So conceptions of the present, of 'now,' in the universe are local, very local. When we look up at the sky, we see stars and galaxies as they were in the past; how far back depends on how far away from us they are.

In the Milky Way galaxy, our presence on Earth is not in any way privileged. For all we know, there may have been a golden age for planetary life in our galaxy, say five billion years ago, but now all that life may long since have been lost. This is a sobering thought. It means that when searching for life in our galaxy we are now confined to those habitable planets which happen to be at least at our stage of technological development. Anything before that would be difficult to detect. However, our technology is proceeding by leaps and bounds and so new instruments like the James Webb Space Telescope, now up and running, will probably be able to detect distant biosignatures, such as atmospheric disequilibrium on exoplanets. Technosignatures, such as artificial radio signals, can be detected by many different telescopes on Earth, but their detection is judged to be much less likely.

4.3 Estimating the Size of the Universe

With improved technology, our view of the universe expanded. By the beginning of the twentieth century astronomers knew that we inhabited a Solar System in a galaxy of countless stars. But one of the most important discoveries was the demonstration by the American astronomer Edwin Hubble (1889–1953) that the smudges in the night sky, then thought to be nebulae of gas and dust, were in fact other galaxies and that there was a vast number of them beyond our own. Hubble also ascertained that the galaxies were receding at rates proportional to their distance from us. Using the large telescope at Mount Wilson Observatory near Los Angeles, he calculated the redshift of the galaxies, demonstrating clearly that they were receding in all directions, meaning that the universe was expanding continuously. This insight changed our conception of the universe radically.

With the advent in 1990 of the Hubble Space Telescope (named after the astronomer), scientists and the general public, after a few initial glitches, were treated to a huge array of stunning pictures of the universe from above the blurring shroud of the Earth's atmosphere. Among the most significant of these is the Hubble Deep Field (a small region of the Ursa Major constellation in the northern sky, 1995), the Hubble Ultra Deep Field (a small region of the Fornax constellation in the southern sky, 2003–2004) and the Hubble Extreme Deep Field (a detail from the centre of the Ultra Deep Field, 2012), offering successively deeper images of patches of sky in regions where interference from starlight in our galaxy is minimal. One of the most distant galaxies ever to be registered is an object with the prosaic label MACS0647-JD, in the constellation Camelopardalis in the northern sky, which has a light travel distance of 13.3 billion years. The galaxy is actually much further away from us than 13.3 billion light years because it has also been receding from our galaxy during this time due to the ever-increasing expansion of the universe. This galaxy was probably formed about 400 million years after the Big Bang and contains about a billion stars, far fewer than later galaxies like our own Milky Way.

The ability to view the very earliest galaxies was greatly increased with the James Webb Space Telescope, which went into operation in mid-2022. One of the first images taken and released by NASA was that of a patch of sky, near the handle of the Plough/Big Dipper constellation, which shows a deep field with much greater resolution than even the Hubble Extreme Deep Field. This is part of ongoing research using the telescope to determine what the earliest galaxies were like: the JADES (James Webb Space Telescope Advanced Deep Extragalactic Survey) project has already examined a section of the Hubble Extreme Deep Field, a part of the Hubble Ultra Deep Field from 2003–2004 and attained by staring at an apparently empty spot in the Fornax constellation for a total of 22 days in 2012; this revealed about 10,000 very distant galaxies. The JADES initial deep field survey (lasting a total of 9 days) has already located about 100,000 galaxies in the same field as the Hubble Extreme Deep Field, given its resolution which is about 15 times greater than that of Hubble.

4.4 The Observable Universe

Because of the finite speed of light (see below) and the finite expansion rate of the universe in the past, the objects in the universe which can be observed on Earth are limited to a comoving distance of approximately 46 billion light years in any direction, which means the observable universe is a spherical volume around us, about 92 billion light years in diameter, the so-called Hubble volume (Figure 4.2). Enormous as this is, it is probably only a small part of the actual universe; objects in this larger envelope will remain forever unobserved on Earth because their distance and rate of recession means that light from them can never reach Earth. A further consequence of this fact and of the accelerating expansion of the universe is that the number of galaxies we can see with our telescopes will decrease steadily as they pass the visibility horizon for us on Earth, a sort of cosmological event horizon for distant galaxies beyond which they disappear.

Groups of galaxies, like our own Local Group, consisting of the Milky Way, Andromeda[8] and some smaller galaxies, are gravitationally bound. The Local Group is about 10 million light years across and our galaxy is a tiny piece on the edge of a much larger one called the Laniakea Supercluster (from the Hawaiian word for 'infinite heaven'), consisting of about 8,000 galaxies. Most of the galaxies in this supercluster are migrating towards a region of space known as the Great Attractor, which is exercising a gravitational pull on these galaxies. Between the superclusters of galaxies there are vast voids which contain no stars and which are increasing in size with the expansion of the universe at an ever-increasing pace due to dark energy, which is constant across the universe.

There will come a time in the far future when the distances between the clusters are so great that light will not be able to cross their divides (see next section) and the sky will look to observers within any cluster like a dark void beyond their cluster. In addition, because gravity is the dominant force within any given cluster, its galaxies will come closer together to merge into increasingly large structures.[9]

8 In the constellation of Andromeda, just left of the centre of the Milky Way, is a small smudge just about visible to the naked eye. That is a massive spiral galaxy, known prosaically as M31 (according to the Messier catalogue). It is much larger than our own galaxy and is hurtling towards us at over 100 kilometres per second. You might think that at that speed it should arrive here in the not-too-distant future. However, it will take about four billion years to travel the two and a half million light years it is distant from our galaxy. And when it arrives there will not be a single gigantic crash. Rather, the galaxy will gradually appear larger in the night sky and then seem to disappear in front of us and we will finally notice some extra stars in the night sky, depending on the precise angle of collision and the perspective from which we view it. Naturally, if vast clouds of dust collide, they may well spark cycles of new star formation and, if the centres come close, there may be some spectacular distortions of star distributions. Andromeda will probably pass through our galaxy only to slow down hundreds of millions of years later and fall back on us still later again, settling down, again hundreds of millions of years later, to a new, much larger spiral shape of the combined galaxies. Of course, the use of 'we' in this paragraph is anachronistic: no inhabitants on Earth, nor the planet itself, will be around in four billion years to witness the collision.

9 The result of the collision of the Milky Way and Andromeda has been given the tongue-in-cheek nickname of Milkdromeda.

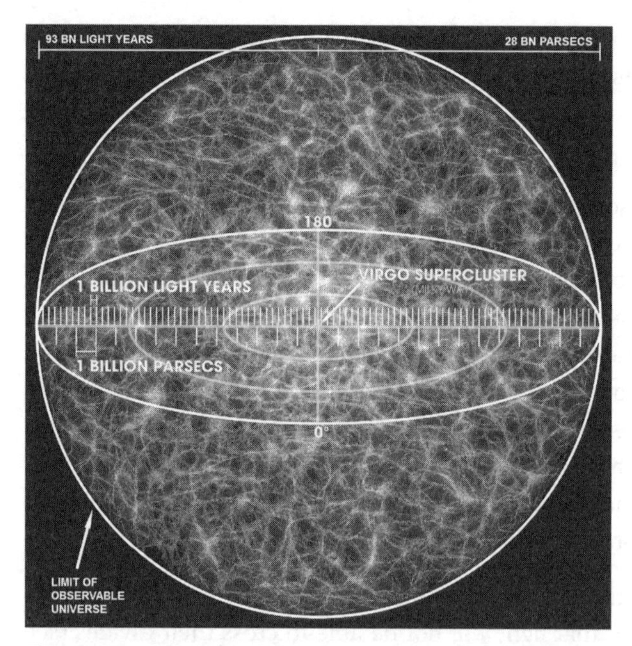

Figure 4.2 The observable universe. The Virgo Supercluster (the dot in centre) is about 55 million light years in diameter and contains over 100 groups of galaxies, including our own Local Group. There are at least 10 million such clusters in the observable universe.

The Speed of Light

At various points in this book reference is made to the speed of light. This is a fundamental value of our universe and it is important to grasp its significance. Its speed is just under 300,000 kilometres per second in a vacuum. This is the fastest speed at which anything in the universe can move – the speed of massless particles, in this case that of the photon, the quantum of the electromagnetic field[10] which we humans perceive as visible light when it is within a specific wavelength spectrum (between about 300 and

10 The electromagnetic force is one of the four fundamental interactions in our universe. The other three are the strong nuclear force, the weak nuclear force and gravity (electromagnetism has been combined with the weak nuclear force in the electroweak theory, an issue not of relevance right now).

700 nanometres in length). Other, practically massless particles, such as the various neutrinos, move at a speed close to that of light.

The speed of photons, that is, of light, is invariant. When photons are emitted, for instance in certain physical reactions, they do not accelerate to 300,000 kilometres per second, they move instantaneously at that speed. As nothing can move faster than a massless particle, its velocity is the maximum speed at which information can travel across space. This upper limit on particle movement applies across the universe, and so would be a base constant on all exoplanets. It also determines the amount of energy released on the annihilation of matter as shown in Einstein's famous equation, $E = mc^2$, that is, the energy released is equal to the mass lost multiplied by the square of the speed of light (a truly enormous figure, hence the power of nuclear bombs and the constant light of the Sun).

Although the speed of light limits the movement of particles, and hence of any conceivable human or exobeing artefact travelling through space, objects such as galaxies can theoretically move faster than light. For an observer on a planet such as Earth, extremely distant galaxies, such as those in the Hubble Extreme Deep Field (see previous section), recede at an increasingly faster rate. At some stage in the future the rate of recession will pass the speed of light, after which these galaxies will disappear from our sight because the light they emit will never reach Earth.

5

Star Formation and Planets

Stars form when vast clouds of gas and dust begin to swirl and collapse with a concentration of matter at the centre. Around the newly born star a proto-planetary disk rotates, out of which planets can later evolve by clumps arising and accumulating ever greater amounts of material. The smaller planets which arise tend to be largely composed of rock, like the first four in our Solar System, whereas the larger ones tend to be gas or ice giants, like the outer four in our system.

The Eagle Nebula in our own Milky Way is an example of a presently active star-forming region, a stellar nursery. It contains a region known as the Pillars of Creation, immortalised in a famous picture by the Hubble Telescope in 1995, which shows long columns of dust in which young hot stars have formed and which are clearing away the dust with their intense solar wind. In other galaxies, similar but much larger regions can be observed, for instance, the (New General Catalogue) NGC 604 region in the (Messier) M33 galaxy (Triangulum Galaxy) contains hundreds of superhot blue stars in a gas cloud some 1,300 light years across. Another active star-forming region in our cosmic neighbourhood is the Tarantula Nebula, 160,000 light years away in the Large Magellanic Cloud (a satellite galaxy to our own). It is about 1,000 light years across and is a veritable star factory, including the most massive stars ever observed in the universe from Earth.

Stars do not always come alone, indeed many stars are located in binary or multiple systems such as our nearest stellar neighbour Alpha Centauri, a mere four and a half light years away and consisting of one large and two smaller stars. Many of the familiar constellations of the night sky contain such groups; for example, Mizar and Alcor in the Plough/Big Dipper constellation make up a sextuplet of stars gravitationally bound to each other. Polaris, the North Star, is a system consisting of three stars.

5.1 Red Dwarfs

One reason why the number of stars in the Milky Way has been revised upwards in recent years is that it contains a huge number of red dwarfs. Some estimates put the amount at about three-quarters of all stars, which are faint when viewed from Earth because of their distance from us. These stars are small and relatively cool compared to our Sun, which is now regarded not as a typical star in our galaxy but rather unusual in its luminosity. Red dwarfs are known to have planetary systems and their habitable zone is closer to their parent stars than in our Solar System. There are dangers for any planet in this zone as red dwarfs frequently emit strong solar flares, jets of high-energy particles flung out from the star's surface into space. Any planet close to the star and with life forms on it would experience considerable damage if struck by such a flare, and frequent repeated flares could lead to atmospheric erosion. But flares are not necessarily all bad news: they could break down oxygen molecules (O_2) and the free oxygen atoms could react with other O_2 molecules to form ozone (O_3), which would in fact form a shield again detrimental ultraviolet radiation from the star further down the road, so to speak.

Because of its close orbit to a parent red dwarf, a planet could well be tidally locked to the star, always pointing the same face to the star, much as our Moon does to Earth.

Red dwarfs are very long-living given that they are completely convective. By this is meant that the helium resulting from thermonuclear fusion of hydrogen in the core is mixed through all layers of the star and is not just confined to the core as it is with our Sun. Such stars can last trillions of years, though we cannot tell exactly as none has reached this age yet. The nearest star to the Sun, Proxima Centauri, is a red dwarf with a planet in the habitable zone, but unfortunately this is subject to mega-flares from the parent star, which have probably long since stripped the planet of its atmosphere. Another well-known star is Gliese 581, about 20 light years away in the Libra constellation, which has a planet in the habitable zone, a mere 33 million kilometres from its sun but with only about 30 per cent of the light which we get from ours.

Red dwarfs can become inactive after several billion years and then release fewer solar flares reducing the threat to possible life arising on them. This is the case with Teegarden's Star, discovered in 2003 in the constellation of Aries about 12 light years away from Earth and approximately 8 billion years old. The star is known to have two planets, Teegarden b and Teegarden c, the former of which is well within the star's habitable zone and just may have a water-carrying atmosphere. There are other quiet red dwarf stars in our cosmic neighbourhood, such as Luyten's Star, again about 8 billion years old and just over 12 light years away from Earth. Another one is Ross 128, over 9 billion years old and about 11 light years from us.

5.2 Brown Dwarfs

The next class does not consist of stars, but of substellar objects, thus classified because they are not massive enough to sustain the thermonuclear fusion of hydrogen in their cores. This means that they are more like massive gas giants, tens of times more massive than Jupiter. The closest brown dwarf to us is Luhman 16 (actually a binary system, discovered in 2013) at a distance of about six and a half light years from Earth. Some planets are known to orbit such objects but whether life could be sustained in this scenario, given the much-reduced energy outflow from a brown dwarf, is unknown, although the issue has been investigated.

5.3 The Life of a Star

Stars derive their enormous energies from fusing light elements into slightly heavier elements in their cores. This process involves the loss of a small amount of matter, which is converted into a huge amount of energy generated in this process, equal to the mass lost multiplied by the speed of light squared: a truly astronomical figure.

The fusion of elements results in increasingly heavier elements arising in the centre of the star, starting with hydrogen, then helium and other elements such as carbon. At each stage of fusion, a small amount of matter is converted into a large amount of energy, which pushes outwards and counteracts the force of gravity drawing inwards in the star. These forces are in approximate balance in our Sun, which explains why it can keep burning its hydrogen fuel for billions of years.

The fusion of elements goes all the way to iron, which has an atomic number of 26.[1] However, because iron fusion actually consumes energy rather than releasing any, this is the end of the line. If a star is massive enough, the iron core collapses and draws in outer layers of the star. This infall is halted and rebounds outwards, with other heavy elements being formed during the process known as explosive nucleosynthesis. The amount of energy released is tens of times greater than what the Sun will release in its entire lifetime of about 10 billion years.

If the core collapse of a star stops at a diameter of about 20 kilometres, which happens with stars less than about 20 times the mass of the Sun, a neutron star is the result, a star which consists of tightly packed neutrons rotating at great speed, something like 30 times a second, and with an enormous magnetic field around it. If the collapse results in a mass that is greater than that for a neutron star, then a black hole is the outcome, a singularity at the

1 Atomic number refers to the number of protons in the nucleus of an atom. This internal structure of the atoms was made explicit in the periodic table of chemical elements, ordered by their atomic number, and first devised by the Russian scientist Dmitri Mendeleev (1834–1907). Atoms may have an unequal number of protons and neutrons, in which case one speaks of isotopes.

centre of which nothing, not even light, can escape because of the magnitude of its gravitational field. Black holes cannot be observed beyond their event horizon, the point of no return for infalling matter and beyond which light cannot be emitted, hence the moniker 'black hole'.[2]

A type 1a supernova involves two stars, one a white dwarf, the extremely dense stellar core of a former star, and the other one another star, somewhere in its lifespan, often at the end as a red giant. The key point here is that the first star starts accreting matter from the second, dragging gas off the second star by its enormous gravitational field. This accumulation of matter in time heats up the first star and, if it reaches the limit of 1.4 solar masses, known as the Chandrasekhar limit, the star can no longer resist the pressure to collapse through gravity and carbon fusion is initiated, which suddenly releases an enormous amount of energy in a runaway core reaction. The result is a supernova, the star explodes. At about five billion times the brightness of our Sun, the peak of luminosity for this kind of supernova is fairly consistent, which makes such stars useful for calculating distances in the universe.

Neutron stars can come in different forms, for example as a pulsar, which emits beams of rays at very stable periodic intervals. A magnetar is a type of neutron star with an enormous magnetic field and which can emit beams of powerful X-rays from the poles of its magnetic field. In 2004, a huge wave of X-rays hit satellites orbiting the Earth coming from a magnetar, called SGR1806-20, which is about 50,000 light years away, and still the blast was stronger than a powerful flare from our own Sun, eight light minutes away.

Gamma-Ray Bursts

The final moments of a star can represent a serious threat for life on planets around any star in its cosmic neighbourhood, say within tens of light years of the exploding star. The intense burst of radiation from the star can be a cause of the mass extinction of life on a nearby planet. One such burst,

2 Black holes nonetheless emit powerful jets of particles at right angles to the plane of their accretion disk, so-called Hawking radiation, named after its discoverer, the English physicist Stephen Hawking (1942–2018). Because of this, black holes may well evaporate completely over vast periods of time.

which our planet has not been subject to, at least not recently, but which could well have affected exoplanets, is a gamma-ray burst. Such bursts are the most powerful events known in our universe and occur when infalling matter of a gravitationally collapsing star sends out highly charged gamma rays in two directions along an axis at right angles to the plane of infalling matter. If either side of this axis is pointing towards Earth, it is possible to pick up the flash of the burst. By the time the astronomers get around to training their telescopes on such a source, the peak of the burst has usually past and they are left examining the afterglow, which typically lasts some hours.[3]

A single gamma-ray photon has the energy of about a million visible light photons. These photons represent ionising radiation, which means they can knock out electrons from their orbits around nuclei, breaking the atomic bonds which contain the electrons in the process. They can also break up the DNA molecules in cells, with dire consequences for the life forms in question.

A gamma-ray burst can release more energy in a few seconds than our Sun will in its entire lifetime. One such burst has been suggested as a source of the early Late Ordivician extinction about 450 million years ago.

5.4 Where Do the Elements Come From?

In the aftermath of the Big Bang, hydrogen, helium and some lithium were created. But there are close to a hundred naturally occurring elements on Earth. So where did all the others come from? The answer is: 'From nuclear fusion in the cores of stars with later ejection at the end of stars'

3 There are actually two types of gamma-ray bursts. The first results from a supernova, leading to a black hole, and yields a long burst lasting about a minute; and the second is triggered by the merging of two neutron stars, with a short burst lasting about a second. The latter is due to the gradual reduction of the respective orbits of the neutron stars, which causes them to get close enough for gravity to collapse them together, resulting in a black hole.

lives'. Some stars end quietly, with a white dwarf or neutron star as final stage, while other, more massive, stars explode as a supernova at the end of their lives. The latter type are the source heavy elements (those heavier than iron). These elements are then ejected into the interstellar medium (the space between stars) and can aggregate and end up on the planet of a star in a following generation, hundreds of millions of years later. Thus it is fair to say that we are made of 'stardust'. Take gold, for example, with the chemical symbol Au (Latin *aurum*) and an atomic number of 79. Gold is shiny yellowish in colour, fairly soft and chemically inert. The gold in the wedding ring you may be wearing, or in the fillings which you may have in your teeth, was formed by nucleosynthesis during a supernova explosion many billions of years ago and probably made its way to Earth during the period of asteroid bombardment about four billion years ago.

5.5 Peering into the Future

We are at the beginning of the digital age. But what about a civilisation which is, say, 1,000 or 10,000 or 100,000 or 10,000,000 years into their digital age? What sort of technologies would they have? To quote an anecdote told by the Swedish astronomer Erik Zackrisson, asking whether a human 10,000 years ago (in the middle of the Stone Age) could have imagined what a smartphone charger is (as the astronomer took one from his pocket), the conclusion was that he certainly could not. So maybe we might not be able to imagine what future technologies would be like. However, there is a levelling off effect in discoveries. For instance, we now know what electricity and atomic power are and how to harness solar energy. There is no 'second electricity', which we know nothing about, and whose potential is hidden from us (unless we learn what dark energy is and manage to harness it to our benefit). Is it possible to imagine how far technologies might go, which are now in their infancy on our planet, or have not even been conceived of? Areas such as genetic engineering, including synthetic life (creating entire genomes via synthetic biology and getting them to replicate and grow into adult organisms) and artificial intelligence (way beyond the faltering steps it has been taking now – see Chapter 18) may well, in thousands, tens of thousands or millions of years, develop to levels which we cannot envision at the moment. Such developments would be part of cultural evolution which is orders of magnitude faster than biological evolution. At the moment, in the early part of the twenty-first century, we have glimpses of the avenues which future research may take, especially in genetics and digital technology. We are moving forward by leaps and bounds, but whether scientific developments yet to come will fundamentally alter the nature of human existence is an unanswered question.

The Future of Our Earth

The past five to six thousand years constitute history, in the sense of time for which there are written documents that go back to attestations of languages such as Sumerian, written in cuneiform, from the area of present-day Iraq during the third millennium BCE. Complex societies with internal organisation, which left architectural remains, go back about 12,000 years. This was the time of the Göbekli Tepe, a temple-like structure, possibly the first example of architecture, which was excavated in the south of Turkey (at the top of the Fertile Crescent), dating to some time before the agricultural revolution began (*c.* 10,000 years ago). To build this, the people will have had a social order with a clear structure, that is to say they will have had a complex society which has up to recently only been assumed for much later groups, such as the Neolithic people who built the first pyramids in Egypt in the third millennium BCE, about 5,000 years ago.

To put this in the perspective of geologic time, the 12,000 years we are talking about here is 0.000264 per cent of the age of the Earth (4.55 billion years). So, if human life is to continue for at least 1 per cent of the age of the Earth, that would give us another 45.5 million years. Given that modern astronomy and digital technology are not even 100 years old, we can well ask what the technology made by humans might be like in, say, 10 or 20 million years time. But will we really survive that long to find out? Considering the lifespans of animals which have gone before us, the answer should be a clear 'Yes'. Look at the dinosaurs who survived over 150 million years on Earth before their fate was sealed by the asteroid which hit the Earth about 66 million years ago. [4]

But many commentators on humanity's probable future do not think that we will survive for millions of years to come. They postulate a so-called Great Filter, which determines just how long we can expect to survive. The assumption that there is such a filter, as put forward by the American scholar Robin Hanson (1959–), rests on the fact that there do not appear to be any digitally capable civilisations in our immediate cosmic surroundings despite the presumed existence of thousands of exoplanets in the habitable zones of neighbouring solar systems. Hanson has looked at extinct civilisations, trying to discover what factors determine the sustainability of terrestrial civilisations. He has posited that climate change, triggered by energy-intensive civilisations, is a key factor. However, the elephant in the room for humanity on Earth is surely overpopulation. We are consuming non-renewable resources at an alarming rate and severely damaging our environment in the process.

But even if we were only to survive on Earth for, say, another 100,000 years, this would be a vast amount of time for us to refine and extend our technology, probably making space exploration more possible than it appears right now and hence improving our chances of survival in other cosmic environments.

However, all of that may well depend on how we behave as collective humanity, such as how we react to global challenges like climate change. And human concern for others, and our common future, is notoriously lacking. We are beings which show a readiness to act for the benefit of our offspring but not for that of others. For example, parents will generally do almost

4 Dinosaurs became the dominant land-based life form after the Triassic–Jurassic extinction event just over 200 million years ago and remained so until the Cretaceous–Palaeogene extinction 66 million years ago, the so-called K–Pg event.

anything for their children, especially when they are young, but not for those of others. Or at least we need to appeal to notions of morality and altruism to get humans to make sacrifices for the common good. There are sound evolutionary reasons for this: as long as each member of a species concerned itself with its own offspring, the species as a whole survived. For instance, a blackbird will vigorously defend its young if attacked by a magpie, but it will not defend the young of another blackbird if this is being attacked. It is true that with cognitively more developed animals altruism seems to kick in, especially when animals live in herds. For instance, elephants will defend the young of others from predatory attacks, for example by lions. And I suppose that most of us would step in to defend someone else's child, if attacked. But how much are we actively prepared to do for the common good, especially when it involves sacrifices on our part? The answer is precious little. Although climate change cannot be denied and is a global danger for our planet, few of us, who are the consumers of resources, are prepared to give up our accustomed comforts – cars, air flights, warm houses in winter – which all consume non-renewable resources and contribute to climate change.

Given that exobeings could only arise through similar processes of evolution as those which led to us, the question is then whether they would show different degrees of altruism than we do. We do not know: their societies could be similar in complexity to ours, with a wide range of collective behavioural patterns. And this could mean that there would probably be at least some groups concerned with the welfare of the exoplanet as a whole and prepared to make sacrifices for its good, but whether that would suffice for such an exoplanet to manage its natural resources would probably be as much an open question for them as it is for us on Earth.[5]

5 One can see this situation in smaller instances, such as on Easter Island/Rapa Nui in the south-east Pacific. Here, the Polynesian population cleared the forests on the island, leading to environmental degradation which in turn critically weakened the island's ability to sustain the local population. The situation was only compounded with the arrival of the first Europeans in 1722, who not just brought hitherto unknown diseases with them, such as smallpox and tuberculosis, but later abducted large numbers of the islanders, keeping them as slaves.

The Future of the Universe

The universe which we observe around us is at a particular stage of its development during which it has expanded greatly and is continuing to do so at an ever-increasing rate. The question naturally poses itself of how this will continue and what scenario is likely for the end of the universe. Earlier models assumed that, after the period of acceleration, the universe would begin to contract again, much like an object thrown up in the air which begins to fall when the upward velocity of the object declines and gravity pulls it back down. [6] However, recent measurements of the accelerating universe have shown that a likely scenario is one where the expansion continues for ever (with an open or flat universe), reaching a stage where there is no free thermodynamic energy, resulting in thermodynamic equilibrium, with the universe remaining in that state indefinitely. Gravity will not be able to pull material back together again. The universe will then be a cold and inconceivably vast region. This may be a dismal prospect but that will not be until many trillions of years from now, if and when it occurs.

6 Since Einstein's general theory of relativity, we know that gravity is the result of space–time curvature caused in our universe by objects with mass. The degree of curvature triggered by a given mass is then the value for gravity.

6

The Likelihood of Life

We know that life has arisen at least once in our universe. This simple fact testifies to an extraordinary feature of the building blocks of the universe – subatomic particles, atoms and molecules – the ability to aggregate to form immensely complex entities, which display a vast array of emergent structural properties. One of the pinnacles of this potential (on Earth) is the human brain, with the consciousness it engenders (see Chapter 17 for a detailed discussion).

When examining the building blocks of the early universe there is no indication whatsoever that these elements – initially hydrogen with some helium – would ever give rise to beings capable of reflecting on the nature of the universe and mulling over its origin and possible future. It would seem that all one needs are suitable conditions and enough time. We have no idea how or why this extraordinary potential slumbers in the building blocks of the universe. After all, it is conceivable that the universe could have developed galaxies from great dust clouds with suns and planetary systems but that, with this, the end of the line would have been reached. Such a universe would have planets in the form of gas giants, rocky planets or water worlds, yes, but nothing more; no biology arising on any of these planets. Then we would have a universe which, to all intents and purposes, would look like what we have now but there would be nobody in this universe to observe it and to reflect on why it is the way it is.

6.1 Basic Preconditions

There are obvious preconditions for life. Just imagine that nuclear fusion did not commence in the centre of proto-stars so they did not ignite.[1] There would be no source of light energy for the planets surrounding stars. Gravitational flexing for planets near a star might produce internal heat, as it probably does in Jupiter's moon Europa (see Section 8.12 below). But that would probably not be enough for a life-supporting energy regime to evolve into complex forms like ourselves or putative exobeings.

Another basic precondition concerns the precise value gravity has in our universe.[2] A slightly greater value would have prevented the necessary expansion of the early universe yielding the space–time in which the galaxies then arose. A slightly lesser value would have prevented the aggregation of dust and gas particles to proto-stars, which provided the framework for later planets.

We know that at the beginning of the universe, as we observe it, the only atoms which were created were hydrogen along with a little helium and tiny amounts of lithium. All the more complex atoms were created in stars. Those with atomic numbers up to iron (26) arose through fusion in the cores of stars and the still heavier elements were created in supernova explosions of dying massive stars. What we observe in our immediate cosmic surroundings are intricate molecules which show that processes of combination have taken place, with commonly occurring elements aggregating into structures of increasing complexity, ultimately leading to life on Earth.

1 This situation actually characterises what are called 'brown dwarfs,' proto-stars several times the mass of Jupiter but where the pressure in their cores is short of what is required for nuclear fusion, see Section 5.3 above.

2 Gravity is identified in science by the gravitational constant, written as capital *G*. Its value is absolutely tiny, about 6.67×10^{-11} m^3 kg^{-1} s^{-2}.

The assumption that similar processes would take place on other planets with similar conditions led to an extension of terrestrial biology to beyond Earth. This is the field of astrobiology, a burgeoning domain of scientific study concerned with the genesis, development and future of life in the universe, building on the insights of biology, chemistry, physics and astronomy. Among other things this relatively new field concentrates on molecular structures involving carbon, the basis of all terrestrial biology, examining proto-planetary disks, other stellar and planetary systems, locating liquid water (keeping to the mantra 'follow the water') and studying all the many processes which could be involved in the genesis of extra-solar life.

For life to get going on Earth, the following preconditions are worth considering here: liquid water and complex organic compounds. Water is an ideal biosolvent and it thus provides a favourable environment for a host of chemical reactions. Water allows complex molecules to remain in suspension rather than falling under gravity to the base of the form containing the water. The specific heat capacity of water is also considerable, which provides stability for the organisms living in it or consisting largely of it, such as we humans.

When the Earth formed about 4.55 billion years ago it was a seething cauldron of molten rock without any water. However, various celestial bodies, notably comets, bombarded Earth and brought water with them. We know that meteorites, small pieces of debris from comets or asteroids, can have water droplets trapped in salt crystals, and indeed amino acids, as for example in the remains of a meteorite which fell on Allende, Mexico on 8 February 1969.

Further proof of this has come from the European Space Agency's space probe, Rosetta, launched in 2004. Along with its lander Philae, the probe examined the comet 67P/Churyumov–Gerasimenko in detail in 2014. Complex organic compounds (those containing carbon usually along with hydrogen, oxygen and nitrogen), which are the building blocks of the amino acids that in turn are the organic precursors of all life-essential proteins, were found on the comet. This means that, while it cannot be shown that life forms arrived on Earth from space, water and organic compounds did, thus providing essential material for the later development of life.

In fact, organic compounds may be present in any given planetary disk out of which a solar system can arise. When the gas in a star-forming region cools, complex, possibly prebiotic molecules called polycyclic aromatic hydrocarbons (abbreviated as PAHs), consisting of carbon and hydrogen in multiple rings, are formed. The Swedish–American scientist Karin Öberg examined the light from proto-planetary disks spectroscopically and found signs of organic molecules in the disk of a nascent exosolar system, showing that the ingredients of life are common at an early stage in any solar system. The carbon-centred molecule she found is methyl cyanide, which has carbon, hydrogen and nitrogen, three of the basic chemical ingredients for life.

The precise compounds which may exist in a proto-planetary disk will vary. The reason why such compounds form is that the dust and gas of the disk are initially at an extremely low temperature, maybe just 20 °C above absolute zero (at which temperature atoms and molecules will easily stick together). Now bear in mind that there will be radiation from the interstellar medium striking off such a disk, such as ultraviolet radiation. This will split molecules and allow them to recombine to form new compounds, utilising combinations of the basic elements carbon, hydrogen, oxygen and nitrogen. The examination of disks at this key stage in the chemistry of early solar systems has been greatly facilitated by the ALMA (Atacama Large Millimeter/submillimeter Array) telescope in the Atacama desert in Chile.

Further organic compounds have been found in dust clouds and stellar systems in the process of formation. One of these is a sugar molecule, called glycolaldehyde, which has been found in a proto-stellar binary star system 400 light years from Earth. It has been postulated that glycolaldehyde plays a role in abiogenesis, given that it is required to form RNA (ribonucleic acid), the evolutionary precursor to DNA.

How Much Carbon and Water Is There?

Water, a simple molecule of two hydrogen atoms and an oxygen atom, is known to be abundant in the universe. Hydrogen was there from the beginning and oxygen (along with carbon) was formed during nuclear fusion already in the first generation of stars, one to two billion years after the Big Bang (indeed maybe much earlier). Several of the planets, discovered beyond our Solar System, are probably water worlds, indeed with far more water than we have on Earth. If they also have an atmosphere, providing weather, and sufficient heat from their parent star, they could well have a water cycle similar to that on Earth. For life, it is probably better for a planet to be a mixture of land and ocean rather than a complete water world as the latter could only harbour marine life. Complete water worlds are 'born' with their water, in the form of ice, because they would likely have formed beyond the frost line in their solar system. The rocky water worlds are like Earth, for example they would have probably started as rocky planets which accrued water through bombardment by icy comets from the outer regions of their solar systems (though there is not entire agreement on this among scientists). They could be water worlds closer to stars if they migrated from outer to inner positions. This type of planetary migration within a solar system is known to happen and is assumed – in the migration hypothesis – to be the reason for 'hot Jupiters', gas planets close to their parent star, which were the first type of exoplanet to be discovered, given their size and short circumstellar orbit. Planetary migration is also held to be responsible for ice giants like Uranus and Neptune, which are much further away from the Sun than where they would be expected to form.

The next question is whether water worlds would also have complex carbon-based structures, which would be an important step on the way to life. Consider first of all the polycyclic aromatic hydrocarbons. These are compounds which consist of several aromatic rings (stable rings of chemical bonds), the latter containing delocalised electrons. These hydrocarbons are now taken to be very common across the universe and to have formed in its early period. Such compounds are furthermore taken to be prebiological and, by extension, prebiotic, and could have provided an early input in the formation of amino acids, which are precursors of proteins, themselves essential to life as we know it. Our bodies contain thousands and thousands of proteins, which perform the most diverse functions, but they are all built from a set of 20 amino acids. These molecular structures would appear to be common across the universe and have been found in meteorites which landed on Earth.

Just how prebiotic structures like amino acids arose concerned scientists during the first half of the twentieth century. Then two scientists conducted a famous experiment in 1952, which now bears their name, the Miller–Urey experiment. They set up an electric current through a solution consisting of water, ammonia, methane and hydrogen and after several days it had become cloudy from the amino acids which had formed. The insight of the experiment is that amino acids could have arisen through the interaction of lightning and water containing common carbon compounds. For its time this experiment was important for the results it yielded, but for life to arise cellular self-replicating DNA would be required, or at least RNA molecules from which DNA would later develop (in the 'RNA world' hypothesis, which is, however, not supported by all scientists).

Carbon can combine with other elements in thousands of different ways to yield very different and diverse molecules (it is tetravalent, meaning that its four outer electrons can form covalent chemical bonds with other elements). It can occur in different shapes in nature and, depending on the manner in which the carbon atoms arrange themselves spatially (in three dimensions), one can have very different manifestations of the element. Coal is a black soot but diamonds are very hard crystals due to the right angles at which the atoms are arranged; this results in a regular lattice consisting of cube-like arrangements of carbon atoms, yielding very tight bonding.

Oxygen is a sign of a carbon-based biology in which plants give off this element. At the beginning of Earth there was no oxygen. This appeared in the atmosphere with oxygen-releasing bacteria - cyanobacteria - which took in carbon dioxide and released oxygen. These bacteria along with green algae freed oxygen into the air and nowadays the process of photosynthesis, which takes place in the chloroplasts of terrestrial green plants, provides us with a constant source of the element. Photosynthesis splits water into hydrogen and oxygen, binding the former into a glucose along with the atoms from carbon dioxide. The net result is that oxygen is released (in what is called a Calvin cycle). This can be represented in simple form as follows:

$$6\,CO_2 + 6\,H_2O + \text{photons} \quad > \quad C_6H_{12}O_6 + 6\,O_2$$

$$\text{carbon dioxide} + \text{water} + \text{sunlight} \quad > \quad \text{glucose} + \text{oxygen}$$

Oxygen is highly reactive and is quickly bound into compounds by oxidation (accepting their electrons to form chemical bonds)[3]. It has been estimated that, if all plant life on Earth ceased producing oxygen, this would be used up within a million years through oxidation.

In the atmosphere of a planet, methane (CH_4) points to life forms. Methane is the main component of natural gas and is released in the anaerobic fermentation of organic matter and by animal digestion (cows release a lot of it, for example).

ORGANIC COMPOUNDS IN SPACE

Complex organic compounds, the building blocks of life, are present throughout our galaxy (and by extension in others), even in the earliest phases of star and later planet formation. This situation greatly increases the likelihood of carbon-based biologies on exoplanets.

The Temperature Question

Temperatures vary enormously throughout the universe. The space between galaxies is as cold as it gets, well below –200 °C (getting close to absolute zero), while stars can be many millions of degrees in their cores. Temperature stability on an exoplanet would be an essential precondition for life. Consider the scale of possible temperatures and the range which we can tolerate on Earth:

(i) –273.15 °C (absolute zero; 0 K on the Kelvin scale). This is the temperature at which solid matter is in its ground state, which means it is at its lowest level of thermal energy. While this temperature cannot be reached by thermodynamic means, experiments have come very close to absolute zero. In the vicinity of 0 K matter exhibits effects such as superfluidity and superconductivity.

3 There are other types of bonds, notably ionic and metallic bonds. However, these would be unsuitable for an entire biology: ionic bonds are too weak and metallic bonds are too rigid to aggregate into forms of life.

(ii) 0 °C is the freezing point of water, the main biosolvent on Earth, the starting point for the Celsius scale.

(iii) 37 °C is the temperature of central organs of the human body, controlled by the hypothalamus located in the base of the brain. If the body's central temperature drops below 35 °C, one begins to shiver, have an increased heart rate and contraction of blood vessels (hypothermia). Further cooling leads to disorientation with confusion, and metabolism begins to slow down and eventually stops, causing death. A few degrees above 37 °C and the body is suffering from hyperthermia, for example with heat stroke or high fever. A further elevation in temperature and the body's functions would no longer work: proteins would denature and coagulate, as an egg does when boiled.

(iv) 100 °C is the boiling point of water on the Celsius temperature scale, a common reference point for terrestrially occurring temperatures.

(v) Millions of degrees Celsius are the typical temperatures of the interior of stars[4]. This varies greatly depending on many factors; the main point is that the temperature of a star is vastly too great to allow the formation of the molecules which might be the building blocks for biology. But stars are the furnaces which supply heat energy to planets and in which nuclear fusion takes place, creating the atoms essential for life. Stars which explode as supernovae create heavy elements in the process.

6.2 Favouring Factors

The universe is capable of giving rise to life, a fact that we are living testimony to. There are conditions on Earth which led to this, but on a more fundamental level there are factors concerning the Earth as a planet which led to it being positioned favourably for life in the first place. Such factors might obtain on other planets as well, so it is worthwhile looking at these before scrutinising the later conditions for life on Earth.

4 The surface temperature of a star can be surprisingly low, for our Sun this is less than 6,000 °C, indeed much less for the periodically recurring dark sunspots. However, the core is about 15 million degrees Celsius and the corona surrounding the Sun can reach temperatures of several million degrees.

Main factors providing the right conditions for life on Earth

1. Earth has an orbit in the habitable zone, the 'Goldilocks' zone, of our Solar System, that is, not too close to the Sun for the temperature to be too high and not too far for it to be too low. Furthermore, the orbit of the Earth is not too elliptical as otherwise the temperature fluctuations would be too great and hence not conducive to life.

   Galaxies have habitable zones as well. Close to the densely packed central region, with its supermassive black hole, there would be intense radiation which would prohibit life as we know it on Earth. Furthermore, the gravitational pull of massive nearby stars would be dangerous as it could cause meteorites or comets to be flung widely around, possibly hitting us, sooner or later. Close to the edge of the galaxy, stars are spread out over greater spaces, which means that the heavier elements (beyond primordial hydrogen and helium) would not necessarily be found in huge dust clouds providing the atomic input to later solar systems like ours.
2. There is an abundant presence of the most important biosolvent, water in liquid form, that latter fact is a consequence of (1).
3. A mixture of land and water can be found on the surface of Earth. This is due to it being a rocky planet on which much water was deposited by comets in its early period.
4. The Moon is comparable to the Earth in size (over a quarter of its size). The gravitational field of the Moon prevents the Earth from spinning too fast and too erratically.
5. A large iron core at the centre provides a stable magnetic field around the Earth.[5] This field deflects high-energy particles emitted by the Sun, so-called 'solar wind', around the Earth and so holding such damaging

5 The magnetic lines actually converge on the poles of the Earth and are responsible for the displays of colour in the night sky known as the Aurora Borealis and the Aurora Australis. The colours come from light emitted by (some of) the particles when they strike the higher layers of the Earth's atmosphere.

particles to the DNA of life forms at bay.[6] The ozone layer in the upper atmosphere is important in this respect as it absorbs damaging ultraviolet rays from the Sun when the ozone (O_3) molecules break down to form O_2 molecules (a more stable form of oxygen).

6. The Earth has an axial tilt (or obliquity) somewhat in excess of 23°. This tilt is the difference between the rotational axis and the orbital axis of the planet. It is what gives us the seasons: the tilt means that part of the Earth is titled towards the Sun for one half of the year (summer) and away from the Sun (winter) for the other half, with the transitions between these positions yielding spring and autumn.
7. The atmosphere of the Earth contains essential elements for life: nitrogen and hydrogen with oxygen being produced by plants and providing a source of energy for aerobic life forms. There is also carbon dioxide in the atmosphere (now just above 400 parts per million); this traps heat and means that the temperature differences between areas in sunlight and those in shade are not great. Without an atmosphere we would sizzle or freeze depending on whether we were facing the Sun or not. On other planets this may be the case, with a difference of over a hundred degrees plus in sunlight (close to a star) and under a hundred degrees minus in the shade.
8. The Earth is a slowly revolving planet, which results in an alternation of light and darkness over some tens of hours. This alternation provides an environment in which life forms can have periods of activity and periods of rest (though some animals do the reverse, they are active at night and rest during the day). This is conducive to the rise of complex forms of life.

---

6 The Earth's magnetic field is due to a geodynamo found in the layers outside the Earth's solid core. These layers consist of molten iron and are continually in motion, with the magnetic material in them acting like the coils in a dynamo. The resulting magnetic field can vary in strength, for instance, there is a South Atlantic anomaly in which the magnetic field is quite weak. The position of the poles is also shifting and a switch of north and south magnetic pole has occurred several times in the Earth's history and is likely to occur again in the future.

6.3 Key Developments and Events

Ideally, to discuss the issue of exoplanets and exolanguages, what we would like are, say, at least a dozen planets, each with a whole range of languages (see the opening discussion in the book). This would allow us to compare and contrast evolutionary paths, present scenarios, special developments, parameters which are similar and those which are different and so reach statements about the likelihood of life beyond our Solar System and the nature of exolanguages. We may well be in this fortunate situation within the present century if the discovery and investigation of exoplanets continues at the pace it is at now. But at the moment, mid-2022, we do not have the necessary information to do this kind of comparative work. So we are left with one Earth, the life which arose on it, the language faculty we have and the languages we humans speak.

We do not know the probability of life arising on an Earth-like planet. But we can examine the developments and events which favoured the rise of intelligent life on Earth. Consider these key developments/events in the history of Earth.[7]

Key developments and events in the history of life

1. The rise of microbial life, probably around four billion years ago, about half a billion years after the formation of the earth. Microbes[8] are single-cell bacteria, usually so-called prokaryotes, with DNA; they duplicate via cell replication. This is the very first step on the path to ourselves. For over a billion and a half years, the only type of life on Earth consisted of simple, prokaryotic cells. Then came the advance from simple to complex cells, the development of eukaryotic cells from prokaryotic cells (mostly bacteria

7 These developments are often regarded, in their totality, as unlikely and this view has been labelled the Hard Steps Model, referring to the difficulties for life to progress relatively unscathed through these stages.

8 Quite literally, microbes are those organisms which are too small to see with the naked eye and which require an optical instrument like a microscope.

and archaea), triggered by a process called endosymbiosis. This happened with the incorporation of one type of prokaryotic cell into another. The one which was incorporated was co-opted as a worker, its DNA modified and it ended up producing bio-energy in the form of ATP (adenosine triphosphate) for the host cell. The incorporated cells are the precursors of the mitochondria which we still have in all our cells.[9] The complex eukaryotic cells developed other organelles (internal structures within a cell involved in self-maintenance) and a membrane for the nucleus containing the DNA.

2. The move out of the sea onto land. Between 400 and 350 million years ago (mya) some types of fish developed a means of extracting oxygen from air and not just water, probably developing an amphibian lifestyle, similar to frogs today, before adapting completely to life on land.[10] These fish also developed their fins into limbs, which were in time to become those of tetrapods (four-legged animals).
3. After the demise of the dinosaurs, the mammals (animals which give birth to live young and breastfeed them) could develop relatively unhindered, with the primates appearing as a group within this class at around this time. Among the most cognitively developed of the primates are the chimpanzees, bonobos and monkeys, as well as orang utans, gibbons and gorillas, all of whom derived from ancestral species at this time.
4. The primates may be traceable to an animal species called *Purgatorius*,[11] which lived roughly between 66 and 63 mya. It had nails, which were

9 Plants also have chloroplasts, which are used to convert sunlight to chemical energy stored in carbohydrates (sugars) synthesised from water and carbon dioxide, thus providing the plant with energy and incidentally releasing oxygen into the atmosphere. The chloroplasts are probably derived from cyanobacteria through endosymbiosis, similar to that which occurred with mitochondria that were integrated into simple cells, creating more complex ones, the eukaryotes.

10 The reverse has also happened: the cetaceans, which include whales, dolphins and porpoises, migrated from land back to the sea probably between 40 and 50 mya (fossil evidence from the area of Pakistan points to this). Cetaceans are mammals, giving birth to live young, providing milk for them and breathing air directly through an airhole; they are also endothermic ('warm-blooded' in everyday parlance).

11 This animal is named after Purgatory Hill in Montana, USA, where the first remains were found.

better for grasping than claws, eyes in sockets and bone plus joints suitable for life in trees. Primates are also known for their increased brain capacity compared to other mammals. There are two suborders, one labelled 'wet-nosed' and one 'dry-nosed', the latter including monkeys and apes. The decline in the olfactory power of the 'dry-nosed' primates is usually linked to improved visual capability.

5. Around 66 mya, after the dinosaurs had been reigning for over 150 million years, Earth was hit by an asteroid (probably about 20 kilometres in diameter), which struck just over the coast of the present-day Yucatán peninsula in Mexico, an event which was confirmed by the existence of the Chicxulub crater there. This heralded a period of life extinction due to fires and tsunamis as well as the prolonged disturbance of the climate given the amount of material spewed high up into the atmosphere. The dinosaurs which did not die as a result of the actual meteor strike did so due to the lack of vegetation in the hundreds of years which followed. Of these huge animals, only the avian dinosaurs (precursors of modern birds) and some reptiles survived, thus opening up the way for the mammals in general and the primates in particular to develop freely as species, especially in Africa, which had arisen as a continent, geographically separate from the Americas, Europe and Asia after the break-up of the original mega-continent Pangaea, around 250 mya.
6. At around 6–7 mya there was a divergence between the line which led to chimpanzees and that which led eventually, not directly, to the genus *Homo* of which we are a part and now the sole survivors. The exact mechanism of this split and its further development is not fully known, but some key features can be listed here:
   (a) the rise of dedicated bipedalism (continuous upright movement on two feet)
   (b) the selective pressures of life on the food-stressed savannah, which promoted intelligent behaviour and social cooperation in an environment with many fluctuations in climate
   (c) the management of fire and the development of cooking.

Life forms on Earth have also survived mass extinctions. There have been at least six such extinctions in history (see Section 6.4 for details) which greatly diminished the amount and variety of animal life on Earth. However, none of these led to complete extinction as, each time, life, however much reduced, recovered and spread to regain more or less its former distribution.

In summary, one can mention a number of issues which have a bearing on the topic of this section and which are central to the question of whether complex life could arise on an exoplanet.

Issues concerning key developments/events

1. Was the appearance of eukaryotic cells a chance development?[12] Maybe it would not take over a billion years on other planets, or maybe it might never happen at all. That is, there could be Earth-like planets in other solar systems which consist of seas full of microbial life, but nothing else. Indeed such microbial life might already exist within our Solar System in the underwater worlds of Europa and Enceladus (moons of Jupiter and Saturn respectively, see Section 8.12 below).
2. Does marine life always move to terrestrial life? This move is potentially possible on rocky planets with water in other solar systems, but does it always occur?
3. If the dinosaurs had not disappeared there might not have been much chance for mammals and then primates to have flourished and thus provide evolutionary precursors to the genus *Homo*.
4. If there had been no divergence between the ancestors of humans and the ancestors of chimpanzees, today's Earth might have some higher primates, including apes and monkeys, but no humans. This could also be true if early human-like species had become extinct, as indeed some of them did.

---

12 The British biologist Nick Lane sees this development as central to the evolution of complex life, see his book *The Vital Question*.

CELLS: A UNIVERSAL OF ALL BIOLOGIES?

All animals and plants are made of cells consisting of membranes enclosing fluid in which various functional parts are to be found: (i) a nucleus which contains DNA, (ii) mitochondria, a source of energy, (iii) ribosomes, little machines to translate RNA strands (transcribed from DNA) into proteins along with other subunits (organelles) inside the cells. The walls of cells (somewhat different in animals and plants) allow nutrients in and send waste out. The purpose of various cells in an organism will vary but they all show the simple ability to divide, triggering growth. It is hard to imagine a different, let alone a better type of structural organisation for exolife forms: vesicles, shielded from the outside, in which chemical reactions and protein synthesis can take place.

6.4 Mass Extinctions in Earth's History

The rise of species on our planet is not a straight ascending line from some distant starting point to now. Rather, the process is characterised by at least six major periods when life almost became extinct. The first has to do with the rise of oxygen, but the others, often dubbed 'the Big Five,' have other possible causes.

The charting of mass extinction events in the world's history goes back to seminal work by the American palaeontologist Jack Sepkoski (1948–1999) who, together with David Raup (1933–2015), studied the fossil records for Earth and discovered that severe reductions in biodiversity occurred at identifiable periods in the planet's history.[13]

These extinction events represent bottlenecks through which life forms had to move to survive. The number assumed for Earth's history varies among geologists and palaeontologists, probably because determining whether an extinction event occurred is done by examining detailed fossil records, and these can be uneven and contain gaps, given that fossils are found in rocks and some regions of the world have fewer fossils than

13 See Raup and Sepkoski (1982) and Raup and Sepkoski (1984).

others. In addition, the event in question may simply have been too old, as with the Great Oxygenation Event, which occurred when there was only microbial life on Earth and for this there is no clear fossil record.

The effects of mass extinction events can vary. Some wiped out an entire branch in earlier life on Earth, such as the late Cretaceous extinction event, with the loss of the non-avian dinosaurs whereas others, such as the Late Ordovician extinction events, led to a thinning out of the entire biosphere.

The likely causes of mass extinctions - asteroid strikes, exposure to harmful ultraviolet radiation from the Sun due to ozone layer[14] depletion and huge volcanic eruptions which triggered global winters for decades, if not for hundreds of years - could just as easily occur on exoplanets so that life there could well have been subject to similar catastrophes. And, of course, what we do not know is on how many planets life got started and afterwards did not survive such mass extinction events.

MASS EXTINCTIONS ON EXOPLANETS

Life is tenacious and from Earth we can see that it will cling on to whatever niche it can occupy no matter how small and threatened. On more than one occasion life has recovered on Earth from major setbacks. But that is not a foregone conclusion. And there may well be exoplanets on which life was indeed extinguished due to one or more catastrophic events like those being discussed here.

6.5 Strikes from Beyond

As discussed in the previous section, 66 million years ago a large asteroid struck the Earth with the force of about 10 billion Hiroshima atomic bombs and approximately 75 per cent of all life disappeared from Earth. This was

14 The ozone layer, at about 30 kilometres above the Earth's surface, protects us by absorbing ultraviolet radiation.

Mass extinction events in Earth's history

1. *Great Oxygenation Event*: Nearly 2.5 billion years ago the levels of molecular oxygen greatly increased in the Earth's atmosphere due to the oxygen released during photosynthesis by cyanobacteria (single-celled organisms, also known as blue-green algae, although this classification is scientifically incorrect). This led to the demise of most anaerobic life forms, those for which free oxygen molecules were toxic.
2. *Late Ordovician extinction events*: A series of events between about 450 and 440 mya at the transition to the Silurian period. It led to the loss of between 60 and 70 per cent of all species.
3. *Late Devonian extinction event*: Occurring approximately 375–360 mya, this extinction lasted about 20 million years and may well have led to the loss of at least 70 per cent of all species.
4. *Late Permian extinction event*: Occurring about 252 mya, at the transition to the Triassic period this resulted in the loss of between 90 and 95 per cent of all species, hence the moniker for the event, the 'Great Dying'. The volcanic eruptions in the Siberian Traps, which released vast amounts of carbon dioxide and which could have lasted for up to two million years, may have played a key role here.
5. *Late Triassic extinction event*: Taking place approximately 200 mya, at the transition to the Jurassic period, this resulted in the loss of about 70–75 per cent of all species.
6. *Late Cretaceous extinction event*: Happening about 66 mya, this event led to the demise of about 75 per cent of all species, including all the non-avian dinosaurs, either directly by incineration at the location of the asteroid strike, or by the huge fires and tsunamis this would have triggered, or more generally in the following decades when plant life declined almost completely due to ash and debris blocking out sunlight entirely. This event was a fortuitous event for us because mammals became a dominant animal form and later evolved into primates, great apes and finally humans.

a serendipitous occurrence for us because it got the dinosaurs out of the way and gave us, the mammals, and later the primates, room to expand.

There have been a number of such asteroid hits. At the beginning of the twentieth century an event occurred which, fortunately for the inhabitants of Earth, happened in a largely uninhabited part of our planet. The Tunguska Event in Siberia refers to a meteor which struck on 30 July 1908, completely destroying an area over 2,000 square kilometres, largely of forest. The force of the blast would have been that of at least 1,000 Hiroshima atomic bombs. The meteor probably exploded a few kilometres off the ground and the force was due to the huge spike in air pressure from the explosion.

On 15 February 2013, a meteor, about 20 metres in diameter, suddenly appeared in the Earth's atmosphere above the city of Chelyabinsk in southern Russia. It exploded at a height of about 23 kilometres off the ground with a force equivalent to roughly 30 Hiroshima atomic bombs, shattering windows and damaging buildings in a radius of many kilometres. Our planet can handle impacts of this size but we have to be prepared for even larger near-Earth objects ('NEOs') which might collide with Earth. To this end NASA has been monitoring our immediate space neighbourhood to recognise a potential threat and perhaps undertake measures against it.

7

Possible Conditions on an Exoplanet

The question here is whether the biochemical processes observable on Earth would be replicated on another planet. Take photosynthesis as an example. This is the means by which plants utilise sunlight in the production of adenosine triphosphate (ATP) and glucose as sources of energy. During this process oxygen is given off and carbon dioxide is absorbed, hence the value of photosynthesis for environments on our planet. The actual process is highly complex and involves electrons going through intricate chemical reactions leading at the end to glucose formation. There is also a kind of reverse process, which involves the release of energy through the oxidation of a chemical derived from carbohydrates, fats and proteins. This is known as the citric acid cycle, an essential metabolic pathway used by aerobic organisms.

We do not know whether the biochemical processes for gaining and releasing energy in plants or organisms on an exoplanet would be exactly as on Earth. But we can say that an exoplanet with exobeings would have an entire range of life and plant forms and that they would have a means of extracting energy from their surroundings and then storing it – if they did not, they would not exist. An obvious source of such energy would be light from the parent star (at the smallest level we see that photons of light are the quanta of energy in the electromagnetic field). The system of electron transport, which we find in plants and organisms on Earth, would apply in principle on an exoplanet, even though the actual steps involved would likely not be exactly the same. Furthermore, electron transport can cause a proton gradient (a positive charge) across a membrane and trigger the formation of ATP. Again, this could in principle be a means of producing an energy 'currency' for organisms on an exoplanet, even though the details of the mechanism would most likely vary compared to that on Earth.

7.1 The Fine-Tuning Problem

The laws and constants of physics and the fundamental forces in our universe have very precise forms and values. This means that, if they were only very slightly different, life would not have been possible. For instance, the precise value of gravity has enabled our universe to arise by permitting the aggregation of dust and gas particles to proto-stars around which planets later came to orbit, including the Earth around the Sun. If the value of the electron had been ever so slightly larger or smaller, chemistry, as we know it, would not have been possible and life, which is based on organic chemistry, could not have started. The universe was not designed for us to evolve, we have no privileged position in the universe; however, the laws and constants of physics allowed advanced life to evolve.

For cosmologists – scientists concerned with how the universe arose – the fine-tuning problem is an issue. Why do the fundamental forces of nature have their precise values and why do the laws of physics behave in just the way they do and not differently? One solution to this question is to maintain that there are a multitude of universes, potentially an infinite number. We happen to live in the one which has the right values and the right elements and which are conducive to complex life arising (this is one version of the 'anthropic principle').

Whether our universe is just one part of a multiverse is contested among scientists. One argument presented to make a multiverse more palatable to sceptics (apart from the fine-tuning problem) is to point out that at key stages in the history of science we have discovered that the universe is much larger than we originally thought. First of all we assumed the Earth was everything, then the Solar System, then the Milky Way, then the hundreds of billions of galaxies, so why not accept that there are billions, if not an infinite number of universes? Supporters of the notion of a multiverse maintain that it is in a process of eternal inflation, with each universe arising through a bubble into which it expands. Some bubbles may not last too long, they might collapse in on themselves because gravity is too strong while others, like ours, can go on for trillions of years, possibly never ending. In this model, each universe would have its own settings for fundamental forces and for laws and constants of physics so that those which obtain in our universe are just one of a multitude of possible settings. These happened to hold here with favourable conditions for the development of galaxies, stars and planets and, at least in the case of Earth, of intelligent life. This, the supporters claim, solves the fine-tuning problem. In an infinity of universes, everything occurs, including our universe with its fine-tuned values.[1]

There are major consequences stemming from the assumption that we are part of a multiverse. One concerns infinity: if there are an unending number of universes, there is no limit to the size of the multiverse. If our universe arose during the process of eternal inflation, which is forever spawning new, variously configured universes, we cannot assume that there was a beginning to the multiverse. The figure of 13.8 billion years, given as the age of our universe, is just when our bubble arose due to eternal inflation. The multiverse has no limit in size, had no beginning and will have no end.

There are different views of the development of the universe and various terms are associated with these. The 'teleological' view (from the Greek word *teleos*, 'goal') sees the development as oriented towards a goal, the rise of humankind, this view being associated with many religious views of cosmology.

This does not sit well with many sceptics or non-theistic thinkers. For them, a better way of thinking of the universe might be as 'ekinological' (from

1 For more information on this matter, see Lewis and Barnes (2016) as well as Barnes and Lewis (2020).

the Greek word *ekkinisi*, 'start, beginning of something'). In this view it is understood in terms of present states being dependent on previous states, or future states being dependent on present ones. It's all about those initial conditions: the Big Bang becomes the important thing to understand in determining our present, and thus our future.

Unanswered questions concerning our universe

1. Does our universe extend infinitely in all directions? If so, is it isotropic, similar in shape and form, in all directions? Martin Rees has pointed out that the structural variation across the observable universe is so slight, less than one part in 100,000, that it is most unlikely that there is no more to it, that is, he believes, like the majority of astronomers and astrophysicists today, that the observable universe, the part which we can observe with our most powerful telescopes, is only a tiny fraction of the actual universe.[2] This would mean that there are areas of our universe, far beyond our ken, which are structurally and organisationally different from our section of the universe. If that is the case, where are the boundaries between different parts of the universe? Do they have the same dimensions of space and time which we have? Might these other parts be among us, just in dimensions which we have no access to?
2. Is our universe just one bubble, which arose through eternal inflation, out of a higher-order multiverse that spawned our Big Bang, along with a multitude of others? If so, are the bubble universes in contact with each other or do they occupy disjointed regions of the multiverse? Is the multiverse, which spawns all the individual universes, in a dimension to which we have no access?
3. If there is a multiverse, how many single universes are there? How could we find out? What exact processes might lead to the spawning of new universes? What fundamental laws govern the multiverse as a whole, are there such laws? How should we envisage this? Is the multiverse logically

2 Figures vary here. Max Tegmark claims that the observable universe is 1×10^{-23} of the size of the universe resulting from the inflationary epoch in the first second after the Big Bang.

separate from the universes it spawns? If each of the 'child universes' is different in its fundamental composition, how many of these have complexity on a sufficiently high level for life to arise, at least in theory?

An intriguing part of our universe is that each time we think we have found a limit and can recognise the universe in its totality, we discover that this is not the case. It looks as if a total grasp of the universe is always receding in front of us: we are almost there but we never quite make it. It's like climbing a mountain: each time you think you are near the summit you realise that there is still another piece to climb, and another and another.

7.2 Small-Scale and Large-Scale Structures

When looking at the various celestial bodies – stars, planets, moons – it is striking that they are all spherical in shape like the balls we know from our own surroundings on Earth. It would seem that to a certain extent small structures are like large-scale structures. Think of galaxies for a moment. One of the most common types is the spiral galaxy – our Milky Way galaxy is a good example. It has a dense core of stars at the centre, with a central supermassive black hole, and spiral arms swirling outward from this. It looks in fact like an eddy in a pool or the surface shape of water when it is being let out of a bath. So many of the shapes on a small scale look uncannily like those on a large scale.

One could ask then why there are no cube-shaped planets or moons shaped like pyramids, consisting of triangular sides with a square base. The answer is simply that if one has an amount of loose matter which is spinning around in space without an outside force pulling on it in some way, you will end up with a body basically shaped like a sphere. To have a cubed-shaped planet you would need some force pulling on the matter in such a way that right-angular corners arise. Furthermore, these forces would have to be constantly maintained to stop a cube-shaped planet from turning into a sphere with time. This means that – *all other things being equal* – loose spinning matter will end up in the shape of a sphere. This is not entirely true because the centrifugal force of rotation causes the matter to push outward somewhat at the centre. This has led to the Earth being a bit elliptical – a feature known as the equatorial bulge, with a value just under 43 kilometres.

These considerations show that the laws of motion and physics are the same on the terrestrial and the cosmic scale. This can be confirmed for near and far objects: far-away galaxies look like our own or those near us (though there are different types); if they do not, we can provide a principled explanation for this; for example, there are irregular galaxies whose shapes have resulted from the merging of separate galaxies or from the gravitational pull of a nearby large galaxy.

If large physical objects are similar across the observable universe, it will hold that small ones are too. Stars in other galaxies appear to be similarly structured to those in our galaxy so we can assume that the planets around those stars will resemble those in our galaxy and our Solar System. This means there will be big gas planets like Jupiter and Saturn, ice giants like Uranus and Neptune, as well as smaller rocky planets like Mercury, Venus, Earth and Mars; some of these rocky planets will also have liquid water. These planets will be spread outwards from their parent stars like ours and there will be a region, the habitable zone, somewhere at a distance from a star, which will provide favourable initial conditions for life.

7.3 The Underlying Basis of Structure

A key assumption of scientists is that complex structures evolve from simpler structures, which have the potential to develop complexity. There is no way to get from simplicity to complexity without taking intervening steps. Take terrestrial biology: this is based on carbon atoms which are tetravalent, with four outer electrons that can form a multitude of combinations with other atoms to yield complex three-dimensionally arranged molecules, typically the proteins of living organisms, including ourselves.

So when we come to human language it will have been the case that the complex languages, which now exist across the world, arose from simpler systems, originally from one such system (assuming that human life arose at only one location on Earth, in East Africa and spread from there).[3] This does not mean that each step in a long development represented an increase in complexity: there are cases of reversal when a system becomes simpler at some point due to a combination of forces, and there are instances where we have 'leaps and bounds,' relatively rapid developments triggered by a sudden push in evolution. But overall, complex systems arise from simple ones, of that we can be sure.

7.4 Emergent Properties

Consider this question: Why does the world which manifests itself around us on the scale we experience look so different from the world of particle physics which underlies it? The answer is: Because of emergence, a type of self-organisation of elements from the bottom up. The collective behaviour of particles, atoms and molecules is implicit, but not obvious, on the atomic level. This brings us to a consideration of just what properties appear on larger scales and how they relate to lower ones. There are basically two types of emergence.

Types of emergence

Weak emergence

All the properties of higher-level scales are inherent in the lower-level scales, though not visible or discernible. This means that the potential for higher-level properties is present at lower levels on which they do not receive any expression, however.

Strong emergence

There are properties which only appear on higher-level scales and which do not appear to derive directly from properties on lower levels.

---

3 Whether central aspects of language structure arose from a single individual is a contested matter in contemporary linguistics; see Section 32.7 below.

The question of emergence is relevant to almost every level which has arisen from a lower one. It applies to the early universe. How did the structure of the present universe arise from very uniform initial conditions? Later on, we had issues such as heavy elements forming in the centres of stars or during supernovae. Did their properties inherently result from the combinations of particles (protons, neutrons, electrons) of which they are constituted? If so, one could make the case for weak emergence; if not, then one is dealing with strong emergence.

There is a criticism of strong emergence: if there are properties which an emergent system has and which are not derived from the fundamental constituents of the system, an external source to explain these higher-level properties needs to be posited.[4] Strong emergence thus conflicts with deterministic views of the universe, but is nonetheless taken to hold for such complex systems as societies and economies, where the dynamics of such systems are not apparently derivable from their constituents.

Furthermore, emergent complexity does not necessarily arise from increasingly complex behaviour of individuals but from the overall increase in the size and organisation of a system. For instance, the complexity of a beehive or termite mound does not arise from bees or termites developing ever more complex behavioural patterns but rather from the precise and specific behaviour of each animal, which, taken together on a larger scale, yields complexity.

Consciousness is another phenomenon which is often seen, for instance by the computer scientist Ray Kurzweil or the physicist Jim Al-Khalili, as emergent: the collective concerted action of billions of neurons and their trillions of connections is taken to trigger the subjective, first-person experience of consciousness, which is not in any way visible or detectable in the physical substrate of the brain (see Chapter 18).

This is a problem for scientists: if one allows that strongly emergent properties can arise, we cease to know the limits of the universe. There may be,

4 Such an external source is often called a *deus ex machina*, lit. 'a god out of a machine', a device used in previous types of drama to solve an impasse which has arisen in the action of the play.

in other parts of the universe, or in the future of our own world, properties due to developments on higher scales of which we now, in our present, have no inkling.

Language can be viewed as an emergent system, which arose from simple beginnings and gradually developed complexity making it increasingly more flexible and powerful. Just how this evolutionary process proceeded is a topic of debate in research concerned with the origins of language and is discussed in detail in Chapter 32.

Limits to Emergent Complexity?

One should not confuse the results of evolution with human artefacts. What we can produce is limited but what biology has resulted in, over billions of years of evolution, is far more sophisticated. There is something to be learned from this consideration, namely that the potential for atoms to form exceedingly complex interacting molecular aggregates has no apparent upper limit. Or, if there is an upper limit, it is determined by environmental or structural factors. This is important when considering the rise of large brains with hominids: there does not seem to have been a predetermined limit to the degree of encephalisation – increase in brain size – which gradually took place through the stages of hominin evolution (see Chapter 13). The limit to brain size was probably determined by factors such as the amount of energy consumed by the brain, already just under a quarter of our intake, the width of the birth canal, putting limits on the size of the skull, and the weight of the head (skull with brain) on our shoulders which, if larger, would cause serious issues of balance and stability of the back for humans, given their upright gait.

7.5 Unintended Side Effects

In the evolution of our universe there have been several incidental side effects. In science these are labelled epiphenomena (singular: epiphenomenon). This is something which occurs unintentionally as the result of something else. A traffic jam is an epiphenomenon: nobody plans one, it results from the density of traffic and the stop-and-go nature of movement in such traffic. It is nonetheless very real, even though it has not been planned.

The linguistic complexity of our world is epiphenomenal. No one set about to split languages up into larger and larger sets, but this did happen. All the 6,000–7,000 languages of the world arose due to the fact that in social groups individuals vary their speech slightly and certain variants are preferred as carriers of social and personal identity. This variation often becomes entrenched with population movements and mixtures and with the interaction of social groups. So the language change which resulted in the rise of more and more languages was not planned but nonetheless occurred.

The Role of Chance

About 4.5 billion years ago, not too long after the Earth had formed from the accumulation of rock out of the original proto-planetary disk of our future Solar System, it was involved in an almost head-on collision with another body about the size of Mars; this small planet is called Theia, in Greek mythology the mother of the goddess of the moon, Selene. The impact was so great than it knocked about a quarter of the mass of the Earth into space. This material was later captured by the gravitational field of the Earth and began to revolve around the Earth, clumping together to then form a relatively smooth spherical body of rock, which we call the Moon. The impact event was the result of chance – the Earth and the other small planet happened to be on a collision course and then the inevitable happened. All of this was before even microbial life arose, but the benefit is with us to this day: because we have such a large moon, relative to the mass of the Earth, we are affected in turn by its gravitational field – witness the tides of the oceans which involve the movement of billions of tons of water, all day every day. The Moon's gravitational field prevents the Earth from either spinning too fast or revolving too erratically – both of these phenomena would impact negatively on life.

For about 135 million years, from the Jurassic period about 200 million years ago to the Cretaceous period about 66 million years ago, the world was ruled by dinosaurs, the stuff of so much science fiction and films. These animals did exist and many will indeed have been as menacing as they are portrayed to be in popular media, going on the fossil remains to be seen in many museums. But their demise came suddenly and is associated with the chance impact of an asteroid (see Section 6.3). The forms of life which survived this catastrophe were then able to flourish in a dinosaur-free world.

The favourable conditions for life on our planet can thus be seen on the one hand to be the result of structural features of the planet along with principles of growth and development, and on the other to derive from a sprinkling of serendipity. Chance events have played an important role. This may be equally true of life on exoplanets, especially evolutionary bottlenecks and hindrances which may be removed by fortuitous events. Here on Earth, evolutionary biologists and other scientists remain unsure of the exact extent of contingency and determinism, that is, whether evolution is dependent on forces which happen to apply at a specific time but which are not repeatable, or whether evolution is dependent on forces which always apply rendering the outcomes predictable (this issue was and is one of the main concerns of evolutionary biologists such as Stephen Jay Gould (1941–2002)).[5] Assessing the relationship of contingency and determinism on Earth will be key to assessing their roles in the origin and development of exoplanets and in the evolution of life forms there.

7.6 Things Which Only Happened Once

The advance from simple (prokaryotic) to complex (eukaryotic) cells happened well over a billion years after the appearance of the first simple cells. It would seem to have occurred accidentally when one cell got

5 In his best-selling book, *Wonderful Life*, he wrote the following oft-quoted lines: 'Wind back the tape of life to the early days of the Burgess Shale [fossil deposits in Canada, dating to over 500 million years ago]; let it play again from an identical starting point, and the chance becomes vanishingly small that anything like human intelligence would grace the replay' (Gould 1990: 45).

incorporated into another and the latter co-opted the guest as a worker rather than digesting it. In time, this led to the cell powerhouses, the mitochondria, which produce ATP, the energy currency of organisms. Further elements within cells continued to develop, so-called organelles, and the DNA of cells came to be enclosed in what became the nucleus. The resultant complex cells formed the basis for all forms of life which evolved afterwards.[6]

Did this development just happen with a single cell, or with a group of cells? If this just happened with a single cell somewhere on Earth, it would have been a chance development, which could just as well not have happened, depending on how the dice of chance fell.

The consequence of this for life beyond Earth is that it would also be subject to a greater degree of chance than we might first assume, along a path from simple bacteria-like cells to more complex cells, which would then expand into a complex biosphere comparable to that we know on Earth. However, for some suitable exoplanets, such serendipitous developments might have taken place earlier; for example, the development of complex cells might have been much more rapid than on Earth.

7.7 What Are the Alternatives?

Given the abundance of carbon in the universe,[7] indeed of amino acids (building blocks for proteins), we would expect exobiologies to be carbon-based, as on Earth. Some other suggestions have been put forward, for instance a silicon-based biology. Silicon is a non-metal, has an atomic number of 14. It occurs in crystalline form and is commonly

6 The evolution of cells and their internal structures was the key research focus of the American evolutionary biologist Lynn Margulis (1938–2011) who elaborated greatly on the notion of symbiosis, which involved organisms interacting to their mutual benefit.

7 The creation of amino acids is thermodynamically downhill and so they are found abundantly in our Solar System, for example in comets and asteroids, and, by legitimate extension, in the rest of the universe.

found in rocks and dust on Earth and on other astronomical bodies. It also combines easily with oxygen, silicon dioxide being particularly abundant, for instance as sand. So how likely is a silicon-based biology? Certainly, it cannot be ruled out, but life is opportunistic, so why go for silicon when carbon is so common and that much better?

8

How and Where to Look for Exolife

Consider that space exploration is not yet even 100 years old nor is digital technology, which is advancing at a breath-taking pace.[1] Assuming that such technology will continue unabated and that there are no negative impacts from other quarters,[2] we can further assume that the ability of humans to probe the universe with increasingly powerful instruments will continue to increase and allow us to discover ever more about the planets around other stars.

The great advances in astronomy in the past century or so were initially theoretical in that they rested on predictions about what the universe is like and how certain phenomena such as light would behave on scales much larger than those on Earth. This could not be measured on a terrestrial scale but with measurements on a cosmic scale it could be observed that light bends, for example when it passes large celestial objects like our Sun and so predictions made by Einstein were confirmed.

1 Moore's law is a prediction, made by Gordon Moore, a co-founder of the American chip-producing firm Intel, which is based on observations in the integrated circuit industry, according to which the numbers of transistors on chips would double every year or at least two years.

2 The amount of interference from terrestrial radio transmission is already compromising radioastronomy. In addition, the amount of space junk and orbital debris is a serious concern for scientists and engineers today. See Deudney (2020) for discussions of these and similar matters.

The second type of advance in astronomy is technological: with great improvements in optical and electronic engineering it has become possible to peer deep into space and gain much more information about our cosmic surroundings. These advances in astronomy have meant that communities of amateur astronomers and interested individuals have arisen, greatly increasing the numbers of people concerning themselves with this branch of science.

8.1 Recent Finds in Our Cosmic Neighbourhood

Progress in astronomy became especially relevant with the discovery, cataloguing and classification of exoplanets. Exoplanets have been found in our little corner of our galaxy. Those in other galaxies, in the Local Group of galaxies, such as the Andromeda galaxy or the nearby Large Magellanic Cloud, can be detected by examining dips in the brightness of X-rays. However, for the topic of this book, detecting planets around stars in, say, the Andromeda galaxy, which is over two million light years away, would be of little benefit. In addition, we will probably not be able to gain detailed information about exoplanets on the opposite side of the galactic plane of the Milky Way because the dense concentration of stars and dust in the central bulge is obstructing our view. But the number of exoplanets which have been discovered in our section of the Milky Way is at the moment – mid-2022 – already over 5,000 (confirmed) with about 50–100 in the habitable zone of their parent star(s).[3] This means, by extrapolation, that there are millions and millions of potentially life-bearing exoplanets just in our galaxy, orbiting about the hundreds of billions of stars of the Milky Way. And it has been estimated that there are several hundred billion galaxies in the observable universe. How many of these harbour intelligent forms of life comparable to human life on Earth, namely exobeings, is unknown. Any estimates can only be made for our corner of our galaxy. And what fraction of these exoplanets can be assumed to be populated by exobeings?

3 For more up-to-date information, see the website *The Extrasolar Planets Encyclopaedia* (exoplanet.eu/team/).

The figures given by different astronomers vary greatly, because we have as yet too little concrete information about exoplanets in the habitable zones of stars and whether they might harbour life.

8.2 Improved Technology

The considerable advances in observational technology over the past few decades can be expected to help in the search for exolife. These have been made possible largely by the American space organisation NASA (National Aeronautics and Space Administration) and the National Science Foundation along with the Canadian Space Agency and the European Space Agency. NASA's Great Observatories program, started in 1990, encompassed four major telescopes, each dedicated to observing a section of the electromagnetic spectrum from infrared (Spitzer) to gamma rays (Compton) as outlined below.

Some major telescopes of recent years

1. *Hubble Space Telescope*: This telescope is still in service having come online in the early 1990s. Launched in 1990 the telescope had a disappointing start (with a flawed mirror) but was quickly repaired and later extended. As it captures light in the visible spectrum it was able to present scientists, and then the world, with pictures of near and distant objects in the universe in remarkable resolution and colour. The Hubble Ultra Deep Field image, made of several images of the furthest objects detectable to the telescope, portrayed a tiny patch of sky with several thousand galaxies proving that the observable universe contains several hundred billion galaxies.
2. *Compton Gamma Ray Observatory*: Detecting gamma rays, the shortest and most powerful rays of the electromagnetic spectrum, this observatory yielded insights into the sources of such rays, very often objects from the very early universe, such as gamma-ray bursts from active galaxies (those which show considerable emissions from their centres not stemming from

stars themselves but probably from a supermassive black hole, which is accreting matter from nearby stars). The observatory was operational from 1991 to 2000.

3. *Chandra X-ray Observatory*: Located between gamma rays and visible light in the electromagnetic spectrum, X-rays allow astronomers to see through clouds of gas and dust which otherwise block light. Thus, this observatory is able to provide information about many objects not otherwise accessible, such as new-born stars, inside clouds where they come into being. The observatory was launched in 1999 and is still operational.
4. *Spitzer Space Telescope*: A space telescope sent into orbit in 2003 and decommissioned in early 2020. It was designed to observe celestial objects in the infrared spectrum with its 0.85-metre primary mirror. This needed to be cooled to −268 °C by liquid helium carried on board. This fuel was used up by 2009 but the telescope has still been collecting data because parts of its camera equipment have continued to work without special cooling.

Additional technology in space, which is dedicated to the search for exoplanets, includes the Kepler Space Telescope and the TESS, which are exclusively for exoplanetary searches, and the James Webb Space Telescope (used for many other purposes as well). Other projects, such as the European Space Agency's Gaia space telescope, billed as 'ESA's billion-star surveyor', are relevant to the detection of exoplanets.

Satellites and observatories dedicated to the search for exoplanets

1. *Kepler Space Telescope*: Launched in spring 2009, the Kepler Space Telescope was designed specifically to track Earth-like exoplanets in a small patch of sky containing about 150,000 stars. On board the observatory is a photometer monitoring stars and checking for dips in their brightness when a planet passes in front of the star from Kepler's line of vision. Detection began in January 2010 and by 2015 over 2,600 exoplanets had been detected and confirmed, despite the fact that components which had outlived their minimal lifetime finally failed in

2012 and 2013, affecting the manoeuvrability of the observatory. In 2014, an extended mission began, called K2 'Second Light', which utilised Kepler's reduced capabilities to examine red-dwarf solar systems in search of exoplanets. It was finally retired in late 2018 (its fuel had been used up by then) after nine years of active and successful service, delivering huge amounts of data.

2. *Transiting Exoplanet Survey Satellite (TESS)*: The purpose of the TESS project is to scan an area of the sky, 400 times greater than that investigated by the Kepler telescope, and examine about 500,000 exoplanets using the transit method over a two-year period. The satellite can mark those candidates worthy of later closer inspection. It is expected to glean information from about 20,000 exoplanets over a two-year period, measuring the mass, size, density, position and orbit of the planets as it does so. This will in turn provide lists of possible candidates for further investigation by the James Webb telescope. TESS was launched in spring 2018 and produced its 'first light' image in August of that year.
3. *James Webb Space Telescope*: This telescope is about 100 times more powerful than the Hubble telescope, whose successor it is. Given this great increase in power, the James Webb Space Telescope is used for cosmological projects, tracking extremely distant objects and offering information about the very early universe and the formation of the first galaxies. Direct imaging of exoplanets is also planned. The telescope consists of 18 hexagonal segments, which combine to realise a mirror 6.5 meters in diameter that unfolded when the telescope reached it orbit in space.[4] The spectrum it covers is visible light at long wavelengths and the infrared range. A sun shield consisting of several layers protects the telescope from the Sun's rays so that its operational temperature can be kept at −222 °C (50 K) to minimise interference in the infrared spectrum. The telescope is located at the Earth–Sun Lagrange point L2, which is a stable position about 1.5 million kilometres out from the Earth where

4 There are plans at NASA for a still more powerful telescope, the Large Ultraviolet Optical Infrared Surveyor (LUVOIR), which would be a general-purpose telescope with hitherto unreached sensitivity (currently proposed in two versions, A and B, with a 15- and an 8-metre segmented mirror, respectively). Another proposal has been for a 12-metre visible-light-range telescope, given the working title of the Carl Sagan Observatory in honour of the renowned astrophysicist.

the telescope remains in the Sun's Earth shadow and hence favourably positioned for maintaining its low operating temperature. The launch of the James Webb Space Telescope was delayed repeatedly; the last delay being due, among other things, to the Covid pandemic. It was finally launched successfully on 25 December 2021 and became fully operational in mid-2022, returning its first spectacular pictures including a 'deep field' image showing, in much higher resolution than Hubble, the galaxies which formed soon after the Big Bang.

In addition to these telescopes there are a number of proposals for future ones, put forward by NASA in 2016. An interesting one of these in the present context is the Habitable Exoplanet Observatory (HabEx) telescope. In the words of the NASA website, this telescope 'is a concept for a mission to directly image planetary systems around Sun-like stars. HabEx will be sensitive to all types of planets; however, its main goal is, for the first time, to directly image Earth-like exoplanets, and characterize their atmospheric content. By measuring the spectra of these planets, HabEx will search for signatures of habitability such as water, and be sensitive to gases in the atmosphere possibly indicative of biological activity, such as oxygen or ozone' (www.jpl.nasa.gov/habex). For this and other projected missions, technologies are being developed to equip future telescopes with sun shields, in the shape of flower petals positioned some distance from the telescope, and/or coronographs, disks in the telescope, which would effectively cut out the star's light and allow better imaging of any planets orbiting these stars.

Another telescope, currently under development at NASA and scheduled for launching in 2027, is the Nancy Grace Roman Space Telescope (formerly known as the Wide-Field Infrared Survey Telescope). This will operate in the visible and near-infrared wavelengths and, importantly, one of its cameras will incorporate novel technology for suppressing starlight, thus rendering it suitable for exoplanet observation.

Mention should also be made of the telescopes being used by scientists engaged in the search for extraterrestrial intelligence (SETI) to search the skies for possible signals from beyond Earth. The Allen Telescope Array in Northern California is used by the SETI institute (www.seti.org/ata); other telescope facilities, such as Green Bank in West Virginia, USA and Murriyang in Australia (formerly the Parkes radio telescope) as well as the Very Large Array (New Mexico) and MeerKat (South Africa) are used for SETI searching at intervals.

Planning for future exoplanet research is ongoing and key documents are regularly published to point to ways forward, such as *Origins, Worlds, and Life. A Decadal Strategy for Planetary Science and Astrobiology 2023–2032* (National Academies Press, Washington, DC). In this document, 12 priority topics are listed, divided into sections which look at (a) origins, (b) worlds and processes and (c) life and habitability.

8.3 Methods for Finding Exoplanets

We can assume by default that stars have planets rotating around them as the original disk of dust and gas which gave rise to a star also contained clumps out from its centre, and these became proto-planets and, with time, actual planets. The problem with proving this fact has been the difficulty of detecting planets around stars beyond our Solar System. Various projects were initiated to improve exoplanet detection, for example the High Accuracy Radial Velocity Planet Searcher (HARPS) at the La Silla Observatory in Chile in 2002 and, above all, the Kepler Space Telescope along with the later Transiting Exoplanet Survey Satellite, greatly augmenting successful detection. There are a number of methods used to track exoplanets, the two main ones are the transit method and the radial velocity method.

Transit Method

This is the most common method by means of which astronomers measure the extent to which the light of a star diminishes when a planet passes in front of it. This value is tiny but can nonetheless be measured and the method was used by the Kepler Space Telescope. Sometimes the method can allow one to estimate the size of the planet and perhaps to examine the elements in its atmosphere via spectroscopy. The transit technique has a number of disadvantages, however, and has led to a number of false positives. So it is used together with other methods for confirmation.

The size of a planet can be calculated from the size of the dip in light, and the mass from the size of the reflex motion it exerts on its parent star (a result of the gravitational pull of the planet on the star). Size and mass allow the calculation of the planet's density so that it can be determined if the planet in question is a gas giant, like Jupiter or Saturn, with a low density, or a rocky planet with metal in its core, like Earth, which has a much higher density.

There is one obvious disadvantage to the transit method and that is that a planet is required to pass in front of its parent star, which it will only do if the solar system in which it resides is more or less parallel to our plane of vision, that is, we are looking edge on. But this will only apply in a fraction of cases; in others, the disk of the exoplanets will be at some other angle to our plane of vision, which means we are looking down/up at the planetary system. Furthermore, if planets are further out from their parent star, their orbital period will be longer, closer to the one year for Earth. This means there will be fewer opportunities to check on a planetary candidate by repeating a transit measurement than with a tidally locked planet orbiting a red dwarf in a matter of days. Nonetheless, the transit method has been responsible for finding over 75 per cent of all exoplanets to date (mid-2022), about four times as many as with the radial velocity method (about 18 per cent).

Radial Velocity Method

The second method involves star 'wobble': when a large planet is close to its parent star it exerts a gravitational pull on the star and causes it to wobble ever so slightly, altering the star's radial velocity. Bear in mind that a star has a small orbit around its own centre. A planet tugging gravitationally on a star causes the centre of orbit to lean slightly towards the planet. The value of this displacement can be measured by considering the Doppler shift in the spectrum of the star; hence the technique is also known as Doppler spectroscopy. Recall that the Doppler effect is the shift in wavelength (light or sound) which occurs when an object emitting waves is either approaching or receding from the observer. So, if the centre of orbit of a planet and the star's centre are not co-terminous, the star is being pulled slightly to one side by the gravity of the orbiting planet and the value of this force can be measured via the Doppler effect. To do this, however, the planet has to be large, a Jupiter-sized body at least. But technical improvements (in the HARPS spectrograph) have led to many less massive exoplanets being detected.

Other Techniques

Some other techniques are used, especially to examine if there are several planets around a star and if they have moons, for instance by measuring variations in the timing and duration of a transit. A further technique, gravitational microlensing, is based on the fact that light from a distant object is bent around a near object in our line of sight, triggering a slight distortion of the distant object's light (used with about 2.6 per cent of the exoplanets found thus far). In theory, a planet moving across its parent star in our line of sight can have the same effect, although the effect of such gravitational microlensing is tiny. Planets from several times the mass of Jupiter down to a couple of Earth masses have been detected using this technique. Direct imaging is unfortunately not normally possible and accounts for only about 1.2 per cent of the exoplanets finds.

8.4 A Planet in the Habitable Zone

The Kepler Space Telescope has photometrically monitored almost 150,000 stars in a small section of the sky in the constellation of Cygnus. Kepler 22b was the first planet to be discovered in the habitable zone of its parent star, and was confirmed in 2010. It is a fairly massive planet with a radius about 2.5 times the size of Earth, a so-called super-Earth, and is located about 587 light years away. With an orbital period of about 10 months, Kepler 22b revolves around its parent star at a distance 15 per cent less than that of Earth's distance from the Sun but its parent star's luminosity is about a quarter less than that of our Sun so it is firmly within the habitable zone (Figure 8.1). Kepler 22b can be classified as an Earth analogue, a planet

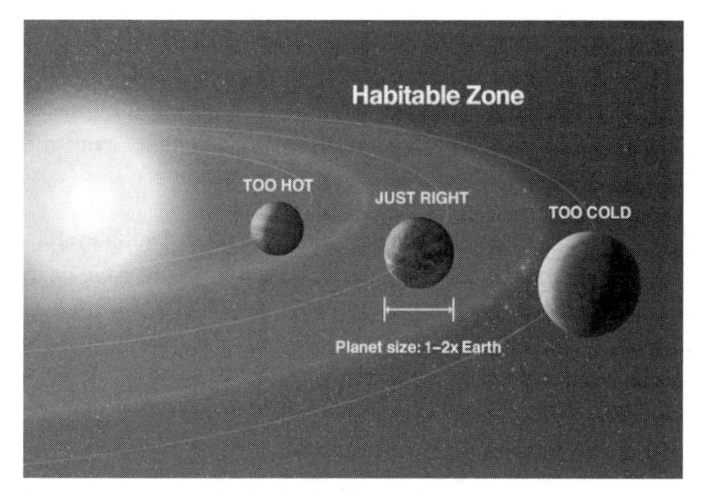

Figure 8.1 The habitable zone of a star

which in many crucial respects resembles the Earth: in the habitable zone of a star which is slightly cooler than our Sun, it probably consists of rock with liquid water on the surface.

There is an upper limit to Earth analogues: the planets cannot be too big: if they are, they end up trapping too much gas (mainly hydrogen and helium) and become gas giants. Earth-sized planets can be expected to trap a certain amount of gas, providing them with an atmosphere much as we have on Earth. It is notable that there appears to be a gap in planet sizes: between 1.5 and 2.0 that of Earth (known as the Fulton gap after a key study), with rocky planets below this and the super-Earths over twice the size of Earth. The atmospheric composition of the small terrestrial planets may well be different from Earth, so they might contain methane rather than hydrogen and nitrogen, but oxygen would be essential for any biology similar to our own.

8.5 What About Rogue Planets?

Our galaxy does not just consist of orderly solar systems with a star and a set of planets. There are many lone stars wandering across the Milky Way, such as Alpha Camelopardalis, which is racing through space at about five million miles per hour and has a hot bow wave in front of it due to the energy it is generating. There are also rogue planets, often single planets wandering around on their own; but they also occur in populations of hundreds or even a few thousand, as announced by scientists from the University of Oklahoma in 2018.

Rogue planets may have deviated from their normal orbit in a solar system and now wander through interstellar space. Such planets could have been catapulted off their circumstellar orbit by the shock wave of a supernova or could have been pulled out of their orbit by the gravity of a passing celestial object such as a star moving close to a solar system or, indeed, by another rogue planet of the sub-brown-dwarf type, which would have a strong enough gravitational field to cause this effect.[5] It is thought that there may be billions of rogue planets in our Galaxy. Could exobeings evolve on rogue planets? Now while it is true that rogue planets might have an internal source of heat, for instance, if there was enough seismic activity under the surface to generate heat, such planets would not have a source of light and hence no bacteria or algae which might produce oxygen for complex and flexible life forms to arise.

8.6 Rare Earth Hypothesis

In astronomy and astrobiology this hypothesis[6] maintains that life as we know it on Earth, with eukaryotes, sexual reproduction and, most of all, primates and humans is extremely improbable, as the precise conditions for this kind of life would be very unlikely to exist elsewhere in our galaxy or even in the universe. The various astrophysical, geological and chance

5 Brown dwarfs are not generally regarded as engaging in the nuclear fusion of hydrogen to helium and beyond as in our Sun and hence they do not emit much energy in the form of light, if any at all. A star's type is determined by its mass, and the smallest of these – brown dwarfs – are under 100 times the mass of Jupiter, the largest planet in our Solar System. Because such stars emit so little light they would probably not be suitable for the harnessing of light by plants, as happens on Earth in the process of photosynthesis. Nonetheless, brown dwarfs may have planetary systems; for instance, the brown dwarf 2M 1207, in the constellation Centauraus about 170 light years away, has a planet five times the mass of Jupiter orbiting it.

6 This term comes from the book *Rare Earth: Why Complex Life is Uncommon in the Universe* (Ward and Brownlee, 2000) .

settings/events in the history of the Earth are also regarded as extremely unlikely and contribute to the view that our Earth is very exceptional in harbouring advanced life. Above all, it is the relative stability of life-favourable conditions on Earth for several billion years which proponents of this hypothesis see as quite unlikely to hold elsewhere (see the extinction possibilities for an exoplanet listed below). And certainly, from research into exoplanets in recent decades, our Solar System, with the Earth a good distance out from its parent star, does not seem to be a typical planetary system (though further research may not uphold this view entirely). So there may be something to the Rare Earth hypothesis, especially when one considers just what could go wrong for an Earth analogue in another solar system.

As planets with life go, we might also be an exception in that only *Homo sapiens* from the different species of *Homo* have survived into the present. The

---

Some extinction possibilities for an exoplanet

1. An exoplanet could be hit by the cosmic rays from a nearby supernova, sterilising all forms of life on the planet.
2. An exoplanet could have its atmosphere repeatedly and seriously impaired by powerful solar flares (not improbable for a planet orbiting close to a red dwarf star).
3. An exoplanet could be moved out of its habitable zone (due to gravitational perturbations from another celestial object), either away from its parent star, making it too cold for life, or towards the star, making it too hot.
4. An exoplanet could lose its protective radiation shield by its magnetic field diminishing and/or by ozone depletion (assuming that either or both of these two factors already applied).
5. An exoplanet might lose its atmosphere by having too little gravity to hold on to it while exposed to intense radiation from its parent star.
6. An exoplanet might suffer a catastrophic extinction event such as a large asteroid impact or a lethal increase in volcanism.
7. An exoplanet could suffer from runaway global warming like our planetary neighbour Venus.

---

last group to go extinct were the Neanderthals, about 35–40,000 years ago. Having different levels of exobeings on an exoplanet could have very negative consequences. What would life be like on a planet on which there were indeed significantly different levels of intelligence and ability, assuming that earlier species survived alongside more cognitively advanced exobeings?

8.7 The Copernican Principle

This term is used to denote that the Earth is nothing special but rather a mainstream type of rocky planet orbiting a typical mainstream star in a common or garden barred spiral galaxy. The reference is to the claim by the Polish astronomer Nikolaus Copernicus (1473–1543) that the Earth is not the centre of the system of planets, but the Sun, demoting the perceived status of the Earth in the process. The implication is that life would be likely on planets like Earth assuming that they show similar environmental conditions.[7] This view has been supported by scientists such as Frank Drake (1930–2022) and Carl Sagan (1934–1996).

To decide whether the Rare Earth hypothesis or the Copernican Principle holds for our planet we have to consider where life is likely to be found, both in our Solar System and beyond.

8.8 Earth Similarity Index and Planetary Habitability Index

In order to compare exoplanets to our Earth, a similarity index was developed and first published in 2011 in the journal *Astrobiology* (Schulze-Makuch et al. 2011). It is used to quantify the importance of key variables for exoplanets, where the variables (1) to (3) have a value of 1.00 (Earth). Mean surface temperature (4) and the chemical composition of the atmosphere (5) are further relevant variables.

7 A similar term for this assumption is the 'Cosmic Zoo Hypothesis' put forward in Schulze-Makuch and Bains (2017).

Earth similarity index (major variables)

1. Radius
2. Density
3. Escape velocity
4. Surface temperature
5. Atmospheric composition

From Schulze-Makuch et al. (2011)

The original proposal also included a planetary habitability index: the authors discuss values for four variables (Schulze-Makuch et al. 2011). In order to avoid 'terracentric bias,' the authors do not specify that (L) must be water but, given its abundance in the universe, there is no reason not to see it as the default value for this variable.

Planetary habitability index (four main variables)

Stable substrate (S)
Solid surface; atmosphere; magnetosphere

Available energy (E)
Light; heat; tidal flexing
Redox chemistry (with energy-yielding reactions)

Appropriate chemistry (C)
Probably carbon-based polymeric chemistry

Liquid solvent (L)
Temperature and pressure values determine the presence of liquids in the atmosphere and on or beneath the surface.[8]

8 A complete water world could support marine life but hardly lead to a digitally aware exocivilisation with which we could communicate.

8.9 Classifying Exoplanets

Planets generally come in systems, not in ones. So the discovery of large Jupiter-sized planets implies that smaller planets are also orbiting the stars with these big planets. The smaller ones could also be rocky planets, possibly with liquid water, in the habitable zone of stars. These assumptions led to a classification of exoplanets according to size (see the four basic types listed below). Note that the label 'terrestrial' refers to rocky exoplanets without specifying whether they might have all the features necessary to bear life. For the latter, the labels 'Earth-like' and 'Earth analogue' are used in this book (see 'Criteria for 'Earth-like' exoplanets' below).

Four basic types of exoplanets

1. Neptune-like: large ice planets
2. Gas giants: usually hot gas giants near their star
3. Super-Earths: rocky but several times the mass of Earth
4. Terrestrial: rocky and less than two Earth masses

This quantitative distribution of the planets according to type is approximately as shown below. It should be stressed that this is just a snapshot of the situation in mid-2022, which is changing and will continue to do so, especially as the James Webb Space Telescope is now fully operational. Furthermore, the classification below is a reflection of detection

---

Distribution of confirmed exoplanets

1. Neptune-like: 1,779
2. Super-Earths: 1,582
3. Gas giants: 1,536
4. Terrestrial: 188
5. Unknown: 5

   Total: 5,090

Source: https://exoplanets.nasa.gov/discovery/discoveries-dashboard/).

---

thresholds and the precision of instruments. The taxonomic distribution of exoplanets is likely to alter with technological advances.

8.10 When is an Exoplanet 'Earth-like'?

The search for exoplanets has been remarkably successful in the past two decades. At the time of completing this book (mid-2022), over 5,000 exoplanets have been confirmed; these are listed in an archive at NASA, (exoplanetarchive.ipac.caltech.edu). The Extrasolar Planets Encyclopaedia (exoplanet.eu) maintained at the Paris Observatory offers a similar service. Among these planets are some which can in principle be compared to Earth. To be classified as 'Earth-like' there are certain criteria that should hold.

It is possible to examine the spectrum of a planet's atmosphere from space. Elements like hydrogen, carbon, oxygen and nitrogen, as well as molecules such as water and methane, absorb light each with a characteristic pattern. The points in the spectrum where the light is absorbed result in spectral lines, dark stripes in an otherwise continuous band of rainbow colours from red to violet, thus spectroscopic analysis is the examination of colour spectra in an atmosphere to detect the presence or absence of key elements. In addition, a planet might have compounds like nitrous oxide,[9] which would disintegrate in the ultraviolet light of the parent star unless it were continually replenished from a source on the planet, whether biotic or not.

Early Finds

The Swiss astronomers Michel Mayor and Didier Queloz,[10] working in Geneva at the Haute-Provence Observatory, discovered an exoplanet, 51 Pegasi b, a hot Jupiter, orbiting its parent Sun-like star only 50 light years

9 Chemically N_2O and commonly known as laughing gas, nitrous oxide is often used as an anaesthetic. It is a potent greenhouse gas and large amounts of it can lead to ozone depletion in the higher reaches of our atmosphere.

10 Recognition of their contribution to the field was given in 2019 when Didier Queloz and Michel Mayor, together with James Peebles, were awarded the Nobel Prize in Physics.

Criteria for 'Earth-like' exoplanets

1. The planet should be within the range of 0.5 to a maximum of 2.0 Earth masses to ensure a tolerable gravity level.
2. The planet should not be tidally locked with its parent star, and therefore rotate on its own axis, providing a night and day cycle.
3. The planet needs to orbit its parent star in the habitable zone and hence not be too hot or too cold.
4. The parent star should be Sun-like or smaller, though cooler stars are subject to high-energy flares and the habitable zone can be too close to the star.
5. The planet should have a core of moving molten iron producing a strong magnetic field, which would act as a shield against solar wind and cosmic radiation, both of which would be detrimental to the rise and maintenance of biological systems.
6. The planet should have a nearly circular orbit (low eccentricity) and not too great a spin on its own axis.
7. The planet should consist of rock with liquid water on the surface.
8. The planet should show a degree of geological activity with plate tectonics and volcanism, these providing a dynamic environment for life forms to evolve in.
9. The planet should preferably have an atmosphere with a constant level of oxygen, continually replenished by an oxygen-emitting biology.
10. The planet should preferably have a weather system which would distribute water across the surface and ensure equable temperatures across the entire planet.

away from us. It has an orbit of 4.2 days and is only 8 million kilometres out from its star (Mercury, by comparison, is 55 million kilometres out from our Sun). The astronomers began by using fibre-optic technology to detect planets. By the end of 1994 they already had variable measurements for several stars. Within two months of the discovery of 51 Pegasi b, two further gas giants (many times the mass of Jupiter) were found circling their stars, 70 Virginis b and 47 Ursae Majoris b. The following paragraphs list some of the more significant exoplanet finds of the past decade or so.

Proxima Centauri b, 1.3 times the size of the Earth orbits Proxima Centauri, the closest star to our Sun. However, the planet is exposed to considerable radiation from its parent star[11] and is tidally locked to this star.

Trappist 1 (discovered in 2016–2017) is a complex planetary system consisting of some seven planets, three of which (Trappist 1e, 1 f, 1 g) are probably in the habitable zone. Their parent star is a cool red dwarf, just under 40 light years away in the constellation of Aquarius, more massive but not much larger than Jupiter.

Gliese 581 g (20 light years away, discovered in 2010) is tidally locked so the borderland between the bright and dark side would probably be the best place to find moderate temperatures. There has been some dispute about its existence with studies producing differing results.

An increasing number of such planets are being discovered, and some, such as Ross 128 b (11 light years away, discovered in late 2017), are actually quite close to us. This planet, is probably in the system's habitable zone. It is a rocky planet, possibly with a water-rich atmosphere, providing cloud cover and a degree of protection from stellar radiation.

Some of these planets are indeed rocky but considerably larger than our planet, so-called 'super-Earths', such as Kepler 22 b (2.4 times larger than Earth). The force of gravity on such planets would be greater than on Earth and any exobeings would have had to develop in accordance with this gravity level. Other factors would be important when considering these planets. The level of greenhouse gases, mainly carbon dioxide, would determine the amount of heat from its parent star which a planet would

11 Our Sun is subject to flares as well: hot plasma moves around inside causing powerful magnetic fields, which can burst out at the surface causing solar storms. In 1859 a solar storm directed at Earth had a direct hit. Its effects were not too serious because power grids and electronics did not exist then, but if they had, it would have caused devastation. In 2012, another major solar storm erupted from the surface of the Sun but fortunately was not directed at Earth. The Sun can also eject significant amounts of its contents, known as coronal mass ejections.

trap. Another key feature would be a planet's eccentricity (degree to which its orbit is elliptical) and whether it always showed the same side to its star, apart from the obvious requirements of liquid water and an abundance of oxygen in the atmosphere. The role an atmosphere plays in temperature regulation should also be borne in mind: it distributes temperatures fairly evenly between day and night in rotating planets like Earth. Many astronauts on spacewalks have remarked on feeling the over 100 °C temperature on their spacesuits in sunshine and the minus 100 °C temperature in the shade. This reminded them how near they were to sizzling and freezing, and how close these two situations were to one another outside the tempering effect of the Earth's atmosphere.

8.11 Potential for Life on Moons

Life does not just have to originate on planets, large moons are also possible candidates. Take those around Jupiter, for example. There are four large moons, known as Galilean moons after their first discoverer, Galileo Galilei, in 1610. They orbit Jupiter with at least 70 others into the bargain, in fact, many more, depending on how one draws the line between a moon and just debris, such as objects from the asteroid belt captured by the gravitational field of Jupiter, or the remnants of an early moon-like object, which broke up into small fragments.

EXOPLANETS: THE ROLE OF ASTRONOMICAL ART

Despite the rigorous research into exoplanets of the past two to three decades, direct imaging of exoplanets has not yielded much visible detail (the James Webb telescope has already provided a fuzzy image of an exoplanet). However, NASA has artists who draw impressions of exoplanets given their scientific knowledge of them. Two such artists at CalTech in California, Robert Huart and Tim Pyle, say that their art is the public face of these objects, forming images in people's minds about exoplanets, and will be for some time to come and so, as they maintain, they carry a responsibility to be as accurate as possible.

A day on Jupiter (one rotation) lasts about 10 hours despite it being about 300 times the mass of Earth and almost 12 times the width of Earth. This is due to its great rotational speed, about 13 kilometres per second. But it is only about a thousandth the mass of the Sun and definitely not big enough for the pressure at its core to initiate hydrogen fusion and begin to glow like a star. Despite being under 800 million kilometres out from the Sun, Jupiter emits more heat than it receives. This is due to its liquid metallic core, which has a temperature of several thousand degrees and which radiates heat outwards, this then escaping from its surface. The much smaller rocky planets on the inner orbits of the Solar System are closer to heat balance, basically emitting the heat which they gain from the Sun, though this is not true of Venus and is becoming less and less so for Earth at present.

8.12 A Lunar Trio

The reason why it is worthwhile examining moons is that many have conditions which could be conducive to life. Take heat, for instance. If a moon has an elliptical orbit around its parent planet, it would be gravitationally flexed (squeezed and stretched) by the planet as it moves closer and then further away from it, this action generating heat within the moon in question. A famous example of this in our Solar System is Io, a moon of Jupiter, which is riddled with very active and powerful volcanoes. But this flexing could be beneficial as well. It could lead to water remaining liquid under a thick coating of ice on a moon despite this being a long way from the Sun as a source of heat.

If a moon has liquid water on its surface, periodic flexing could also lead to hydrothermal vents arising at underwater cracks in its submarine surface (these are common at the joints of tectonic plates and at volcanic hotspots on Earth). Mixtures of hot gas, water and lava, rich in iron and sulphur and at temperatures of several hundred degrees, can be released into the ocean at the vents. Life forms can arise at these points, consisting of bacteria which feed on the emissions and then form the beginning of a biological chain, with tube worms and shrimp-like creatures feeding on these. These ecosystems are very fragile and disappear if the vent emissions stop. However, they are postulated to be a source of early life forms on Earth and may well exist in the water worlds[12] of moons such as those to be discussed in the following sections (for more information, see the Planetary Science and Astrobiology Decadal Survey 2023–2032, which has six panels in which the exploration of planets and small Solar System bodies has priority).

Europa

Europa is a moon of Jupiter. It is the smallest of the Galilean moons, slightly smaller than our Moon, and is made of silicate rock with an iron–nickel core. Europa is covered entirely by water, frozen at the surface but with more water underneath its surface than all the water on Earth. From on top, the kilometre-thick crust of ice appears relatively smooth with long dark streaks on it. At the bottom of the subglacial ocean is a rock floor (Figure 8.2), which could have added salts to the water, along with carbon compounds, providing substances which could be useful for microbial life forming. Water-vapour plumes have been detected on the surface, similar to those found for Enceladus (see below). These are believed to be caused by cryogeysers, volcanoes of ice and bits of other matter. The European Space Agency, in cooperation with Airbus Defence and Space is preparing a mission – the Jupiter Icy Moons Explorer ('JUICE') – to the Jovian moons Ganymede, Calisto and Europa, launched in April 2023 and planned to arrive at Jupiter by 2031. The mission will not attempt to bore through the ice of Europa but will examine its surface, estimate its minimal thickness and search for organic molecules and thus assess the likelihood of Europa being able to support submarine life.

Another mission of interest is the Europa Clipper being developed by NASA in connection with their Ocean Worlds Exploration Program. It has been designed to complement the Jupiter Icy Moons Explorer and is due

12 The surface of Europa is subject to considerable radiation from Jupiter, which would be damaging to any life forms, but the ice cover is sufficient to shield the subglacial water from this harmful radiation.

Figure 8.2 Jupiter's moon Europa (cross-section showing internal structure)

for launch in late 2024. The spacecraft will study Europa in a series of some 45 fly-bys while in orbit around Jupiter. Initially, it was planned to have the spacecraft orbit around Europa but the intense radiation from Jupiter was regarded as a potential danger and it will now adopt an elliptical orbit, coming close to Jupiter only when it is to fly by Europa.

Speculative questions about possible life on Europa

1. Is there microbial life on Europa? Did this arise there or was it transported to this moon by the comets which supplied all its water in the early days of our Solar System? If the former, abiogenesis (the beginning of life) should be common in the universe where the conditions are right. If the latter, we still do not know where the microbes came from; they may well have been in the water-supplying comets striking Europa during its infancy.
2. Is there complex life on Europa? To answer this question positively it would be enough to have small crab-like or worm-like creatures hovering around the black smokers on the floor of Europa's subglacial ocean. That fact alone would show conclusively that evolution occurs whenever the conditions are right, in this case the existence of minerals and heat from the black smokers providing an energy source.

3. How advanced could life be in the underwater world of Europa? Would there be not only crabs and worms but larger forms, which we would recognise as fish? Such creatures might be similar to deep-sea fish on Earth, which have evolved in darkness and under great pressure due to the depth of water. The underwater pressure on Europa would be far greater than that in the oceans of the Earth. True, the force of gravity on Europa is nothing like as strong as here on Earth, but the ice sheet is about 30 kilometres thick and that would raise the pressure on the liquid water underneath considerably. Such fish might have gill-like structures for extracting dissolved oxygen from the water. Life on Europa or other possible water worlds would be chemosynthetic (using energy from chemicals, not light, as is the case around black smokers on the floors of our oceans) because there is complete darkness in the water world under the ice covering of the moon. If Europa has fish in its ocean, another question would be how they propagate. Would they be sexually distinct, with males and females providing input DNA to the eggs from which young fish develop? If that were the case, a number of further essential questions concerning the origin of life would be answered; above all, that the principle of sexual reproduction for forms of life, as they evolve with increasing complexity, is universal, given the right conditions.
4. Could civilisations develop under water? This is difficult to imagine, though doubtless high levels of intelligence could in theory be reached, and have been, with mammals that live under water, such as whales and dolphins, though these stem from land creatures which returned to the water. Yet again, there are many water worlds throughout our galaxy. Many are covered by thick sheets of ice, like Europa, so any creatures they might harbour could never venture to the surface of their own moon/planet, let alone leave it on a spacecraft. And even if exobeings did exist in extrasolar waterworlds, for space travel they would have to carry their environment with them, much as human beings do with molecular oxygen in gas form. However, travelling through space in a water tank would present a unique set of challenges.

How could one get under the ice sheet of Europa to answer some of the speculative questions concerning this moon? Drilling might sound like an option, but the mechanical challenges would probably be insurmountable. Another option would be to melt through the ice: a radioactive rod could maintain a constant high temperature of several hundred degrees and melt its way down through the ice being drawn towards the centre by the moon's gravity. However, that would involve the unwanted effect of contamination. And then how would a probe under the ice relay data back to Earth? Perhaps via a parent device which would remain on the moon's surface but maintain contact with the subglacial probe.

Enceladus

Enceladus, a small moon of Saturn about 500 kilometres across, is a much smaller icy moon than Europa, with a surface temperature of nearly −200 °C. The surface is very smooth in some parts, suggesting that it was resurfaced in its history by some kind of geological activity. Active water geysers at the south pole of Enceladus, photographed by the Cassini

EXPLORATION AND POSSIBLE CONTAMINATION

Contamination is an issue in space exploration and can take either of two forms: (i) forward contamination, which would mean introducing microorganisms (bacteria/viruses) or radioactivity from Earth to an extraterrestrial environment, such as Europa, were we to investigate its water world; and (ii) back contamination, which would occur if microorganisms were introduced to Earth from spacecraft returning from a mission to an extraterrestrial object. This was a serious concern during the first Apollo missions to our Moon and the first astronauts were kept in quarantine to ensure that they did not bring back microorganisms which would have been undesirable on Earth. In the event this did not happen because the Moon is dry and sterile, but that might not be the case with spacecraft returning from worlds which could potentially harbour life.

spacecraft orbiting Saturn in 2005, have confirmed the presence of under-ice water. When flying straight through the plumes of these geysers, Cassini detected organic molecules in the water clouds emanating from the geysers.

Titan

Titan, the largest of Saturn's moons at 5,170 kilometres in diameter, is the second biggest moon in the Solar System after Jupiter's Ganymede and is larger than, though not half as massive as the planet Mercury. Titan has an atmosphere which is denser than the air on Earth. This is mostly nitrogen, along with hydrogen and methane (CH_4), and has a temperature of about −180 °C. There are lakes of methane near its poles, which change their shape, showing that they are active and move. These lakes are the only evidence of surface liquid found anywhere in the Solar System apart from Earth. The planet has an active weather system, based on methane and nitrogen, which has formed its surface features.

Titan was discovered by the Dutch astronomer Christiaan Huygens (1629–1695) in 1655, and the probe which landed there on 14 January 2005 was called after him. This was the first landing on a celestial body beyond Mars. The mission was successful, with the probe, carried by the spacecraft Cassini, transmitting data for about 90 minutes after landing. This data was transmitted back to Cassini, which had let the probe down onto the surface of Titan. The data included information about the atmosphere of Titan and even a picture of the surface.[13]

At present there is a further project from NASA – Dragonfly – designed to land on Titan and explore its topography and geology. Dragonfly, due for launch in 2027, will seek to determine if life forms on Titan could exist, explore both possible subsurface oceans, former surface water in impact craters and the methane lakes already observed.

13 The Cassini spacecraft was launched in 1997, arrived at Saturn in 2004 and completed its mission in 2017 when it was directed into Saturn's atmosphere, where it disintegrated on entry.

8.13 Microbial Life on a Moon: What Could It Tell Us?

What conclusions might we draw if we found microbial life on Europa or Enceladus? That would be a great step forward in determining how common the type of biology we have on Earth would be across the universe, assuming that such life was an instance of 'second genesis', with a different source from life on Earth.

Possible conclusions if microbial life, not deriving from the same source as life on Earth, were found on water world moons

1. Self-maintaining and self-replicating life forms can exist and proliferate. They would probably have an origin in material becoming trapped in fatty acid vesicles (enclosed structures) forming primitive cells. These cells would later divide and differentiate, forming very basic forms of life with further differentiation into various types of microbes.
2. If the microbial life on a moon were eukaryotic, this would suggest a progression from prokaryotes to eukaryotes on the assumption that simple cells evolved before complex ones with internal structures (organelles). There was such an advance on Earth but it took nearly two billion years to happen.
3. Carbon-based biology in water as a solvent is a common evolutionary path for life.
4. The evolution from simple cellular organisms to more complex forms is a natural development which could, in principle, lead to much more complex organisms comparable to mammals and primates.

Exomoons: Habitats for Exobeings?

Going on what we know from the planets of our Solar System it is probable that exoplanets will also have moons. But how likely are they to be locations not just for microbial life but for exobeings? This would depend on a number of factors not least of which are the conditions which would similarly apply to exoplanets themselves: location in the habitable zone of a star, availability of liquid water, a long period of stability in the planetary and solar system, etc. However, there could be important issues with exomoons: if they orbit large planets, as do the moons of Jupiter and Saturn, the radiation emitted by the parent planet might be too great for vulnerable life forms to evolve; a magnetic field around the exomoon could provide a shield from damaging radiation. Another consideration is that a large parent planet might cause an exomoon to be in the shade of the parent star for long periods, causing extreme cold which would not be conducive to life.

8.14 Where Are We at Present?

Many scientists have proposed that the period of significant impact of humans on the environment, say about the last 200–300 years (since the beginning of the Industrial Revolution), be recognised as a particular epoch of geologic time – the Anthropocene – given the extreme reduction in biodiversity and the deterioration in natural environment and climate which is now taking place. This is due to a number of factors, such as the encroachment by humans on the habitats of various animals, the use of fertilisers in agricultural industry, which has drastically reduced insect and other populations, and global warming, which has negatively impacted animals of the polar regions and ecosystems such as coral reefs.

However, if humanity does not succeed in orchestrating its own demise,[14] through overpopulation, depletion of natural resources and global warming or even worse through a nuclear conflagration, it may continue for millions of years to come. Technology will advance accordingly so that

14 See Martin Rees' book *Our Final Century?* (Rees, 2003) for more on this type of scenario.

computers and exploration devices, like telescopes and space observatories, will continue to improve in an exponential manner, if the present is anything to go by. But the question remains whether we can really extrapolate from the present, maybe there are too many wildcards and jokers in the predictions, for instance just how artificial intelligence will pan out in the near future.

9

The Limits of Exploration

The options for space travel will determine how much of our corner of the Milky Way we might explore in the future. Whether exobeings will have crossed the frontiers we recognise in this field now is an open question. The speed of light will be the same absolute barrier for them as it is for us. Indeed, reaching a significant fraction of this speed will represent an immense technological challenge. This means that, for all practical purposes, the search for life elsewhere is, and will be in at least the near future, limited to our corner of the Milky Way galaxy. Life can only arise on planets (leaving moons aside for a moment) and these are relatively small compared to stars. And, of course, they only reflect light from the latter so that detecting planets in other galaxies is presently out of the question, despite advances in technology. This means that when considering the question of whether or not we are alone we should remember that we live in an island universe, our galaxy. We can assume that the amount of life in the Milky Way (if we find any) will be roughly the same across the hundreds of billions of galaxies of the universe, as our galaxy is in no way special with respect to the possible conditions for life. But we will never know, the distances are simply too vast. In a way it is a sad realisation that there are probably exocivilisations scattered across the universe and we will never know where they are, let alone get in contact with any of them. You might regret this fact and wonder why we cannot contact such exocivilisations, why the universe did not pan out differently enabling us to have such contact. But the universe did not develop for us, we developed because the circumstances just happened to be right on our planet. And the great distances between the stars and galaxies work in our favour: potentially threatening events, such as an exploding star, a supernova, occur about once every hundred years in our galaxy – one was registered by Chinese astronomers in 1054 and its remains, the Crab Nebula, can be seen in the constellation Taurus today. It is about 6,500 light years from Earth, 11 light years in diameter and is still expanding rapidly, at about 1,500 kilometres per second, pushing an enormous shock wave out into space around it. But given the distances just within our galaxy alone, we have not, to our knowledge, been negatively affected by such events before.

The non-predetermined origin of life on Earth is something of a damper for those individuals who regarded the rise of human life on Earth as something which was destined to happen all along. There is no scientific evidence that this is the case and we have no privileged or protected position in the universe. If we were unlucky enough to have a supernova go off within 30 light years of us, that would probably spell the end of planet Earth. Fortunately, there are no likely candidates within that distance of our Solar System.

9.1 Getting Around the Universe

As yet, all rockets which have lifted off the Earth have used chemical fuel which enables them to reach the velocity necessary to escape the Earth's gravitational field. There is a problem here: rockets have to quickly achieve a high speed after starting, their so-called escape velocity. This depends on the weight of the rocket, of course. To reach this velocity the rocket needs a lot of fuel, which in turn makes it heavier and it is thus more difficult to reach the escape velocity. One of the solutions to this conundrum is to jettison fuel tanks after take-off, when their fuel has been burned. It helps somewhat to solve the problem but one is left with a relation between rocket weight and the payload, for instance the satellite which it is trying to bring into a geostable orbit around the Earth, of roughly 10 to 1.

For interstellar travel, an option might be for the spacecraft not to carry its fuel along. So how could a spacecraft propel itself through space? There are a number of suggestions, one of them being to use a laser beam from which a spacecraft with a light-sensitive sail could harness the energy in the photons of this beam. The maximum calculated speed for such a space sailcraft would be about 10 per cent of the speed of light. A spacecraft would also need to expend energy in slowing down when approaching a destination, given that there is no material in space capable of decelerating a spacecraft by friction.

Another proposal would involve a rocket in which explosions in its rear would drive it up out of the Earth's gravitational field and then through space. Theoretically controlled nuclear fission explosions could be used to propel the craft at possible speeds of up to five per cent of the speed of light through interstellar space.

Incidentally, using the natural light of the Sun as an energy source would not be much use because, when we move away from the Sun, the pressure of photons decreases and hence the source of energy for any theoretical spacecraft would diminish.

Yet another proposal would be to use nuclear fusion as a source of energy. What one would need is deuterium (heavy hydrogen with a nucleus of one proton and one neutron), the atoms of which could be made to fuse together in a field of high-energy electrons to form helium (with an atom consisting of two protons and two neutrons), releasing a great amount of energy in the process. A nuclear fusion spacecraft could replenish its heavy hydrogen supply by scooping up hydrogen as it moves through space (from dust clouds or the atmosphere of gas-giant planets like our own Jupiter). Such a spacecraft might reach about 15 per cent of the speed of light.

Other proposals further from reality include constructing a spacecraft which would generate antimatter during its voyage. The collision of this antimatter with normal matter would lead to the annihilation of both and the release of a great deal of energy, including gamma rays, which are the most powerful known type of radiation. Such a spacecraft could reach a speed which would be a significant fraction of the speed of light.

The practical problems involved with such a device would be immense. For one thing, the astronauts would have to be shielded from the lethal gamma rays emitted in the matter–antimatter collisions. For another, hitting a piece of space debris, even larger dust particles, at just a fraction of the speed of light would automatically destroy the spacecraft, unless it had some type of shield to surmount this difficulty.

A further source of fuel for space travel might be the dark energy which is an intrinsic property of space itself and drives it to expand. It accounts for over 70 per cent of the energy of the universe and it might just be that some exocivilisation has developed a method of harnessing this energy and employing it for spacecraft propulsion. Earthlings are nowhere near anything like that – we do not even know what dark energy is, just that it exists.

Much interest in recent years has been directed at the option of asteroid mining as a source of water and fuel energy for future space missions, but only within the Solar System. The two types of asteroids which are of interest are (i) carbonaceous asteroids, which may likely contain water, and (ii) metallic asteroids, which may contain metals like iron (and possibly gold) that could be used as construction and repair materials for spacecraft. Various new techniques are being developed for mining as traditional techniques involve drilling or blasting, neither of which would be possible on an asteroid with virtually no gravitational field. Among the innovative methods are the use of sunlight focusing (using the magnifying-glass principle) to split parts of an asteroid by heat pressure and collect the debris for processing.

And then there is the warp drive: this would remain stationary in a warp bubble but compress space–time before it and expand space–time behind it, thus moving through space, apparently without any of the problems just described for more conventional spacecraft. A serious proposal for a warp drive was made by the Mexican theoretical physicist Miguel Alcubierre in 1994, based on one possible solution to Einstein's field equations in general relativity.

Similarly radical proposals for space travel also exist, such as wormholes, a kind of hypothetical tunnel connecting two distant points of space–time, or even two different universes. The point about a wormhole is that the distance inside the hole – which would be traversed at a subluminal speed – is much less than that outside the hole.

9.2 Sending Out Probes

It might make more sense, for both us and possible exocivilisations, to send out robotic probes to exoplanets as these would not be affected unduly by radiation (as long as there are shields around their sensitive electronic parts) and the time factor would not be so crucial. By the time we have developed the technology, we might have intelligent machines on board such probes, which would be capable of taking decisions independently of any mission control centre on Earth.

Even if we or an exocivilisation were to construct a spacecraft in which all the problems outlined here were solved, such a craft would still be an artefact of humans/exobeings and, like all such artefacts, subject to human/exobeing error in both their construction and operation.

Now assume that a spacecraft from an exoplanet managed to reach our Solar System; this would mean that they would have solved a whole raft of scientific problems, many of which are apparently insuperable barriers to us, mainly the speed of space travel problem.

Self-replicating probes could be a solution to the speed and distance issue, colonising a fraction of a galaxy. If an exocivilisation had a head-start on technology over us of some few million years, this could have already happened. To date, no evidence for this has been found (despite a few false positives) but admittedly we have hardly looked.

9.3 Getting Here After We Are Gone

Considering the obstacles to space travel leads quickly to a reality check. Imagine that an exocivilisation discovers that Earth exists and decides to send out a few probes to investigate our planet. They might take millions of years to arrive by which time we may have disappeared and the only traces we will have left will be contained in a very thin stratum of the Earth's crust. Nothing of us will remain on the surface of the Earth as this is far too active, with far too much erosion, volcanic activity and decomposition due to weather and plant life.

Animals will come to re-inhabit ecologies which were previously exclusively human, such as the megalopolis of present-day Earth. Insects will proliferate. Rodents of various kinds, rats, mice will abound. Wild dogs, pigs and goats will spread quickly. We know this from their introduction to island ecologies where there did not exist before, such as the feral dogs on the Galapagos Islands. These animals will roam the continents unimpeded by humans.

Maybe the best chance for human artefacts to survive will be beneath water by not being exposed to the elements on the surface. If there were buildings in the sea, they would be covered in barnacles like a shipwreck on the sea floor. But even they would disintegrate and, in a few million years, become a mere geological stratum.

9.4 A Feeling for Distance

Nothing is more sobering in the context of space travel than the distance issue. Consider that Voyager I is the space probe which has gone the farthest from Earth. It was launched back in 1977 and has been hurtling through space at about 60,000 kilometres per hour for 45 years now (mid-2022); it has only just left our Solar System a few years back. It makes sense to pause for a moment and consider the size of our immediate vicinity in the Milky Way and the challenges posed to us by space travel beyond the Solar System.

A speed of 60,000 kilometres per hour is the typical maximum velocity for a modern spacecraft, like the New Horizons spacecraft currently at Pluto (actually this varies between 50,000 and 80,000 kilometres per hour). A higher speed than that would present huge problems: if a spacecraft travelling at, say, 500,000 kilometres per hour was hit by a piece of rock the size of a pea, this would rip a huge hole in the craft making it unusable, ultimately causing the death of any crew onboard. So we will keep with the figure of 60,000 kilometres per hour. Now, there are 8,765.81 hours in a year so our hypothetical spacecraft could cover a distance of about 526 million kilometres in one year. This might seem like a lot, but at this pitifully slow speed the spacecraft would require 18,000 years to travel a single light year. Given that light travels about 9,467,082,000,000 kilometres in a year and the nearest star is 4.21 light years from us, it is about 39,856,415,220,000 kilometres away (just under 40 trillion kilometres). It would take our craft 6,642,735,87 hours to get there at 60,000 kilometres per hour. This means it would require 75,780 years for a present-day spacecraft to get to Proxima Centauri, the nearest star to our Sun; hence, it would take over 75,000 years to get to the first star beyond our Solar System, the first of the several hundred billion stars 'out there' in our Milky Way galaxy. These are sobering figures; we could contact a nearby planet perhaps, using radio waves, but visiting an exoplanet is currently way beyond our most advanced technological options.

THE GOLDEN RECORDS ON VOYAGERS I AND II

The Golden Records are discs attached to the sides of the spacecraft Voyager I and II. They were launched in 1977 and are currently beyond the Solar System, heading out into interstellar space. The records have grooves on one side with greetings in 55 different languages with the label 'The Sounds of Earth. United States of America, Planet Earth'. No exobeing could understand this, indeed it would be doubtful whether they would at first realise that this is the written representation of a human language (an exolanguage for them) – but given some ingenuity on their part they might work it out. There are drawings giving pictorial information about how to play the records, where Earth is located, what radio waves are, etc.

10

Assessing Probabilities

10.1 Considering the Fermi Paradox

Ever since the beginning of our space-faring age scientists have wondered about the likelihood that intelligent life could be found on planets outside our Solar System. At present there are no indications in our cosmic neighbourhood that there are any exocivilisations on exoplanets. The Italian-American physicist Enrico Fermi (1901–1954) was among those wondering about why we have no evidence of any life beyond our Solar System. He considered this with some astronomer colleagues in 1950, and after that this situation came to be known as the 'Fermi paradox'[1] and is still discussed widely.

There are probably hundreds of millions of more or less Earth-like planets in our galaxy but no sign so far that intelligent life exists on any of them, apart from our own planet, has been detected. The Earth formed about nine billion years after the Big Bang. So throughout the universe, and across the Milky Way, there could be millions of planets on which

1 Jill Tarter, former director of the SETI institute, maintains that there is no paradox because, given that we have not looked very far yet, the absence of data cannot be considered significant.

technologically advanced civilisations exist or existed. Would these have made themselves known throughout their part of their galaxy, at least through radio-wave leakage? If so, then there should be tell-tale signs of such civilisations across our Milky Way. But we have not reliably detected any such signal. This means that the Fermi paradox is still in need of an answer and a number of solutions, some of which are related, have been suggested.

Proposed solutions to the Fermi paradox

1. Intelligent life in the universe is rare indeed, maybe only once or twice per galaxy, separated by great distances. Furthermore, life in other galaxies could never get in contact with us in the Milky Way, given the vast space separating galaxies. If there were, for instance, an Earth analogue with exobeings in Andromeda, our closest major galactic neighbour, establishing contact with it would be impossible as Andromeda is currently over two million light years away from us.
2. The time factor – we could only contact exocivilisations which are more or less at our stage of development, for instance with digital technology, but have not run their course and declined, died away or been destroyed.
3. Neocatastrophism – we are just lucky that we have not been wiped out by a nearby supernova or a gamma-ray burst in our direction.
4. Sustainability – civilisations in general either exhaust themselves quickly or move very slowly, in which case it would take them a long time to reach a level of technology at which they could contact others.
5. The zoo hypothesis – exobeings exist but hide their presence from us, for a variety of reasons. They might fear attack from us or contamination of their planet or just think we are not worth investigating.
6. Aggression on exoplanets is turned inwards and hence exobeings have no time for space exploration. Any technology they have is used exclusively by the military.
7. Exobeings are everywhere but are invisible to us: it could be a question of scale. Maybe they are very much smaller than we are and we have not as yet noticed them. This is an intriguing consideration as it involves the

question of how small an organism can be with a brain comparable to ours in cognitive power.

8. They use (unspecified) means of communication which we cannot detect. We generally use electromagnetic waves (radio signals) to communicate and we assume exobeings would do the same given the possibility of modulating these waves to carry information at the speed of light. Recently, laser beams have been considered (and tried out) as a means of communicating our presence to possible others in our corner of the galaxy.
9. The exobeings live in virtual reality; our universe is only a simulation.
10. There is no Fermi paradox, we are alone in the universe (not just in our galaxy).

Speculation about how intelligent life in a galaxy might behave is a matter which concerned the Russian astrophysicist Nikolai Kardashev (1932–2019), who in 1964 proposed three basic types according to the amount of energy which a given galactic civilisation might harness. A Type I civilisation utilises just the energy which reaches it from its parent star (essentially the situation for our Earth). A Type II civilisation is able to harness all the energy released by the parent star. How this is supposed to happen is unclear, but there have been suggestions how it might be realised partially, for example by means of large structures in space to capture more of the energy output of a star than falls on the planet's surface. However, serious consideration of this option would need to address a variety of questions: (i) Where would the material come from to construct such structures? Bear in mind that about 98 per cent of the entire mass of the Solar System is contained in the Sun and, for exoplanets, the relationship would be similar. (ii) What materials could one produce to withstand the intense heat of a star outside a planet's atmosphere? (iii) How would one transfer the energy from a such structure back to an exoplanet to use for its local energy needs? The Type III civilisation that Kardashev proposed would be one which can utilise energy on the scale of the entire host galaxy, though how that is supposed to happen is anyone's guess. Other scenarios where exocivilisations would be evident are conceivable, for instance if they sent out self-replicating probes, so-called von Neumann probes, to populate the galaxy.[2]

Advanced civilisations might have already created networks of interstellar communication, which self-replicate and distribute themselves across a galaxy, according to the American science writer Timothy Ferris, although we cannot see any signs of this. The source civilisation could upload information about itself to such nodes and then document itself for later eons even if the source civilisation disappeared. However, given the lack of evidence of any such situation, this can only be regarded as pure speculation.

10.2 Looking at the Drake Equation

In 1961, the American astronomer Frank Drake (1930–2022) devised an equation to get scientists to think about how we could estimate the occurrence of intelligent life in the Milky Way. According to Drake, the likelihood of civilisations beyond our Solar System depends on the values for a set of seven parameters. The equation was written to estimate the number (N) of exocivilisations and has the following variables and form:

$$N = R^* \times f_p \times \eta_e \times f_l \times f_i \times f_c \times L$$

As Frank Drake has himself readily admitted, we do not know the values of the parameters in the equation and so its predictive power is slight. Nonetheless, Drake has said that there may be anything up to 10,000 civilisations in the Milky Way. While other scientists, such as Paul Davies, regard the lack of signals from exoplanets as an 'eerie silence'. Drake believed this is not the case given that our search for extraterrestrial intelligence is still very much in its infancy. Hence, the value of his equation lies in the impetus it provided to scientists to reflect on the issue of exocivilisations.

2 There is a project, called Breakthrough Starshot (one of the Breakthrough Initiatives funded by a number of wealthy philanthropists), which has as its goal to send a fleet of lightsail probes to Proxima Centauri b (the planet in the habitable zone of the nearby star Proxima Centauri, only 4.27 light years from Earth).

Breakdown of the Drake equation

$N = R^* \times f_p \times \eta_e \times f_l \times f_i \times f_c \times L$

1. R^*, the average rate of star formation, in our galaxy
2. f_p, the fraction of formed stars that have planets
3. η_e, for stars that have planets, the average number of planets that can potentially support life
4. f_l, the fraction of those planets that actually develop life
5. f_i, the fraction of planets bearing life on which intelligent life, comparable to human life on Earth, has developed
6. f_c, the fraction of these civilisations that have developed digital technologies and means of communication, emitting detectable signs into space
7. L, the length of time over which such civilisations release detectable signals

For scientists involved in exoplanetology, the key parameter is η_e (read 'eta Earth,' the first character is the Greek letter eta). The value has been discussed in great detail, both before and after the Kepler mission (see Chapter 8). This has had estimates which are widely different among the scientists involved but the sobering result from the Kepler mission is that very few, if any, planets it discovered match this parameter, that is, can be classified as Earth-like. However, this does not mean that there are no Earth-like planets orbiting stars in our immediate cosmic neighbourhood, it has more to do with the limits of current instrumentation, which is why advances in technology in the future could change the picture here.

Part III

Our Story on Earth

Exobeings can only have arisen through the long, slow process of Darwinian evolution by natural selection no matter how advanced any present situation for them might be. Any suggestions about how evolution could have panned out for them can only be made by considering key steps in the evolution of organisms leading to humans on Earth.

11

The Slow Path of Evolution

Our story begins with the formation of our Earth about 4.55 billion years ago from the swirling disk of dust and gas, at the centre of which was the young Sun. The latter was formed from the large concentration of material in the middle of this disk. Other concentrations had begun to emerge outside the centre and these grew with time, attracting increasing amounts of material by their growing gravity. The more matter gathered in these concentrations the greater the gravity they had, this in turn causing some of them to steadily increase in size. These concentrations yielded the eight planets we know in our Solar System, with many smaller fragments forming asteroids in the region between Mars and Jupiter and other objects, far beyond the planets, in the Kuiper Belt and the even more distant Oort Cloud.

To begin with there was a lot of loose material flying around and much of this rained down on the early Earth along with comets from the outer reaches of the embryonic Solar System. An actual collision of Earth with a smaller planet, given the name Theia, led to a significant fraction of Earth being flung out into space, this then orbiting the Earth, with the fragments finally clumping together to form our Moon (see 'The Role of Chance' in Section 7.5). The comets were the carriers of water and their continual bombardment of Earth led to enough of it reaching our planet for oceans to result later.[1] This is an essential feature of our planet: it consists of land and sea. But life began in the latter as water: the legacy of life's beginnings in the sea can still be recognised in the fact that our bodies consist mostly of water (about 60 per cent).

Four and a half billion years is a long time and geologists divide this into different sections, which together form the geologic timescale. The chronological units are in order of decreasing size: eons, eras, periods, epochs and ages. The first era is known as the Hadean Era and was characterised by heavy bombardment of Earth. By four billion years ago this had given way to the Archaean Era, which lasted until about 2.5 billion years ago. Exactly when the first life forms in the sea appeared (simple cells, prokaryotes) is controversial, but they may have arisen shortly after four billion years ago, when the period known as the Late Heavy Bombardment had come to an end.

Initially, our Earth did not have molecular oxygen (O_2), in its atmosphere. This probably appeared about 3.2 billion years ago when photosynthesis arose, allowing the production of energy-rich glucose from carbon dioxide, water and sunlight, giving off oxygen in the process. This led in turn to the rise in oxygen levels and the demise of the majority of anaerobic life forms (though there are some of these still in Earth's biosphere). Cyanobacteria and algae (eukaryotic organisms) were initially responsible for this rise in oxygen. Despite these major changes for nearly two billion years life only existed as single-celled prokaryotes until a chance event (or events) led to one type of cell being engulfed by another and, instead of its just being digested, it was co-opted into producing the energy currency of our cells, adenosine triphosphate (ATP). With that, the road was open for complex, differentiated cells and ultimately for complex life forms to arise. About 200 million years ago (mya), mammals first appeared on Earth. Initially, their progress was kept in check by the dinosaurs, but the latter became extinct after the serious asteroid strike 66 million years ago, known as the K–Pg extinction event (see Section 6.3

1 There has been a recent suggestion that Theia, like Pluto, formed in the outer Solar System and subsequently migrated inwards to the orbit of Earth. This view could make Theia responsible for a large amount of the water found on Earth.

above), ending their 150-million-year reign on Earth. When life recovered from this, nearly all the dinosaurs were gone, with modern birds[2] and reptiles surviving as their descendents. In the over 3.5 billion years[3] from the rise of life to the extinction of the dinosaurs, 99 per cent of all species on Earth died out. There were periods of flourishing and then of mass extinction (see Section 6.4) but, through these cycles, life managed to survive and to later flourish again.

11.1 Just What Is Life?

Although most people have an intuition about what constitutes life, this turns out to be a difficult term to define. There are various definitions, which depend on what perspective one adopts. Consider what the Austrian physicist Erwin Schrödinger (1887–1961) had to say: he ascertained that all living things strive to avoid decay into disorder and energy equilibrium (entropy). Hence for him life is characterised by the resistance to entropy. This makes sense: your life is the aggregate of all the thousands and thousands of chemical processes which are constantly taking place in your body to keep you alive. You are in a constant state of self-maintenance and self-repair, which overall can be characterised as homeostasis: the state of internal balance of the body's chemistry, its temperature, its energy intake, etc. Schrödinger's ideas have been echoed by later scholars, maintaining that there is a continuous tension in the body between the inherent tendency towards decay and hence equilibrium (entropy) on the one hand and that towards self-organisation and self-maintenance on the other.[4]

2 This is known from the fossil remains of transitional creatures such as *Archaeopteryx*, which was a bird-like dinosaur.

3 Stromatolites, a by-product of microbes, are large structures (like giant black mushrooms) which are about 3.5 billion years old and can still be found on the shores of Western Australia.

4 The English physicist Paul Davies has claimed that there must be laws which determine how organisms are to organise themselves. In his opinion there is a level of self-organisation which separates biology from chemistry.

Self-maintenance in biology also involves organisms reacting appropriately to their environment so as not to be subject to external threat. This in turn can concern the operation of antagonistic systems within the organism. For instance, our autonomic nervous system has two major sets of pathways, the sympathetic pathways (responsible for the 'fight or flight' response to critical situations) and the parasympathetic pathways responsible for the opposite set of actions (the 'rest and digest' attitude adopted when danger is not present). From this, one can see that self-maintenance in animals is both behavioural and biochemical.

Focusing on the latter perspective leads to the definition of life given by NASA scientists: 'Life is any self-sustaining chemical system capable of Darwinian evolution.' You can see this when someone dies, the body immediately begins to fall into multiple states of disorder and decay. The NASA definition also adds the dimension of evolution. Life forms strive to replicate themselves and to continue through the generations. This is achieved by cell replication controlled by genes which are present in our DNA. All the cells in our body have the same DNA, our individual version of this universal genetic code, and all forms of life have DNA (bar some RNA viruses). The different cells in our body, constructed by parts of our genome (the entirety of our DNA), result from the activity of certain protein-encoding genes, which build the particular type of cell in question.

Genes come in various kinds and there is at least a division into two main types: (i) structural genes, which encode proteins; and (ii) regulatory genes, which control the expression of the structural genes.[5] Thus, genes enable reproduction and growth for all forms of life. There are, of course, many factors involved in the expression of genes; there may be other genes which determine if certain genes can be expressed, a phenomenon called epistasis. Then there are epigenetic factors like histone modification, which concerns the manner in which DNA strands are wrapped around histone proteins, which in turn determines if the genes on that stretch of DNA can be

5 There is also a third type which appears to have no function. A very large portion of our DNA is just such non-coding DNA. Collectively, it was previously labelled 'junk DNA,' but more recent views see its presence as functional, even if it is not known in what way and to what extent.

read and processed or not. There is also DNA methylation, a biological process by which methyl groups are added to the DNA molecule. Methylation can change the activity of a DNA segment without changing the sequence. When located in a gene promoter, DNA methylation typically acts to repress gene transcription. This methylation is essential for normal development, with demethylation and remethylation occurring during embryogenesis.

11.2 Energy Regime of the Body

Self-maintenance and self-repair require energy. Consider that we must maintain our internal organs at a constant temperature of 37 °C and regulate water concentration and relative acidity in our bodies, the latter yielding a constant pH value of about 7.4 (very slightly alkaline or basic, on a scale of 0 to 14), which is essential for enzymes (proteins for accelerating or slowing down chemical reactions) to operate efficiently. The acid-alkaline balance is maintained by metabolic means via the kidneys and by respiratory means via the lungs.

In the evolution of our species the access to meat (muscles, tongue) and bone marrow of prey led to a great increase in protein intake and to a greater level of metabolism. While plants gain their glucose by photosynthesis – technically they are autotrophic – animals create their glucose by digesting food – they are heterotrophic. Furthermore, cellular respiration is necessary to break down glucose and to obtain ATP, the body's energy currency. With the improvement in our food intake, especially with cooking, our energy regime reached levels not possible for other animals.

11.3 Finding Out How Life Works

The discovery of how life is manifested in cells which replicate using information encoded in DNA occurred in various steps, with each generation building on the discoveries of its predecessors. The story of this discovery involves at least the following major scientists, who moved the field forward by their innovative foundational research.

WHERE WOULD EXOBEINGS GET THEIR ENERGY FROM?

Among higher animals there are two types of energy intake: one is breathing, the intake of oxygen and the related exhalation of carbon dioxide; and one is the intake of water and food. Oxygen intake is constant: interrupting it for more than a few minutes leads to death. Water intake is essential for life and without it we could only last a few days at the most. Food intake is less critical as we can also utilise body fat (animals which have brown fat can go without food for much longer periods, e.g. during hibernation in the winter). Depending on energy demands and our metabolism we can go without food for a week at least. To maintain their metabolism and the essential functions of their major organisms, exobeings would need an external source of energy. Whether they would have the split between breathing and eating as we have is uncertain. If they have carbon-based biologies with plant and animal life, oxygen is likely to be a key element in their atmospheres and one they would tap into for energy. If their metabolism produced carbon dioxide as a waste product, this would need to be released from the body, though by what means is uncertain. Food intake could show great variation compared to humans. It could also lay the foundation for cultural developments and sociality arising around the procurement/production of food (hunting, farming), its preparation (cooking) and consumption (eating together), as it did on Earth.

Major scientists in the history of evolution and genetics

1. Charles Darwin (1809–1882) proposed the theory of natural selection, first laid out in his ground-breaking publication *The Origin of Species* (1859).
2. Gregor Mendel (1822–1884) discovered the laws of heredity (during the 1860s, but his work was not appreciated until 1900).[6]
3. Thomas Hunt Morgan (1866–1945) showed that genes, the carriers of genetic material, are found in chromosomes contained in cells.

6 The English geneticist and statistician Ronald Fisher (1890–1962) managed to combine Darwinian natural selection with Mendelian laws of inheritance, which he expounded in his book *The Genetical Theory of Natural Selection* (1930).

4. Francis Crick (1916–2004), Rosalind Franklin (1920–1958), James Watson (1928-) and Maurice Wilkins (1916–2004) all played a role in discovering the DNA double helix and determining how it works during the early 1950s.[7]

These discoveries replaced older ideas, such as orthogenesis, an outdated view that biological evolution moved towards a certain goal, typically the development of modern *Homo sapiens*. This view, along with Lamarckism, the notion that characteristics acquired during the lifetime of an organism could be passed on to their offspring, were rejected in the modern synthesis which arose in the early twentieth century and which sought to unite Darwinian natural selection, Mendelian inheritance and more recent models of population genetics.

11.4 Our Restless World

While catastrophic setbacks represented major hurdles for life on Earth, a certain amount of volubility on a planet is beneficial for life. For instance, volcanic activity is advantageous as it causes material to well up from below the Earth's crust, which, in the long run, leads to greater fertility on the planet. But there is always a tension between the dynamism of a planet and events which cause species extinction. One such event was the Toba eruption on Sumatra, a massive volcanic explosion about 75,000 years ago.[8]

7 Francis Crick and James Watson created their famous double helix model of DNA, which they presented to the world in 1953. Their work was, however, based on that by molecular biologist Maurice Wilkins and, in particular, Rosalind Franklin, whose distinguished career as a chemist, and possibly later as a molecular biologist, came to a sudden end with her early death at the age of 37.

8 Lake Toba is contained in the caldera of the volcano, measuring about 100 × 30 kilometres. This volcano is not to be confused with the volcanic eruption of Mount Tambora (Sumbawa Island, Indonesia) in April 1815, which led to huge amounts of volcanic ash being spewed high into the atmosphere and blocking out the Sun for well over a year. It caused the 'Year without a Summer' in 1816 in Western Europe and parts of the eastern USA, with freezing temperatures during the summer months.

Another source of dynamism on Earth is continental drift. When tectonic plates press against each other the sea floor can well up and form islands out of the sea – Iceland arose in this manner, rising up over the surface of the sea from the Mid-Atlantic Ridge, which is moving apart at about 2.5 centimetres per year in the process known as sea-floor spreading. Subduction is a process which occurs when tectonic plates are pushed under one another and can lead to volcanoes on the surface, such as Mount Fuji in Japan. Tectonic plate activity would probably occur on all rocky planets, leading to continual changes in the planets' ecologies over time. This activity also leads to major climate change over geological time, providing a stimulus for new life forms to develop. It is assumed that all the land mass of Earth was originally a single block, called Pangaea, which began to split up about 250 mya, with the modern continents arising and still drifting apart to this day.[9] The break-up led to different winds blowing in different directions and, with the smaller continents which resulted, moisture in the atmosphere could more easily reach the interior of continents in the form of rain, leading to ecologies more favourable to the genesis of life.

11.5 Energy Gradients

An energy gradient is a kind of slope from a source of high energy to a target of lower energy. The flow of energy can be beneficial for life forms, as they can tap into this. Stars illustrate a type of energy gradient: their cores have temperatures of millions of degrees, so there is movement of energy from the stars to elsewhere around them, such as onto planets in their habitable zones. There are also intrinsic energy gradients on planets. The plate tectonics just mentioned would feed the dynamics of planets, providing new energy gradients via volcanoes, which could promote evolutionary diversity, though sometimes retarding this, as it did on Earth. Another energy gradient can be seen in the hydrothermal vents, so-called 'black smokers',

9 Evidence for continental drift is to be seen in fossils found in different parts of the world, now separated by water, but assumed to have been once part of the same land mass.

on the ocean floor where hot mineral-rich smoke is emitted into cold water on the floor of a deep ocean. Around such energy sources life forms can flourish, tapping into the minerals and heat emitted from the smokers. Of course, if such sources disappear, life will soon follow the same path.

11.6 Life Getting Under Way

When looking at the development of our planet across geological time we should recall that for nearly two billion years the only life forms were simple-cell bacteria and archaea. Viewed across the habitable-zone planets of our galaxy, that period might be unusually long, it might be typical or it might indeed be short. There has been discussion recently about whether comparison of exoplanets with our Earth would yield the right insights concerning the potentiality for life. Some scientists, such as René Heller and John Armstrong (2014), have posited a further category, superhabitable planets, on the assumption that the conditions on our Earth are good, but not optimal for the development of life. A slightly higher density, with an attendant stronger gravitational field, would keep more oxygen and carbon dioxide closer to the surface, providing a greater source of energy (oxygen) and warmth (the greenhouse gas, carbon dioxide). Superhabitable planets would orbit so-called K-type stars, slightly less luminous and less massive than our Sun (a G-type main-sequence star) but brighter than M-type stars (including red dwarfs). K-type stars could well have an average lifespan of around 50 billion years, many times that of our own Sun.

The move to complex-cell life forms might be an extremely rare development. As always with such questions, we do not know the answer as yet and hope to find answers in the coming decades of space probing and exploration. What one can say is that, for an exoplanet with exobeings, there will have been a similar step forward leading to a great diversity in cell types, and ultimately in life forms, no matter how long it took for this step to be completed.

For the development of more complex forms of life on Earth, the process really got underway with the so-called Cambrian 'explosion,' which probably took place around 540 mya. During the 15–25 million years of this event a great diversification of marine life occurred and all the major animal phyla arose (these can be found in the fossil record). Before this period most animals (in the sea, of course) were probably filter feeders, engaged in a minimum of movement, waiting for food to come their way, for example in the form of plankton or other tiny organisms. Another reason for the relatively sudden diversity among vertebrates in or around this period is the duplication of whole genomes within the DNA of cells, and not just of single genes, which would require far more time to trigger the level of diversity we observe.

About 370 mya (in the middle of the Devonian period), the first amphibians moved onto land. Animal fossils found in Canada and Greenland over the past 140 years show downward-pointing, bony fins which can be interpreted as early forms of arms and legs, allowing the creatures to move on land to a limited extent. The most famous find in this area is the fossil of an animal known as *Tiktaalik* (an Inuit word used for the assumed animal found on Ellesmere Island in Nunavut, in the far north of Canada, by the palaeontologist Neil Shubin). This animal belonged to a genus of lobe-finned fish which formed a transition to land animals. It had a flat head and fins pointed downwards, with sturdy bones which would have allowed it to move somewhat on land, as if it had fin-legs. Although the *Tiktaalik* genus still had gills, later animals developed the ability to extract oxygen from the air (usually via lungs but sometimes through the skin also, as frogs do when under water) and not just from water as fish do with their gills. The proto-lungs were probably the swim bladders which many fish use to trap air, helping them to remain buoyant in water (The swim bladders themselves may have evolved from an even earlier lung-type structure). These bladders would have had networks of blood vessels allowing them to remove the oxygen from the air.

Early amphibians developed their leg-like limbs from bony fins which allowed them to move on land; those which developed front and rear fins in this way became the first tetrapods (animals with four legs). They also developed necks, which improved their ability to scan their environment visually (by moving their heads) and use their mouths and teeth at various angles. A further essential development was the rise of mammals who, as the name implies, give birth to live young who are fed by the mother via milk glands (breasts). Carrying young in the body before birth put a limitation on the size of animals. Admittedly, the variation is considerable, witness the size of mammals like elephants and giraffes, but these pale in comparison to some of the dinosaurs, such as *Brachiosaurus*, which could be up to 20 metres long and weigh over 50 tons because it laid eggs out of which the young hatched.[10]

WOULD EXOLIFE FIRST ARISE IN WATER AND THEN MOVE TO LAND?

Given the abundance of water in the universe, and its ideal nature as a biosolvent, it seems likely that exolife would arise in water first. There are, however, limitations on how far life could develop under water. Without limbs comparable to our hands, underwater beings would not have any equivalent to our manual dexterity. Furthermore, it is not probable that their marine environment would promote particular intelligence. In the oceans of Earth, the most intelligent animals are air-breathing mammals, which moved from land back to the sea.

LOCOMOTION AND LIMBS WITH EXOLIFE FORMS

The ability to move around freely in their environment is key to all complex life forms. On an exoplanet, this would be equally essential. But whether exobeings would have four leg-like limbs to move on land, like terrrestrial quadrupeds, or two for motion, freeing up another two for activities comparable to those we use our hands for, is speculation. Other combinations are possible, such as four legs and two or four hands. However, hand-like limbs would be necessary to construct artefacts and ultimately digital technology for interstellar communication.

10 Other physical limitations are the thickness of bones relative to tissue mass. The large dinosaurs had very thick bones to avoid cracking under the weight of their bodies. Birds have air cavities in their bones, which reduces their weight and gives them buoyancy in air.

The ability to breathe air arose independently in a number of fish, so it is not a feature of evolution which was dependent on a single species or a small group of animals. This is quite different from the situation with the transition from prokaryotic to eukaryotic cells, which would seem to have occurred with a small group, or even just a tiny number, of cells that managed to engulf other cells by co-opting their DNA to build a source of energy with these cells – the mitochondria – thus giving a great push to the evolution of life. In addition, the drop in oxygen levels in the water (due to oxygen-consuming algae) added an additional motivation to fish to start to breathe air and so keep their oxygen levels up to what they required.

11.7 Functional Principle and Realisation

Life on Earth uses oxygen as a source of energy. For the simplest forms of life, like sponges or jellyfish, oxygen can diffuse from their surface to the body's cells. More complex life forms, however, all have a method to transport oxygen from outside to the cells inside the body and then rid the body of carbon dioxide (a waste product). So the basic principle we are talking about here is transportation. But how is this realised in different life forms? This is where the variation is apparent. Oxygen may enter an animal's body via receptor cells in lungs (for land animals and sea mammals) or gills (for fish) or it may diffuse through the skin (as with frogs under water). The system of transportation we humans have is blood, which contains macromolecules called haemoglobin, a complex three-dimensionally arranged protein that contains iron, which can hold oxygen and release it to the cells around the body when necessary. The haemoglobin can furthermore carry carbon dioxide away from cells back to the lungs for release out of the body. There are different mechanisms for oxygen and carbon dioxide transportation and different types of molecules which carry out this function in different animals. For instance, haemocyanin contains copper, which can reversibly bind oxygen and thus be used for its transportation to the cells of an organism's body. Such is the case with arthropods like today's insects, which have an outer skeleton, called an exoskeleton, rather than an internal system of bones like vertebrates (an endoskeleton).

This situation is found throughout the natural world: there are functional principles which are variously realised. It is the variation which results in the differences in phenotype (outward appearance). For exobeings, the same would probably hold. Assuming an oxygen-rich atmosphere on an exoplanet, exobeings would utilise this source. But just how they would intake oxygen and transport it to different parts of their bodies would be another matter. The sum of functional principles and their realisations would thus result in phenotypes which would be very different from us humans but nonetheless similar on the level of general biology. Furthermore, we could expect that exobeings would use certain organisational principles of their bodies for more than one function as we do. Our blood is not only used for carrying oxygen to cells and carrying carbon dioxide away, but also to carry many other substances, such as diverse nutrients (those deriving from the food we eat) to cells. In fact, our blood is a universal transport system in our body. Terrestrial biological evolution would lead us to suppose that exobeings would also have such a system though its realisation would probably not be identical to ours.

11.8 The Rise of Predators

The Cambrian period is noted for the appearance of predators – fish who hunted other fish. The act of hunting involves planning ahead, sensing where prey is, ambushing or chasing this and making the final grab at the prey – the animals must imagine prey and, when hunting, they must realise the prey is still around when out of sight. This is cognitively much more complex than simple filter feeding where fish suck in and eat tiny animals or where land animals consume foliage or grass. Large, sophisticated predators all have brains with the ability to plan and execute cognitively challenging tasks, such as scanning the ground from a height (as birds of prey do) or diving into the water to catch fish (as seen with cormorants, for instance).

The rise of predation led to a number of strategies appearing among animals who were potential prey for others. In nature, there are basically three ways to defend yourself: (1) fight, if you have claws and teeth, or are just bigger than your foe; (2) flee, if you are a quick runner; or (3) camouflage yourself, if neither (1) nor (2) applies. Cephalopods – octopus, squid and cuttlefish – are pre-Cambrian animals who still exist today availing of option (3) above: in less than a few seconds these animals can change their skin to merge in with the surroundings they happen to be in. This includes not just changing colour, but also adopting highly complex surface features on their skin, which can make it like sand or a rock covered with barnacles. This ability has meant that the neurons of these animals are found distributed across their body and not solely concentrated in a large controlling brain, as with mammals.

Predation and reactionary defence mechanisms are regarded by evolutionary biologists, such as the English scientist Richard Dawkins, as a key step in the increase of intelligence. Furthermore, predation is a relative phenomenon: the animals which are predators for some can be prey for other, somewhat larger animals, and so up along a line which leads to apex predators. An additional factor is the social element to predation: animals which hunt animals larger than themselves generally do this in packs, like wolves. When a single animal would be a match for the hunter, like a water buffalo for a lion, a pack is necessary to kill the prey. This demands a high level of cooperation and the animals of a pack must be able to communicate with each other.

Early hominins were not only predators but also preyed upon. The fact that in the savannah of eastern and southern Africa they represented prey for many animals, such as large cats, hyenas and even large birds like giant eagles, provided stimulus to their cognitive development. Many of the fossil skeletons of early hominins show marks made by the claws, talons or sabre teeth of predators during an attack.[11]

Those hominins who cooperated in groups, and so enhanced their chances of survival, were generally successful. This cooperation was dependent on

11 The notion that aggression within the genus *Homo* was a driving force in evolution has been supported for a long time, for instance by the well-known palaeoanthropologist Raymond Dart. But there are two sides to this coin: early *Homo* species were both predators and prey for others. So the avoidance of predation is equally a significant factor in our evolution.

a theory of mind, the ability to work out what others were thinking and feeling and so cooperate with them. Such a situation would have meant that the more intelligent hominins survived, thus providing considerable impetus for our evolution towards the genus *Homo*. As time passed, intelligence was our chief quality, not physical prowess, like so many other animals.

Modern humans are among the apex predators of the animal world. We are at the top of a long line of hunters and hunted because there are no animals who eat us to survive. But we are different from, say, lions or sharks, other apex predators. Lions, for instance, live in the savannah of eastern and southern Africa where they can hunt for prey, such as zebra or impalas, or even a baby elephant if they can get past the adults shielding it. But they are confined to this environment; they could not survive in a tropical forest or a desert or a temperate zone for lack of suitable prey. We humans, on the other hand, control our environment and hence are flexible in where we live: with housing, temperature control and food management, we can live anywhere between the ice and snow of the far north or south and the heat of a desert or in any environment in between. As opposed to animals, we usually create the habitats we occupy. This flexibility of an environment comes at a price: the amount of space occupied per human is enormous compared to most animals, we have houses and usually work in buildings. Our carbon footprint is also a major consideration: we heat or cool our buildings, depending on climate, and we use various modes of transport which consume fossil fuels.

11.9 Different Kinds of Evolution

Before delving into the details of how evolution works, a few words are required on the kinds of evolution and the timescales involved. Darwin himself realised that hundreds of millions of years were necessary for natural selection to occur on a scale which would produce very different species on Earth. We now know that genetic mutations occur fairly frequently, but these need to also result in a change in gene expression, and hence protein encoding, for natural selection to take effect. For instance, if a mutation occurs which renders an animal's fur more light-coloured and this animal lives in a cold habitat with much snow, that mutation confers a selectional advantage on this animal. The long-term cumulative effect of such mutations are ultimately responsible for the white fur of the arctic fox or the mountain hare, for example.

The fastest type of evolution, cultural evolution, can take place within some thousands of years. Such changes are not as clearly connected to enhanced survival as are instances of the slowest type, macroevolution, where a deleterious change in the gene pool of a population can have very serious consequences. The reason for this is that cultural evolution is accompanied by cultural buffering in which developments in a society can reduce negative effects of our environment (see Section 14.9) Modern medicine is the best example of cultural buffering as it protects us from a whole variety of pathogens which could severely impact on our survival. However, deleterious genetic mutations can cause major problems as they are often only expressed in later generations.

Major types of evolution, listed by relative timescale

Macroevolution

Slowest kind, relies on the operation of natural selection and the results of favourable mutations.

Microevolution

Relatively slow, but gene flow between populations with subsequent separation can speed up speciation, as can genetic drift within a population.

Cultural evolution

Fairly rapid, depends on developments in communities and societies leading to biological changes in humans, such as the rise of adult lactose tolerance with pastoral farming.

Can evolution be observed today? With humans, this is not the case, which is the main reason why scientists disagree about the rate at which we are currently evolving, if at all. However, with intensive breeding of animals, such as various types of dogs, the results of selection, here driven by human choice, can be seen within a very short time span.

11.10 Genes and Phenotypes

Genes are functional stretches of the DNA molecules in the nuclei of a body's cells; they form the units of heredity in our DNA. In humans, there are about 20,00 to 25,000 genes, which vary from a few hundred to up to two million base pairs on the DNA double helix (a full strand of DNA contains about three billion base pairs). Not all genes code for proteins (complex molecules which perform specific functions in the body). Not only that, there is, as yet, no way of identifying the genes which are responsible for building an organ like the stomach or the heart, let alone those involved in the generation of the language faculty in the brain.

The word 'phenotype' refers to the outward appearance of an organism or its constituent organs. For instance, one could say that, phenotypically, chimpanzees and gorillas are similar, but in evolutionary terms they belong to different genera.

The differences between the genetic and the phenotypic levels are very important.[12] Consider humans, chimpanzees and gorillas. Humans and chimpanzees are more closely related to each other than either are to gorillas, sharing well over 90 per cent of their genetic material. But if you consider their phenotypes, that is, what they actually look like on the scale of their entire bodies, then chimpanzees are much more like gorillas. In the words of the palaeoanthropologist Daniel Lieberman, a gorilla is like

12 In their innovative textbook on evolutionary linguistics, McMahon and McMahon (2013: 11) make a three-way distinction in the use of the term 'evolution': '... we will ... be making [a] ... three-way distinction between evolution at the genetic level, physical phenotypic consequences at the structural level, and the variable behavioural systems these structures enable.'

a pumped-up chimpanzee. Phenotypically, humans are very far removed from both chimpanzees and gorillas.

One could ask an obvious question: why does virtually identical DNA not produce virtually identical phenotypes? And does that few per cent contain the blueprint for all that makes us different from chimpanzees, for instance human language? The answer has to do with gene expression. Genes do not translate directly into physical features. Genes are expressed, meaning they trigger complicated physical processes that involve many stages, in which other genes can play a role in enhancing or inhibiting the development of physical features. Genes are used to build function-specific proteins in pieces of biological machinery inside our cells called ribosomes. These synthesised proteins then go on to perform myriad functions, including cell growth itself, and this is where the differences become apparent, including those in communicative ability, given that our stem cells trigger the growth of vastly more neurons than do those of chimps.

11.11 Control from Above or Below?

When we consider an organism as complex as the human body we might think that there must be control from above which insures that the whole thing works properly. Take the example of the liver. If a scientist tells us that the liver is a bundle of atoms and molecules which interact in a certain way, we may feel inclined to counter that the liver is an organ, which works so well and performs its functions so efficiently that it must have some level of control or supervision above the level of its lowest-level components.

The motivation behind such a view probably derives from our experience of human artefacts. Think of some large complex object made by humans, an electric power plant, for example. We recognise the necessity for a level of supervision – provided by a team of engineers – to ensure that all the individual parts of the power plant work correctly and, should a problem arise, that it is solved satisfactorily. So we have a notion of 'control from above' for complex objects. Now we seem to transfer this thinking to natural organisms. When we think of animals or humans, we tend to think that there is a level of supervision or control above that of the individual atoms and molecules which these beings consist of. Why? Well, we might say, someone must look after all the different parts of the being and ensure that they work properly and someone must step in if something goes wrong and guarantee that the being as a complex system continues to function. To take the example of the power plant again: the supervisors must ensure that electricity is continuously generated. We might then think that the 'supervisor' in the body ensures that it remains healthy so that it can survive for as long as possible. For example, if an atom or molecule fails to react the way we expect it to, someone must step in and do something, otherwise there are consequences for the functioning of the entire organism.

Going back to the power plant: if the cooling system fails, the engineers must do something, like switching on the auxiliary cooling system while they repair the primary system. But atoms and molecules do not 'fail' in the way that human artefacts do. If the conditions change, their behaviour may alter, but then other atoms and molecules may kick in under these changed conditions, restoring the original state and hence ensuring that the first group of atoms and molecules continue to behave as they did before the conditions changed temporarily.

Of course, it is an enormous simplification to talk of just 'atoms and molecules,' what we have in practice are very complex and finely tuned structures, consisting of myriad building blocks which behave in predictable ways. Take, for instance, the reaction of the body to a change in temperature. If it decreases you may start to shiver. This consists of involuntary muscular contractions – caused by chemical changes at the relevant points in the muscles – which generate heat in the body. If the temperature increases above its normal level, the pores of your skin widen and your sweat glands become active. The sweat, which is secreted onto the skin, evaporates and the body loses heat in the process.

The genesis of organic structure is always from below and human artefacts arise through organisation from above. So do we need a level of organisation or control from above for the human body? Not if all organic structure arises from below, and ultimately from atomic and molecular constituents. There are, in an organism as complex as the human body, different layers of organisation. Moving up from the simplest constituents we recognise that structures appear which afford control over the maintenance and repair of the organism. The organs of the body, which perform dedicated functions, are found on a higher level and the immune system becomes evident, a system which consists of elements with specific functions when fighting off pathogens (see Section 32.8). But all these higher levels arise from below, from a level of organisation 'further down' towards the ultimate constituents of the organism.

All of this means that we do not need to posit a level of organisation from above, that is to say we do not need a creator to 'make' the things we have in the universe, including ourselves. What remains is the absolutely staggering ability of atoms to aggregate into exceedingly complex molecules, such as the proteins in our bodies, and construct from below structures of extraordinary intricacy and efficiency. Just think of differences in consistency for a moment: atoms/molecules can form substances with differing degrees of rigidity and consistency. Consider the major components of your body from this perspective for a moment: your bones are hard, your tendons are tough but pliable, while your tissue is soft and your blood is liquid, but can solidify via coagulation if required, for instance at a wound. If all atoms/molecules displayed the same consistency, the differentiation of structure and function on the level of our bodies would not be possible.

11.12 'Design' from Below

When design from below fulfils its function, there is no further motivation for it to continue developing. Take the human throat as an example. If someone were to design an animal, the very last thing they would build into the creature is a common pipe section for both air and food – that's plainly ridiculous. But evolution works differently. For animals like mammals, and then the primates, the throat is short. For instance, in the chimpanzee the uvula at the back of the mouth makes direct contact with the epiglottis, which drops down, closing off the trachea (windpipe) when the animal is swallowing food. However, with humans, due to our lowered and elongated larynx, the epiglottis is quite far down in the throat and the danger that food could enter the windpipe is greater. But our epiglottis does the job of closing off the trachea and so there was no evolutionary pressure to develop some other mechanism to deal with the bifurcation of wind and food pipes in the larynx. Once swallowing is initiated, the swallowing reflex is triggered. This is under involuntary neuromuscular control and consists of four distinct but closely coordinated movements: (i) the tongue blocks off the oral cavity to ensure that food does not flow back into the mouth, (ii) the uvula closes off the nasal passage to ensure that food does not go up the nose, (iii) the vocal folds close off the lungs and (iv) the epiglottis closes off the windpipe to ensure that the food, made soft with saliva into bolus, glides down from the pharynx through the larynx into the food pipe (oesophagus), where peristalsis (muscle contractions in the walls of the food pipe) ensures that it is pushed further down into the stomach. So although we live with the constant danger of pulmonary aspiration of food and then choking from airway obstruction, this rarely happens.

12

How Does the Whole Work?

As humans we are confronted with devices which are supposed to work and often do not. Just think of all the domestic appliances you have at home. Do all of them work? I am sure that you can remember the time when the toaster gave up the ghost or the torch in the garden shed did not work. We have a notion of device and we have an expectation that it will work. But when we say a device does not work, what do we mean? Generally, we mean that it does not perform the function we expect of it. If you put sliced bread into the toaster and press down the lever at the side and nothing happens you utter a sigh of frustration because the device is not working. If you go out to the garden shed on a wet winter's night and find that the torch on the shelf as you enter does not work, what you mean is that the light does not come on when you press the switch.

What you really mean is that the declared purpose of the device is not being fulfilled. The components still have the same behavioural properties they always had. However, something else might have happened which prevents the intended purpose of the device from being realised. In the case of the torch, the contacts inside it have probably oxidised from being left in a damp shed; in the case of the toaster, the wire of the heating element may have broken at some point due to a surge in electricity or to brittleness which arose through age (most electrical devices cease to function for some such reason).

The point being made here is that the notion of working and not working is peculiar to human artefacts, to the devices which we build. On the level of physical properties and laws, nothing ceases to work in the sense of showing unpredictably altered behaviour. Electricity will only flow through a medium if this medium allows an electric charge to propagate through it. Air does not conduct electricity (at small currents, lightning is a different matter) so the broken wire in the toaster will stop the element from heating and making toast. Metal oxides do not generally conduct electricity in appliances either, so contacts which have oxidised will prevent the flow of a charge and so that from the battery will not reach the light bulb in your torch.

12.1 Devices and Organisms

Now think of the human body. It is certainly highly complex and appears to function always (at least for young individuals in good health). But the functioning is not due to extrinsic design, as with a torch or a toaster, but due to the fact that the myriad of atoms and molecules behave in very specific ways. The human body was not constructed like a machine made by humans, rather it evolved. In the course of evolution, the specific behavioural properties of atoms and molecules in the body led to certain structures arising and to certain functions being fulfilled by the body. In the course of evolution, repair mechanisms also arose so that, when we break a bone, the ends at the break will normally grow back together again. By comparison, the oxidised contacts of your garden shed torch will not clean themselves for the torch to regain its original functionality.

You might think the repair mechanisms in the body are sensible and you can imagine that if you were asked to design a human organism you would certainly include such self-repair strategies. If the human body shows the types of properties which you would imagine including in a designer organism, this is a coincidental parallel. In some ways, the body does not have functions which it would be very useful for it to have. For instance, the human body is practically incapable of repairing damaged nerves, which is why damage to the spinal cord has such devastating effects for accident victims. It is easily imaginable that the body might have strategies for allowing nerves to grow back completely again and regain their original functionality but this just did not happen in the course of the evolution of humans and so we do not have this ability today. Repair mechanisms also differ across the animal world: some animals, called polyphydonts, regrow teeth in adulthood, such as reptiles like crocodiles, but humans do not so we have to make do with what we have and resort to dentists when our teeth start causing problems.

12.2 Evolution and Design

There is strong evidence that the body has evolved and not been designed. What sort of evidence? Well, take left over bits from earlier stages of evolution, so-called vestigial structures. At the base of our backs, we have a coccyx, a tailbone, consisting of three to five vertebrae at the end of the spine. It is a remnant from the time when our distant ancestors had tails. Attached to it are some tendons and ligaments, so it serves a function in stabilising these, but not as the tail which it once was.

There are also parts of the body which are less than optimal in their present manifestation. For instance, your eyes have a blind spot where the optic nerve is connected to the retina (the concave back of the eye with the millions of nerve cells, rods and cones, whose function it is to convert the image coming through the lens into electrical signals which are transmitted to the visual cortex at the back of the brain). If the eye were designed 'from above', the designer would most likely not have left the blind spot (unless as an oversight).

Another example concerns the female reproductive system. The two ovaries in the female body are each covered by a bell-shaped structure at the end of the fallopian tubes (the connection with the uterus), known as the fimbriae. Just before ovulation these swell with blood and then massage the ovaries gently to move an egg into either of the fallopian tubes, where it can be fertilised by a male sperm. In some very rare cases, a fertilised egg can escape between the fimbriae and the ovaries resulting in an abdominal pregnancy (an unusual type of ectopic pregnancy). If the area of the ovaries and fallopian tubes were designed 'from above', surely the ovaries would be contained within the fallopian tubes to avoid the chance of a fertilised egg escaping into the abdominal cavity. But because the present system works in the overwhelming majority of cases there was no evolutionary pressure for any other system to arise and replace it.

12.3 Do the Parts Know the Whole?

If the functionality of the body is due to the interactions of billions and billions of atoms and molecules, one is justified in asking whether all the functions of the body conspire to keep the body going? That is, does the body have a notion of itself? When we think of our bodies, we do very much have the feeling that the human body behaves as a whole and has at least two major aims: to stay alive as long as possible and to propagate itself. The fact is that, under normal circumstances, our bodies stay alive for several decades and have a sexual drive, which we can interpret as an indirect but usually effective means of getting us to propagate. But the question is: Does the body behave in such a way *in order to* stay alive and propagate?

If you take a bottom-up approach and look at all the biochemical processes in the body, then it is hard to maintain that each process takes place *in order* that the body survives. To claim this would be tantamount to saying that biochemical processes 'know' what they are doing; say, that the acid and enzymes in our stomach 'know' that they are preparing life-essential nourishment for intake by the intestines. Chemicals in the body do not 'know' anything so this assumption is untenable. The evolutionary biologist would maintain that the interactions which take place in our bodies have the serendipitous effect of letting us survive for several decades and propagate ourselves and so we, as a species, are still here.

For us as thinking beings, it may be difficult to accept that organisms as complex as ourselves have resulted just because it happened that our different parts came together, or arose, and now work harmoniously as a unit. There is one other possible explanation. This is that the brain has a conception of the entire body and makes sure that the right biochemical processes take place to ensure that the organism survives. But this turns out to be a weak argument. It is true that the brain is very firmly embedded in the body, but it does not control many of the processes which take place, or only very indirectly. Anyhow, if it turned out that the brain controlled the body *in order* for it to survive, there would still be the further question to answer: Where does the brain get this notion of the body from?

One reason for thinking that the body knows itself could rest in the phenomenon of proprioception – the ability of the body to perceive itself, feel its parts and register its movements. Proprioception is essential to our successful negotiation of our surroundings and to avoiding danger by feeling pain, for instance. But that does still not mean that the body knows itself, rather that feedback from the nerves around the body is processed by the brain. The notion that the body knows itself is an epiphenomenon of proprioception, something which our consciousness makes us think. On a conscious level, we have a notion of body, but this is a side effect of the way in which we conceptualise the world we live in, including our own bodies – but I am getting ahead here – the topic of consciousness will be dealt with in a later chapter.

12.4 A Question of Scale

We exist on a scale which is very roughly about halfway between atoms/molecules and stars. This is the scale of the world we know: anything smaller than what we can see with our bare eyes is beyond our world of normal experience and we need a microscope or at least a magnifying glass to perceive it. Anything larger than the Earth is beyond our cognitive grasp unless we study astronomy.

So why do we exist on the scale we do? Well, consider that vertebrate animals with skeletons are suitable for moving around on land at a height of about one to two metres. Some animals, such as giraffes, reach up to five metres, but this is very much the exception. Animals with a shell-like outer casing (an exoskeleton), such as insects, are severely limited, given that the size of the casing and its internal volume have strict physical limitations. These limitations have played a role in our evolution as well. Take our heads, for instance. They are quite heavy with our large brains and big skulls around them – a typical adult head weighs between 4.5 and 5 kilograms. This weight sits on our neck and shoulders, and is kept upright and movable by a series of muscles. In fact, the head is balanced on the atlas, the top cervical vertebra of our spine (C1). Which is located under the skull in the middle of the head. Given this delicate arrangement, a much larger and heavier head would not have been favoured by evolution. Of course, we could have developed different types of heads, but with our current anatomy a significant increase in the size and weight of the head would not be feasible.

12.5 When Do Cells Become an Organism?

The large-scale appearance of a life form, called its phenotype, is made up of billions and billions of individual cells. So these cells are the starting point. For a complex life form, like humans, to exist, some 220 different cell types come together into a single, interacting and functioning whole. Contrast this with slime moulds, for instance. These consist, at a minimum, of single complex (eukaryotic) cells, which can aggregate to form clumps but do not form a single organism as the cell aggregations are not divided into separate functional sections.

Cells have walls based on fat molecules, called lipid bilayers, in which two layers are positioned together with a hydrophilic (water attracting) head, consisting of a fatty acid, pointing outwards, both on the outside and the inside of the cell membrane, and a hydrophobic (water repelling) tail pointing inwards away from the edge. Such membranes provide an environment for reactions to occur isolated from their surroundings. The walls of cells have special qualities: they allow certain substances to pass through, either taking in material which they convert into energy or giving off material it has generated and which is required outside the cell or which is waste to be disposed of.

There are minute machines operating tirelessly in all cells: the ribosomes are constantly reading off the messenger RNA (produced from the genetic code DNA in the cell nucleus), taking any group of the 20 amino acids and then combining them in specific three-dimensional configurations to form any of the hundreds of thousands of complex, task-specific proteins used in your body.

Cells react to their environment and, importantly, they divide, producing replicas of themselves. At this scale, being alive is the sum of all these processes taking place at any one time in the cells of your body. Of course, some cells have priority on the higher level of the organs of which they are composed. For instance, you can do without a leg or an arm but obviously you cannot do without a heart or a brain.

The multi-part cells in our bodies and other life forms can only have arisen through evolution. No biological organism is an artefact of another organism, that is to say, no organism has been constructed consciously by any other.

12.6 Sexual Reproduction

Plants and animals use various mechanisms to reproduce. Budding is the simplest form: a piece of an organism breaks off and forms a new one as with hydra, a group of small organisms not larger than one centimetre, which live in fresh water. Sporogenesis is where a fungus, such as a mushroom, releases spores into the environment, as a little cloud of cells, and these can form new fungi if they find a suitable substrate to grow on. Some plants reproduce via runners, roots which move out horizontally and sprout somewhat away from the parent plant, with the daughter plant later severing the connection with the former. Pollination is a further common method of reproducing; for instance, a plant's flower has an anther and a stigma, with the former producing pollen (with male sex cells) and the latter receiving it. This is usually by a visiting insect, such as a bee, inadvertently transferring the pollen from one flower to the stigma of another to be fertilised with the egg cells this contains. Wind pollination is another method found, for example, with flowering trees, which blossom before their foliage appears.

The reproduction types just mentioned are typical of simpler forms of life. However, most animals, and all mammals, reproduce sexually and are distinct in that their sexual organs are different, with the male providing the sperm and the female the egg and then bearing the offspring (with mammals) or laying the egg (with birds or reptiles, for instance).[1] In addition, many animals are sexually dimorphic in phenotype with the males larger and stronger than the females and more flamboyant, such as showing more colourful plumage as with pheasants or peacocks. The flamboyancy is related to mating: the males need to impress the females with a show of plumage to convince them to mate with them.

Going on fossil evidence, sexual reproduction arose about 1 to 1.2 billion years ago. Essentially, this is a process in which genetic material from two individuals, different in sex, combines to form a new individual, the offspring of the original two. At the beginning of sexual reproduction is the making of gametes - sex cells - which in males are sperm and in females eggs. A process called meiosis[2] occurs in which the number of chromosomes of the parents is halved (from 46 to 23), going from a diploid number to a haploid number in the gametes which result. During fertilisation, two gametes, from different sources, the mother and father, unify to form a diploid zygote, which then contains twice the number of chromosomes, and so the original number of 46 is restored. In all mammals and primates, the female gamete is the egg and the male gamete is the sperm, the resulting zygote is the fertilised egg, which develops into the embryo and then the fetus, with 50% of its chromosomes from the mother and 50% from the father.

There are two alternatives to sexual reproduction. The first is asexual reproduction in which the offspring stem from a single organism without the fusion of gametes. The second is hermaphroditism in which animals

1 There are other mechanisms for producing young; for instance, marsupials, such as kangaroos or koalas, keep the joeys in a pouch in front of their belly and nurture them via milk glands until they have grown sufficiently and are ready to leave the pouch.

2 Not to be confused with *mitosis*, which is the duplication of cells, a process which takes place continually in our body to replace dead cells or create new ones as part of growing.

have more or less an entire reproductive system and can therefore act as male or female, producing gametes normally associated with only one of the sexes. At first sight either of these processes would seem easiest. So why did nature come to favour sexual reproduction for more complex forms of life? This is a much discussed issue and the general opinion is that two advantages are salient.

Putative advantages of sexual reproduction

1. The combination of chromosomes from two parents has the advantage of reshuffling their genetic code and reducing the effect of any damaging mutation which might have arisen in either parent. Any disadvantageous mutation in an organism without sexual reproduction is automatically passed on to the offspring.
2. Because females can choose sexual partners, reproduction can result in adaptation to the environment of an animal, assuming the female chooses a partner who suits her habitat well and can meet its demands.

The Cost of Sexual Reproduction

In the animal world, and not infrequently among humans, there are dangerous conflicts between rival males in competition for a female. Indeed among some animals a single male serves as partner for a whole group of females, a situation which can be quite exhausting for the male in question, such as a male walrus during the breeding season. Females can be choosey with males who must go to considerable lengths to demonstrate their suitability as partners; for instance, some birds perform complex dance-like rituals to convince a female that they are just the right partner to have. And some female animals have several male partners (a system called polyandry). Other animals, such as whales, undertake great journeys in the pursuit of a suitable environment to rear their young. Still other animals die after mating (called semelparity), such as the marsupial antechinus, or after reproduction, such as salmon after migrating up rivers to a spawning ground. There are also animals which undertake long journeys to reach suitable breeding grounds; for example, some birds migrate great distances, probably making use of the Earth's magnetic field and possibly use of quantum effects to follow the extremely weak magnetic field lines of the Earth. An example would be a subpopulation of the European robin which migrates from the Mediterranean where it spends the winter, to northern Scandinavia, and even further north to Svalbard, in the spring, the breeding season for these birds.

The relative success of males and females in mating can and does vary. Females invest more time and effort in striving for successful mating (and producing offspring), given that pregnancy involves considerable time and effort along with the commitment as primary carer to any offspring born. Males, on the other hand, generally invest less time and produce millions of sperm on a daily basis. The number of offspring a female can bear is furthermore limited by factors such as duration of pregnancy and time span of fertility in the life of an animal whereas for males there are greater possibilities to generate offspring. This means that variability in reproductive success is greater in males than in females, a phenomenon known as Bateman's principle after the English geneticist John Bateman (1919–1996). This principle seeks to explain why females are more selective and cautious in their selection of a male to mate with. The principle also specifies – a not uncontested claim – that female mating and breeding is the limiting factor for a species.

All animals which reproduce sexually, including humans, have a strong drive to do so, which implies that this means of reproduction has been favoured by evolution and, despite its obvious costs to many species, has been maintained in those cases where it has arisen.

WOULD EXOBEINGS REPRODUCE SEXUALLY?

This is surely one of the most intriguing questions concerning complex life forms beyond Earth. Would they then also have sexual desire on the physical level and affection and love on the emotional level? This fact would not be relevant to language as there are no basic differences in the manner in which the two sexes on Earth acquire language, though their use of it later is another matter.

12.7 Variety is the Spice of Life

There is a common saying, 'Variety is the spice of life.' This could well be adapted to 'Variety is the spice of evolution,' the gradual change in flora and fauna over many, many generations, which can only occur if there is variety inherent in the reproduction system between generations.

At some initial stage the variation would cluster around a mean, the input mean. If this mean shifts, as it is likely to do over time, the variation becomes vectorised, meaning that it has a direction. The question then for the investigating scientist is: Why do we have the observed direction and not some other one? There might be limiting factors which would exclude other directions for the mean and/or there might be some adaptive advantage in the mean moving in the direction it takes: it may provide a benefit in a specific environment, as with birds of prey which developed extra sharp eyes and silent flight, both very useful for locating and catching prey.

12.8 A Quirk in Meiosis

There are different points where variety can occur between generations. One is the mixing of the haploid sex cells from each parent to yield diploid cells with the full set of 46 chromosomes (in humans). But another point can be seen in a process called 'crossing over,' which takes place during prophase I in meiosis. The latter is the label given to the creation of sex cells by division and reduplication. The input cells have 46 chromosomes, with two chromatids on each, in a stage called interphase in which they prepare for mitosis (cell duplication) or meiosis (making of gametes, i.e. sex cells). There are four main stages in meiosis known as prophase, metaphase, anaphase and telephase and these go through two cycles, the net result being four cells with 23 chromosomes; on fertilisation two gametes come together, forming a zygote with 46 chromosomes. In the present context the important stage is prophase I, where the chromosomes come close together and the chromatids (the tightly wound strands of DNA) begin to exchange genes, leading to new chromatids with novel mixtures of genes not present originally (this is the 'crossing over'). It is like shuffling two packs of cards together and then splitting the combined pack in half again. The two new halves are different from the two packs before shuffling. This crossing over is the main reason why children look like their parents but are not exact replicas of either and can look like grandparents as well (because the parents themselves have chromosomes which arose from mixtures of their parents' genes).

Crossing over is essential in realising variety among offspring from any set of parents and hence provides points at which evolution can kick in when some variants have an inherent advantage in a given environment.

12.9 Genetic Mutation

An organism is continually making copies of DNA in the myriad cells of its body. Some of these copies contain mutations. To label these 'random' is misleading as they are often the result of external factors like stress for the organism. Most mutations are harmless but some of the mutations are deleterious and are repaired by the body while others are not. Mutations can be the source of inherited disease. They may result from changes to stretches of genes or in some cases to a change in a single gene. An example of the latter is sickle-cell anaemia, which is caused by a defective copy of the ß-globin gene on chromosome 11, inherited from both parents (when from one parent it is usually innocuous). The phenotypic result of this is that the red blood cells, in which the haemoglobin for transporting oxygen in the body is abnormal, take on a sickle shape, which can, however, offer resistance to malaria.

A change in just one 'rung' of the DNA 'ladder' is called a single nucleotide polymorphism (abbreviated to SNP), for instance when there is cytosine–guanine instead of adenine–thymine in the base pair (in one 'rung'), technically called an allele at this position. Such SNPs occur in about one in a thousand base pairs in the human genome. With about five billion base pairs on human DNA, this means that any two individuals will differ in their genome by approximately six million base pairs. As only about 2% of our genome codes for proteins, which exercise some function in the body, there is a lot of space for SNPs which are apparently not of relevance to the individual in which they occur.

Looking for patterns and regularities in the SNPs of human DNA helps scientists group individuals according to the part of the world they and their immediate ancestors come from, on the assumption that sets of SNPs cluster geographically.

Health and Disease

Apart from genetic mutation, we humans are susceptible to infectious diseases. Whether and to what extent this would be true of exobeings is unknown, but their evolution would presuppose a vast kingdom of bacteria, and possibly viruses, just as on Earth. Exobeings could well live symbiotically with some of these (as we do with our intestinal flora, for instance) while other bacteria could potentially damage an exobeing by interfering with its normal biochemical functioning as some bacteria do with us on Earth.

12.10 Divergent Evolution

By divergent evolution is meant that changes in the environment of a species lead in time to changes in traits. The standard example here are the finches on the Galapagos Islands, named after Darwin who first investigated them in detail. There are about 18 of these finches, all from a common source on the South American mainland. Over millions of years, some became insect eaters, some woodpecker types, some seed and nut eaters, depending on the environment of the island they happened to live on. Another example comes from the Grand Canyon area where two groups of squirrels are found. On the north side of the Grand Canyon, on the Kaibab Plateau, there are now the Kaibab squirrels, which are grey-brown in colour. These originated from the canyon region and are originally a subspecies of the Abert's squirrel, which are grey-black in colour, and which occupy a much larger area to the south of the canyon, including the niches vacated by the ancestors of the Kaibab squirrels when they moved to higher ground to avail of nuts in the forests of the Kaibab Plateau.

The relevance of these considerations for this book is as follows. Assuming that life on exoplanets starts as single-celled organisms and evolves from there, the environments in which the species would then develop will determine their later phenotypes. For instance, if the planet orbited a mature, stable red dwarf, the life forms might have eyes with vision adapted to infrared wavelength, something which would be different from our vision, geared to perception approximately in the range of 350 to 800 nanometers (depending on light conditions). This could mean that, in theory, if we gained enough experience of exoplanets and life on them, scientists could devise a metric for divergent evolution at these locations: the extent to which the environments diverge from Earth's environments could be used to predict the kinds of evolution one might find there.

12.11 Convergent Evolution

Convergent evolution, on the other hand, is determined by common environments with similar selectional preferences. It is interesting in that it shows how evolution can move along common paths, given similarities in the surroundings in which species develop. This is true of the major divisions in animal habitats, as shown below.

Life in Water: The Development of Hydrodynamic Bodies

There are many animals in the sea, including mammals, such as whales and dolphins, whose ancestors migrated from land to sea. These evolved to have a shape like fish with smooth skin, fins and a hydrodynamic body form. But so do penguins, who fish in the sea for food and walk on land, but have feathering which is extremely fine, allowing them to move in water effortlessly. Penguins are flightless: their wings have been adapted to work like flippers when under water.

Animals partially or wholly at home in the sea

1. Fish (largest group by far)
2. Sea mammals (originally land animals)
3. Penguins (birds)

Life in the Air: The Development of Wing-Like Structures

Animals which use air to move around have wings or wing-like structures. Just like fish are the largest group of sea creatures, birds, derived from avian dinosaurs, are the largest group in the air. But there are others. Flying squirrels and sugar gliders have developed membranes between their fore- and hind-limbs, technically called patagia, allowing them to glide, that is move passively at an angle through air, and thus get from one high point, such as on a tree, to another without first descending to the ground and then ascending to the next high point.

Animals which engage in flight or gliding

1. Birds (largest group by far, from avian dinosaurs)
2. Flying insects (insects)
3. Flying squirrels (rodents)
4. Sugar gliders (marsupials)
5. Flying lizards (reptiles)
6. Flying frogs (amphibians)
7. Flying fish (fish)
8. Bats (mammals)

Here we can see how flight/gliding operated by convergence: some aspect of the anatomy of an animal was co-opted for moving through air. With birds, which have had a much longer period of evolution, wings have developed along with a reduction in weight, and a large breast muscle to render flight possible. With flying lizards, the membranes are actually between their ribs, which can be distended to provide the enlarged surface area of the membranes to permit gliding. With flying frogs, the membranes are webs between their fingers and toes. With bats the membranes are between extremities which correspond in evolutionary terms to the fingers of our hands and link up with the animal's abdomen. Bats are functionally between flying squirrels/sugar gliders on the one hand and birds/flying insects (butterflies, bees, etc.) on the other: they can fly (not only glide) but not as well as birds and only to hunt for insects at dusk and at night, using echo location.

The effects of an environment can be seen in adapted life forms. For instance, birds which dive for fish, such as cormorants or gannets, pull in their wings and streamline their bodies into an elongated oval shape, diving vertically into the water to surprise their prey. With gannets, the nostrils are located in the mouth and they have air sacs around their face and chest to cushion the impact of diving into water from heights of tens of metres. Thus, one can see what adaptation animals have undergone in their evolution to perform a certain activity.

Using the Options of Hard Tissue

There are many different types of tissue and some are very hard, like bones, claws, nails and teeth, and serve distinct purposes. Beaks are a good example. These are bony protusions from the upper and lower mandibles (jaw bones), which have developed independently in birds, but also in some types of cetaceans, and in fish and turtles, for instance. Beaks are used basically for feeding, which includes extracting prey from the ground or crevices in plants and trees, grasping and killing larger prey, feeding of young and, in some cases, such as New Caledonian crows, for manipulating objects. Beaks are also excellent evidence of environmental adaptation as their forms have evolved in many different ways depending on the long-term habitats of the animals which have them, for instance short, powerful beaks with animals which feed on seeds or nuts and long, narrow beaks for animals which extract their food from mud and silt.

Hard tissue has evolved independently as protective shells in animals, such as turtles, tortoises, snails, shellfish, molluscs, etc. Shells have developed in many different species, either as the exoskeleton of an invertebrate or as a growth deriving from the ribs of an animal, as in the case of a turtle or tortoise. Shells serve the purpose of protecting an organism from the outside world, most often from the predators which may be found in its environment.

12.12 Analogous and Homologous Structures

Structures in organisms which have developed independently but have similar functions, analogous structures, can be assessed as more likely to develop on exoplanets given that they have arisen multiple times on Earth. For instance, the flippers in whales and penguins both derive from forelimbs (more obvious with penguins), but they did so independently. Another example is the heart, which, as a simple pump for blood in animals with a circulatory system, developed about 600–700 million years ago at several separate points in the animal world. The same is true of eyes, which evolved from primitive light-sensitive cells; the earliest evidence of eyes appears in the fossil record about 500–600 million years ago. The following is an example related to human phonetic capabilities: toothed whales (odontocetes) have two pairs of phonic lips (all except the sperm whale) in the nasal complex and use these to produce sound, so they are analogous to the vocal folds in the human larynx. The phonic lips can be controlled with a sensitivity comparable to human vocal folds.

Homologous structures are those which derive from a single primitive structure, which is conserved through speciation over long periods of time. For instance, there is one pair of original extremities which evolved into wings, fins and arms in different species. Basically, they consist of three-joint structures, with one bone, two bones and many bones in the first, second and third section (from the torso to the edge of the body).

Terminology might be confusing when considering common or independent evolution. For example, the giant panda and the red panda are not related. The giant panda is indeed a bear but the red panda is descended from a raccoon.

Comparison of homologous and analogous structures

Homologous	**Analogous**
Different in function but from same ancestral source	Similar function but not from same evolutionary input
wings of a bat, fins, arms; claws, nails	bird wings and insect wings; human vocal folds and whale phonic lips

12.13 Epilogue: Profusion in Nature

Nature (and the universe) is profuse at the initial stage of a development and this maximises the chances of success. Imagine it as a cross-country race with many obstacles along the course. The runners line up at the beginning; they all want to reach the goal of the race. But nature cannot make the race easier to ensure all the runners are successful (the course of the race was not designed by nature, it just evolved). So what happens? Nature increases the number of runners to start with, thus improving the chances that at least some of them make it to the goal, indeed in some cases just one suffices (see human conception below).

Think of a pool somewhere in the countryside in the spring. It is full of tadpoles, but only a few will mature into adult frogs. So to ensure that frogs survive, a system developed whereby thousands of tadpoles are hatched from frog spawn, thus providing a reasonable chance that a certain number will make it to be frogs and ensuring that the next generation survives and can propagate. Those who do not make it are usually eaten by other animals (birds and fish) and so they actually contribute to the chance that other species are successful in their own race for survival. And our biosphere is healthy if the various species vying for survival are in balance.

Examples of this kind of situation abound in nature: profuseness at the beginning of a development to ensure success. Just think of all the fruit a tree will bear in the hope that one new tree (or a few, if it's lucky) will result. Another example would be male sperm. A young healthy male can produce up to 200 million per millilitre but only one of these will be successful in fertilising the egg of a human female.

Quantity of Planets and the Chances for Advanced Life

Now switch to planets. The situation here is not the same as with organisms because planets and their parent stars do not develop in order to propagate life, but just bear with me for a moment. Assuming that there is, on average, at least one planet per star (most likely more, given that our Solar System has eight) then there are hundreds of billions of planets in our own galaxy alone with an unknown number of 'rogue planets' not orbiting a star (any more). How many of these will be successful in the long development from proto-planet to exobeing-bearing Earth analogue? Probably very few, but those few also probably represent a figure to be reckoned with. If the chances are one in 300–400 billion, there is only one planet in the Milky Way with intelligent life at a digital age, our Earth. However, if the chances are better than one in 300–400 billion, say one in a billion, the search is worthwhile because there would be planets 'out there' with exobeings we could in theory communicate with (several hundred in our galaxy at the extremely modest estimate of one in a billion).

13

The Road to *Homo sapiens*

The long pathway from simple cells to intelligent beings, capable of developing digital technology, will have been traversed on any planet with such beings. So it makes sense to consider how the development panned out on Earth. Up to about seven million years ago (7 mya), about 99.85 per cent of the age of our planet, there were no animals walking the Earth which looked like us. Yes, there was great biodiversity on land and in the sea. But there were no species which were in any way capable of reflecting on the larger world in which they lived, beyond their own environment, let alone reflecting on Earth as a planet and its admittedly very modest role in the cosmic scheme of things. If any exobeings had by chance taken a closer look at our planet, they would not have been motivated to investigate it in any detail.

But then there was a split in the lineage of great apes between chimpanzees on the one hand and initially similar primates on the other hand, which were to form the lineage out of which we humans evolved. All the species after this split are known as hominins. These are a subgrouping within the hominids, which consist of the four genera (plural of genus) (i) *Gorilla* (gorillas), (ii) *Pongo* (orangutans), (iii) *Pan* (chimpanzees [*Pan troglodytes*] and bonobos [*Pan paniscus*]) and (iv) *Homo* (different species leading ultimately to modern humans).

The split from the chimpanzees led first to australopithecines (from Latin, meaning 'southern apes') and in time to the genus *Homo* and to the species *Homo sapiens*, that of anatomically modern humans, the one to which all humans on Earth today belong and which is the only surviving species of this genus (see details below).

13.1 The Pitfall of Compressing the Past

The further we look back in time the greater the tendency not to perceive time spans like those closer to us: in our perception we compress time the further it is away from us. For instance, in palaeoanthropology when estimating the ages of Neanderthals and *Homo sapiens* one finds different estimates, anything from 300,000 to 800,000 years ago. So 100,000 years either way is hardly worth talking about. But stop and think about just how long 100,000 years is. Think of the time from the Roman Empire to today, consider how dramatically life on this planet has changed in these 2,000 years. Now try to think of 50 blocks of such 2,000-year units of time strung together, that will hopefully convey to you *subjectively* how long 100,000 years is, the length of time which palaeoanthropologists do not usually discuss because it is so short in their scheme of things. When considering when the genus *Homo* and the genus *Pan* (which led to chimpanzees and bonobos) diverged, you will find the figure of 6–7 mya (i.e., 6,000,000–7,000,000 years ago) for the last common ancestor of the two genera. This time span is negligible in terms of the Earth's evolution, but one million years is nonetheless equivalent to about 500 blocks of the period of time from the Roman Empire to today.

Across the Generations

At about 25 years per generation,[1] 1,000 years has 40 generations, a million years (i.e., 1,000,000 years), has 40,000 generations. By comparison it is

1 Recent research has shown that the intergenerational time interval is somewhat longer than what is intuitively accepted, i.e. 20–25 years, with the lower figure taken to apply to pre-industrial cultures, see Tremblay and Vézina (2000).

only 80 generations since the beginning of the Common Era, that is, since the height of the Roman Empire.

A generation refers to the time period from birth to the average age at which a species has offspring, not the average lifespan of a species. This average is calculated by taking the average age at which a species starts and finishes having offspring, across an entire population, hence the figure of not more than 25 years for pre-industrial societies.

Now the last common ancestor of humans and chimpanzees, that is to say since the point in time where the human lineage diverged from that of chimpanzees lies about 6–7 mya, so that somewhere in the region of 250,000–300,000 generations separate us today from the last common ancestor (often abbreviated to LCA). So it took something over a quarter of a million generations to get from common ancestors of chimpanzees and *Homo* species to language-speaking modern humans on a separate developmental line.

13.2 Palaeoanthropology: Reaching Back in Time

Fossils occur when minerals replace organic matter – typically an organism's body parts, usually bones – turning them into stone. It is not always possible to date fossils directly but one can date the strata of rock they are found in.

In 1959, Mary Leakey, with her husband Louis Leakey, discovered fossil remains in the Olduvai Gorge in northern Tanzania of an early hominin which they called *Zinjanthropus*, now usually designated as *Paranthropus boisei*, dating to some 1.76 mya. The age was determined by geochemists examining the sediment layer it was found in. The following year (1960) they discovered another fossil, labelled Hominid No. 7.[2] They concluded on the basis of major morphological differences that there were at least two different lineages of early humans alive at

2 At that time the label 'hominid' was used for species of the genus *Homo*. Nowadays this refers to the larger group containing humans and great apes, as outlined above.

this time. Hominid No. 7 had a large cranium and a smaller face, which suggests that its lineage was that which produced the tools also found in Olduvai Gorge.

The Great Rift Valley region of Eastern Africa (Figure 13.1) was subject to repeated volcanic eruptions, which spewed out ash with radioactive elements that helped scientists accurately date the layers in which early *Homo* fossils were found. In the Hadar region of Northern Ethiopia, the American palaeoanthropologist Donald Johanson (1943–) discovered the australopithecine called Lucy in the Afar Triangle in November 1974. The scientific designation for this specimen is *Australopithecus afarensis*. This incidentally confirmed the view of the South African palaeoanthropologist Raymond Dart (1893–1988) that australopithecines were bipedal.[3] While already bipedal they were still arboreal, at least during the nighttime to gain protection from predators like sabre-toothed large cats.

These hominins dominated during the Pliocene epoch, which lasted from about 5.3 to 2.5 mya. The Pleistocene epoch which followed, and lasted from about 2.5 mya to 11.7 thousand years ago (kya), was characterised by recurring glacial periods during which *Homo habilis* 'handy man' (*c.* 2.3–1.65 mya) and *Homo erectus* 'upright man' (*c.* 2 mya–115 kya) evolved from the foregoing australopithecines. When these finally ended we have the transition to the Holocene epoch (*c* 11.7 kya–), which leads down to the present. There is considerable overlap between species in different parts of the world. *Homo sapiens* dates back to several hundred thousand years ago during the Middle Pleistocene (or Chibanian age as it is now known), but the very last *Homo erectus* appears to have been Solo Man from Java, which died out about 115 kya. These different species are recognisable by clear morphological features of the skull, for example the heavy brow ridge in the pre-*sapiens* species.

3 Bipedalism is an example of an analogous feature as it evolved separately in different species, such as birds and many marsupials; it was also a feature of many dinosaurs. Traits which are inherited from common ancestors by two later species are termed homologous.

Figure 13.1 The Great Rift Valley (East Africa)

13.3 Rummaging Around in Caves

Some of the major palaeoanthropological finds have been made in the open, due to exposure of bones. But many were made in caves where early hominins sought shelter from inclement weather and predators of various kinds. Some of the caves are large and open structures, others are not. The Dinaledi chamber ('chamber of stars') in the Rising Star cave complex near Johannesburg has yielded plentiful fossil remains. Access to the inner parts of the cave is not easy: it involves crawling flat on the ground through a bottleneck not more than 25 centimetres (10 inches) high before getting into the first chamber, then up a cliff face called the Dragon's Back into the chamber with the fossils.

Sites in the Levant have been particularly rewarding for palaeoanthropological researchers, given that this region was a corridor through which migrating early humans from Africa moved northwards into Europe and Asia Minor. Mislaya Cave in Israel has provided the earliest evidence of modern humans outside Africa, at between 177 kya and 194 kya (a row of teeth was found there). Skhul Cave has remains of at least 10 adult humans and dates from 100 kya to 135 kya (excavated in the 1930s). Other countries have valuable sites such as the Fuyan cave complex in Hunan Province, China, where 47 teeth, all from modern humans, were found and dated to between 80 kya and 120 kya or the Zhoukoudian site in Beijing where the fossils known as Peking Man (*Homo erectus*) were found in the early 1920s, dating to between 500 kya and 660 kya. The cave on the island Flores (Indonesia) and the Denisova cave (southern Siberia, Russian Federation) are more recent significant fossil sites (more on these below). Fossil sites are often revisited, given improvements in dating techniques and computer simulation, which have prompted scientists to reassess fossil finds, such as those at Jebel Irhoud in Marocco (more on this below).

13.4 The Out of Africa Hypothesis

Today there is consensus among palaeoanthroplogists that the species of the genus *Homo* for which we have fossil records all arose in Africa and spread out from there later (Figure 13.2). The dispersals were probably triggered by the search for food. Climate fluctuations, periods of aridity, would have forced early *Homo* species to move to areas where plant and animal life was more abundant. It is important, when using hindsight, not to imagine that early species were explorers like the colonial Europeans during the Age of Discovery (*c.* 1500–1900). Early hominins followed sources of food, which led to them radiating out from an original source over thousands of years.[4]

4 The older view of *polycentric evolution*, the notion that humans arose at various points around the Earth, such as China, was favoured by the German anthropologist Franz Weidenreich (1873–1948) and some others but nowadays it is not supported.

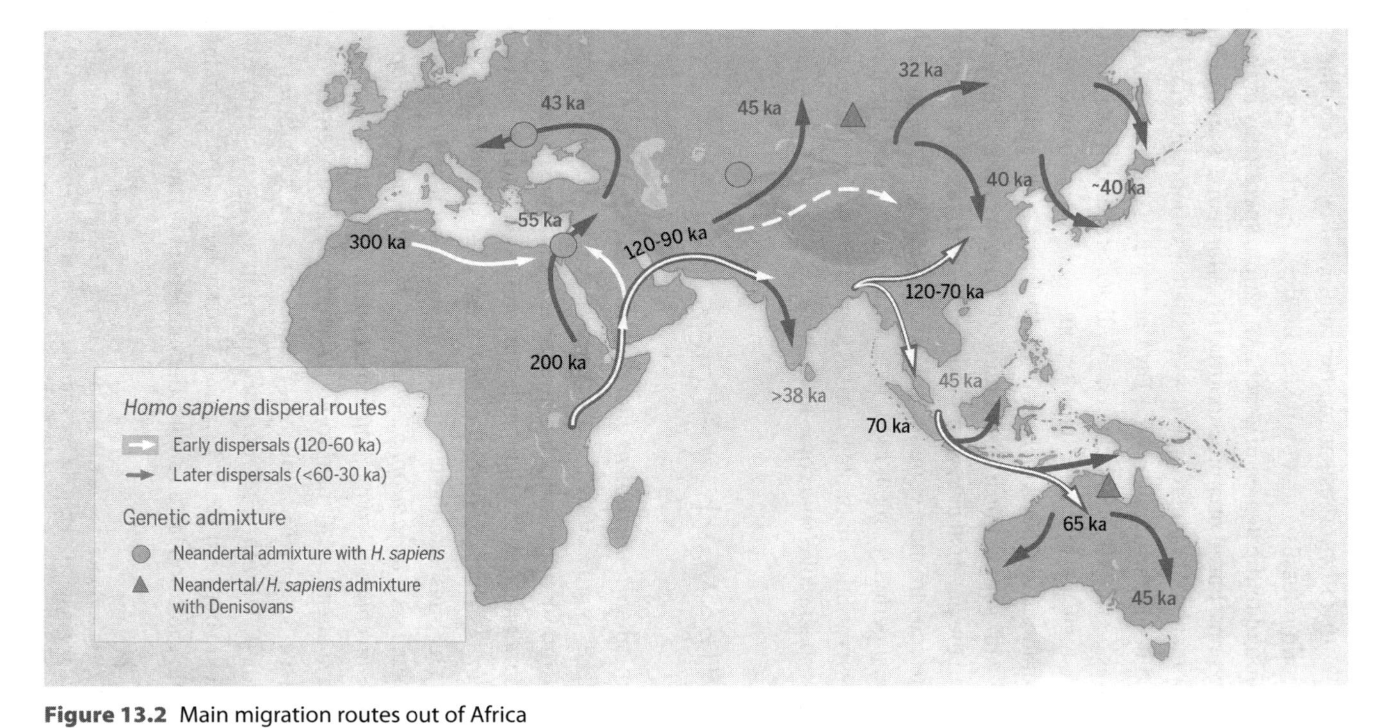

Figure 13.2 Main migration routes out of Africa

The Out of Africa hypothesis is mainly supported by the fact that all the oldest hominin fossils come from sites in Africa. Further supportive evidence is found in the degree of genetic variation, specifically in the range of haplogroups, based on mitochondrial DNA for females and Y chromosomes for males. A haplogroup is a set of alleles (variants of a gene) which occur at different locations on the DNA and which are usually inherited together as a group (smaller clusters of alleles are known as haplotypes) through a single parent. These haplogroups came to vary over time, due to slight alterations of DNA which crept in. Different haplogroups generally correlate with geographical distribution of human groups. For instance, the R1b (in the paternal line, with subbranches)[5] and the H haplogroup (in the female line) are most common in (Western) Europe and spread during the colonial period (c. 1600–1900) to North America and Australasia. But the greatest variation in haplogroups is found in Africa which suggest that the oldest species of *Homo* come from there.

13.5 'The March of Progress'

The evolution of *Homo sapiens* after the divergence from the ancestors of chimpanzees is not a straight line as often shown in two-dimensional line drawings, the so-called 'march of progress', with chimpanzees on the left and *Homo sapiens* on the right, and with supposed intermediary species in between. Such representations are great over-simplifications. There were parallel lines, like the genus *Paranthropus*, with various postulated species, such as *Paranthropus boisei* and *Paranthropus aethiopicus* or *Kenyathropus platyops*, which did not continue, all of these existed about 3.5–3.2 mya or earlier. *Kenyathropus platyops* may have been the first user of tools (see Section 13.15 below). A skull cap of a specimen from this genus, called *Paranthropus robustus*, found in a cave in South Africa in 1938, dates to between 1.8 and 1.5 mya, and shows that these parallel lines of hominin evolution lasted a considerable length of time.

5 Both the R1a and most of the R1b haplogroups are assumed to have spread from the region of the Pontic-Caspian Steppe with the dispersal of the original Indo-European groups, who spoke early forms of the later languages from this family, around 3,000 BCE.

Spectacular fossil finds (of at least 15 individuals) in the Rising Star cave complex in the Cradle of Humankind area north-west of Johannesburg in South Africa in 2013, under the leadership of the American–South African palaeoanthropologist Lee Berger, led in 2015 to the positing of a hitherto unknown *Homo* species, *Homo naledi*. This species probably derived from *Homo erectus*, but had become arboreal once more and had a reduced brain size. Theirs was a mosaic anatomy, in part similar to other *Homo* species (in their hands and feet and probably their dentition as well), in part still reminiscent of earlier *Australopithecus* (in their small brain size). These brains, with an average range of around 550 centimetres, were not much different from those of chimpanzees; they were on average under 160 centimetres in height (less than 5 feet). Whether they had an identifiable culture, involving the burial of the dead, is disputed, but there are strong arguments in favour of this view: only *Homo naledi* fossils are found in the cave, there are no signs of predation on the bones and the innermost chamber, the Dinaledi Chamber, is very difficult to access, so the fossils are extremely unlikely to have ended up there by chance. The interesting fact about the newly discovered fossils is that they date back to not more than 335 kya, a time at which *Homo sapiens* was already living on the plains of East Africa and possibly elsewhere. *Homo naledi* illustrates clearly that the development of the genus *Homo* involved many different species co-existing for longer periods of time.

And then there is *Homo floresiensis* from the Liang Bua Cave on the Indonesian island of Flores, popularly labelled the 'hobbit' because of the small size of the main complete fossil (LB1), just over one metre high, which survived on the island before anatomically modern humans arrived in the area about 50 kya. This species was likely subject to 'island dwarfism,' a development in which animals that have no or few natural enemies dwindle in size as they do not need to fight with others for food and survival. Incidentally, the opposite can also happen: if a species is not contained by (usually larger) predators, it may increase in size, leading to island gigantism. It should be said that there is disagreement in the scientific community about whether *Homo floresiensis* is really a separate species or, indeed, whether LB1 suffered from a medical condition similar to microcephaly, which would account for its small brain (*c.* 400 cubic centimetres (cc)), only about half that of its predecessor *Homo erectus*. Even accounting for its brain-to-body mass ratio, its brain was remarkably small.

13.6 The Context of the Genus *Homo*

The following paragraphs summarise the essential classifications in the development of humans from their primate ancestors. This should not be interpreted as a linear development. Palaeoanthropologists today emphasise that many species of *Homo* (Table 13.1) co-existed over several million years, some discontinuing but possibly contributing, through gene flow, some aspects to the species that survived. The image of a braided river, with several parallel channels, which flow apart and rejoin later, is one which is often used.

Hominidae/hominids (great apes): The family of primates which consists of the following genera: *Pan* (chimpanzees and bonobos), *Gorilla* (gorillas), *Pongo* (orangutans) and *Homo* (humans and our near relatives, such as the Neanderthals).

Table 13.1 A top-down view of *Homo sapiens* in its larger evolutionary context

Taxonomy	Label
Superorder	Mammals
Order	Primates
Superfamily	Hominoids
Family	Hominids
Tribe	Hominins
Genus	*Paranthropus, Australopithecus, Homo*
Species (main)	[*Homo*] *habilis, erectus, naledi, floresiensis* (?), *neanderthalensis (denisovans), sapiens*

Hominini/hominins: In anthropological taxonomy, hominini is the name given to a tribe which contains the three genera listed below, that is, all hominids after the divergence from *Pan*, so all genera and species after they diverged from the ancestors of chimpanzees:[6]

1. ***Paranthropus* (Greek 'beside man'):** Evidence for this genus comes from a few fragmentary fossils such as *Sahelanthropus tchadensis* (7.2–6.8 mya) which was discovered in 2001–2002. It consists of part of a skull, but no part of the body which could confirm possible bipedality. However, the situation with *Orrorin tugenensis* (6 mya) discovered in 2000 in Northern Kenya is different. Grooves on the hip ball at the top of the femur suggest that this species was already bipedal, confirming that this feature is ancient in our lineage (though not uncontested).
2. ***Australopithecus* (Latin 'southern ape'):** This genus arose in Eastern Africa at around 4.4 mya going on the fossil *Ardipithecus*, found by Tim White and his colleagues in the Afar Depression of Ethiopia in the early 1990s. This find was almost a million years older than the fossil named 'Lucy' found at Hadar in the Awash Valley, Ethiopia in 1974 (see above) and dated to 3.2 mya.

   An early find of an australopithecine was that of the Taung Child discovered by Raymond Dart in 1924 – but not recognised by the wider scholarly community until decades later – showing that 2.8-million-year-old *Australopithecus africanus* was indeed present in Africa, a precursor to the genus *Homo* and the first evidence to support Darwin's idea that humans originated in Africa. The Taung Child was probably the victim of a large eagle-type bird, as scholars concluded after a re-examination of the skeleton in 1995 which showed traumata from talons in the skull. The site where it was found contained the remains of baboons and the shells of the eggs of large birds. Even as late as the period of Neanderthals, humans, often children because of their relatively small size, were hunted by large birds of prey.

6 There is not complete agreement on this: some anthropologists prefer to include the genus *Pan* in the tribe of hominini. For this book, the label 'hominini' is taken to encompass all genera after the split from chimpanzees.

The cut-off point from *Australopithecus* to *Homo* is a brain size of about 500–600 cc. The transition is also seen in the limbs: the legs of *Homo* are long, the arms shorter, the shoulders further back with the ability to swing the arms around. The shape of the teeth is telling: smaller in shape, especially the molars, pointing to a diet of meat, but also (much later) of cooked meat which does not require the great amount of chewing needed to break down the tough fibre of so many plants.

3. ***Homo* (Latin 'man')** The genus *Homo* derived from the genus *Australopithecus,* over 2 mya, though no precise dating for the separation of the two genera is possible and there may well have been considerable overlap. Whether there are species intermediate between *Australopithecus* and *Homo habilis* is a matter of debate. Scientists disagree about what features should be cited as documenting the split of the two genera: shorter upper limbs/longer lower limbs, human-like torso (with barrel-shaped ribcage and not cone-shaped, suggestive of a large belly), brain size, teeth size, manual dexterity (hands with an opposable thumb/index finger) and the production of tools, etc. (see McGinn 2015 for a broader discussion of this).

13.7 Divisions within the Genus *Homo*

Homo habilis

Homo habilis is an ancient species of the genus *Homo*, generally regarded as the first to follow the late australopithecines. It lived in East Africa from about 2.3–1.6 mya. The name means 'handy man,' a reference to the fact that it made tools, though it is now thought that it was not the first to do this. Nonetheless, it must have been manually dexterous and have had the precision grip typically of all later *Homo* species. This ability probably arose gradually with the development of the *Homo* lineages and there are precursors among the other non-human primates, such as bonobos and chimpanzees, which can use sticks as primitive tools, holding them between their thumb and index finger, but without the distal pad-to-pad contact at the tips which is characteristic of humans enabling us to exercise fine motor control.

Early remains of *Homo habilis* were found in the Olduvai Gorge in northern Tanzania by the palaeoanthroplogists Louis and Mary Leakey in 1955 (see above). It overlaps in time with the next species. Going on the fossil record the big jump in terms of hominin size is from *Homo habilis* to *Homo erectus*. *Homo habilis* was about one and half metres in height with a weight of about 50 kilograms.

Homo erectus

Homo erectus 'upright man' appears in the fossil record about 2 mya in Africa. It had a cranial capacity of between 550 and 1,250 cc and stood between 145 and 185 centimetres tall, depending on population. It was the first species of the genus *Homo* to disperse out of eastern–southern Africa to inhabit a broad band of territory from western Europe though the Middle East and central Asia to northern present-day China and south-eastwards into present-day Indonesia, the last confirmed population being that of Solo Man, less than 120 kya on the island of Java. This means that it was, in terms of time span, the most successful species of *Homo*.

Homo erectus is known for its use of advanced tools, so-called Acheulian tools of the Lower Palaeolithic period (*c.* 3 mya to 300 kya) with well-crafted flint wedges surpassing the Oldowan tools used by *Homo habilis*.

Among the most interesting sites where *Homo erectus* was found is that of Dmanisi in the present-day Republic of Georgia. Five well-preserved skulls were found there, all with a cranial capacity of between 500 and 600 cc. It is also notable that there is no evidence of the use of fire at Dmanisi, which dates back to about 1.8 mya. An even older site with *Homo erectus* is Shangchen in north-central China (Shanxi province), first discovered in 1964, showing stone tools which, according to a study published in 2018, can be dated as far back as 2.1 mya.[7] This would imply that *Homo*

7 The mechanism used to establish this date is palaeomagnetism, which examines the Earth's magnetic field, on a geologic timescale, as evidenced by having been trapped in rock and sediments from previous ages. The direction of the trapped magnetism can then be compared to known historical fluctuations in the Earth's magnetic field and an accurate dating can be arrived at.

erectus began wandering out of Africa almost immediately after its rise as a distinct species, or, as the authors of the study assume, the makers of the artefacts at Shangchen were a separate species preceding *Homo erectus* and possibly an australopithecine.

After *Homo erectus* a number of other species have been identified, their classification resting on particular mixes of archaic and more recent features: they show, for example, cranial capacity and shape, brow ridges, tooth size, jaw structure along with other parts of the anatomy, such as length of arms, shape of hands and feet, shape of hips and back. A prominent one is *Homo heidelbergensis*, which is either a follower of an earlier *Homo antecessor* or just *Homo erectus* and occupied both eastern–southern Africa and the circum Mediterranean and northern regions of Europe. But evidence from the Sima de los Huesos ('pit of bones') site in the Sierra de Atapuerca, north-central Spain, with the remains of some 28 individuals, suggests that *Homo heidelbergensis* is part of a larger Neanderthal lineage and so the split between these and *Homo sapiens* would be pushed back, perhaps to 600–800 kya, given that the Neanderthal finds at this site have been dated to 430 kya (Meyer et al., 2016) and more recently – in 2019 – to much further back than that again.

Homo neanderthalensis

Of all the species of *Homo* it is this one which has undergone an extensive revision in the past few decades. The Neanderthals[8] are a closely related species to ourselves, the split between them and modern humans (*Homo sapiens*) took place at least 400 kya, possibly as much as 800 kya. The last Neanderthals probably died out about 35–40,000 years ago. The understanding of our relationship to this species has changed due to what we have found out about it in recent years.

8 Named after the Neander Valley near Düsseldorf, Germany, where the first specimen was found and recognised in 1856 by a schoolteacher, Johan Carl Fuhlrott (1803–1877). There were earlier finds in Belgium and Gibraltar but these were not recognised for what they were.

The Neanderthals split off from *Homo sapiens* in Africa and then spread over large tracts of Europe and Asia, with fossils found at locations as far apart as Gibraltar at the entrance to the Mediterranean and Okladnikov Cave in the Altai Mountains of Siberia. They were slightly smaller than us, with a robust, stocky build which probably made them more suitable for the harsh northern climate of the Middle Palaeolithic (Stone Age). Their relatively short arms and legs together with their barrel-shaped torso meant that they lost less body heat that comparable *Homo* species in Africa, with longer limbs and a narrow torso which permitted relatively rapid heat loss, essential in a hot climate and for persistence hunting. The Neanderthals seemed to have lived less long than we do, at least nowadays. Most died before 40 and matured more quickly, perhaps leaving less time for learning. In the opinion of researchers, such as Steven Mithen, the Neanderthals had a domain-specific mentality in contrast to the cognitive fluidity of *Homo sapiens*. They seem not to have integrated their technical and natural history, along with their social intelligence, in any creative manner which would have led to innovations and hence better survival chances for them.

It is known from the analysis of the Neanderthal genome (recovered from bones) of individuals stemming from different locations across Eurasia that a small amount, typically about two per cent, of our DNA derives from a certain amount of interbreeding with Neanderthals in common spaces across northern latitudes, after *Homo sapiens* began its last dispersal out of Africa about 70 kya.[9]

Much has been written about the probable causes of the Neanderthals' disappearance (Wragg Sykes 2020). The proposals range from extinction through violent conflict with *Homo sapiens* to inbreeding in small

9 Evidence from Mislaya Cave in Israel and the recent examination of two cranial finds in Apidima Cave in the extreme south of mainland Greece imply that there was an early *Homo sapiens* dispersal out of Africa about 180 kya, with interbreeding between modern humans and Neanderthals. In fact, the Eastern Mediterranean is now regarded by many scholars as a contact zone between Neanderthals and modern humans. It is tempting to see this early movement as an earlier, less well-documented dispersal followed by the larger and more highly attested one less than 100,000 years later.

populations, which would have meant less genetic diversity and greater susceptibility to unfavourable environmental conditions and to disease.

With climate change, that is periods of cooling, the Neanderthals did not appear to adapt, but were reduced in population and retreated to core areas. This is the opposite of what humans did and do. It also meant that by the time *Homo sapiens* was spreading some of the regions they moved into had already been vacated by Neanderthals. But exactly why the Neanderthals died out has still not been determined with certainty.

Denisovans

In 2010, tiny fragments of a human were found in the Denisova Cave in the Altai Mountains of Siberia which shows signs of occupation perhaps going back to around 200 kya. The bone fragments (of a tooth from an adult male and a little finger from an adolescent female) dating from around 30 kya to perhaps over 100 kya, were analysed at the Max Planck Institute for Evolutionary Anthropology in Leipzig by Svante Pääbo, Johannes Krause and their colleagues. Mitochondrial DNA could be extracted from the little finger (some additional fossil material was discovered somewhat later, including a mandible (jawbone) at the Baishiya Karst Cave in Xiahe, Tibet) and a molar from Cobra Cave in Northern Laos, showing that this species of *Homo* was different from Neanderthals and genetically related to Tibetans and populations in South-East Asia, for instance on Papua New Guinea, and even in Australia. It may well have been that Neanderthals and Denisovans co-habited this cave and other locations; they share an amount of DNA in double percentage figures. The Denisovans would appear to have diverged from the rest of the genus *Homo* over 1 mya.

Homo longi (?)

The well-preserved skull of a member of an assumed new species of *Homo,* called *Homo longi,* was found in Harbin, north-eastern China, in 1933 but has only been examined by palaeoanthropologists in the past few years (Ni et al. 2021). Also labelled more colourfully as Dragon Man, the find shows that this species was closely related to humans and possibly the same as, or directly related to the Denisovans. The skull fossil is between 146,000 and 309,000 years old; it is that of an adult male and shows a marked brow ridge and an oblong shape with the rear of the head bulging out in a manner similar, though not completely identical to Neanderthals.

Homo sapiens

The above sections show that the road from chimpanzees to humans is by no means a single straight line, rather it is a complex, braided pathway with stages of divergence and again of convergence. And the movements out of Africa probably consisted of overlapping routes, often going in both directions, with intermingling of difference species along the way. This makes the fact astounding that *Homo sapiens*[10] is the only extant species of the genus *Homo* on Earth today. The journey out of Africa began about 70 kya, probably by a very small population, maybe around 10,000. From this original group all modern humans are derived. This fact finds corroboration in genetic similarity across the present-day populations of the world but also in the fact that all human languages are similar in their structures and principles of organisation, despite the differences in pronunciation, grammar and vocabulary which make them mutually incomprehensible.

13.8 The Progression of Consciousness

We know close to nothing about how consciousness developed during the evolution of *Homo* species. At what stage did human primates begin having a concrete notion of past and future and engage in complex, abstract thought? We can get at least a partial handle on the latter development: by the earliest attestation of art, for instance, on the walls of cave walls using red ochre. The earliest examples of this art are generally dated to 40–50

10 This term, meaning 'wise man' in Latin, was first used in 1758 by the Swedish botanist Carl Linnaeus (1707–1778), who introduced the binomial system of nomenclature, the first part indicating the genus and the second the species.

kya, though paintings from the Iberian Peninsula (Cueva de la Pasiega, Puente Viesgo, Cantabria; Cueva de Maltravieso in Cáceres, Extremadura; Cueva de doña Trinidad, Ardales, Andalucia) have been dated (with the uranium-thorium method) to at least 64 kya. A hashtag-like marking with red ochre on a small stone found in Blombos Cave in South Africa has been dated to 73 kya. Important in the current context is that this art, apart from including animal representations and hand stencils,[11] contains ideomorphic pieces, abstract representations which document the capability of the artists for symbolic thought.

However, all this is way after the divergence of chimpanzees and early *Homo* about 7 mya. So we cannot tell what level of consciousness *Homo erectus* enjoyed, or indeed *Homo habilis* before that. But given that *Homo habilis* (2.4–1.6 mya) was able to make primitive tools (Oldowan or Mode I type) with a brain capacity of 500–900 cc we can assume that manual dexterity preceded the cognitive abilities we associate with *Homo sapiens*.

13.9 Defining *Homo sapiens* Anatomically

The phrase 'anatomically modern humans' is used to refer to *Homo sapiens*. There are a number of features which delineate such individuals from others which preceded them.

Salient features of anatomically modern humans

1. Large prefrontal cortex with a brain size of about 1,200–1,400 cc
2. A round head without a bulge at the back
3. Small mouth, teeth and jaw muscles, suggesting the consumption of cooked food

11 The most famous example of hand stencils on a cave wall is from the Cueva de las Manos (Cave of the Hands) in Santa Cruz province in Argentina. It contains umpteen stencils in an irregular pattern across a stretch of wall. These date to between 9 and 13 kya and so are quite recent in the context of early cave art.

4. Long legs, somewhat shorter arms, with wide-range joints on the torso; longer neck
5. Flexible tongue muscle capable of realising different configurations of the oral cavity.[12] Lowered larynx with hyoid bone positioned for precise muscular movements to produce sounds. Agile vocal folds for generating voice, essential for vowels and many consonants.

Teeth are archaic features of virtually all animals. Already at the period of the Cambrian radiation, about 540 mya, jawed fish with strong teeth are to be found as early examples of predators. We no longer use our teeth to catch prey, as do great cats like lions and tigers who still have sabre teeth used to bite into the throat of an animal and suffocate it. Our smaller teeth point to easy-to-obtain high-protein food (not primarily leaves, shoots and tubers). Hence we came to lose the great molars to grind fibrous plants and grass, which early hominins had. Teeth are also useful in palaeoanthropology as they do not decay like soft tissue and given their small size commonly survive intact, as with the molar of a Denisovan which helped to identify its bearer as a member of a separate lineage of the genus *Homo*.

Anatomically modern humans are defined by a small face and a globular brain case (Figure 13.3). So what evidence is there for this in the fossil record? The findings at Jebel Irhoud in Marocco, made initially in the early 1960s, have been re-evaluated for their age using thermoluminescence.[13] Flints found there were dated to approximately 315 kya and a tooth to approximately 285 kya. The fossil finds at this site show (i) a small face, but (ii) an oval brain case (viewed from the side), rather than the modern

12 As Mu and Sanders (2010: 771) note: 'the innervation of the human tongue has specializations not reported in other mammalian tongues, including non-human primates. These specializations appear to allow for fine motor control of tongue shape'.

13 This technique is based on the fact that crystals of certain substances, such as quartz, absorb ionising radiation from their surroundings, which in turn leads to electrons being 'trapped' in the crystals of the material. When heated to a high temperature these electrons are freed and a faint light is emitted. This can be measured and, by means of certain other calibrated factors, the time at which the electrons were initially trapped can be determined.

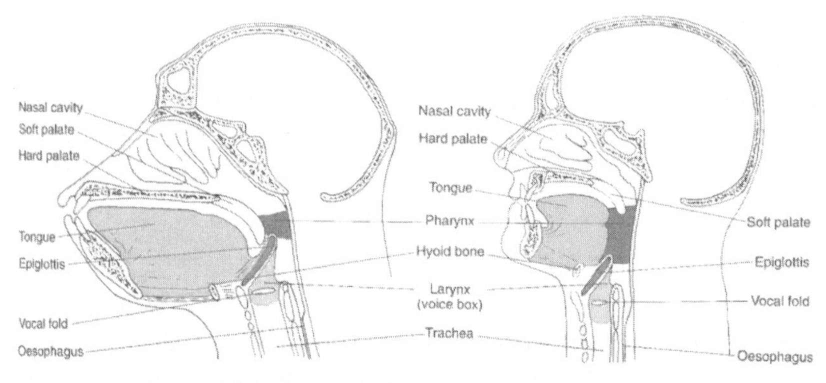

Figure 13.3 Head of chimpanzee (left) and human (right)

round case. This in turn would imply that anatomically modern humans were not a sudden development 100 kya as scientists had been led to think going on earlier examinations of the fossil record. Rather, they show a gradual transition from archaic *Homo* to modern *Homo sapiens*, which was already underway 300 kya.

13.10 Energy Intake

We humans have two main sources of energy, which we take in quite differently. The first is oxygen, which we inhale with the air we breathe and use for cellular respiration, above all in our brains. The act of breathing is constant all our lives long and the consequences of interrupting it are dire: it quickly leads to unconsciousness, when the supply of oxygen is cut off from the brain, and suffocation if the interruption lasts more than a very few minutes. If someone experiences a shock in which breathing stops for any length of time it can lead to irreparable brain damage; this can also happen during birth if the supply of the mother's oxygen is cut off from the baby before it is born.

Breathing has two aspects to it: (i) the inhalation of oxygen and (ii) the exhalation of carbon dioxide, a waste product which arises in cells. Our lungs are constructed in such a way that they can take in oxygen and give off carbon dioxide. These two gases need to be maintained in our blood within a fairly narrow range of pressures for each. If the relationship between the two changes to any appreciable degree in one direction or the other it leads to distress in the body (hyperventilation – too great an intake of oxygen and hypoventilation – too little intake of oxygen).

Not all forms of life require oxygen. Certain simpler forms (anaerobic organisms whose metabolism uses some other source such as hydrogen) can not only exist without oxygen, but are damaged by its presence. But all more complex forms of life, including all vertebrate animals, need a constant supply of oxygen, which is gained by inhaling air from the atmosphere or filtering oxygen out of water in the case of fish.

The second source of energy is our food intake, which is periodic. Modern humans, who can afford it, have major food intake at least three times daily, with breakfast, lunch and dinner and bits in between in the form of tea/coffee breaks. But we can survive for relatively long periods without food, at least without solid food; the intake of water is different, that must occur at least daily or the body quickly gets distressed, suffering from desiccation. If, in extreme circumstances, there is little or no solid food available, the body can survive on its own fat for quite some time, days, if not weeks. Solid food is important for a number of bodily functions, such as maintaining the constant temperature of the internal organs at 37 °C

EXOBEINGS AND OXYGEN

Would exobeings use oxygen as their primary, constant source of energy? The chances are that they would. Organic material, such as trees and certain bacteria, releases oxygen into the atmosphere. Assuming that exobeings arose on planets through evolution from plant life through animal life then, given that life is opportunistic, it would come to exploit the presence of oxygen as a supply of energy for its own metabolism. This is a tacit assumption in the search for Earth-life exoplanets: the spectroscopic signature for oxygen in an atmosphere is regarded as an important biosignature pointing to the existence of life on an exoplanet.

or providing energy when we are involved in muscular exertion. It is also important during growth periods and in cell replacement procedures during our entire lifetime.

13.11 Narrow Range of Values

There are many parameters in our body which show a very narrow range of values. A well-known example is the pH balance in our blood, a measure of relative acidity and alkalinity, more precisely of the extracellular fluid which needs to be in balance (technically called homeostasis) with the intracellular fluid. The normal value is around 7.4 in arterial blood, with very little tolerance in either direction. The body undertakes corrective measures should the level vary ever so slightly, for instance by initiating acid or base release by the kidneys, using chemical buffers to maintain the right balance or provoking a greater release of carbon dioxide from the blood via the lungs to reduce acidity. In fact, if you hold your breath, the sudden uncomfortable feeling you experience, is the drop in pH value in your blood, that is, the increase in acidity. Equally, if you hyperventilate, you will get dizzy because of the increase in alkalinity.

Another sensitive parameter is that of glucose in our blood. The average is about 80–100 milligrams (mg) per tenth of a litre (decilitre (dL)) before eating and rising an hour or two after a meal to anywhere between 100 and 140 mg/dL or more. During longer periods without eating, blood glucose can drop to 70 mg/dL or even 60 mg/dL. Control of the sugar level is realised by food intake and by the action of the liver, which can break down glycogen (the stored form of sugar) into glucose that the body can use. Persistently high sugar levels are dangerous and can cause a host of issues, the irregular production of insulin by the pancreas being the one people with diabetes are most aware of. Low blood sugar (hypoglycaemia) can have serious consequences, sudden and severe or prolonged, if unchecked.

13.12 Brain Size

Palaeoanthropologists used to believe that what they first needed to find was the fossil of a creature with a big brain, one about twice the size of a chimp, say 700–800 centimetres, rather than the 400 centimetres of the chimpanzees today, in order to document the evolution of our species. This concentration on brain size led to many false starts and even to such academic disgraces as the Piltdown Man Hoax.[14] Various methods were employed, for example Louis Leakey used a method of filling a cranium with small glass beads and then measuring these in a calibrated cylindrical glass. But it became apparent that brain size was not the only factor driving evolution forward: *Homo erectus* was successful at making Acheulian tools and spread into the Caucasus and across South Asia. And yet, such examples as the famous skulls of the Dmanisi fossil site in southern Georgia show, on average, a cranial capacity of about 600 cc, a rather small value given that the site goes back to about 1.8 mya.

Homo habilis appeared between about 2.3 and 2.5 mya, preceding *Homo erectus* which appeared between about 1.9 and 2 mya. So there was a big jump in physical size in less than one million years. Furthermore, a continuous weakening of the jaw muscle in *Homo habilis* through *Homo erectus* allowed the brain case to be more flexible and expand, with the jaws, mouth and teeth being proportionately reduced in size.

Our species, *Homo sapiens*, appeared somewhere at least 400 kya, possibly as much as 800 kya, so that the development of the brain from the 700 cc of the *Homo erectus* to the approximate 1350 cc of *Homo sapiens* took place over the course of approximately 1.5 million years (Figure 13.4 and Table 13.2). However, absolute brain size is not the defining factor for

14 The scholarly community, certainly in England, was blinded by the apparent finding in 1912 by one Charles Dawson of a large-brained precursor of *Homo*, called Piltdown Man, named after the site where it was found in East Sussex. The finding was a hoax, a complete fake, and was not exposed as such for 41 years, a time lapse due not least to the willingness of members of the Geological Society to believe in the appearance of a 'missing link' between ape and man on English soil.

cognitive power. On average, the Neanderthals had a slightly larger brain that we have, but with us the largest part is the prefrontal cortex, where all higher-order cognitive functions are located.

What matters is the brain-to-body-weight ratio, technically called the encephalisation quotient. For instance, this is 0.84 per cent for a chimpanzee. For elephants, who have brains weighing nearly 6 kilograms, the ratio is less than 0.15 per cent. For us humans, the ratio is 1.85 per cent, assuming an average brain weight of 1.35 kilograms and an average body weight of a male between 80 and 90 kilograms. In the present-day animal world, dolphins are second only to humans.

When it comes to the number of neurons, elephants, with 250 billion, win hands down, nearly three times our *c.* 85–90 billion. But, as just mentioned, it is the size of the prefrontal cortex which is important, as it is here that our cognitive abilities reside. These are found in the associative cortices, the regions of the cerebrum which are not directly engaged in perception and motor control. Humans have almost three times the absolute number of neurons in the cerebrum than do elephants.

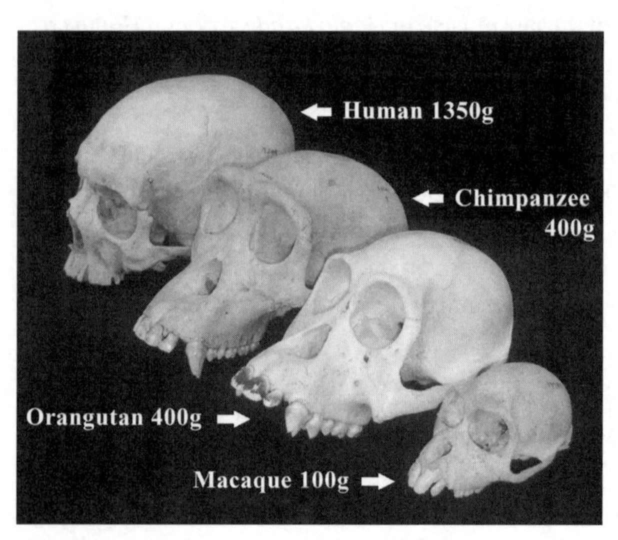

Figure 13.4 Comparative brain sizes of humans and some primates

Table 13.2 Increase in brain size in hominins over time

Genus + species	Brain size (in cc)	Time (mya)
(Chimpanzee/gorilla	400–500)	
Sahelanthropus tchadensis	>300	*c.* 7
Orrorin tugenensis	??	6.1–5.7
Ardipithecus ramidus	300–350	5.5–4.4
Australopithecus afarensis	450–550	3.8–2.9
Homo habilis	600–700	2.4–1.6
Homo ergaster	850	2–1.25
Homo erectus	550–1,250	2–0.12
Homo heidelbergensis	1,250	0.6–0.4
Homo neanderthalensis	1,450	0.6–0.04
Homo sapiens	1,350	0.6–present

Note. *Sahelanthropus* and *Orrorin* probably represent separate lineages related to chimpanzees but not continuing into later periods.

Our big brains come at a price. They consume almost a quarter of our energy intake, chiefly glucose and oxygen. The brain poses a danger to women during childbirth, given its size with a newborn relative to the birth canal. And it is vulnerable to accident and can be susceptible to various kinds of serious diseases, like Alzheimer's disease or meningitis, an inflammation of the membrane covering the brain and spinal cord called the meninges.

A Sudden Leap?

Between the last common ancestor of chimpanzees/humans and the end of the *Homo habilis* line lie at least five million years. This means that, in all that time, brain size was not developing to any appreciable extent. *Homo habilis* did not have an elevated forehead like *Homo sapiens* and the skull shape behind the eye orbitals narrowed somewhat, both features indicating that the frontal part of the brain, the embryonic prefrontal cortex of later humans, was quite modest in size, implying much more limited cognitive abilities compared to later *Homo* species.

One theory nowadays sees the fluctuations of climate along the Great Rift Valley in East Africa, with periods of drought followed by tropical wet periods as an essential environmental trigger for the relatively rapid increase in brain size from *Homo habilis* to *Homo erectus*.[15] What we do know is that the impetus toward increasing brain size was not cooking, as the increase began with *Homo erectus* before there is evidence for consistent cooking. But the consumption of meat could be connected with this development, providing a high-protein diet for *Homo erectus*. The advent of cooking meat with fire, not earlier than 1 mya, could well have provided a positive feedback loop for early humans by reducing the amount of time actually spent eating and hence allowing for more time to spend in social interaction, often indeed in the context of eating. It is known from observations of the animal world that the most intelligent animals are those who engage in intensive group interaction, such as dolphins in the sea and elephants on land.

13.13 Evolution of Our Anatomy and Physiology

Bipedality: In and Down from the Trees

Footprints in volcanic ash at Laetoli on the Serengeti Plain of northern Tanzania betray the shape of the foot of the creature which walked across this ash. Above all, the big toe was no longer of a grasping type (as with chimpanzees) but flat and parallel to the other toes, so helping the creature to balance on two feet. Going on this evidence, dated to about 3.7 mya, it

15 Deep-sea sediment probes reveal when the climate on the adjacent land was dry and when wet (the probes can be dated by examining the shells of tiny sea creatures embedded in the sediment).

is clear that the hominin species *Australopithecus afarensis* was similarly bipedal to modern humans, with feet with longitudinal and transverse arches. With that, it became clear that bipedality preceded large brain size, something which was thought to be the other way around well into the twentieth century.

It may well be that bipedalism began to develop around or just after the last common ancestors of chimpanzees and later humans, about 6–7 mya. Bipedalism might have arisen as a strategy for climbing thinner tree branches, leaving both hands for grasping other branches (Figure 13.5); it may also have been, in part, a postural development used when feeding. Now while bipedality probably arose in trees, we lost our prehensile toe (with the ability to grasp) and, after some time, it straightened out to be parallel to the other toes. When we moved down from the trees we lost our prehensile tails, previously used to grasp branches and provide stability when in an arboreal environment. It is this fact which indicates that we had left our tree life behind us. Hands became useful for bipedal hominins to hold their young (with quadrupeds, the young usually cling to the back of the mother, in the manner of a piggy-back).

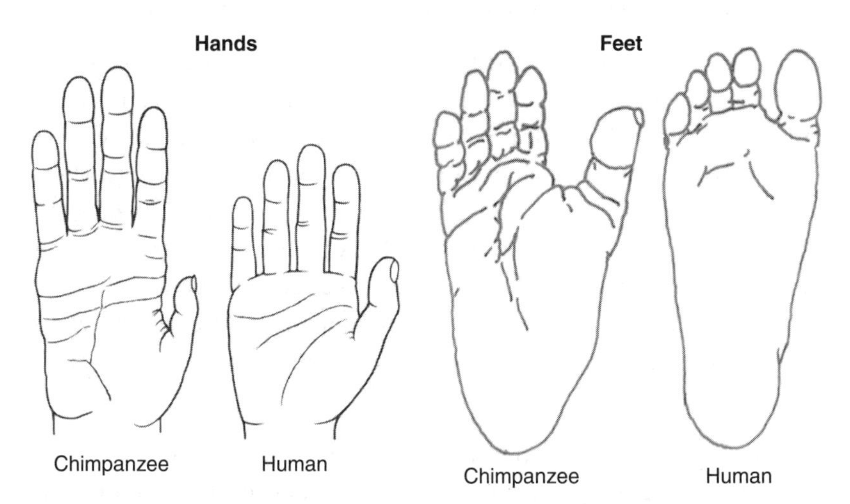

Figure 13.5 Chimpanzee and human hands and feet. Note the handlike structure of the chimpanzee foot, which allows the animals to grasp branches but not to walk upright like humans

The fossil known as *Orrorin tugenensis* dated to about 6 mya, seems to indicate a very early date for bipedality. The main part of the fossil is a femur including the ball at the top, which fits into the hip socket. This ball shows grooves from wear of the type found in upright bipedals. Later fossils, such as *Ardipithecus ramidus*, found in sediment between two layers of rock both dated to a little over 4.4 mya, show that hominins by that time had become committedly bipedal.

Our hips are quite different from those of other higher primates. Chimpanzees have a pelvis which is long and reaches up the animal's back. Our hips are shorter and stockier, wrapping around the hip area in a ring-like shape to hold the muscles which control pelvic tilt while we walk. We also have an inward-pointing femur (the bone between the hip joint and the knee) as opposed to the straight femur of chimps. Furthermore, the spinal cord and brain stem are perpendicular to the ground with those early hominins who moved on two feet and faced forward. We can also consider the foramen magnum, a large oval hole in the base of the skull through which the spinal cord reaches from the back into the lower part of the brain, the medulla oblongata. With humans, the foramen is at the bottom of the skull, as humans walk upright, but towards the back in a quadruped like a dog or cat.

It was originally thought that with bipedality we lost our agility in trees but that is not necessarily the case. Walking on tree branches has advantages and hence many palaeoanthroplogists today believe that bipedality developed among australopithecines while they were still leading arboreal lives. Nonetheless, when our early ancestors became committed bipedals, this meant a change in lifestyle away from that of other primates like orangutans who mostly suspend themselves in trees using their hands and feet. Whether or not bipedality arose while our ancestors were still in trees is, however, not a central question here. The fact is that they came to move more on the ground than hang out in trees.

Habitual bipedalism[16] is important in our evolution for another reason: it led to the decoupling of gait and breathing and hence facilitated the flexible use of a pulmonic airstream for the production of speech.

16 Many species are bipedal, such as kangaroos and birds, the latter deriving from bipedal dinosaurs, which go back at least 200 million years.

Our Hands (i): Nails versus Claws

A key feature of primate anatomy is the presence of nails instead of claws at the ends of their hands and feet. Claws are useful for predators, like great cats, as are talons with birds of prey. They are practical when clutching prey, climbing or holding on to branches and the like. But animals with claws do not show motor control of individual claws. True, claws can be retracted or distended or can be moved sideways to expand an animal's grasp but they tend to act as a unit. Digits with nails, as we find with primates, are more flexible and normally show a degree of individual control. In the evolution of the genus *Homo* this was advantageous when it came to making tools or constructing artefacts which require a high degree of dexterity and accuracy.

Our Hands (ii): Power Grip and Precision Grip

The development of hands rather than claws - extremities - which had nails on fingers instead of sharp curved tooth-like structures, has been associated with a life in trees for primates like modern monkeys. Long thin fingers (Figure 13.5) would be good at holding onto branch endings, which the primates could then exploit for grasping flowers and fruit, giving them a niche in the ecology of their time.

Nothartcus (45–50 mya) is among the first genus of primates with long fingers, nails and a thumb at an angle outwards from the other fingers, so potentially opposable. It was a small primate similar in outward

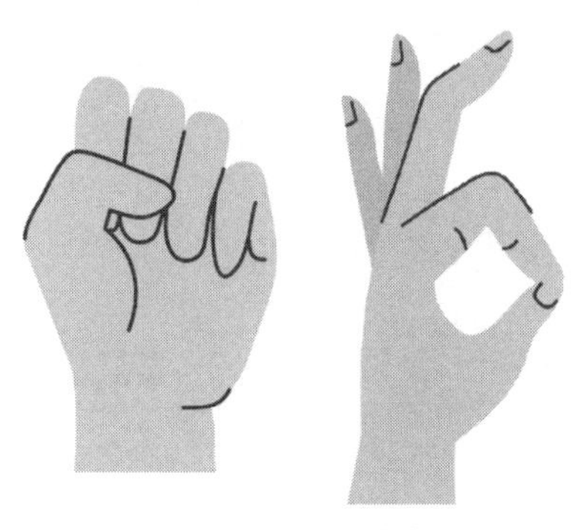

Figure 13.6 Power and precision grip

appearance to a modern rat and had a body length of about 40 centimetres with a long tail.

We modern humans have a power grip and a precision grip. The former is seen when we use our entire fist, as when rowing in a boat. or pulling a rope. The latter is when we bring the tip of the thumb and the tip of the index finger together (Figure 13.6) to finely hold some object and manipulate something with precise movements such as tying our shoelaces or putting on a watch on one's wrist. These are not things you could do with your fist and any animals with claws could not perform comparable actions either.

Our Ears

Our ears consist of three parts; outer, middle and inner ear (Figure 13.7). Many animals have large outer ears, which channel sound signals into the

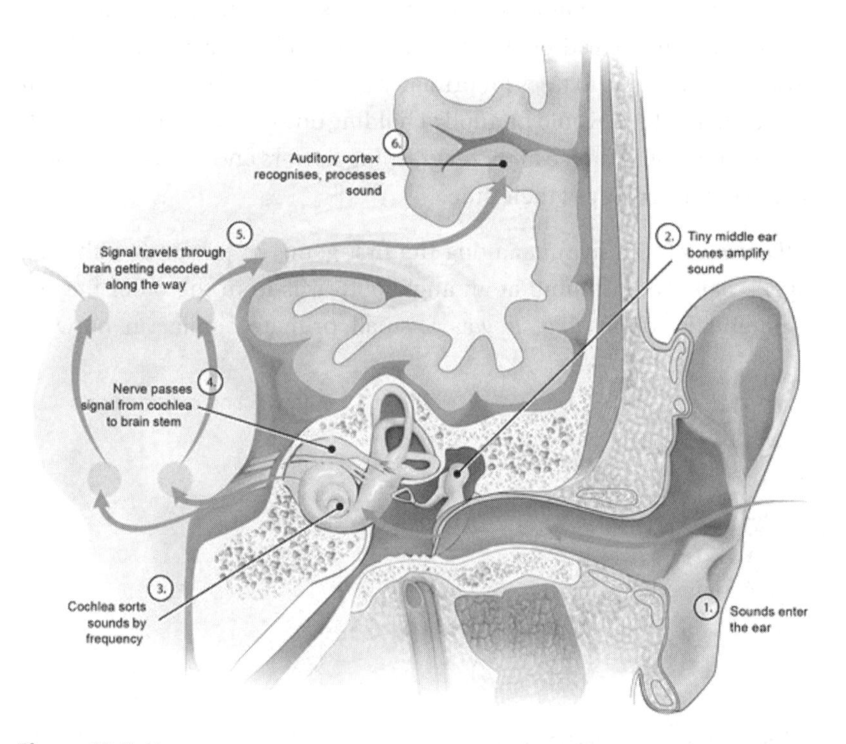

Figure 13.7 Human ear

inner ear. With others, the outer ear, visible as the auricle, is missing, for instance with fish or birds for whom an auricle would be a hydrodynamic or aerodynamic disadvantage, respectively.

Mammals have outer ears and a middle ear with three important bones, the malleus, incus and stapes (the hammer, anvil and stirrup, collectively called the ossicula auditus, or middle ossicles, which contrast with the single stapes bone of older reptiles). These bones serve to amplify the sounds coming in through the outer ear and eardrum. The additional bones in the mammalian ear resulted from the refunctioning of the jaw joint bones, which were reduced in size in later reptiles when the jaw bone itself became so large as to attach directly to the skull. The former jaw joint bones migrated towards the ear to then form the incus and malleus of the modern mammalian inner ear.

13.14 Defining *Homo sapiens* Culturally

Genetic evolution has led over millions of years to the defining features of human anatomy just listed. But humans have also developed culturally, a process which is much quicker as it is not dependent on the natural selection of genetic variants, on gene flow, chance mutations or genetic drift in a species. Standing back for a moment, one sees that humans did (and do) things which other primates did (and do) not.

Specifically human activities

1. Tool production (*c.* 3.4/2.6 mya)
2. Controlling fire (*c.* 1.5 mya)
3. Cooking (*c.* 0.8 mya)
4. Wearing clothes (*c.* 0.3 mya)

None of these features are found with chimpanzees and bonobos, our closest relatives. Chimpanzees are not really capable of learning from others, they do not observe what others do in the hope of shortcutting to some new ability (this is a kind of visual theft, as the observer does not have to go through the process of acquiring this knowledge from scratch). Chimpanzees do not pass knowledge down through the generations. We humans, on the other hand, chose to share ideas and cooperate with others in our groups. This led to the accumulation of knowledge which was transmitted across the generations, leading to considerable aggregation of skills.

13.15 Tool Making, Cognition and Communication

The first of the above features, tool production, is in its manifestation among humans, quite unique. True, there are animals like otters which use stones to crack nuts, and New Caledonian crows have been well studied for their ability to use twigs to poke insects and larvae out of holes. While such activity presupposes a degree of forward planning and hence implies the cognitive ability for this, no animal tool making can compare with the human ability to produce tools for a whole range of purposes. This ability has shown increasing sophistication through the ages. Various types of tools are recognised and used for classification. Note that this tool production was connected with hunting and the procurement of food (perhaps through scavenging in its earlier stages) and is different from the invention of other devices used for building and transportation, such as the lever or the wheel, which were developed very much later.

The oldest archaeological site offering evidence of tool making is Lomekwi 3 in northern Kenya, where stone tools were discovered which date to about 3.3 mya. Thus, they are by far the oldest tools made by a hominin.

Tool use is associated with the various species of *Homo* in the millions of years of their evolution. But tool use may in fact predate the rise of *Homo*, as there is some evidence that *Australopithecus africanus* and *Australopithecus sediba* used tools, and it has also been assumed for *Australopithecus afarensis* (*c.* 3.4 mya) which apparently butchered animals using sharp stones (but none have been found).

Early hominin tool technology

1. Oldowan or Mode I (named after the Olduvai Gorge in Tanzania) refers to simple stone chipped with another stone to produce a sharp edge (scrapers used when butchering dead animals). Date from 2.6 mya.
2. Acheulean or Mode II (named after finds in 1859 at a site in Saint Acheul in Amiens in north-east France), characterised by sharp bifaced flints used in axes and spears. Date from 1.7 mya.
3. Mousterian or Mode III (named after the site at Le Moustier in the Dordogne in France), involved the production of flints from the core of a stone by knocking off flakes and exposing a sharp pointed surface. This group can be taken to also include Lavallois tools (names after a site on the outskirts of Paris). Date from 300 kya.

The most complex tools were those in Mode III. It is a moot point whether early humans would have been taught tool making from their elders. Would it have been enough for them to conceive of a finished product (which they had seen sometime before) to then take a rough stone and craft a flint from it? But even then, they would have had a conception of the finished product to strive for that goal in the first place. And in the later periods of tool making, sharp flints were gained by knapping, a process which involved shaping a stone by striking it with another, harder stone and this would certainly have entailed a vision of the finished product.

13.16 Making Flint Tools

Flint is a form of the mineral quartz, which is found in the sediments of rivers or the sea. To produce a sharp tool, for instance to butcher meat, chop fruit, pare wood, etc. the early hominins would have had to find a block of flint, knock off pieces of it, with a hard stone, to get at the core of the rock and produce a so-called rough-out, rather like a thick axe-head. This was then further refined, often using the antler bone of a deer, to produce a thin piece of flint with razor-sharp edges, which then could be held in the hand and later fixed to a piece of wood, either at the tip of a long stick to produce a spear or on a shorter piece at right angles to yield an axe. Such flint tools date back to somewhere between 1.5 and 1.75 mya, for instance, from the West Turkana region in northern Kenya.

Working the rough-out to a spear- or arrow-head involves detailed knowledge of how to hit the stone so that the centre piece comes off and the result is a sliver of stone with a centre-line running right around the stone, edge-on.

13.17 The Management of Fire

Humans are unique in the animal world in that they have learned to control fire. It is hard to overestimate the significance of this fact in our evolution. Fire had three main advantages for early humans: (i) it provided warmth in colder climates; (ii) it kept predators at bay, as they feared it; and (iii) it allowed them to cook food. This use probably began with naturally occurring fire, for example after a lightning strike. There is evidence of this from burnt stones in a region called Koobi Fora, on the eastern shore of Lake Turkana in Kenya; this dates back to about 1.5 mya. Some time after that, humans came to control fire by stoking it and extinguishing it, as required. At a still later stage humans learned to create fire, probably by heating tinder through friction to the point where it ignited, thus providing the seed for a fire. An alternative is to generate sparks by striking stone and thus cause the ignition of dried twigs and leaves.

All animals have an instinctive fear of fire and avoid it whereas we learned to manage fire. Camps of hominins in the savannah plains of eastern and southern Africa then became safe. Keeping fire going during the night would have successfully kept away night hunters and scavengers, like lions or hyenas, from human camps.

13.18 The Advent of Cooking

Food is an essential source of energy. It provides us with three macronutrients, which we require in our bodies.

Nutrient sources in food

1. Proteins occur in thousands of different types. They are essential for building muscle and tissue and are found in a whole range of bodily functions, such as the immune system, DNA replication, various stages of metabolism. The building blocks of human proteins are some 20 amino acids.
2. Carbohydrates, such as starch, are molecules consisting of carbon, hydrogen and oxygen and have a high energy yield. The monomers (building blocks) for carbohydrates are monosaccharides (simple sugars).
3. Lipids are a group of biomolecules, the most well-known of which are fats, useful for storing energy in the body for later use and in the formation of cell membranes (lipid bilayers).

In addition to the four elements carbon, hydrogen, oxygen and nitrogen, which account for over 95 per cent of the body's mass, our bodies require micronutrients, mainly vitamins and minerals. The latter include calcium, phosphorus, potassium, sodium, magnesium and minor minerals (in terms of the quantity needed) such as iron, sulphur, zinc, iodine, etc. (trace elements). Vitamins are a heterogeneous group of micronutrients required by the body in small quantities and are generally available in uncooked plants and vegetables.

The intake of food by various species in our evolution shows distinct stages, providing a transition to the cooked food consumed by *Homo sapiens*.

Food intake in hominins

1. *Australopithecus*: raw food
2. *Homo habilis*: pounded food
3. *Homo erectus* (late): cooked food[17]

The above progression led to an increasing ease of digestion. Pounding food is entirely mechanical while cooking is chemical, in that it leads to changes in the molecular structure of the food and to its oxidation. Cooking is a form of external pre-digesting of food. When we eat cooked food, it enters our stomachs with a soft, paste-like consistency, which allows us to derive maximum nutritional benefit from what we eat.

Once we managed to control fire, we could use it for our own needs. The rise of cooking was probably accidental: some meat got burnt at a fire and humans realised it was quite tasty.[18] Cooked food has two main advantages: tough fibrous proteins are broken down during cooking and harmful pathogens are rendered innocuous. Heat denatures proteins, as when an egg is boiled and the white goes from transparent to white, and makes food more digestible. The proteins unfold and many toxins in plants are made harmless in this process. Their biomass is actually digested and utilised by the consumer of the food through metabolic processes. How far cooking goes back cannot be said for certain. But the Wonderwerk Cave in South Africa offers evidence of such activity, about 1 mya.

An Unresolved Question

There is an unresolved question here. The increased calorie intake due to consuming cooked meat would have provided a basis for the

17 The British anthropologist Richard Wrangham has repeatedly stressed the significance of cooking in the development of our species, see his *Catching Fire: How Cooking Made Us Human* (2009), where his ideas are laid out in detail.

18 Cooked meat may well be responsible for the enhancement of umami taste receptors in humans, which promoted the further consumption of meat treated by fire.

expansion of our brains. But the increase in brain size with hominins can be seen before 1 mya, which is before the earliest date for regular cooking. In anthropological terms, cooking arose somewhere between *Homo erectus* and *Homo sapiens*, after the expansion in brain size which began between *Homo habilis* and *Homo erectus*. So, although cooking played a role in the positive feedback loop of augmented protein consumption and increasing brain size, it is not the trigger for this development. Furthermore, it is not the case that only hominins with increased brain size began to spread out of Africa. Those who inhabited the Dmanisi site in Georgia about 1.8 mya (see Section 13.7 above) had brains of only about 600 cc and did not use fire, so they did not know cooking either.

When plants and meat are cooked the fibre[19] in their cell walls is broken down, releasing nutrients. Cooking thus softens food and there is less chewing involved, which meant, in evolution, that the teeth and jaws of our ancestors got smaller. This in turn allowed the braincase to expand in size by the opening up of the skull at the top and, by a movement downward, of the reduced jaw. All of this meant that *Homo erectus*, by about 1 mya, had access to much higher levels of energy via cooked food. While this extra energy was not the trigger for increasing brain size it did feed into this process, quite literally, leading ultimately to the modern brain, which consumes about a fifth of our energy intake.

A recent study (Fellows Yates et al. 2021) has demonstrated that both Neanderthal and *Homo sapiens* showed evidence of consuming starch-rich food (nuts, seeds, tubers) alongside the meat they obtained from hunting. The evidence was provided by examining the oral microbiome of the two *Homo* species as well as chimpanzees, gorillas and howler monkeys (more distantly related to hominins). Specifically, the analysis of dental calculus (tartar) showed that early humans were consuming food that provided them with glucose, which would be important for

19 As the Brazilian-American neuroscientist Suzana Herculano-Houzel has pointed out, because cooked food is so readily available we overeat and what was once a solution to our nutritional problem has itself become an issue and – ironically – the new solution is to return to eating raw food with lots of fibre.

maintaining their large brains. Whether the consumption of glucose through plants played a role in the increasing encephalisation which set in during the two-million-year evolution of *Homo erectus* (brain size: *c*. 550-1,250 cc) is uncertain. The authors of the study state that there are "possible implications" of dietary starch in the "*Homo*-associated encephalization".

Meeting at the Cooking Place

A key aspect of cooking is that it involves a degree of centralisation: people take food from where they procure it to a single habitual place – campfire or kitchen – where it is cooked, usually in the presence and to the benefit of others. So cooking strengthened social bonds no end. And to this day, the most common way for humans to get together socially is for a meal, whether it is everyday in the work canteen, at home in the kitchen or in a restaurant on a special occasion.

Gaining Spare Time

There is a further advantage to cooking, which could well have played a role in our evolution: it resulted in our having much more time on our hands, because we did not have to chew fibrous plants or leaves for hours on end, as do panda bears and koalas, or on tough raw meat for that matter.

FOOD, COOKING AND EATING ON AN EXOPLANET

If animal life on an exoplanet were to live off fibrous material like leaves and shoots, they would probably have to spend about eight hours a day eating as do gorillas and orang-utans. But they might have combined diets, like chimps who eat fruit but also hunt and eat monkeys when available. Or they could have some common, high-energy foodstuff, which could cover their daily calorie needs without spending too much time eating.

13.19 Wearing Clothes

A somewhat whimsical feature of humans, when compared to animals, is that we wear clothes which have become an integral and differentiated aspect of human cultures. Animals do not need clothes because they have skin and/or fur which suits their environment. Great cats, like lions or leopards, have little or no fur (apart from the manes of lions, an indicator of male prowess) because they live in warm climates. Polar and grizzly bears have thick fur because colder northern regions are their habitat. Chimpanzees and bonobos are somewhere in between, with a medium cover of fur. In the development of our genus *Homo* we lost this coating of fur.

The Loss of Fur and Its Consequences

Why natural selection should have favoured early hominins who began to shed their fur is something which scientists still puzzle about. Maybe there was a sexual advantage to this when females came to prefer individuals with less fur. However, its loss brought disadvantages with it: fur is useful for camouflage and for keeping us warm in cold weather. Nonetheless, with its loss came a new mechanism for humans to rid the body of excess heat: we came to sweat through glands in our skin. This seemingly minor development had two huge advantages for us. (i) It allowed us to engage in persistence hunting by chasing animals over considerable distances through the midday heat until they collapsed from heat exhaustion – that does not happen to us to anything like the same extent because we sweat through our skin and the ensuing evaporation, cooling over a large surface area, can reduce our internal body temperature. (ii) With the switch to sweat glands in the surface of the body,[20] the tongue was free from the function of giving off excess heat through panting (as dogs do, for instance). Then the tongue could go from a flat, limp shape to a bunched and agile muscle, which could produce the sounds of human language.

20 This is not something which began after our splitting off from the chimpanzees. The latter have axillary sweat glands (in the armpits) as humans do and do not rely on panting for thermoregulation as do many other animals.

A further development after the loss of fur was the rise of subcutaneous fat to prevent heat loss in the cold with the now absent fur (just consider how bony sheep look after being shorn). This fat was also useful for stabilising our bodies after the rise of bipedalism.

Clothes to cover the exposed body may date as far back as 300 kya. Given our upright gait, our genitals were more exposed than with quadrupedal animals, where they are largely concealed and protected between their hind legs (for males), with a tail providing additional protection, especially for females. This could have meant that, initially, clothing (around the loins) was probably just protective, as it still is for some non-industrialised groups living in the warm tropics. For those living further away from the equator the hides of animals reduced exposure of the skin to cold. With time, wearing clothes became the default situation for humans and a lack of clothing in public became a source of embarrassment.

The loss of fur had a number of other consequences for us. We have increased our facial expressivity as we cannot raise our body hair for expressive purposes as many hair-covered mammals do. This latter behaviour is called pyloerection (bristling hairs) and is yet another example of secondary functionalisation, like using surface blood vessels in huge ears, in elephants, for instance, to cool blood.

The exposure of larger surfaces of skin led to the cultural practice of marking our skin with tattoos. This has a long history: the famous Ötzi skeleton, over 5,000 years old from the Tyrolean Alps, found in 1991, had deliberate body marks (on his ankle).

Making and Wearing Shoes

The American palaeoanthropologist Erik Trinkaus (Washington University, Missouri) thinks that humans began making and wearing shoes about 40 kya, pointing out that humans then began to develop weaker bones in the four toes beside the big toe as a consequence of wearing shoes/boots.

WOULD EXOBEINGS WEAR CLOTHES?

This might seem like a trivial question but the answer might well colour our reaction to exobeings should we ever see any. The temperature range on an exoplanet would play a role here as would thick bodily hair (fur), which exobeings might have.

13.20 Setbacks in Our Evolution

The outline of our evolution so far seems like a direct line in which things got better and better. But that is by no means the case. First of all, there were many species of *Homo* around in the past few million years with sidelines which were discontinued. And for those who survived it was not all plain sailing. Many events led to a severe reduction in the numbers of *Homo* and threatened the extinction of the entire genus. For instance, we know of the Toba supervolcano, which erupted some 75,000 years ago where present-day Lake Toba is to be found (in Sumatra, Indonesia). It is one of the Earth's largest known eruptions. Such a megaeruption would have caused a volcanic winter which could have lasted for up to a decade, with disastrous consequences for flora and fauna throughout the globe.

Another setback for humanity, which we know from history, is the plague which struck in the eastern Mediterranean in the mid-sixth century and again in the mid-fourteenth century throughout Europe. In each case it caused a loss of anything between a quarter and a third of the population at the time.

These are known setbacks for humanity. But before them there may well have been others which are not documented and not in evidence in the geological or fossil records. In addition to disease and activity in the Earth's crust there have been many fluctuations in our climate, producing periods of great cold, dryness or warmth to which life had to adapt or go extinct.

Periodic variations in our climate are known as Milankovich cycles after the Serbian scientist and climatologist Milutin Milankovich (1879–1958), who explained fluctuations in the Earth's climate on the basis of the following factors.

Factors affecting climate fluctuations on Earth

1. Eccentricity (degree of elliptical orbit around the Sun)
2. Axial tilt (degree to which the Earth is 'lying on its side')
3. Precession (change in the orientation of the rotational axis of the Earth)
4. Shape of the continents
5. Volcanic activity (leading to increases in carbon emission)
6. Amount of ice cover (leading to increased reflection of sunlight)
7. Greenhouse gas emissions (extreme at present)

These factors determine the amount of solar radiation reaching the surface of the Earth and being absorbed by it. Originally, Milankovich thought the cycles of climatic variation lasted about 40,000 years, the time span between ice ages on Earth, but scientists today regard 100,000 years as a more realistic figure. Increased technology for working out climate cycles in the past, drilling ice cores in the Antarctic or the ocean floor and studying rock layers has documented patterns of climate change over some millions of years. During ice ages there are glacial maxima when the Earth's surface temperature plummets and hence ice cover is at its greatest (sometimes producing almost complete cover, leading to 'Snowball Earth'). These regular cold snaps during geological history would have provided a challenge for life forms which depended on an abundance of temperate vegetation. On the other hand, warming of the atmosphere led to lush ecologies at the poles and consistently high ocean temperatures in between, which again was harmful to most forms of life, interrupting food chains involving organisms such as plankton that were consumed by larger organisms.

Currently, Earth is within the Quaternary glaciation, which had its last glacial maximum about 21,000 years ago, and since then the ice sheets over the north and south poles have been steadily retreating, with this process accelerated greatly due to the effect of greenhouse gas emission caused by the lifestyle of present-day humans. Whether we are in an interglacial period, with another glacial stage in the near-to-medium future, is uncertain. But the millions of tons of carbon dioxide (and methane) being pumped into the atmosphere at the moment makes this seem unlikely.

Paradoxically, fluctuations in climate and the resulting see-sawing of local environments, may well have been beneficial for the evolution of the genus *Homo*. Fluctuations in aridity and humidity in the Great Rift Valley of East Africa can be recognised in the geological record of the past few million years and these ecological challenges may well have stimulated the survival of robust and cognitively advanced species of *Homo*.

13.21 Hominins: The Big Picture Once More

In discussions of palaeoanthropology, the stages of hominid developments are debated and analysed. But in a book about possible life beyond Earth, the central question is: How likely is what happened on our Earth to have also occurred on an exoplanet?

CLIMATE ON EXOPLANETS

On exoplanets, temperature and climate fluctuations would be as important as on Earth: long-term stability would be essential for the evolution of exobeings, although a degree of climatic variation might be a beneficial stimulant. The factors determining or at least affecting climate might be stacked against the evolution of life in other solar systems. For instance, a perfectly circular orbit around a star is more the exception than the rule and many planets show considerable eccentricity in their orbits, coming dangerously close to their star at some part of their orbit or moving away too far at another. Whatever the irregularity of a planet's orbit, it would have to be contained with the habitable zone of its solar system to offer a suitable scenario for the evolution of life.

We all started in the sea and then went through an amphibious stage, moved onto land and continued our evolution there. The animal known as *Tiktaalik* (see Section 11.6 above) began about 375 mya to spend time in shallow water and on adjacent land, leading to the amphibians who spent part of their time in the water and part on the land, such as modern-day frogs, who have lungs as adults but gills as tadpoles. Mammals in the sea, such as whales and dolphins, resulted from some of the ancestors of these animals returning to the sea. These animals bear live young, suckle them and breathe oxygen directly into lungs via an air hole on the top of the head.

Would exobeings have begun their long evolutionary path in water? Most probably, as water is an ideal biosolvent. It is an excellent carrier medium and only reacts chemically with the molecules it carries under specific conditions.

13.22 A Unique Species and the Great Cognitive Gap

A curious fact of human evolution on Earth is that there is only one species of *Homo* left. All others, especially the Neanderthals and the Denisovans, died out for reasons which are not entirely clear. The many intermediary species and 'side branches' of our evolution are also long since gone. There will have been various conditions, chiefly in Africa, but also in Europe and Asia, which were unfavourable to the survival of different *Homo* species, with only one group managing to 'hang on,' *Homo sapiens*. There may well have been severe reductions in population size, for instance during periods of environmental stress as with prolonged drought. This would have caused population bottlenecks through which only a small number of hominins managed to endure. One such bottleneck occurred between about 130 kya and 195 kya when, according to the American archaeologist Curtis Marean, only a few thousand *Homo sapiens* eked out an existence of the exposed shores (the Agulhas Bank) of South Africa.

GEOGRAPHICAL ORIGIN OF EXOBEINGS

An Earth analogue would be a rocky planet with oceans of water and hence a distinct geography. This would mean that continents would exist, which in turn would imply that different habitats and climatic conditions would obtain at different locations on the planet. Would exobeings have arisen at one location on their planet and spread out from there as we did with the dispersal of *Homo* species from Africa? This would depend on what the local geological and ecological conditions were like on an exoplanet. It would be possible for advanced life to arise at different sites, maybe at different times. Furthermore, depending on the size of an exoplanet and the shape and distribution of land masses, certain parts might not be in contact with others, leading to very disparate ecologies arising.

The individuals who survived may well have been those with language, which allowed cooperation in groups and hence led to improved survival rates. A bottleneck view of recent *Homo sapiens* survival would also account for the relative uniformity of the species and its language faculty.

Whatever the reasons for our survival, the fact is that there is a gap between us and our nearest relatives (in evolutionary terms), the chimpanzees and bonobos. There is also a cognitive gap between ourselves and other relatively intelligent animals, such as dolphins or elephants, despite the sophistication of their communication systems. Nowhere is this cognitive gap more evident that in the fact that we have complex language and advanced technologies and societies based on this.

What could the situation be like on an exoplanet? We cannot assume that exobeings would be the only species on their planet with a significant difference in cognitive power between them and the animals of their planet. Our Earth may well be unique in this respect. And what might the consequences be like if cognitive powers were more spread out among living beings on an exoplanet?

14

The Rise of Human Societies

As exobeings can only arise through evolution the roots of their sociality would lie in earlier stages of their biological development. From Earth, we know that animals bond and form communities in different ways, with the common purpose being the survival of the species. This is the basis of their very divergent kinds of behaviour. There is no reason why this should not be the case on an exoplanet as well.

Animals differ widely in the extent to which they interact with each other. Some animals, like certain types of birds, such as albatrosses, remain together as couples to rear their young and return to each other every year; others do not, with the male leaving once mating has occurred. However, if there are pressures from the environment, as with extreme cold, animals may form colonies as do king penguins in the South Atlantic, on Georgia Island, for instance, or in Antarctica. In other cases, animals may congregate in particularly suitable feeding places, such as Lake Nakaru, Kenya, where thousands of flamingoes flock to feed. Such large assemblies of animals consist of breeding pairs and their offspring but they do not generally show an internal vertical order which might be regarded as proto-social.

The situation is different with smaller groups, which usually form because of a scarcity of food or difficulties in obtaining it. Animals must then cooperate and adjust their behaviour to that of other animals. A good example is hunting. Wolves typically live in cold inhospitable climates and so it was those early wolves, who began to hunt in groups, who had a better chance of obtaining sufficient food and survived better to produce offspring. Among these offspring, those who fine-tuned the practice of hunting were in their turn more successful at producing offspring, and so the hunting wolves came to dominate to such an extent that group hunting became a defining characteristic of wolves' behaviour, which in turn made demands on their cognitive abilities, creating a positive feedback loop.

14.1 In the Beginning Was the Group

Small cooperative groups, which formed to hunt prey and also to avoid predation, are probably the origin of societies. Those animals which were not as strong as others not only learned to cooperate for mutual benefit but also to use their cognition for survival. This definitely would have been the case with early *Homo* species, who were not particularly strong and did not have canine teeth like the many big cats in their environment. Group formation was the recipe for survival in the wildernesses of Africa, not just to have a means of combating and countering predation but also to come through periods of environmental stress, such as prolonged droughts. The 'groupishness' of humans, similar to the behaviour of chimpanzees in their communities but on a greater scale, became a prominent, even a defining feature of our behaviour and has remained so to this day.

14.2 Humans, the Great Extenders

We humans show a marked propensity to extend traits and abilities we have far beyond their original purpose and range. The early groups, which served the goal of surviving in inhospitable environments, led to increasingly far-reaching social networks. We began to establish and maintain contact with other humans who lived outside our immediate surroundings. These networks were exploited for resources and represent the beginning of trade relations among groups. Both *Homo sapiens* and Neanderthals engaged in these activities, for instance, getting stone for tools from regions well beyond their locales.

14.3 The Origins of the Leader

With such groups there will be natural differences in the size of animals, with the largest, strongest male usually dominating. This is a key feature of all social groups: the members vary in many ways. Coming to terms with internal differences and finding one's niche in the group promoted intelligence. In any social group there will always be a tension between competition and collaboration, between power and solidarity. Manoeuvring one's way between these poles requires planning of one's actions, working out how others are likely to react and behave, assessing situations in the group – these are all processes which demand cognitive abilities.

A further aspect of small groups of intelligent animals is cooperation among non-kin. This is a distinguishing feature of advanced forms of life: elephants in a herd will protect any calves from attacks by lions, but in a colony of seals or walruses (both semi-acquatic pinnipeds) no member will help another to ward off an attack by a polar bear, for instance.

Cooperation among non-kin is more likely than not the origin of altruism. If an animal invests time and energy in the protection of another, non-related animal, there is a chance that it might be the recipient of similar help at a future date. Such behaviour presupposes that the animals in question have a certain level of cognition, which enables them to imagine a situation that might arise, this providing the motivation for their action in the here and now. In-group cooperation is the opposite of competition where animals simply fight each other for a partner or food; such behaviour is not indicative of higher cognition, unless it involves complex deception patterns.

Sociality increases as we view increasingly complex forms of life. And we humans are clearly social animals. We need continual contact with others and deprivation of this, as with solitary confinement, leads to serious psychological disturbance with any individuals subject to it. High levels of social cooperation among early species of *Homo* had the additional advantage of allowing them to vary their habitat as environmental factors restricted them less and less, as opposed to other animals which were trapped in their habitat and often went extinct when this changed too much for them to adapt in time.

With human groups, cooperation continued to increase. It became entrenched as typical behaviour of our species. This can be seen with young children who share information and regard cooperation as the default mode. If it is flouted for some reason, children are aware of this. Sharing information with others is a powerful mechanism for technological advancement. Anthropologists, like the American Dietrich Stout, have stressed that the crafting of tools is an essentially social activity in non-industrialised groups, such as indigenous peoples in Papua New Guinea. It goes further than that, as another American anthropologist Augustín Fuentes has stressed: the sharing of information with others involves imbuing our actions with symbolic significance. This in turn can lead to agreement about customs and practices within a culture, such as the nature of work, the division of time, the use of money, the practice of schooling, marriage and caring for the elderly. From a very early age, children participate in their culture and through this participation absorb knowledge and become part of a larger community. In history this led to the rise of social norms which are necessary to organise collaboration beneficially for all involved. These norms vary across cultures, of course, and, depending on how they pan out, they result in sometimes very diverse cultural practices.

Our ability to plan for the future allows us to conceive of resources and then sharing can become an issue, especially if there is strong competition from others outside our group for the same resources. Since the advent of sedentary populations (as most hunter gatherers gradually became agriculturalists), vertical differences in proto-societies arose. These differences very quickly resulted in societies with kings and servants, the latter divided into many vertically organised levels. Paradoxically, although the development of farming led to these differences arising, it is the farmers and similar labourers who formed the lowest level in early societies. Kings and leaders profited from their supposed closeness to deities, in fact in many cultures, for instance the early Nile Valley culture, one of the early river civilisations, the kings were regarded as deities. While there were disadvantages to the rise of socially stratified societies, they furthered the practice of social learning and innovation, through the specialisation of roles for individuals in society. This greatly increased the scope of an individual's activity and extended the collective behaviour of societies.

Compare this to the situation with even advanced animals: they can discover things and have basic forms of knowledge (as opposed to instincts, like the ability of a spider to spin a cobweb). But in general they do not share this knowledge with their fellows, though some higher animals will pick up things by observing others. The cognitive scope of animals is limited: anything that they have not seen does not exist for them. We can learn about things which have happened and which we have not seen, either in the past or in a part of the world far removed from us.

What makes human knowledge so powerful is that we share it with others, we teach others to do things, we organise and structure our knowledge and we pass it on to following generations. Our cultural evolution gave rise to schools, formal institutions which all children are expected to attend in order to benefit from the accumulated knowledge of the past. For us humans, knowledge is what we share in our communities and across the generations. In fact, it is the drive to do this which led to systems of writing being invented, essential for the intergenerational transmission of language.

14.4 Societies on Exoplanets

For exobeings, what has just been said would also apply. They could only invent advanced technology if they had first developed complex cooperative social groups. These would consist of varying individuals so that social hierarchies with dominating and dominated subgroups, yielding a vertical structure, would be built into the fabric of exosocieties. Societies of this kind would also imply concentrations of non-nomadic populations such as those which formed the first cities built by humans: Çatalhöyük in present-day Turkey, for example, which flourished about 9,000 years ago, and the later Sumerian city-states from about 6,000 years ago, such as Ur in present-day Iraq. The dense infrastructure and interaction, typical of all cities, would only be possible if their inhabitants had language.

An advanced exosociety might also have gone through stages of social development in which the notion of a single sovereign at the pinnacle of a society was replaced by the idea of a representative body from this society forming an institution, which decided on rules and regulations for communal behaviour. Such a society would probably have developed the view that some of the freedoms of the individual would have to be curtailed for the greater good and that the state would have the right to devise laws which would determine the permissible actions of individuals within that state. However, considering this level of detail for an exosociety is at present mere speculation.

If it were true that exobeings reproduced sexually, with one sex bearing the offspring, there would be families which – in principle – would be comparable to those on Earth. A broader question is whether their social organisation would rest on the nuclear family, as it did traditionally in European societies at least, or if it would show the diversity of personal and sexual preferences found in twenty-first century Western societies. This question is tied up to how much freedom exobeings would possess in their societies. What or who would provide them guidance? How would they derive their sense of belonging? How would they deal with weakness and uncertainty? We can only surmise about that, but it would be an interesting comparison to make, nonetheless.

14.5 The Question of Violence

Physical violence among humans has existed for as long as we have records and it would, in principle, seem to be a continuation of violence in the animal world. However, with animals the contexts in which it occurs are clearly defined. Animals will defend themselves using violence, if necessary. They will fight for the right to mate with females. Animals also engage in infanticide: a lion which is attempting to gain a lioness as mate is likely to kill any cubs she may already have. This is to ensure that his genes survive and not those of another lion. The readiness to kill offspring is uncannily widespread in the animal world: a dominant female meerkat will kill the offspring of other females in her group and then force these to wet-nurse her pups. This behaviour is disturbing for humans, but it is highly contextualised. While infanticide among humans is regarded as horrific, there are nonetheless far too many adults ready to kill other adults, something which is not widespread as intraspecies behaviour in the animal world. But our nearest hominid relatives, the chimpanzees, who used to be regarded as peaceful and cooperative, can engage and have engaged in inter-group warfare and occasional in-group killings, notably recorded in Gombe, Tanzania in the mid-1970s, this clearly documenting the violence which higher primates are capable of (see Goodall 1990).

There is evidence from work such as that by the ethologist Anne Pusey that our patterns of violence have their evolutionary roots in the behaviour of chimpanzees, who show strong in-group bonding among males (unless in competition for monopolisable females) along with inter-group aggression to increase their territories. Female chimpanzees do not exhibit anything like the same levels of violence as males. Now consider that human violence on a physical level is overwhelmingly an issue involving males – the briefest of glances at rogue states around the world today will confirm this. In the United States, over 93 per cent of federal prisoners are male and five out of six murderers are male. On an exoplanet with sexual reproduction among males and females (whatever their phenotypical equivalents) it would be interesting to know in which group violence would be more common.

Violence appears to be deeply rooted in human history. Many Neanderthals may have been killed by *Homo sapiens*, though the view that they were exterminated by the latter is not supported today. The 5,000-year-old human remains found in the Ötzi valley in Austria were of an individual who was murdered. But on the other hand there are cases like the Neanderthal Shanidar I, an individual from a site in northern Iraq with severe injuries and who lived for several years after sustaining these due to his community, in which caring for the weak was obviously practised. And early hunter-gatherer societies also tended to be egalitarian: acting the bully, striving to be dominant within the group was (and is) frowned upon.

Some of the reasons for warfare are shared with animals. Raiding – taking away resources from another group (food caches, potential mates, etc.) – is very common. Competition for resources can cause population pressure in a habitat, but in the animal world this usually results in fewer offspring rather than animals killing each other, in contrast to human competition in such contexts.

In one key respect humans differ from the animal world: we have developed tools for warfare. Swords, axes, clubs, spears, bows and arrows – weapons have a long history. Furthermore, virtually every scientific discovery has been misappropriated by the military to serve in war. In the twentieth century alone we have the production of highly toxic mustard gas, the construction of armoured cars and tanks (already for World War I) and, not least, the making of the atomic bomb, all instances where scientific and technological knowledge was abused for the purposes of war.

So our higher cognitive and manual abilities have been put to many uses, some of which sadly involve aggression by humans against other humans. In addition, our ability to reflect on our lives, on events we were involved in or connected with, has led to the urge for revenge or retaliation which we experience when we feel we have been wronged. Many acts of revenge are directly coded into cultures, such blood revenge for the death of a relative/group member which is generally motivated by a perceived loss of honour and respect in society.

Violence in both the human and the animal world is a multifaceted phenomenon with many motivations and realisations. However, there is one thing which is of relevance to the present book as it stands out clearly with humans: the perpetration of violence towards those outside one's own group – family, extended family or regional grouping. To be realistic, there is a danger that exobeings might react violently to any attempt at contact by humans. However, any contact we might achieve in the forseeable future will most likely be very indirect, for example via some kind of electromagnetic signals, so that any fear of attack by exobeings would appear unfounded, but not all scientists agree about this.

14.6 Evidence for Social Organisation

Some *Homo* fossil bones show traumata such as teeth bites from predators or bone fractures. The fact that individuals of *Homo* species lived with such injuries would imply that they were cared for by the group while they were recuperating even if they did not recover their prior state of health fully. These are cases of reciprocal altruism, whereby the carer exercises care in the expectation of being the recipient in future, should the necessity arise. Only higher primates do this. With other animals, if a member of a group is injured or weak it gets left behind and is eaten by a predator or its cadaver by vultures or some other scavengers.

Caring for members of one's group is a behavioural trait which was already in evidence with the Neanderthals, as just mentioned. It shows a degree of attachment between members of a group and is similar to grieving. Cognitively advanced animals are known to grieve. For example, elephants can remain with a dead animal for some days, defending it from scavengers, smelling and touching it. A chimpanzee mother is known to carry her dead baby with her for days after its death.

The upshot of these considerations is that higher forms of life show concern and attachment for their offspring but also for others in their group. This behaviour developed further with humans such that we can maintain social relationships with quite a large number of individuals in the social group to which we belong. The Oxford anthropologist and evolutionary psychologist Robin Dunbar is well known for his belief that the upper limit on the number of stable social contacts which humans can maintain is about 150 – this has become known as Dunbar's number.

14.7 The Advent of Farming

The hunter–gatherer lifestyle was continued by *Homo sapiens* for much of its existence until about 12,000 years ago when, in the Fertile Crescent (an area covering most of present-day Iraq, southern Turkey, Syria and Lebanon), humans began to grow crops by planting seed in spring and harvesting in autumn. This heralded the agricultural revolution which was one in a long line of cultural innovations made by humans. Farming involves planning in spring for autumn, several months away. So it was in principle different from hunting and gathering, which was locked in the here and now. Probably at the same time, or perhaps somewhat later, rice came to be domesticated in the Yangtze River basin in south-central China and was soon followed (independently?) by the cultivation of rice for human consumption in India.

The advent of farming led to population growth due to the greater reliability of food sources and also contributed to short inter-birth intervals. Sociality also increased as groups became larger and contact across groups increased given their spatial proximity. Some populations pursued a nomadic lifestyle, which had more in common with previous hunting and gathering and were often forced to engage in this living pattern because of seasonal fluctuations in feeding grounds for their animals.

It is hard to overestimate the importance of farming for modern humans. It led to the domestication of numerous animals, many of which themselves acted as food sources for humans, such as goats and cattle, as well as supplying food in the form of milk and, later, cheese. But perhaps the most significant domestication was that of horses around 5,500 years ago, in the steppes of central Asia (the area of present-day Uzbekistan and Kazakhstan). As a work animal for humans, horses (and to a lesser extent their near relatives, donkeys and mules, as well as oxen) maintained their position right up to the early twentieth century only to be replaced by the tractor in the field and the motor car on the street.

The Industrial Revolution in late-eighteenth-century Europe gradually introduced machinery to farms. The early twentieth century also saw a further, seminal innovation: the introduction of artificial fertilisers to increase plant growth by adding elements like nitrogen, phosphorus, potassium or calcium, to the soil. It is no exaggeration to say that the present world population of nearly eight billion can only survive because fertilisers did away with the need for farmers to rotate their crops and leave fields fallow to recover (using the so-called three-field system of crop rotation).

The mechanisation of agriculture made it less labour-intensive and, with increasing industrialisation during the nineteenth century, larger sections of the population came to work in manufacturing, in the factories which sprang up in Europe and America. The late-twentieth and twenty-first century has seen a decline in the manufacturing workforce, with increasing automatisation of production processes due to digitisation. This has led to an increase in the services area of Western economies, so much so that in many countries the majority of the workforce is engaged in services of one kind or another and no longer in manufacture. The same situation could apply to an exoplanet with an advanced digital culture, given that one of the main goals of such technology is to automate production and thus reduce the use of manual labour.

HORSES AND WHEELS

Humans have domesticated many animals that have proved of value to them. Among these is the horse which came to be used over 5,000 years ago in the steppes of Central Asia, allowing people to cover much greater distances. Roughly at this period, wheels were developed, which together with an axle yielded horse-drawn wagons, revolutionising transportation. In addition to other devices, such as the lever or the inclined plane, humans could lift and move objects far heavier than themselves.

Whether exobeings would exploit animals on their planet for their own purposes is unknown, though not unlikely. Whether they would have wheels for various devices is quite probable given that the circle is a basic geometrical form recognisable in many shapes and movements in our three-dimensional universe.

14.8 Culture and Human Evolution

In the deep past of humans, natural selection along with random mutations, genetic drift and gene flow, were the main factors determining the path of evolution. However, with the rise of human societies, aspects of these societies came to play a role. From that point onwards we speak of cultural evolution.[1]

Humans may (inadvertently) guide their own evolution through cultural development. Take the example of lactose tolerance among people with pastoral herders in their ancestral background. The ability to digest the sugar lactose found in dairy products, above all in fresh milk, is usually restricted to childhood, with this ability decreasing as people enter their teens and early adulthood. But in middle, northern and north-western Europe there is a high incidence of adults who still produce the enzyme lactase necessary to digest the sugar lactose in the small intestine. The abundance of bovine milk in the farming environment of early populations in countries like Ireland, Sweden or Poland stimulated the continued production of lactase into adulthood and led to the persistent tolerance of lactose. Many Americans whose ancestors came from these countries are lactose tolerant as a result. On the contrary, countries like those in East and South-East Asia as well as in southern Africa have a very low tolerance of lactose because of the virtual absence of dairy farming in their histories.

14.9 Cultural Buffering

Developments in culture, rather than through evolution, often offer protection to humans. The development of housing, which protects us from the environment, particularly extreme cold and possible attack by predators

1 The Israeli historian Yuval Noah Harari makes much of the point that our sudden cultural evolution about 70,000 years ago was too quick for our environment to adapt to this. Our leap to the top of the food chain meant that we overtook all other animals. (RH: probably in conjunction with the control of fire.) We were not accustomed to this powerful position and it engendered anxiety in us. This was not true of animals like lions, who had carved out a niche for themselves in the African savannah and were not worried about this position.

is an example. Cultural buffering is connected with the evolution of societies, which in turn implies the advancement of science, but the level of this science would depend on a variety of factors, not just essential discoveries but the attitude of social entities, such as institutional religions, to the value of science. Furthermore, the advancement of science is not necessarily a single straight line from a beginning to the present. There have been good starts which led nowhere[2] or scientific achievements which were unnoticed and were not picked up by others in a society or culture.[3]

Turning to the subject of health, medicine would also have advanced among exobeings to ward off the pathogens which would have evolved with them on their planet. Consider such advances in medicine on Earth: individuals with serious diseases, inherited or acquired, such as diabetes (Type I, which is about 30 per cent inheritable) or heart conditions, now have a greater chance of survival and thus of having children, perhaps passing on a tendency towards these diseases to their offspring. In previous centuries, these people would have died early in life, thus reducing the incidence of these diseases in a given population.

So, in a way, there is cultural buffering against the negative effects of diseases and medical conditions, but there is also the danger of some of these continuing across generations, which would not have survived in previous centuries. But on the whole the benefits of modern medicine outweigh any negative genetic consequences it might have.

2 Surely, the most obvious example of this is the view of the ancient Greek philosopher Parmenides (who flourished in the late fifth century BCE) that the world was spherical. It took the best part of two millennia for this view to be accepted in Western societies.

3 There are a number of cases where artefacts from earlier ages have given us cause to rethink our views. Perhaps the most famous is the Antikythera Mechanism, a sophisticated device designed to model the movements of the Moon and Sun and thus predict lunar and solar eclipses (regarded as portents of misfortune in the culture of the time). The mechanism consisted of a very intricate set of bronze cogs and gears, which moved together and had hands that predicted the movement of several celestial bodies many years in advance with astounding accuracy by pointing at future times drawn on dials on the outside of the mechanism. In that respect the mechanism was a true computer with genuine predictive power. It was built in ancient Greece around 100–200 BCE and predates devices of similar sophistication from Western Europe by about 1,500 years. The device was retrieved in 1901 from a ship off the island of Antikythera, north-west of Crete, which had probably sunk in the area around 70 BCE.

MEDICINE AND GENETICS ON AN EXOPLANET

An exoplanet which was much farther advanced technologically than our Earth might well have options which we can only barely imagine. For instance, if they mastered artificial protein synthesis and the implantation of engineered DNA into carrier cells, they might have possibilities in medicine, and genetics in general, which would allow them to tailor life forms to their perceived needs.

14.10 Would Exosocieties Have Money?

This might seem an irrelevant question, but it is far from that. While many people see money, and especially the institutions which manage it, as a source of distress and injustice in our world, the phenomenon of money permits societies to have a division of labour into very specialised disciplines and activities, and allows them to manage their resources effectively. With money, people are renumerated for their services and can use the money they receive to purchase goods and services from other sectors of society. If we only had a system of barter the range of work activities across a society would be very limited indeed.

The existence of money and technical advancement might well imply a social system of capitalism like that which arose in the Western world with industrialisation. If engineering and technology were only found in certain parts of an exoplanet, where initial conditions favoured their development, there would likely be an unequal distribution of wealth between parts of the planet, much as there is on Earth.

14.11 And Would They Have Art?

Early art forms (over 600 paintings) have been found in the Lascaux Cave in the Dordogne (France) which date to about 17,000 years ago. Even earlier cave paintings have been discovered in Spain and have been attributed to Neanderthals. The development of art proceeded unabated throughout the Stone Age and beyond. For instance, there are decorative grave goods found at the Bush Barrow site (near Stonehenge, Salisbury, England) which date from the early Bronze Age (around 4,000 years ago). The intricate decoration is to be seen on the worked gold objects (a lozenge and a belt buckle) recovered at the site. The societies which created such works had an internal vertical structure and the artistic artefacts were most likely produced for people of high standing.

For all of human history art has been present. It predates farming, writing and mechanisation of all sorts. This is understandable considering that all one needs for art are natural colour pigments such as red ochre (consisting of iron oxide with clay and some sand). The walls of caves served as a canvas for early exercises in art. Furthermore, early art is usually an expression of concern with the animal world (horses, bulls, boar are common motifs), along with natural phenomena, such as sunlight, water, soil, plants, trees, etc., and different forms of human activity, like hunting.

The early appearance of cave art at multiple locations is testimony to the capability for abstract thinking which humans have. Abstract thought is one of the prerequisites for the rise of societies and civilisations and so, if exoplanets possessed this, their inhabitants would, in all likelihood, also engage in representational art based on experiences in their environment.

The Year 1610

To substantiate the tentative answer given to the question in the heading of this section, let us move forward in time. Consider for a moment the year 1610, a momentous year in European history. It occurred during the Age of Discovery when Europeans were exploring the world outside their own continent. In this year the English adventurer John Guy set sail for Newfoundland. Henry Hudson crossed the huge bay in north-central Canada, which now bears his name, in search of the Northwest Passage, allowing Europeans access to the Pacific from the north. Further south, on the Atlantic coast, the colony of Jamestown, Virginia, the first permanent English settlement in North America, started to flourish after its founding in 1607.

At the beginning of 1610, on January 7, the Italian mathematician and astronomer Galileo Galilei discovered the four major moons orbiting Jupiter with his newly constructed telescope, heralding a new era in optics and providing clinching evidence that the Earth was not the centre of everything. That was a scientific discovery.

But in 1610 there were also great artistic achievements. In Galileo's native Italy, the composer Claudio Monteverdi (1567–1643) published his *Vespro della Beate Vergine*, a masterpiece of Western music. He was born in Cremona, the city which also brought forth great makers of early string instruments such as the Amati and Stradivari families. For these activities, exquisite craftsmanship rather than advanced technology was required.

In Spain, Miguel de Cervantes was writing the second part of his world-famous novel *Don Quixote* and, in England, William Shakespeare wrote *The Tempest*, the last play he composed alone.[4] And the High Renaissance in Italian art with Leonardo da Vinci (1452–1519), Michelangelo (1475–1564), Raphael (1483–1520), Botticelli (1445–1510), Titian (c. 1488–1576) and others had begun already and continued in the late-sixteenth and early-seventeenth century, with painters like Michelangelo Caravaggio (1571–1610), whose early death occurred in the year being discussed here.

So while scientific instrumentation necessary for astronomy was getting on its feet, literature, art and music had already attained great heights. The moral of this story is that art preceded the advanced technology necessary for the instruments of science. There are at least three reasons which one can put forward for this.

Prerequisites for scientific development

1. Science needs mathematics, a symbolic number system which can be used to uncover patterns in systems and to construct predictive models of the natural world. For the study of numbers (arithmetic), mathematics needed the Hindu–Arabic numeral system (though Greek geometry and

4 In fact, it is a sad reflection on humanity that in Shakespeare's plays the only bit of technology which actors needed were daggers and swords to kill each other with.

astronomy managed to advance nonetheless), with single digits for the numbers 1–9 and, importantly, the number zero, 0, neither of which were found in the Roman counting system, though there were precursors in the Eastern Mediterranean. Significant sub-branches of mathematics are (i) algebra, which is concerned with the manipulation of variables without predetermined values (also an Arabic invention) and (ii) calculus, which is involved with the study of change in systems, co-invented in its modern form by Isaac Newton and Gottfried Leibniz in the late-seventeenth century and (iii) geometry, the study of two- and three-dimensional shapes and sizes, which can be traced back to the Babylonian and Egyptian civilisations of the second millennium BCE.

2. Science needs instruments to make observations about the natural world which are not possible with our bare eyes. With the telescope, and later the microscope,[5] great strides were made in the natural sciences. Furthermore, instruments used by science frequently need electricity for their operation so the harnessing of this natural resource would have to precede the making of instruments and machines for use in science.
3. Institutionalised religion is often suspicious of science because it can expose many of the former's assumptions as false, for instance that the Earth was the centre of the universe or that the world is only a few thousand years old. Nonetheless, in history many major scientific figures worked within a religious framework, indeed saw their discoveries as evidence of a divine order. This view of the relationship of science and religion is now very much a minority view.

A conclusion from the above considerations is that beings on an exoplanet might have reached great artistic heights but not have the technology to communicate with others across space. Such an exoplanet would be an unremarkable celestial body which could well go unnoticed by astronomical instruments built by humans.

5 There is some dispute about who originally invented, or rather developed, the microscope. A number of prominent contenders are Dutch. One of these was Cornelis Drebbel (1572–1633), who built compound microscopes involving two lenses and which provided considerable magnification. In 1624, he gave a demonstration in Rome, which Galileo attended. The latter

14.12 The View from Science Fiction

Science fiction often involves the projection of dystopian visions onto conjectured worlds beyond our own. The language used is frequently reduced and restricted, similar to the fictional language Newspeak of George Orwell's *1984*, which reflects the totalitarian control of a one-party state over its population.[6] This has nothing to with the kind of communication system exobeings might use on exoplanets, unless one of these happens to be ruled by an Orwellian superstate. Equally, mind control, a common trope in dystopian science fiction, is not likely to occur on an exoplanet. Or at least it could only be a theoretical possibility for an exocivilisation so technologically advanced that we cannot imagine what options for such manipulation it might have, maybe some means of decoding neuronal activity in others' brains to divine their thoughts. For American science-fiction films, another, much exploited subject is alien invasion, following on a long tradition in blockbuster movies of maximum violence and spectacular special effects. The film *Independence Day* (1996) and its sequel *Independence Day: Resurgence* (2016) are typical examples of this genre as is the entire *Star Wars* franchise.

There is a broader issue here, which has to do with what degree of scientific accuracy one can expect of science fiction. True, in fictional literature (and films based on this) the 'suspension of disbelief' is often demanded of the reader/viewer. But what is the extent of such disbelief? To illustrate this question, consider the film *Arrival* (2016), which is based on a literary work called *Story of your Life* (1998) by the American science-fiction writer Ted Chiang (1967–). The main character Louise Banks is a linguist

later built his own compound microscope after Drebbel's model. In England, microscopes were greatly promoted by Robert Hooke (1635–1703). He used them to study organisms only scarcely visible to the naked eye, such as fleas, and produced a ground-breaking work *Micrographia* (1665), which contained his illustrations of microscopic organisms and structures. Hooke also coined the term *cell* for the basic unit of organic biology.

6 *We* (1924) is an earlier dystopian novel by the Russian writer Yevgeny Zamyatin (1884–1937), which depicts a totalitarian state in which all individuality and personality differences were done away with, conformity then being the most salient characteristic of humanity.

commissioned by the US military to decipher the communication system which a group of 'aliens,' who have arrived at various places on Earth in several spacecraft, appear to have. Her personal fate, including her ability to see into her own future, forms the central theme of the film. But neither the story nor the film adaptation are really about exobeings. These are vaguely and blurringly represented as heptapods (creatures with seven feet) behind a scarcely transparent screen. Both the story and the film are more about the human motivations and tensions evident in the actions of the characters. As the author of the original story has said in an interview, "Science fiction is about using speculative scenarios as a lens to examine the human condition". Another important point about the film is the following: the heptapods use ink on a glass screen to convey their language to the humans. No animal would use such a short-supply resource as ink for primary communication,[7] unless they were using this ink as a kind of secondary medium like writing. But in the film there is no indication that the ink is coming from some sort of pen or cartridge which the heptapods were holding in their tentacles.

Quite different is the novel *Contact* (1985) by American astronomer Carl Sagan, which was adapted into a like-named film (the medium for which it was originally conceived in the late 1970s) and finally released in 1997. It tells the story of a scientist Ellie Arroway, who joins the SETI (Search for Extraterrestrial Intelligence) program and picks up a non-natural signal coming from the Vega star system in the constellation Lyra (some 26 light years from Earth). Another character, Hadden, manages to decipher the signal, which apparently contains instructions to build a machine for a single occupant, which could travel faster than light via a series of wormholes. Much of the novel and film is concerned with the terrestrial politics surrounding the signal's discovery, which eventually leads to the construction of the spacecraft in the USA in order to travel in the direction of this signal. However, the craft is sabotaged and destroyed by a terrorist. Arroway is told by Hadden that there is a secret second craft in Japan and she is selected to be the single astronaut on the mission. She experiences

7 Some types of squid emit a cloud of ink to hide themselves from predators, but not as a means of regular communication with other squid.

a journey through wormholes lasting some 18 hours. During this time she meets an 'alien,' who has adopted the appearance of her deceased father, on a beach on an apparently distant planet. During this brief encounter, her 'father' reccounts to her how different his existence and stellar environment is and that she must return to Earth, which she then does, assuring her that this meeting was but the first step in more enduring contact with 'alien' civilisations. Much of the rest of the novel/film has to do with the reactions to her story by various people back on Earth. Because the 'alien,' who Arroway meets, adopted the guise of her father (and his American English), there is no speculation about what type of language/ communication systems such 'aliens' might have in their non-disguised form. So, despite the many scientifically credible details in *Contact*, it does not offer any information relevant to the topic of the current book.

In any story/film involving 'aliens' there are obvious questions which one can pose, first and foremost how 'alien' creatures could build radio equipment and indeed interstellar spacecraft. Would they have evolved on their home planet to have a precision grip with an opposable thumb and index finger (or functional equivalents) like we have? If not, they could not build anything, let alone spacecraft. Would such 'aliens' have advanced cognition which is at least comparable to ours? And lastly, would they have a language faculty similar to that we all possess, which enables us to acquire and later use any human language productively?

Part IV

The Runaway Brain

More than any other species on Earth we have brains which are far more powerful than is necessary for survival in our natural environment. This excess of cognitive power has enabled us to spread across the globe and control our surroundings unlike any other animal. Why our brains should have expanded to this extent, despite the obvious cost, for example in terms of energy consumption, is a crucial issue in human evolution and would be a necessary precondition on other planets for exobeings with advanced technology.

15

The Brain-to-Body Relationship

If you look at the animal kingdom you will see a bewildering array of life forms, with an even more astounding variety of no longer extant species in the past. Among these life forms we find the class of mammals in which the relation of brain size to body mass is greater than in other groups of animals. This is particularly true for cetaceans (sea mammals like whales and dolphins) and for elephants. However, there is one species which stands out from all others: the genus *Homo*, specifically the species *Homo sapiens*, the only surviving species of this genus. We are characterised by our large brains in proportion to our body mass and the prominent cortex (outer layer of the brain), especially at the front of the head. This immediately throws up the question of whether the development which led to our brain size and structure was a fluke of evolution or a perfectly natural development, given the circumstances in our immediate evolutionary past. The answer is not a simple one because the circumstances in question are complex and involve a number of unrelated matters which favoured our development in the animal kingdom, some six to seven million years ago.

15.1 Wallace's Puzzle

In his review of Charles Lyell's *Principles of Geology* and his *Elements of Geology*, published in the April 1869 issue of the *Quarterly Review*, Alfred Russel Wallace (1823–1913), evolutionary biologist and colleague of Charles Darwin, posed the following rhetorical question: 'How ... was an organ [the brain, RH] developed so far beyond the needs of its possessor?'

We could dub this 'Wallace's puzzle': Why do we have such large brains with an expanded cortex which can handle science, language, art, literature and music? To survive fully adequately in our environment, it would be enough to have a brain just like a chimpanzee. The majority of marine and terrestrial animals have survived in their respective environments for millions of years but yet seem not to have developed any intelligence which is significantly beyond what is needed to hunt prey successfully (often in groups) and to ensure the survival of their species. Of course, we do not know what the cognitive experience of a zebra or a trout is, but there is no indication in their behaviour that it is highly differentiated, involving a significant level of abstract thought.

So there is a question to answer here: Why did evolution favour a great rise in intelligence when it was seemingly not necessary in our early hominin environment? And does evolution ever 'favour' anything? Reformulated, the question can be asked: Why did natural selection, in our ancestral past, lead to the rise of intelligence on this level? It obviously started relatively slowly after the divergence of the lines which led to chimpanzees on the one hand and humans on the other.

The Role of Feedback Loops

In our early evolution a positive feedback loop was initiated, placing us on a trajectory which we followed for several million years, but by no means in a straight line. It led to increased intelligence, advantageous in the stressful environment of East Africa at the time of early *Homo*. Other things appeared to have happened when we embarked on this evolutionary path.

We traded brawn for brain, becoming less strong than the chimpanzees but more flexible, agile and capable of adapting to changes, such as climate fluctuations in our surroundings.

This kind of development appears to have happened only once on Earth, certainly when one considers that we have also developed language.[1] That suggests that our brains with our intelligence could be due to a freak runaway development in our lineage but one which is not common in evolutionary systems across the universe – we do not know the answer to this, but it would be a crucial issue for any exoplanet with a biosphere favourable to the evolution of life.

One angle of approach to this unanswered question would be to consider the effect of positive feedback loops. Consider for a moment the three sets of animals in Table 15.1.

The second group in Table 15.1, the cetaceans, are known for their large brains and their intelligence, especially (bottle-nosed) dolphins, which exhibit self-recognition, playfulness, curiosity, social behaviour and non-beneficial interaction with humans, all hallmarks of cognitive power. Cetaceans evolved from small, deer-like animals around 50 million years

Table 15.1 Three monophyletic groups in comparison

Clade	Members	Proposed ancestor(s)	When they evolved (million years ago)
Hominins	Humans	Last common ancestor	8–6
Cetaceans	Whales, dolphins	Pakicetidae	50
Pinnipeds	Seals, walruses	Puijila	Over 20

1 Other intelligent animals, like cetaceans and elephants, do have large brains (relative to body mass) and sophisticated means of communication, but not forms comparable to human language. Dolphins evolved from a family of terrestrial animals, Pakicetidae, which probably resembled small deer, about 50 million years ago. The large brains of dolphins would appear to have developed after their transition to a dedicated marine life. See Marino et al. (2004) and Montgomery et al. (2013).

ago, which first became amphibians and later dedicated aquatic animals while largely retaining their mammalian physiology. The third group in Table 15.1, the pinnipeds, have a similar evolutionary history. They evolved from even-toed ungulates (hoofed animals), which we see today as deer, buffalo, pigs, sheep, cattle, etc. However, the pinnipeds have cognitive abilities roughly comparable to those of their terrestrial ancestors. So some development started in the evolution of cetaceans in sea, which initiated a positive feedback loop that in time led to their large brains and cognitive powers. But no such development occurred with the pinnipeds. True, seals show a limited degree of curiosity, but otherwise there are no indications that their cognitive level is comparable to that of cetaceans.[2]

Now compare cetaceans to humans. They developed their original front legs into fins for swimming in water (with either vestigial hind legs within the body, with whales, or arrested bud development during gestation, with dolphins). This meant that manual dexerity, characteristic of humans and an essential part of the positive feedback loop promoting intelligence, was absent with cetaceans. Furthermore, the acquatic environment of cetaceans placed limits on any adaptability to different and contrasting environments which humans did attain, again promoting intelligence.

The conclusion from these considerations is that the evolution of intelligence on exoplanets would require an array of favourable conditions, not just to initiate a feedback loop in the direction of greater intelligence among forms of exolife, but to continually support this development and to reach a level where language could arise.

Predation and Environments

It is true that we are apex predators, like lions or sharks. We feed on animals (most of us, anyway) but we are not fed on by others (now). However, the important difference between us and other apex predators is that we

2 The Israeli zoologist Arik Kershenbaum (personal communication) believes that the level of sociality among cetaceans is responsible for their remarkable cognitive powers, though the question of what came first and/or whether these factors influenced each other mutually during the evolution of this order of marine mammals remains unclear.

control our environment and have constructed instruments, weapons, which ensure that we are not preyed upon by others. Furthermore, we can switch our environment at will; animals can not. Sharks live in the oceans, lions hunt on savannah plains and do not occupy other environments. It is environmental flexibility, rather than just apex predation, which characterises humans and probably promoted intelligence in their evolution (see previous section).

15.2 Are Brains Necessary for Life?

For the simplest forms of life, the answer is 'No.' Jellyfish do not have brains, but a net of nerve cells around their bodies. Other invertebrates do have brains, like octopuses,[3] which have large concentrations of nerve cells in their tentacles, making them essentially different from vertebrates, which even in their simplest forms have bundles of nerve cells at the head end of the spinal cord, where the developed brain is located in more complex forms of life. Octopus neurology has received considerable attention in recent years as an alternative to one large brain, and it just might correspond more to arrangements found with life forms on exoplanets than vertebrate models do.

Any exobeings will have gone through a process of evolution over at least hundreds of millions of years, from simple microbial life to complex intelligent beings. There is no other way for any beings, on any planet, to get to an advanced stage of cognition. So for exobeings, the stages of their evolution on their planet would be evident in the geological and palaeoanthropological records, maybe not as complete as they might like, with gaps and holes, just like our records.

3 There has been much discussion about just how intelligent octopuses are and if they have consciousness and of what kind. Certainly, they show remarkable abilities; for instance, the mimic octopus, found in the Indian and Pacific oceans, can mimic other animals or objects. They do this by altering the colour and texture of their skin, for example by making it look like a piece of rock with an irregular, multi-coloured crust. They achieve this by manipulating pigment sacs on their skin called chromatophores.

So if there is one definitive statement one can make about exobeings it is that they would have functional biological equivalents to our brains, with equal, or possibly greater, power than ours.

15.3 Structure of the Human Brain

Our brains are the products of millions of years of evolution during which certain parts were developed for certain tasks, collaborating together to yield the functional unit of a brain. An essential revision in the past century of neurological research is that the functions of the brain are often distributed across the brain. It is true that certain parts are associated with certain functions, for instance Broca's area behind the left temple is largely responsible for speech production and Wernicke's area behind the left ear is, in the main, responsible for understanding speech. These are approximations which were initially discovered (in the nineteenth century) by examining patients with specific injuries to the head. Recent research avoids these traditional designations and sees Broca's area as part of the inferior frontal gyrus (sometimes referred to as Brodmann areas 44 and 45 following an earlier classification). The reason for this is that the older terms imply a binary split between two areas, for production and reception, respectively, but investigations have shown that neuronal organisation is much more complex.

It is difficult to pinpoint aspects of brain activity or storage to exact regions of the brain, not least because the greater the level of detail one examines the more individual differences come into play. For instance, it is not known where the vocabulary an individual possesses is stored in the brain. Even more important in this context is the fact that consciousness cannot be pinned down to a specific region of the brain (see Section 15.4 below).

The division of the brain into two halves is deeply rooted in animal evolution. The distribution of functions across the two halves (called 'cerebral hemispheres') does show distinct patterning, but not exclusively. Especially if damage occurs to a part of the brain in childhood (before the lateralisation of the brain just ahead of puberty) the functions carried out by the damaged part can be largely reconstructed in the corresponding part of the brain in the other half.

The two halves of the brain are connected by the corpus callosum, a large complex bundle of some 200 million nerves forming a neuronal bridge. The nerves are differentiated with thick, heavily myelinated nerve fibres connecting parts of the brain responsible for motor coordination and quick physical actions, like running during flight. Thinner fibres towards the front of the corpus callosum are connected with prefrontal association areas in both halves of the brain.

Furthermore, the human brain is contralateral by which is meant that, by and large, each hemisphere controls the opposite side of the body. This is clearly obvious when someone suffers a stroke to the left half of the brain because then the motor control of the right half of the body is impaired. Contralaterality is an evolutionarily deep feature present in all vertebrates but not in invertebrates.

The brain (Figure 15.1) can be divided into different parts,[4] which have associated functionalities roughly as follows.

1. *Brain stem*. Going from the spinal cord up into the head, the first part is the brain stem, consisting of the medulla oblongata (Latin 'longish, innermost part of an organ'), the pons (Latin 'bridge') and the midbrain or mesencephalon, the uppermost part. In terms of evolution, the brain stem is the oldest part and represents an extension of the spinal cord. It controls such essential involuntary physical actions as breathing and the beating of the heart.
2. *Cerebellum* (Latin 'little brain'). Behind the pons towards the back of the neck is the cerebellum, a structure somewhat resembling a small cauliflower at

4 A basic division of the brain, popular in some circles in the late-twentieth century, is the triune brain proposed by Paul D. MacLean. He viewed the human brain as divided into three parts: the reptilian (or lizard) system (the most ancient and responsible for the regulation of essential bodily functions such as heartbeat or blood pressure), the palaeomammalian system (consisting of the limbic system and responsible for core emotions) and the neomammalian system (consisting of the neocortex in humans and responsible for abstract thought and language).

the lower back of the head, largely responsible for voluntary actions like getting dressed and acquired abilities such as playing a musical instrument. With the medulla and the pons (but not the midbrain), the cerebellum forms the hindbrain.

3. *Basal ganglia*. This is a label for a group of neuronal nuclei deep in the brain, under the cortex, which integrate various cortical signals, for example from the primary sensory and motor cortices on the top of the brain, to produce a singular behavioural output by an individual. This can involve motoric actions but also cognition and emotion. There are different pathways, which promote and inhibit signals respectively, to produce a fluid, coordinated output, such as when dancing, performing sports or just walking along.
4. *Limbic system*. This is a ring-shaped part of the brain under the cortex, closely associated with emotions and with very basic aspects of human existence such as hunger, pain, the decision to fight or flee as well as controlling essential hormones in the body (via the pituitary gland). The limbic system is well developed in non-human animals, which suggests that its origins lie deep in their evolutionary past. The following five structures are essential parts of the limbic system.
   (a) *Cingulate cortex*. This is a belt-shaped area of the brain above the corpus callosum and under the cortex. It is involved in emotion, learning and memory and is implicated in the motivation for behaviour and its outcomes.
   (b) *Thalamus* (Latin from Greek 'chamber'). Located in the centre of the brain, on top of the brain stem, the thalamus is a bilateral, midline-symmetrical structure responsible for many involuntary actions such as digestion and the rhythms of sleep and wakefulness. The inputs to the thalamus, such as sensory stimuli, are associated with specific thalamic nuclei which transmit signals upwards into other 'higher' parts of the brain in the cortex. The visual cortex, located at the back of the head, passes on information it receives via the optic nerve to the thalamus for confirmation of interpretation. The flow of information between the thalamus and the cortex is bidirectional, the latter providing feedback to the former.

(c) *Hypothalamus* (Latin from Greek 'under-chamber'). This is a small structure (about the size of an almond), again located at the base of the brain, which links the nervous to the endocrine system. The hypothalamus controls a variety of bodily functions such as hunger and thirst, tiredness, sleep and our circadian rhythm.

(d) *Hippocampus* (Latin from Greek 'sea-horse'). This is a bilateral structure at the base of the brain. It plays an essential role in sorting information input to the brain and thus is involved in the formation of both short- and long-term memories as is the amygdala to which is it attached.

(e) *Amygdala* (Latin from Greek 'almond'). This consists of a pair of neuronal nuclei which are strongly implicated in memory functions, especially the consolidation of long-term memory. The amygdala is also key to emotional memory and to feelings of aggression, fear and anxiety. Olfactory input occurs directly into the amygdala, which is perhaps why smell is so strongly evaluative, either good or bad.

5. *Cerebrum* (Latin 'brain', adjective: *cerebral*). This is by far the largest part of the brain and that which is popularly identified with this organ. It is in two halves like a giant walnut, connected in the middle by a dense bundle of nerve fibres, the corpus callosum (Latin 'hard body'), see above. The left hemisphere is normally associated with language and it contains the Broca area (for speech) and the Wernicke area (for comprehension). The cerebral cortex is the outer surface of the cerebrum and has the appearance of ridges – Latin *gyrus* (sg.), *gyri* (pl.) – and grooves – Latin *sulcus* (sg.), *sulci* (pl.) – which greatly increases the surface area of the cortex. This is conventionally divided into lobes, with the occipital lobe at the rear, the parietal lobe across the top, frontal lobe towards the front and the temporal lobe at the side, all in pairs, one part per half of the brain. These lobes can be further divided depending on what functional subsections they contain, for instance the visual cortex is located in the occipital lobe at the back of the head, the motor and sensory cortices are located in the parietal lobes on the top of the head.

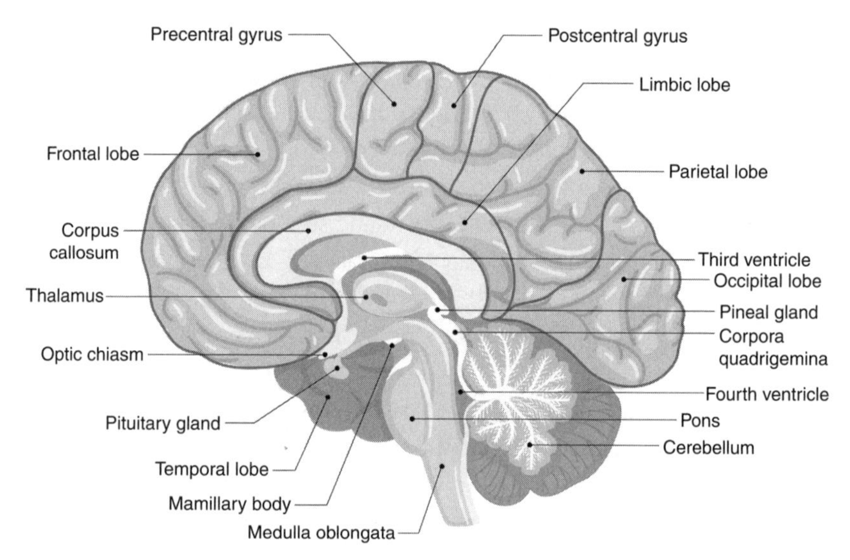

Figure 15.1 Cutaway view of the human brain

15.4 Characteristics of the Human Brain

A human brain consists of about 85–90 billion neurons, all linked together by axons and dendrites resulting in trillions of connections between these neurons.[5] To give you a rough idea of what that figures means, consider that you have about 12 times as many neurons inside your head as there are people in the world today (just under eight billion). The totality of connections between the neurons of your brain is called its connectome. Nerve signals can travel along axons and dendrites (Figure 15.2) at over a hundred metres per second. Whether it will ever be possible to map these connections onto some other substrate, like computer memory or external storage, is very doubtful.

5 The Spanish neuroscientist Santiago Ramón y Cajal (1852–1934) began producing drawings in the late 1880s and 1890s of the dense thickets of neurons and their myriad connections to be found in the brain and the spinal cord. These drawings were the first visual representations of these structures. Together with his Italian counterpart Camillo Golgi (1843–1926) he shared the Nobel Prize in Physiology and Medicine in 1906, although the two were locked in bitter professional rivalry.

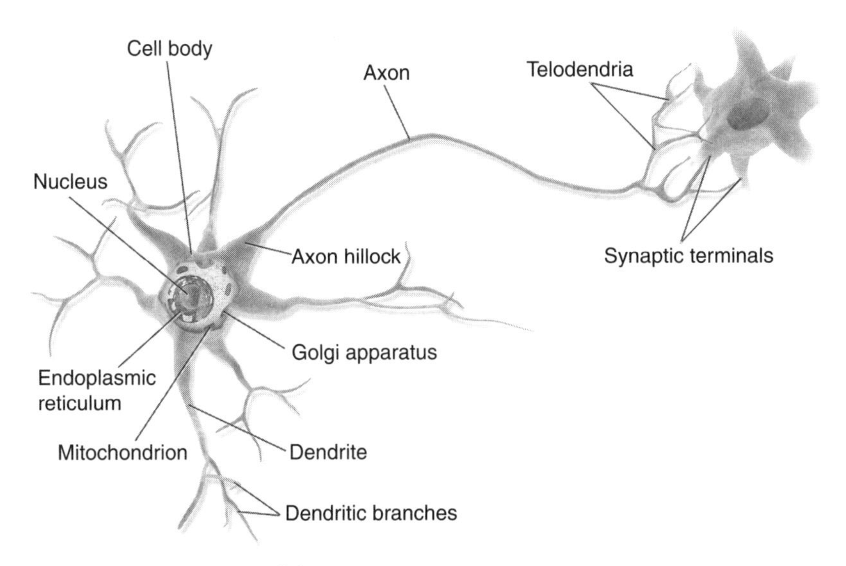

Figure 15.2 Neurons and their connections

About 80 per cent of our neurons are in the cerebellum (about 70 billion) and only 20 per cent (about 16 billion) in the cerebral cortex, with a mere 0.8 per cent (< 1 billion) in the subcortical nuclei. While in the womb, we have the greatest number of neurons in the final phase of pregnancy. After we are born, many neurons are lost and those which are used establish a vast and intricate network of connections via axons and dendrites. In the early years of life, the outer layer of the brain, the cerebral cortex, undergoes folding into grooves (*sulci*) and ridges (*gyri*), which greatly increases the surface area of the cortex.

In addition, the brain has about an equal number of glia, support cells for the neurons. These cells supply the neurons with oxygen and nutrients and remove waste. Your brain is washed during sleep: the cerebrospinal fluid removes waste from the brain, is transported out of the skull and replaced by fresh fluid.

A human brain consists of about 60 per cent white matter, contained in the central nervous system. It is composed in the main of myelinated axons – the high fat content of these structures gives them a light colour, which is usually pale pink because of the blood supply in this tissue. About 40 per cent is grey matter, that part of the central nervous system which consists largely of the neurons, glia and other related elements. It may be somewhat yellow or pale pink in colour, again due to capillary blood vessels.

The brain uses between a fifth and a quarter of the energy intake of the body, above all oxygen and glucose (sugar), although it represents only a few per cent of the body's mass. The power consumption of the brain is about 15–20 watts. Its demands for energy are so crucial that if the blood supply to the brain is cut off we would lose consciousness within seconds.

Special protective measures are necessary for such a vital organ. The brain is protected from the outside world by a skull of bone. And it is internally protected from harmful substances by means of a blood–brain barrier, which prevents many harmful substances from reaching the brain. Nonetheless, certain (psychedelic) drugs do cross this barrier and can affect our consciousness.

The brain does not contain nerve endings which could register pain, so your brain cannot be painful like a gash or burn on the skin. Headaches are usually due to tension on the outside of the brain or to abnormal pressure of the cerebrospinal fluid in either direction, up or down.

There are brain waves, which have specific wavelengths produced by the synchronised electrical pulses from masses of neurons firing together. While awake, there is typically a background oscillation in the cerebral cortex. Beta waves (*c.* 12–30 Hz) are found when we are involved in general thought. During problem solving and when making decisions, gamma waves (over 40 Hz) are generated.

Waves of around 4 Hz are typical of deep dreamless sleep or generally unconscious states (delta waves). During deep sleep neuronal activity is highly local, with little connectivity across regions of the brain. This would seem to imply that there is a minimal level of cross-regional neuronal activity, which is necessary to trigger consciousness, something which is confirmed by coma patients who recover normal consciousness. Similarly, when you wake up, you begin to process the input from your senses activating various regions of your brain in the process. The possible connection of brain waves to language is discussed in Section 32.12 below.

Lastly, one can mention that our brains are eminently flexible. Senses can be strengthened and their scope increased to act as a substitute, if necessary. For instance, blind people utilise their hearing, touch and smell to a much greater extent than do seeing people. Indeed, there is a very active area of neuroscientific research called 'sensory substitution' in which a certain sense modality can be enhanced when another is impaired, for example by loss of hearing or vision or indeed genetic abnormalities such as colour blindness. Such enhancements are achieved through sophisticated technologies which provide inputs to the brain, corresponding to the stimuli of the impaired sense. The brain is then in a position to interpret this new input as equivalent to that which has been lost or critically reduced; for instance, tactile input, which can serve as an alternative to visual input.

WOULD EXOBEINGS SLEEP?

To answer this question, consider planetary motion first. All planets rotate around their parent star (or stars). But not all of them rotate on their own axis as Earth does. If an exoplanet is very close to its star, like a relatively cool red dwarf, it could be tidally locked and only show one face permanently to its star as the Moon does to our Earth. In this case there would be no daily cycle of light and darkness and, assuming that darkness would be associated with reduced activity and thus be a natural choice of period of the day to rest, there would be no given period for sleep. There are other options, of course. Exobeings could live in the twilight zone of their planet, on the border of light and darkness. They might have two periods of sleep, first and second, like humans used to have in previous centuries. They could retire to darker places for rest. They could show bouts of torpor (reduced body temperature and metabolism) during their day as do many animals on Earth. But perhaps the most intriguing question is whether they would need sleep as much as humans do, to detox their brains, organise memories from the day by strengthening neuronal connections, recuperate from physical exertion, etc. And would they dream like Earthlings?

15.5 Windows on the World: The Human Senses

It is obvious that the brain resides in the dark of the skull. Information from the world outside is transmitted to the brain via the senses (Table 15.2). True, only a fraction of the input which reaches our sensory organs, eyes or ears for example, is actually processed but that is the basis of sensory awareness.

For each sense there is a process of sensory transduction by which input signals are converted into electrochemical signals, which travel along axons from neuron to neuron and which ultimately lead to the conscious experience of the sensory input in question.[6] The brain is continually trying to make sense (literally) of all this input and, importantly, by deciding what parts of the barrage of input to ignore and what to focus on; just think of how much visual information enters our eyes and how much of that we actually notice. Furthermore, the brain makes predictions about what is real in the world by evaluating input. For successful evolution, correct assessment of sensory input is essential. The recognition of a predator, or maybe just its shadow, smell or slight sound, could spell the difference between life and death. Hence natural selection favoured those members of a species which were successful in interpreting sensory input. This would most likely be the case on exoplanets with their life forms as well.

Table 15.2 Five human senses

Sense	Organ	Medium
Sight	Eyes	Light waves (photons)
Hearing	Ears	Sound waves
Touch	Skin	Nerve receptors sensitive to pressure and heat
Smell	Nose	Nerve receptors sensitive to odours
Taste	Mouth	Nerve receptors sensitive to flavours

6 Senses can be cross-wired, resulting in a phenomenon called synaesthesia. Some composers, such as the Finn Jean Sibelius (1865–1957), were known to have definite colour associations with certain musical keys.

Sight

At the back of each of our eyes is a convex-curved retina onto which the image entering the pupil of our eye is cast. The retinas are covered with rods (*c.* 91 million) and cones (*c.* 4.5 million). Rods are responsible for vision at low light levels, so-called scotopic vision, and are not involved in colour recognition. Cones are active at higher levels, resulting in so-called photopic vision, and are responsible for colour recognition and the sharpness of vision. There are three types of cones, which are responsive to light at different wavelengths. People with less than three cannot distinguish colours of different wavelengths, such as red and green, and are labelled 'colour blind.'[7]

In the centre of the retina is a small blind spot, called the optic disk, which forms the head of the optic nerve. It is 'blind' because it does not contain photoreceptors. The optic nerve, consisting of over a million nerve fibres, leaves the eye at the back and transfers the electrical signals generated on the retina to the visual cortex at the back of the head. We have two eyes, which allows for telescopic vision: the angles at which light strikes each eye is slightly different and this allows us to interpret the incoming light signals as stemming from three-dimensional space and to assess distances.

Eyes developed very early on in the history of life. Light-sensitive cells became more and more specialised, developing a lens and a cavity with nerve cells on its rear wall, leading to the eyes of later life forms.

All higher forms of life have eyes, though habitually nocturnal animals cannot generally see well and rely on other mechanisms for spatial orientation, such as bats with ultrasound, which they use for echo location.

7 There are extreme cases where whole groups of people are colour blind: on the Micronesian island of Pingelap, because of a much-reduced gene pool, a specific kind of colour blindness, called *maskun* 'not-see' in the local language, came to affect more than 10 per cent of the population of a few hundred. The gene pool was reduced because of a typhoon in 1775, after which only about 20 of the inhabitants were left alive.

Deep-sea fish which inhabit layers of the ocean in perpetual darkness have small, often non-functional eyes.

Hearing

Audio waves between about 20 Hz and 15,000 Hz can be heard by humans. Our ears are delicate organs, which amplify sound entering our external ear, or auricle, first by striking the ear drum and leading to vibrations being conducted and amplified by small bones – malleus, incus and stapes – onto the cochlea; this is a coiled shell filled with fluid and tiny hairs, with stereocilia on top, in little bundles which are attached to nerves. The hairs with stereocilia waft in the vibrations of the cochlea fluid and ions move to the top in turn, causing chemicals (neurotransmitters) to be released at the bottom of the hair cells where they are in contact with auditory nerves generating electrical signals, which are passed onto the brain for processing. There is a range of different hair cells, which are responsible for detecting vibrations stemming from sounds of different frequencies (from low to high).

Hearing and sound production are linked in a species. If an animal can produce sounds at certain frequencies, it can also hear them. The larger the animal, the greater the amplitude of the sounds it is capable of and the lower the frequency it can attain. For instance, blue whales can produce sounds up to about 180 decibels (painful for humans) and elephants can emit sounds in the infrasonic range, lower than 20 Hz.

Hearing and Predation

Does the human voice sound range occupy a privileged section of the sound spectrum? No, not in any absolute physical sense, but this range corresponds to that which can be produced by the vocal folds in the larynx and to the spectrum of common sounds in nature.

The sounds we can hear could spell danger for an animal in the wild, hence natural selection would have favoured those animals which could best register sounds in this range. Ever since the Cambrian 'explosion,' 540 million years ago, predation has been a central characteristic of animal behaviour. Hearing in the central band of our audio signals is essential to avoiding predators: the rustling of leaves as a predator moves up on an animal, the sound of hooves on the ground, the noise of water moving by a predator stepping through it – all of these need to be perceived by animals trying to avoid being eaten. Echo location, on the other hand, is no good for avoiding predators and the animals which use it are either like whales, which do not really have natural enemies, or like bats who hunt for insects at night when there are not any predators about. There are also a few instances of cave-dwelling birds who use echo location in the dark.

Touch

Our sense of touch relies on pressure-sensitive and heat-sensitive nerves on our skin. It is important when negotiating our environment that we receive continuous input. Otherwise we might injure ourselves, something which happens to individuals who have no pain receptors (nociceptors). The touch nerves are not distributed evenly across our bodies. Those on the tips of our fingers are highly sensitive while those on the base of our back are least sensitive.

Touch has a strong emotional element and augments the attachment between mother and child, suggesting that it has an origin in deep time. Among many animals, such as chimpanzees and elephants, touch strengthens bonds in a group, as does the touch-based act of grooming. In Western-style societies, there is a general taboo on touching outside one's immediate relatives or partners, though cultures do vary in this respect.

Smell

There is a small area of tissue at the back of the nasal cavity which contains millions of olfactory receptors, which react to various molecules from chemical compounds entering the nose in low concentrations on a molecular lock-and-key principle.[8] These then cause the olfactory nerve

8 Recent, not uncontroversial work by the biophycist Luca Turin, suggests that molecular docking is not enough to account for how smell works and that quantum effects in the bonds of the molecules are responsible for our perception of smell.

to generate an electrical signal, which is passed back to the brain and perceived by us in our conscious state as smell. Humans have an average sense of smell – many animals, such as dogs, have a much more powerful sense. Nonetheless, we can perceive a wide range of smells, both pleasant and unpleasant in our evaluation of them. Smell is probably a very old sense in the animal kingdom and is important in assessing whether a potential food source is safe and – as a secondary function – in recognising other animals, either of one's own species or of another.

Taste

Similarly to smell, taste is used in assessing potential food and is important is determining whether it is fresh, given that putrid food smells and tastes unpleasantly for humans and animals. The latter are especially careful in assessing solid and liquid food, including water to drink. Animals will reject food they evaluate negatively. Both taste and smell can trigger disgust, which according to Darwin is an ancient protective feature in the animal world, continued in the human lineage.

Our sense of taste is realised via taste buds, of which we have many thousands, distributed in our mouth, above all on our tongue, in different regions. According to a traditional view, buds sensitive to sweet and salty food are located in the front of the tongue, with sweetness perceived at the tip and saltiness around the front and mid-to-rear edge; buds for bitterness are found in the back, while the buds for savoriness (technically known as umami, a Japanese word) are broadly located in the centre of the tongue. However, this is a misconception of gustatory receptors. Smell and taste work together, which is why your food does not really taste when you have a cold and your nose is blocked.

The Binding Problem

The physical pathways of the senses are separate from one another as are the parts of the brain where the input is processed. Nonetheless, the diverse sensory inputs get sewn together and we experience a unified, single picture of the external world in our brain. But explaining how this happens is difficult and is referred to as the binding problem. Somehow the conscious 'whole' is assembled for us to experience our environment as we do. However, one cannot say that the parts of the brain responsible for different senses pass on their processed information to one central control unit, which then presents these different kinds of information as seamless consciousness. Furthermore, there does not seem to be a locus for consciousness beyond the feeling that it is inside our heads, behind our eyes somehow. This and other issues will be revisited below in Chapter 18 on consciousness.

15.6 The Cost of Our Brain

A human brain is about 1,350 g in weight while that of a chimpanzee is approximately 400 g. This extra kilogram contains all the cognitive abilities by which we are distinguished from our closest common ancestor in the family of hominins, though these abilities are not in a one-to-one relation to size. But the size of the human brain comes at a cost: the brain presents difficulties, if not dangers, for mothers at childbirth. Although the brain is only about 3 per cent of body mass it consumes about 20–25 per cent of energy intake (glucose and oxygen). Human babies need to have about 15 per cent body fat to ensure that they can continuously feed their hungry brain. By comparison, baby baboons have about 4 per cent body fat. By 9–10 months, human babies need more than 600 calories per day, which is more energy than their mothers can provide via breastfeeding and so food supplements are required.

As adults we need about 6 calories per billion neurons per day, which means about 500 calories for our brains alone, with an average body requirement of 2,000–2,500 per day (for males, somewhat less for females).

16

How Brains Develop

Humans are mammals, a group of vertebrate animals with backbones, an internal skeleton and a nervous system controlled by a brain. However, the essential feature of mammals is that they give birth to live young, as opposed to other animals[1] like reptiles and birds, from which we split off about 300 million years ago (mya), and which lay eggs from which their young later hatch. The word mammal derives from Latin *mamma* 'breast' and refers to the fact that the females of these animals breastfeed their young to begin with. This system is typical of cognitively advanced animals and there is probably a causal connection here. Live birth and breastfeeding result is a greater attachment between mother and young than does egg laying, adding an increased emotional dimension to the lives of the animals in question and thus providing a positive feedback loop for further cognitive development.

The gestation period for mammals tends to be longer than with other animals and there is an apparent increase with greater cognitive skills; for example elephant cows bear their young for two years before giving birth,

1 There are intermediate situations, for instance female sharks and some snakes produce eggs, but these hatch inside the body so that the young are born alive; such animals are termed 'ovoviviparous,' a Latinate word meaning approximately 'egg-live-bearing,' a general label to cover a number of different configurations.

dolphins and whales require a year or more. Longer gestation periods have advantages and disadvantages. On the one hand, the young are shielded from outside dangers and arrive in the world at a fairly developed stage. On the other hand, a longer pregnancy can be dangerous for mother and young. Complications can arise within the mother's body and the fetus may die before birth. Furthermore, longer gestation periods mean fewer births per unit of time and hence greater pressure on those who are born to survive.

Humans go through a period of development from birth to adulthood which shows key phases with certain traits. This development of the individual is known as ontogeny and parallels have been drawn between the development of the individual and that of the species, known technically as phylogeny. Whether ontogeny (the development of the individual) replicates phylogeny (the development of the species) is a disputed issue[2] There are clearly correlations, but any too strict parallelism is unlikely to be accurate. Nonetheless, the consideration of this issue can be fruitful, in language studies as elsewhere.

16.1 Embryogenesis and the Brain

In the beginning is the zygote, the female egg fertilised by the male sperm, in which chromosomes replicate to form separate nuclei, then cells, and so the process of growth is initiated. To begin with we speak of an embryo, which after the ninth week of fertilisation is known as a fetus.

Very early on, three primary germ layers are formed and of these the outer layer, the ectoderm, is relevant here as part of it thickens and forms the flat neural plate. This in turn dips in the middle and begins to fold with the top sides of the fold joining and the epidermis closing over, leaving a neural tube underneath (complete by the third week of pregnancy). On top of this is the neural crest, a temporary layer of cells which develops

2 The German naturalist and zoologist Ernst Haeckel (1834–1919) believed that the development of the individual recapitulates the evolution of the group, a view which, in any narrow form, has only limited, if any, support today.

into various other cells, including some types of neurons and glial cells. At one end of the neural tube, the spinal cord will develop and, at the other end, the brain, where the fore-, mid- and hindbrain form first (Figure 16.1) and later divide, with neuron production proceeding at a very rapid rate: at peak production during the third trimester of pregnancy this can reach 250,000 neurons per minute.

The developing brain at the front end of the neural tube is like an expanding bulb. During pregnancy neurons need to move from their place of production, lower down in the head, to their final position in the brain. This involves a process of migration by which individual neurons move along radial glial strands to reach their final 'parking place', during which younger neurons can pass older neurons already in place and so expand the brain in the process of corticogenesis. Connections between neurons already begin to form, a process which lasts the entire lifetime of an individual. Interneuronal connections can arise and decay, depending on how the brain has or has not use for them. However, some early connections are primary and permanent, such as those linking the optic nerve with the

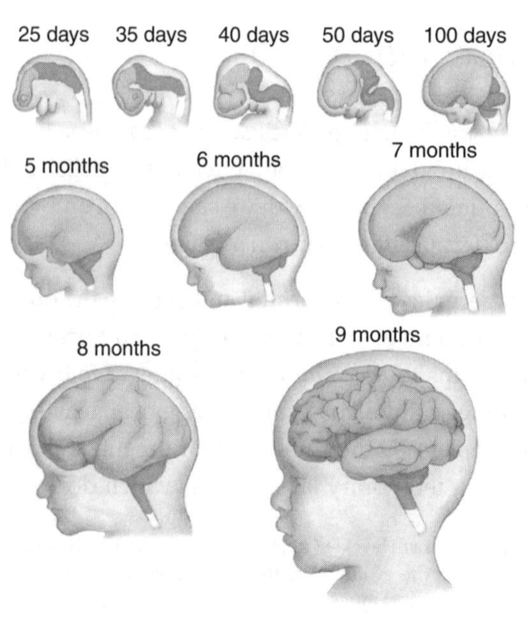

Figure 16.1 Prenatal development of the human brain

visual cortex at the back of the head or those linking the inner ear with the auditory cortex or those in the sensorimotor cortex, which controls body movements.

Brain size and cognitive function are not strictly correlated, as neuroscientists always stress. For instance, a child born with a brain well under normal size (suffering from the complex condition of microcephaly) may have a brain not much larger than a chimpanzee but can still perform much better cognitively than the latter.

16.2 The Proliferation of Neurons

Research by the German neuroscientist Wieland Huttner and his team in Dresden has demonstrated that there is a human-specific gene – *ARHGAP11B* – which shows the highest level of specificity among human cortical stem cells (not present in chimpanzees, with which human stem cells were compared). It arose in evolution about five million years ago, a little later than the last common ancestor of chimpanzees and humans, through a partial duplication of a ubiquitous gene labelled *ARHGAP11A*.

But why does *ARHGAP11B* lead to a proliferation of neurons? The answer is that this gene contains a human-specific sequence which arises by a so-called reading frame shift: 55 nucleotides of the DNA are not transcribed when producing the RNA blueprint for later protein synthesis. This RNA then causes the production of a protein which amplifies basal stem cells, leading to an increase in stem-cell division, which in turn results in an increase in the number of neurons in the human brain. This is seen by Huttner as a major cause for the large brain of hominins developing.

There are two types of stem cells involved in the growth of the cortical cells of the brain: (i) apical radial glia and (ii) basal radial glia. These two types of cells contain about 300 genes, 56 of which are human-specific and not found in the control population of mice investigated. These were highly expressed in the relevant stem cells. The gene, which was expressed to the highest degree in the basal stem cells but not in the (post-mitotic) neurons, was *ARHGAP11B*. It was then shown not to exist in chimpanzees. In the evolution of the hominin lineage, *ARHGAP11B* must have arisen after chimpanzees but before the genus *Homo sapiens* as it is known to have existed in Neanderthals and probably in Denisovans as well.

For a significance test, this gene was introduced deliberately into a mouse genome and the proliferation of neurons was observable, just as with humans, by apical radial glia inducing basal radial glia, not just to complete one round of cell division (leading to two neurons) but to multiple rounds. Furthermore, in the embryonic development of the mouse the brain began to fold like a human brain.

The genetic research which provided these insights was concerned with determining how neocortical folding in humans takes place, yielding the large human brain. However, if the proliferation of neurons during human pregnancy is due to the expression of the *ARHGAP11B* gene, one could ask whether phylogenetically (in the development of our species) this gene is responsible for our greater cognitive powers compared to other primates or whether our cognition grew, triggering the activity of *ARHGAP11B*.

16.3 Childhood and Puberty

Our childhoods are noticeably longer than those of other animals, including those we are most closely related to, like chimpanzees, who reach sexual maturity at around seven years. In early childhood, the billions of neurons in our brains establish complex sets of connections between each other. These lead to the personalisation of the brain, rendering later behavioural patterns of an individual predictable or at least more likely.

Why do we have such long childhoods? One reason could be to complete the very complex process of language acquisition. While it is true that we have acquired the basics of our native language by the age of five, that is to say we are fully fluent and can understand everything, we still need a number of years to complete our acquisition of grammatical details and to build up our vocabulary.

Childhood in the Evolution of *Homo* species

Dating of the teeth of the fossil known as Turkana Boy (formerly labelled Nariokotome Boy), who lived about 1.5 mya near Lake Turkana in Kenya, shows that he was only about eight years old at the time of death. But he was 1.6 metres in height so was almost as tall as the average male of his time. This implies that the childhood of this species was quite short compared to that of later *Homo sapiens*, which in turn implies that he would not have had advanced cognitive abilities and certainly not language, in any modern sense, which would require a longer childhood to be acquired.

In 1993, the remains of an eight-year-old Neanderthal boy were discovered at the Scaldina Cave along the River Meuse in central Belgium. These consist of a mandible (lower jaw) from which DNA was successfully extracted at the Max Planck Institute for Evolutionary Anthropology in Leipzig. The development of the teeth, estimated to be about 127,000 years old, could be analysed and suggested – going on the eruption of the molars at the back of the mouth – that the childhood of the Neanderthals was intermediary between the *Homo erectus* (illustrated by the Turkana Boy, just mentioned) and modern *Homo sapiens* like ourselves, where the onset of puberty and sexual maturity occurs around 12–13 years of age. A specimen of *Homo antecessor* from Sierra de Atapuerca in northern Spain (dated to 0.9 mya) had tooth development (eruption and growth) closer to modern humans than to early *Homo*, again showing the stretching of childhood as we move towards modern humans.

Puberty

Puberty is the period of transition from childhood to adulthood. It is characterised by hormonal changes in the gonads, ovaries in girls and testes in boys, which lead them to become capable of sexual reproduction. To a large extent, though not completely, we lose childhood plasticity at puberty. In purely biological terms, this is fine: with sexual maturity we can produce offspring and so the species can continue.

Just before puberty a further change takes place in humans known as the lateralisation of the brain. This term refers to the fact that neural functions and their associated cognitive processes became more or less fixed in parts of the brain. This development was previously held responsible for the decline in the ability to acquire languages natively after puberty. The so-called Critical Period Hypothesis, introduced by Eric Lenneberg in 1967, is no longer supported in its original, simple form. Rather, present-day scholars believe that there are periods in childhood characterised by particular sensitivity towards certain linguistic abilities, for example the acquisition of phonetics, such as a pronunciation like a native speaker, which depends crucially on acquisition in the very early period of life.

Assuming that exobeings have advanced systems of communication, prolonged childhoods could also be characteristic of such civilisations, allowing them to acquire such systems. The plasticity of childhood is necessary for the acquisition (unconscious learning) of complex cognitive skills. However, exobeings might have other means of attaining these: they might show high levels of plasticity throughout their entire lives and not have a rigid division into two periods, as humans do with childhood and adulthood.

16.4 Lifespan and Aging

Humans are notable for living increasingly longer after puberty, 70–80 years is no exception nowadays. It is our behaviour in this period which makes the human world so different from the animal world. In the latter, lifespan is generally dependent on size, population and on the chances for reproduction. Fruit flies live about 30 days, but turtles can live to over 100 years without any of the deterioration in substance and function, technically called senescence, which other animals show in great age. Bowhead whales can live to up to 200 years. As a very rough rule of thumb, the bigger the organism, the slower the metabolism, the longer the life.

A longer lifespan induces social behaviour. Probably because animals which have difficulty finding food (in an inhospitable environment) tend to live longer (experimentally confirmed by putting small rodents like rats and mice on a shorter supply of food than they would normally have), probably in order for them to have more opportunity to reproduce. Large animals – elephants, giraffes, rhinoceros – live longer not least because reproduction takes time (longer pregnancies). If there is a plentiful supply of food and mates (as with insects), reproduction is quick and easy, but life is short.

Aging is typical of humans and cognitively advanced animals. Some very simple forms of life, like the freshwater organism hydra, appear not to age at all. There are various theories about why and how humans age. The accumulation of errors and damage in our DNA is one favoured reason. This includes telomere shortening. Telomeres are the ends of chromosomes, like the ends of shoelaces, there to prevent them becoming splayed. With repeated DNA duplications over the years, these telomeres shorten increasingly, which has a bearing on the number of years we have left to live.

In the natural world, aging is not so obvious because animals that begin to age and become weak are in great danger. The smallest sign of weakness can signal the end for an animal. This is obvious in herds of animals which are preyed upon by large cats, such as zebra or antelope, which are hunted by lions or leopards on the savannah plains of Africa. A member of the group which is not strong and cannot keep up with the rest, or strays from the others, is likely to be taken out by a group of lions hunting together. An accident can have fatal consequences for an animal. A hunter which breaks a leg will starve in a week or so. For this reason, it is very unusual to see old animals in the wild, though we do have old pets, like cats and dogs, who go grey at the mouth. It is true at the opposite end of the scale. The very young are also at risk: elephant babies form a preferred prey for lions because they can more successfully attack them, given their small size and inexperience in self-defence.

However, humans have managed to buffer the effects of aging by advances in medicine. We actively combat weakness and disease to prolong our lives. This also means that certain degenerative diseases, like Alzheimer's and Parkinson's, are more in evidence in human populations nowadays, because we live to an age where these are more likely to occur.

Depending on how long an individual lives, their latter years may involve a decline in their physical and cognitive powers. For instance, memory performance diminishes with age. The older you get, the more you must train your memory by continually repeating items of information, such as words or names, which you might wish to retain. In advanced age, it is normal to lose brain substance. With the degenerative Alzheimer's disease, the percentage of neurons lost can be considerable, but with age-related decline there is nonetheless enough brain reserve to ensure that all normal faculties are fully available. In addition to this there is cognitive reserve, which refers to the accumulated experience of using our brains over a lifetime. If we are used to availing of different ways to find things in our brains, this can stand us well in later life. Together, brain reserve (physical) and cognitive reserve (functional) form a neurological reserve. When this is exhausted, however, deficits become apparent.

Language and Aging

The hippocampus, which is centrally involved in the establishment of memory, loses about 5 per cent of its neurons in the normal aging process. This affects memory performance (dependent on other factors as well, such as dendritic connections between neurons). However, the ability to construct sentences – a feature of syntax – does not decline noticeably with increasing age nor does the correct formation of grammatical words – technically called inflectional morphology – which implies that these cognitive abilities are very basal and differently stored and retrieved than lexical information, the words of our language. The continual use of syntax and morphology whenever we speak ensures that the connectivity required for the correct generation and comprehension of morphosyntactic structures is maintained in our brains throughout our lives.

Beside normal aging there are various pathological conditions which can affect one's language. They are all subsumed under the global label aphasia and will be dealt with in Sections 24.3 and 24.4 below.

LIFESPANS FOR EXOBEINGS

On an exoplanet, animals would have offspring, otherwise life could not continue through time. When it comes to the lifespan of exobeings, can we assume this would be comparable to our own? For evolution via natural selection to occur, a relatively quick turnover of the generations is required, so exobeings could hardly live for, say 1,000 years each. Furthermore, assuming that they went through a normal procedure of natural evolution, they would have a biology in which their bodies and the processes in them, which keep the bodies functioning satisfactorily, would be subject to degradation as are human bodies, which age inexorably.

Could exobeings have managed to stop aging? An exocivilisation thousands of years more advanced than ours might achieve that, or at least slow aging down. But this could have grave consequences, for example overpopulation, as new exobeings would be born, but none or few would die. In the long term, that would stifle evolution. What is more likely is that exobeings might develop means of more successfully mitigating the effects of aging by targeting specific factors.

17

Our Cognition

By cognition is meant mental power, the performance of the brain. This varies among individuals but we can see when considering humans as a whole that there is a certain level which is characteristic of all humans and separates us from other animals. For instance, we can plan for the future, utilise past experiences, teach ourselves a wide variety of skills and engage in myriad activities which have nothing to do with our survival as a species.

Considering the high end of human cognitive achievement for a moment, we recognise that it is represented by our best scientists and among these there are, and have been, a small number of individuals who have furthered our scientific knowledge to an inordinate degree. Just think of the great names from the golden age of physics in the early twentieth century, of which Albert Einstein or Max Planck are among the best known to the general public. Isaac Newton, Gottfried Leibniz, Charles Darwin, Michael Faraday, James Clerk Maxwell are among the great lights of previous centuries.

17.1 The Limits of Cognition

Do we know the limits of cognition? Is what we have as good as it can get? Bear in mind that there is not a strict linear correlation between brain size and cognitive power. We know that human brains can vary (downwards) by a few hundred cubic centimetres. What is important is the size of the prefrontal cortex, where language production and logical thought processes occur as well as the connection networks in the brain. As evolution is still occurring, we can ask if, given enough time, our brains could get bigger and bigger and our cognition improve accordingly? Well, maybe, but a larger brain for us would mean a larger skull, which weighs more and would cause mechanical and physiological difficulties for us, quite apart from making childbirth even more difficult.

Would this apply to exobeings as well? Would there be an upper limit on the size of their brains determined by similar factors? This question might have to do with the issue of scale on their planet. If they were much larger than we are, with more stable bodies and a thicker skeleton, they might be able to handle brains of, say, three kilograms, with 200 billion neurons, or maybe if gravity was less strong on the exoplanet, in which case there would be more neurons per unit of weight than on Earth. What this would mean for their cognitive power is an intriguing question.

Connected with cognition is the manner in which exobeings might perceive the universe around them. Perception has been and is an issue for philosophers. Externalist positions maintain that the physical world (and universe) exist independently of our perception of them; for instance the American philosopher Hilary Putnam stressed that cognition is not just an individual matter and that the world has a real existence, independent of the observer. But many of our words refer to observational phenomena, and do not have to correspond to scientific classifications which operate on an analytical level. We can see this with the word 'fish,' which is often used to refer to whales and dolphins. But they are mammals who bear live young and breathe air directly into their lungs via a blow-hole (their ancestors in evolution were animals which returned from land to the sea). This kind of thing happens because we use external features on our scale of human experience to classify objects and beings, like when we call the furry four-limb clawed animals found in Australia 'koala bears' although they are not bears, but marsupials.

Resulting from the above considerations are two essential questions concerning exobeings and cognition: (i) whether exobeings would have greater cognitive powers that we have (some people like the British philosopher Colin McGinn think this would be the case) and (ii) how they would perceive their environment: whether they would classify its contents in manners similar to ours, relying on how they experience their world on the scale of their bodies.

Working Memory

The British psychologist Alan Baddely (1931–) has maintained that there is a central executive which directs focus and attention in working memory. Executive control consists of these parts: (i) working memory with attention control, (ii) cognitive flexibility and (iii) inhibitory control. The ability to consciously control our mental activity is particularly human and develops in the first years of our lives. During this period we undergo socialisation, meaning we grow into the society in which we are embedded and unconsciously adopt its norms and customs.

Our working memory can survive attention distraction, which does not appear to be the case with non-human hominins. This retention across distraction would suggest that we have an episodic buffer from which we can retrieve information which was the focus before the distraction.

COGNITION AND COMMUNICATION

Our cognition is based on concepts. For instance, can you remember the first time you heard the word 'infrastructure'? I presume you can use it with ease and have no difficulty when others use the term. But just what does it mean? The point is we do not have to know *exactly* what something means. In fact, there may be no exact meaning to a word, but rather a conventional agreement about what it means approximately. Successful communication is based on consensus among speakers about what things mean *roughly*.

With early humans, we find evidence of advanced tools, sewing clothes, the making of watercraft and constructing dwellings, which are examples of cognitively complex tasks that would have involved planning some time before execution and would thus have involved a well-developed working memory.

The Notion of Object

This arises from our perception of the world. Objects are delimited from their environment, can normally be moved and have a different shape or colour, or taste, when dealing with food, and behaviour, where animals and humans are involved. An apple is easily perceived as an object given these characteristics. Animals would seem to recognise objects as well; for example, when a chimpanzee is offered a banana by a keeper, it recognises it as a piece of the fruit which it likes eating. What we do not know is whether chimpanzees have a notion of class of objects: can a chimpanzee visualise a banana with its eyes closed? We do not know. But humans can. We can talk about bananas even when none are in sight, we have an idea of a banana.

The extension from many instances of an object to the class of objects is a small but important step and one which children make at a very early age. However, there is a certain amount of evidence for concept-based memory with animals who bury food to eat it later. For instance, scrub jays hide food and can later find the location to retrieve the food. They at least have a notion of the food, even when out of sight, and they can remember where they hid or buried a cache of food and retrieve it. This does indeed imply a degree of abstract thinking.

17.2 Theory of Mind and the Notion of Self

As the British philosopher Simon Blackburn has pointed out, there is an epistemological problem here: How do we know that others have minds like us with consciousness like ours? The answer a neuroscientist would give is that we develop a theory of mind in early childhood, which enables us to guess, largely successfully, what others are thinking. It is true that we can never be absolutely sure what thoughts another person is presently having, but the high degree of predictability of others in situations we are acquainted with, for instance when they are tired, hungry, angry, wakeful, satisfied, compassionate, allow us to conclude that we know approximately what their inner state of mind is probably like. This is essential for cooperative action among humans, from hunting to sports, and indeed in all the shared activities in communities of practice.

By the age of four or five, children have developed a theory of mind. It has to be learned, we are not born with it. But the instinct to develop it may well be innate. Working out agency, motivation and causation is essential for us to make our way through the world. If we assumed that everything we observe was random, we would remain ignorant and we would not get very far. Thus, mind-reading, which is at the centre of a theory of mind, is important for survival in complex human groups.[1] As one can only surmise from indirect evidence what someone else is thinking and what their intentions are, it is necessary for all human children to quickly develop a theory of mind which helps them to guess others' thoughts and so play a successful role in the society into which they are growing.

In general, animals are poor at theory of mind and at self-recognition, qualities so typical of humans. They engage in forms of instinctive behaviour like breeding and caring for their young. But there is great variation in the degree of interaction between members of a species and the extent to which male and female remain together and are responsible for rearing their young. Seals meet just for the act of mating, and never again, while certain birds, like swans and albatrosses, form pairs and bond for their entire lives. It would seem that, in general, the more complex life forms are, the more social elements come into play.[2] Already with animals, like

1 The evolution of social thinking has also been investigated for non-human primates, for example in groups of baboons in Botswana, by the American ethologist Dorothy L. Cheney (1950–2018).

2 Is it merely coincidence that those birds which congregate in groups before nightfall to roost, such as crows, are among the most intelligent of their species?

horses or dogs, we see a desire to be together which goes beyond mere survival. Cognitively higher animals, like primates or elephants or dolphins, demonstrate considerable social skills in their complex modes of interaction.

A theory of mind is intimately connected with a notion of self. Because we are so sure that we exist as a single, unique thinking and reflecting individual, we assign this property to others. The notion of self is much debated in philosophy, psychology and neuroscience. It is generally seen as a construct of the mind. A sense of self gives our existence a continuity, a feeling of permanence through the many changes we experience during our lives. Our sense of self is formed by the memories of our unique lives and is also bound to our personalities – the amalgam of behavioural and attitudinal patterns which characterise us. This makes us identifiable and predictable for others and hence allows us to adopt clear roles in groups, from the family to society.

Looking Out on the World

Do you ever have the feeling that you are sitting in a chair, in your head, looking out through your eyes? This feeling is what the American philosopher Daniel Dennett has called 'the Cartesian theatre.' But there is no looking out of the eyes, the only information processing which takes place is that of light coming into the eyes. Because you can focus your attention on objects in your visible field, you imagine that you are looking out. But the focusing takes place inside your head. The retina of your eye can only take in information, it cannot 'look out' at the world.

But such feelings are yet another aspect of our cognition which contributes to our sense of self. The apparent continuity we experience is also strengthened by the continual maintenance of certain bodily functions, such as the precise temperature of you inner organs (37 °C), your heart rate, your breathing and also your digestion (though less critical to your survival). So the continuity of the body provides the brain with a physical scaffolding on which the continuity of our cognition and the sense of self rests.

17.3 Internalisation of the World We Perceive

We construct an internal representation of the world we perceive and develop concepts which we can articulate in communication with others. To illustrate what is meant, consider the question, 'What is Dublin?' We have a concept of the city and we can refer to it in conversation: 'I was born in Dublin'; 'The Dublin of my youth'; 'We were in Dublin last year'; 'Dublin is the capital of Ireland.' But there is no external object 'Dublin.' And you can show this with a thought experiment: If you could replace all the buildings in Dublin with new ones, would it still be the same city?[3] In fact, there are real cases of this: during Word War II, the cities of Hamburg and Dresden were so badly damaged that virtually nothing was left unscathed; the record for this destruction is held by Düren, near Aachen, which was completely destroyed. The cities were rebuilt, partly by reconstruction but also by razing ruins to the ground and starting from scratch. Are these cities the same as the ones before the war? The answer is that there never was a city Hamburg, Dresden or Düren, nor is there a city Dublin, London, New York, Rio de Janeiro or Cape Town, that is, delimited objects, independent of our perception of them. What we have are tacit agreements by speakers to use certain names (words with specific sound shapes and spellings) for certain collections of things (buildings, streets, parks, etc.) which we regard as entities.[4] This might become clearer with another example. We call our nearest major galactic neighbour Andromeda. But that is what we Earthlings call the galaxy. Among astronomers it is also known as Messier 31, or just M31, after its position in the famous catalogue of faint objects in the sky by the French astronomer Charles Messier (1730–1817). If there are exobeings on planets around stars in Andromeda, one thing is certain: they do not call their galaxy Andromeda (nor would any exobeings in our

3 This is like the famous ship of Theseus: in Greek writing, the story is handed down of a hero called Theseus who kept his battleship in a harbour and, as it aged, pieces of the ship were replaced until finally there was no original piece left. Was it still the same ship?

4 During communism in Eastern Europe in the twentieth century, many cities had different names: Chemnitz was Karl-Marx-Stadt (Germany); St Petersburg was Leningrad, Nizhny Novgorod was Gorky, Volgograd was Stalingrad (Russia); later, the original names were restored (except Volgograd, which was previously Tsaritsyn).

galaxy either). The name, from that of a beautiful woman in Greek mythology, does not exist beyond the convention by humans on Earth to use the name for the purpose of referring to the galaxy.

The key feature of words, such as those just mentioned, is that there is broad agreement among those who avail of them about what is meant by them, namely, what they refer to. This is true of the physical world. But with abstract notions the matter can become more difficult: just consider the different views, in any society, of what constitutes justice or equality. This book is not the place to discuss this particular issue but it does show that the figurative and metaphorical use of language[5] is further evidence of how human cognition moves away from the literal and creates features and structures on a more abstract internal level. Indeed we can even have words for things which do not exist at all: neither a dragon, nor a unicorn, nor an angel exist. But we have words for such putative beings and we can list their features: dragons are like huge caterpillars which spew fire, unicorns are like horses with a twisted cone on their forehead, angels are those androgynous beings with wings sprouting from their shoulders who hover above saints and other holy persons in religious pictures.

The agreement in reference about concepts is one of the strongest pieces of evidence for the existence of minds in other people. After all, this only works if one assumes that entities, like those discussed in the previous paragraphs, correspond to internal concepts which all of us share.

17.4 The Tiger in the Bush: Our Love of Patterns

Are mathematicians geniuses? It might seem counterintuitive or peevish to say they are not. It is true that there is a general admiration for people who engage in mathematics because it is difficult and far removed from our daily lives. The last statement betrays an important fact: mathematicians

5 This is a major area of linguistic investigation and one which has immediate relevance to our use of language, see the book by George Lakoff and Mark Johnson (2003 [1980]), *Metaphors We Live By*, for a classic treatment of the matter.

possess a relatively rare ability, namely to juggle numbers easily and to engage in highly complex abstract thought. If we all did this, there would be no admiration of or wonder at mathematicians. The real question is the following: Why is mathematical ability rare? The answer is that it was not favoured by evolution. There is no situation in the outside world where significant mathematical ability would be necessary for the survival of our species.

But the ability of everyone to extrapolate things and situations from fragmentary sensory input is quite an achievement. I'll dub this ability the 'tiger in the bush' phenomenon. Imagine the following situation: A few early humans are out in the wild searching for food. They are on their guard for animals which might attack them. While scanning their surroundings they notice something behind a bush. What they see could be an ear and another part could be a paw or a bit of a tail. They immediately flee to be out of the range of the predator they suspect behind the bush. Those individuals survive to live another day. The ability they have is to recognise an animal from a few visual fragments. In real time, in their brains, they conclude that these fragments correspond to a predator. Their minds connect the dots, fill in the blanks and they reach a snap decision to flee (other non-visual information would also be recruited for this decision: the type of environment the individuals were in, their knowledge that predators often lurk behind foliage to surprise their prey, the time of day they go on the hunt, etc.).[6]

The computation necessary to match some visual scraps to a complete mental image is not trivial by any means and involves some intense vector graphic calculations, which are performed in our brains on the spot, and equal the achievements of the most qualified mathematicians. The moral of this short story is that we humans engage continuously in sophisticated pattern recognition, which enables us to successfully

6 This ability can be set in a broader context. We tend to connect things which may in fact be entirely unrelated; for instance, we associate star positions with events and personalities (astrology), we attribute special status to certain numbers (numerology). As Michael Shermer describes convincingly in his book *The Believing Brain*, this tendency of humans is the source of superstition and our view that coincidences have special meaning.

negotiate our way through the world around us. This pattern recognition is central to facial recognition (from a partial, side view, for instance), in judging possible obstacles in the paths in front of us when moving forward and in general in assessing situations of possible danger in today's complex, material world.

18

Consciousness

> How it is that anything so remarkable as a state of consciousness comes about by irritating nervous tissue, is just as unaccountable as the appearance of the Djinn [Genie] when Alladin rubbed his lamp.
>
> Thomas Henry Huxley (*The Elements of Physiology and Hygiene* 1868)

> But this [research into consciousness] is still a science of correlations not of explanations.
>
> David Chalmers, TED talk, 2014

Consciousness is the seamless inner subjective state which accompanies you in every moment of your wakeful life and which no-one else is privy to. It is a non-physical experience, which cannot be observed by examining the brain. In attempting to define consciousness, various scientists have strived to specify its necessary and sufficient properties or at least to narrow these down so as to get a handle on it. This is where the difficulties arise. While we all have consciousness and recognise it as an experience, it is difficult to pinpoint it in the form of a definition.[1] And how would one go about doing this? One can give an operational definition: consciousness is when we show awareness and when we react to external stimuli. But it is much more than that, it is our inner world which we experience even when there are no external stimuli. Consciousness is where our thoughts are, where we get our ideas. While it is connected to the outside world via sensory input from various sources, consciousness is also something uniquely internal, locked inside our bodies, which we do not share with others, no matter how emotionally close we may be to them. Furthermore, there is nothing else like consciousness. As the English neuroscientist Susan Greenfield has highlighted, the fact that consciousness cannot be defined as a member of a higher set, through a relation of hyponymy as linguists would say, makes it of its own, *sui generis* to use the Latin term. For instance, a violin is a member of the string instruments, these are part of an orchestra. What is consciousness a member or part of? Consciousness is a set consisting of one member, it does not form a word field of related phenomena, it is unique.

When we engage in language, we often use single words for phenomena which are not unitary things, for instance we use words like 'happiness', 'love' or 'boredom', which are concepts, as if they referred to single things. Doing this is technically called 'reification' from Latin *rēs* 'thing'. But is consciousness a thing? Some scientists regard it as a process consisting of various parts. Perhaps it might be best to refer to it as a continuous state. Except for the time we are asleep, and the few occasions we may have been under anaesthetic or otherwise unconscious, we all have a feeling of continuous existence extending back to our early childhood. This of course ties in with memory and gives us the feeling of being ourselves, of being unique individuals.

But apart from defining consciousness, one can ask the legitimate question: What would an explanation of how consciousness arises be like? How could one explain the presence of something non-physical in purely physical terms? The physical and the non-physical are different domains and

1 One could define consciousness negatively. But that is unsatisfactory. What features would you wish to mention which necessarily and sufficiently define consciousness negatively?

it is not possible to describe and explain phenomena from one domain in terms of the other. For instance, in chemistry one can explain how certain reactions occur, for example by describing the elements and molecules involved and how various chemical bonds form. But when looking at two different domains all one can do is register correlations between changes of state in one domain (the physical brain) and parallel changes in the other domain (the conscious mind). And one cannot reverse engineer consciousness: we cannot break it down into its component parts, so to speak, and then determine where each of these comes from and somehow find the areas of the brain responsible for them.

This is a serious difficulty in neuroscience, but some scientists do not recognise it as such. Parallels are often drawn between consciousness and a computer: the desktop is consciousness, the interface with the outside world, and the chips inside are like the physical brain. But such analogies fail because the computer chips are physical matter and the desktop contains software that is fully determined by the commands of the operating system, whereas our consciousness is a fluid, subjective, non-physical experience. Furthermore, consciousness is clearly connected with our feelings, it goes hand in hand with all the sensations which we have.

For exobeings, we can assume that they would be self-aware, conscious individuals, making deliberate decisions and interacting constantly with their environment. The relationship of this consciousness with the physical substrate (brain analogue) they possess, and which is responsible for its generation, would be as crucial to grasping their cognition as it is for understanding our own.

18.1 The Role of Emotions

The reference to sensations is central here: we are feeling beings who experience emotions constantly. Some scientists, such as the American Joseph LeDoux and the Portuguese-American Antonio Damasio, have examined the relationship of rational thought to emotions. What happens in our brains to make us feel fear, love, hate, anger, joy? Do we control our emotions, or do they control us? Do animals have emotions? How can experiences in early childhood influence adult behaviour, even though we often have no conscious memory of them? In *The Emotional Brain*, Joseph LeDoux (1996) investigates the origins of human emotions and explains that many exist as part of complex neural systems that evolved to enable us to survive. LeDoux has stressed that emotions can take over the brain and prevent all other processes, including physical processes such as digestion and saliva production. Furthermore, emotions are contagious: anxiety and fear are easily transferable across individuals, indeed across whole countries or continents.

18.2 The Origin of Emotions

Emotions enhance the likelihood of survival for creatures. They help us avoid dangers (by engendering fear or disgust) or they can allow us to turn a situation to our own benefit (by producing joy or desire in someone, like a potential partner). Emotions assist us in reacting to situations quickly and are a highly conserved feature of our evolution. Emotions are generally localised in the limbic system, a set of complex brain structures under the cortex (see Section 15.3 above). This system is well developed in animals, including those which do not have substantial cognitive abilities.

In the relevant neuroscientific research, for instance by Antonio Damasio, a distinction is made between emotions and feelings. An emotion is a set of changes (physical movements, release of hormones into the bloodstream, release of ions by nerves on trauma to a part of the body) whereas a feeling is the conscious perception of the emotion. The manner in which a creature experiences an emotion, that is, the type of feeling it has, depends on its cognitive abilities.

There is a temporal gap between an emotion and a feeling. Think of touching (by accident) a hot plate on a cooker or a live electric wire. The emotion of pain, manifest in the retraction of your hand, happens before you register the pain of heat or electricity on a conscious level. If you had to wait for the conscious registration and then decide to pull your hand back you would be badly burned or electrocuted. Emotions force us to react appropriately to a given situation rather than just thinking it through cognitively and then making a decision.

It is the feeling which is the pleasant or the unpleasant sensation. This is on the conscious level (within your brain). The intensity of the feeling can depend on a variety of factors influencing your brain. If you are at the dentist and know that the pain of the drill is necessary to remove the caries in your tooth, you are more likely to accept it than if you suspect it is pointless because the tooth must probably be extracted anyway given an abscess in the root. The attitude to pain determines its subjective intensity.

Feelings can remain in memory and we can reflect on them. Sometimes our preoccupation with a feeling is deleterious to our health, as with stress, a permanent low-level state of alertness or worry, or brooding over a perceived wrong which has been done to us. There is no evidence that animals have this level of introspection about their emotions and feelings (inasmuch as they have the latter in the sense being employed here).

18.3 The Hard Problem

When considering the relationship between consciousness and sensation we are confronted with what the Australian philosopher David Chalmers has aptly labelled 'the hard problem': the difficulty of linking the physical brain with the subjective experience of consciousness. It centrally addresses what are termed 'qualia': specific instances of subjective conscious experience deriving from sensory input. Qualia, as the name implies, seem to have a quality which is not in the original signals passed on to the brain as electrical current from the particular sense organ. A few commonly quoted examples of qualia are the smell of cheese, the taste of a peach, the view of a sunset, perception of colours, such as the red of a rose or the yellow of daffodils in spring. The arts quintessentially trigger qualia: listening to a sonata, reading a great poem, dwelling on a famous painting, are all occasions when we experience an emotion not contained in the input signal which triggers a subjective sensation in the brain. It is true that there has been criticism of the notion of qualia, but by using this special term a phenomenon seems to have been recognised, which is in need of explanation. There is no doubt that an explanatory gap between the physical, observable brain and the subjective experience of consciousness does exist, no matter what terminology one prefers.[2]

While it is difficult to define consciousness satisfactorily, it is important to stress that it is not identical with the mind, after all the mind continues to exist when full consciousness is temporarily suspended as in sleep, when one is under anaesthesia, when one faints or is 'unconscious' for some reason. In an attempt to introduce clarity into proposed explanations of consciousness, Chalmers offers a threefold taxonomy.

Chalmers' proposed explanations for consciousness

Type A: Hard-line materialism. There are only physical brain states. Consciousness is an illusion, it has no real existence, but is accounted for by the behaviour and dynamics of the brain (Daniel Dennett).

Type B: There is consciousness and there is a gap between what we observe physically and what we experience subjectively. But it is a matter of perspective: from the outside we see the physical brain, from the inside we experience consciousness. So we have two ways of thinking about the same underlying reality. This view maintains that consciousness and the brain are one thing in nature.

Type C: The gap is only one of knowledge and will be resolved sooner or later, scientists have just not arrived at the solution to this difficulty yet.

The American philosopher John Searle (1932–) has stated that consciousness has an irreducible first-person subjective ontology. He rejects

2 See Pinto et al. (2017).

panpsychism, the view that mental states are universal across all beings and objects, because it does not account for how such consciousness could arise. The brain generates consciousness so, in principle, if we knew down to the very last detail how the brain is constructed and how it functions and could build a new brain, consciousness should arise, but this is a task we are light years away from managing. For Searle, consciousness is the product of special biological processes, which take place in complex neuronal structures of the type that we have in our brains. Its defining property is the subjective, first-person quality we experience. Without that we are not talking about consciousness.

Searle believes that we suffer more from our traditions than we are aware of. He distinguishes two traditions in particular: Tradition 1 says consciousness is the window on our soul, on God, on the metaphysical, on the supernatural realm, which is separate from the physical universe. Tradition 2 says that consciousness is not part of materialism and is not a legitimate subject for scientific inquiry. Both deny the fact that consciousness arises from neuronal activity.

But despite Searle's deterministic and materialistic view of consciousness, Chalmers maintains that something beyond our current physics is needed to account for consciousness because we cannot link it causally to any process in the brain, although we can point to an ever-increasing number of correlations between brain states and conscious experience.

Correlations and Explanations

The Pinnacle Point excavation site near Mossel Bay on the southern coast of South Africa has provided evidence that Middle Stone Age *Homo sapiens* occupied the location from about 170,000 to 40,000 years ago. There is evidence in one of the caves (Cave 13B) that the dwellers there fed on shellfish and used fire in the production of stone tools. The marine animals on which they fed would have had to be caught at low tide, which would require knowledge of the moon phases and their connection with tides. For instance, spring tides, with extreme high and low tides, occur when the Earth, Moon and Sun are aligned. All the early inhabitants of the region would have had to do was to recognise the correlation; they would obviously have had no knowledge of the gravity the Moon exercises on the Earth, moving the water of the oceans, nor of how the Sun comes into play during the new moon phase, adding its gravity to that of the Moon.

18.4 The Sense of Self Again

The sense of self which consciousness engenders in us is often described as an illusion. But the word 'illusion' (Frankish ed., 2017) is not appropriate here as it carries connotations of deception and unrealness. Consciousness is not the physical brain cheating us, but generating a level of awareness which is to our great advantage. A more neutral term than 'illusion' would be 'construction': the brain constructs consciousness, which provides us with a sense of control over our actions. And anyway, as Searle has pointed out, to think about whether consciousness is an illusion or not is already proof of its existence. If you are in a position to doubt consciousness, you are already conscious.

Consciousness is uniquely private and personal and causes one to develop a sense of identity, of being 'me' and no one else, of an agent acting with free will. Whether we have free will is a hotly debated issue, in both neuroscience and philosophy, with scholars from the latter field now taking on board insights from the former.

Quantum Effects, Determinism and Free Will

For many philosophers, and even more so for theologians, the notion of free will is sacrosanct. A purely deterministic view of the human brain states that if the entire configuration of the brain, at any point in time, were known then the following state would be entirely predictable. This would mean that, when making a decision, the feeling that we are free to decide as we wish is indeed an illusion. The outcome of our considering pros and cons, for instance, for a certain of action, is not a matter of conscious choice: the decision is already determined by the state of our brain at the time of considering what to do.

To maintain that free will really exists is would be necessary to show that states of the brain show at least a degree of randomness. This has led many scientists to think about whether quantum effects might be operative here. The British physicist and Nobel laureate Roger Penrose has claimed that there is a gap between the indeterminate quantum level and the deterministic classical level, which we see around us on our accustomed human scale, with cause and effect operating. Penrose believes it is in this area that the non-computational activity – randomness – of the human brain resides (see Section 18.9 below).

The American neuroscientist Peter Ulric Tse has maintained that there may be indeterministic elements to neurons firing, which then add 'noise' to the system and render outcomes on a higher level of consciousness random, or at least add a degree of randomness. However, as Sam Harris has pointed out, any randomness which might result from quantum processes would not account for free will. This rests on the stubborn conviction which humans have that, for any decision they have made, they could have decided otherwise. And that, for any decision they are about to make, they are free to choose whichever course of action they wish to.

Consciousness and Information Integration

The American–Italian scientist Giulio Tononi is known for devising the integrated information theory (known as IIT), which posits that consciousness is determined by the degree to which a system like the human brain integrates large amounts of information. Using the Greek symbol Φ as a measure of this integration, Tononi's theory assumes that consciousness exists to varying degrees with various physical substrates. The theory classifies various characteristics, such as composition, which refers to how subsets within the system interact in a cause-and-effect manner to contribute to consciousness. The theory does not, however, deal with the specific behaviour of the human brain and does not offer a principled account of the relationship between consciousness and unconsciousness. It accepts that information integration on other substrates, such as supercomputers, could in principle result in consciousness. Despite the scepticism shown by many neuroscientists to Tononi's theory, the notion that brain-wide information sharing is the neuronal basis for consciousness is widely accepted, for example by the French neuroscientist Stanislas Dehaene.

Having an Idea

The jury is still out on randomness, with many scientists steadfastly maintaining that quantum effects, characteristic of the subatomic level, are not operative on the level of neuron firing, where they would need to be to add randomness to consciousness and hence invalidate complete determinism. However much determinism chimes with the physics and biochemistry of neuroscience today, there are aspects of the brain which await explanation. Consider having an idea. Say you are concentrating on a problem, trying to find a solution to a situation you are stuck in, in your work, your daily life or whatever. And maybe when you focused on the issue or were just walking along you suddenly had an idea about how to solve it. The idea just popped into your consciousness. Where did it come from and how was it triggered? One could say that ideas are generated and then 'enter' your consciousness. But what is the process by which they are generated? Is this process in itself random or fully deterministic on an unconscious level? And what is the precise relationship between the unconscious and conscious levels of our brain?

If you reflect on having thoughts and ideas, you realise that it happens all the time, in fact there are fixed expressions in language for it, such as: 'It just struck me'; 'It came into my mind'; 'It entered my head'; 'It suddenly dawned on me.' These all refer to the sudden appearance in one's consciousness of thoughts and ideas. It is not known how this happens, but it does, the whole time, with all of us. Indeed, certain individuals seem to have high-quality thoughts, which we traditionally called inspiration – a word which does not, however, explain what it is referring to. The essential feature of sudden thoughts or ideas is that they normally do not require an external stimulus. Thoughts and ideas popping into consciousness are very much part of the hard problem of consciousness.

Consciousness and Personal Identity

It is fair to say that we all experience a relatively persistent sense of self through time. Is the notion of an 'enduring self', which lasts a lifetime and is based on memories, experiences and personality, an epiphenomenon, that is to say that the brain was not designed to create such a feeling, but it just arose from the way the brain operates? There is no simple answer to this question and scholars disagree about how to address, let alone how to solve it. Personality is a key aspect of the sense of self and would seem to be dependent on functioning cortical parts of the brain. For instance, one can remove the cerebellum (small brain at the lower back of the head) and an individual's personality remains largely intact. But if there is a major lesion in the cortex or the thalamus then the individual will be severely impaired behaviourally.

Consciousness and Metaphysical Reality

Does consciousness offer a window on a non-physical universe, which exists parallel to but independent of the physical one? The major religions of the world would appear to believe this and use the notion of soul to explain the link to some metaphysical reality. For now, this is a matter of speculation and will most likely remain so into the foreseeable future. What remains a mystery is how this electrochemical activity of the brain generates a subjective non-physical experience. As Susan Greenfield, the well-known English neurobiologist, has said when considering the relationship of the brain's physical activity and the consciousness it engenders, we do not really know what an answer to how consciousness arises would look like. How do you causally relate the flow of small electrical currents and the release of certain chemicals in and around neurons to the all-embracing first-person sensation of human consciousness? There must be a relationship because changes to the physical substrate of the brain can and do effect consciousness. But how would one even begin to start describing that relationship causally?

Metaphors for Consciousness

To appropriately characterise consciousness, various scientists have resorted to metaphors which we can easily relate to. The American neuroscientist David Eagleman has compared consciousness to the chief executive officer (CEO) of a company. It is just concerned with actions which have consequences for the entire organisation but not with all the neurological details of the actions in our daily lives. This is certainly true: the vast majority of actions we perform do not require conscious decisions. Say you are preparing the breakfast in the kitchen and talking to someone at the same time. You are probably not aware of every single movement you make, putting cups and plates on the table, fetching the sugar bowl and the milk jug, putting the kettle on, etc.

Another metaphor for consciousness is that of an orchestra. The English philosopher Julian Baggini has used this comparison: an orchestra consists of many parts but creates a unitary sensation, as when a symphony is played by many musicians together as a group. So, in the brain, there is a cumulative effect of several neuronal areas and pathways, like the sound of an orchestra is the sum of all instruments playing. This metaphor highlights what is known as the binding problem (see Section 15.5 above). Consciousness probably consists of an amalgam of different physical processes which the brain can use to generate consciousness because (i) we cannot pinpoint consciousness to a certain location in the brain and (ii) it still remains even when brains are split or severely impaired through disease or injury.

So just how do the individual parts of the brain involved in consciousness generate the unitary feeling we experience? How does the brain unify all the information from different senses into a singular subjective experience? One answer is that there would seem to be pathways and subsystems in the brain which show a neural oscillation at 40 hertz or above – called gamma waves. These oscillations (colloquially termed 'brain waves') are associated with focus of attention and working memory and may represent a physical baseline for consciousness (see Section 32.12 for further discussion).

Consciousness and Awareness

Consciousness is intricately connected to awareness. Again, take the example of an orchestra, where one's attention is drawn to certain instruments, say wind or brass, during the course of an orchestral work. Our attention is focused on some part of consciousness under certain circumstances, say one's hearing when one is listening carefully, one's sight when one is straining to see something,[3] one's sensation of pain when one has hit one's big toe off the corner of the cupboard in the bathroom.

Awareness would seem to sit on top of our subconscious registration of sensory input. For instance, when an image enters our eye, say of a car in a street, a complex electrical signal from the rods and cones on the eye's retina is passed back to the visual cortex at the rear of the head. But when we are aware of the car, say noting the colour, its make, if someone is sitting inside, etc., neuronal electricity activity is triggered in the top and front parts of our brains.

The Nobel laureate Austrian-American neuroscientist Eric R. Kandel (1929–) headed the final chapter of his recent book *The Disordered Mind. What Unusual Brains Tell Us About Ourselves* (Kandel 2018), 'Consciousness: The great remaining mystery of the brain'. There, he confirmed that much sensory information, such as that from eyes or ears, is not necessarily broadcast up to the prefrontal cortex for decision making, but processed locally near the dedicated cortices for these senses.

The Origin of Consciousness: Unlimited Associative Learning?

The rise of consciousness during evolution was obviously gradual and various levels of consciousness can be recognised in the animal world today. But what could have been the origin of this unique non-physical quality of so many life forms? One notable proposal is unlimited associative learning

3 Hearing and sight are phenomenologically very different: hearing is based on differences in time while vision is grounded in differences in space. These are two different modalities.

(Ginsburg and Jablonka 2019, 2022; Birch et al. 2020). Its main tenet is that animals (and humans) can remember the connection between a stimulus and a response, storing a memory of that experience that then guides later behaviour. For instance, an animal can take note of the bitterness of unripe fruit and avoid eating it in future. Or it can realise that eating alone is less stressful than in the company of other, possibly larger animals, which might be competition for the same food.

Unlimited associative learning goes far beyond automatic reflexes, released instantaneously by some external trigger. It assumes that the animal engaged in this learning can store the reaction and apply it later to comparable though not identical situations. The animal inherently evaluates stimuli and orders them hierarchically attaching relative importance to them and storing them in memory as a network of interrelated experiences. Such learned patterns can be adapted and modified later, should new instances of the stimulus require this.

Looking for an evolutionary marker for the transition to consciousness, Eva Jablonka and her colleague Simona Ginsburg see unlimited associative learning as central. An animal can learn even if there is a time gap between the initial stimulus and the result as when hunting prey. This requires memory and maybe a degree of self-reflection, but at any rate it posits the mental capacity for object permanence in the animal chasing prey.

There are further aspects to this model such as sensorimotor binding: joining acoustic with visual cues or tactile with olfactory cues or different combinations of these. This leads to mental representations which, in a continuing positive feedback loop, can result in abstract thought. Once this has occurred such representations can be retrieved from memory without the external stimulus being present anymore.

The key role of predation in the rise of goal-directed behaviour and experiential learning has been recognised by many researchers, not just by the current model of the origin of consciousness. But it would be wrong to see this origin as causally dependent on the rise of predation at the beginning of the Cambrian period. Rather, increasing consciousness was a gradual process, linked to the evolution of species and their brains, a process which indeed exhibited certain thresholds, as with the rise of predation, which resulted in a relatively sudden impetus for the further development of consciousness.

18.5 Size of the Brain and Consciousness

How much brain, what number of neurons, are necessary to generate consciousness? We can only tell that indirectly by looking at what the cerebral anatomy of animals are like. The part of chief interest here is the cortex, which is present in all mammals to varying degrees. What is also important is the basic brain-to-body mass ratio, technically called the encephalisation quotient.

This is calculated via the allometry of animals (shape, size, anatomy and physiology taken together in relative terms). At about 7.5, we have the highest ratio of all mammals followed by species of dolphins such as the tucuxi (a freshwater dolphin of the Amazon basin), at 4.56, and the bottled-nosed dolphin, at 4.14. The chimpanzee, our closest evolutionary relative, has an encephalisation quotient of between 2 and 2.5. The quotient for elephants is between 1.75 and 2.35 while that for dogs is 1.2 (about the same for octopuses) and for cats just 1.

These figures do not, however, tell us anything about how various animals subjectively experience consciousness. There are a few additional tests, such as the mirror text used to determine if animals recognise themselves in a mirror and thus can be said to have a 'theory of mind,' which would imply a reflecting self. While the diagnostic value of this test is contested it is true that only higher mammals pass it, so it does have some bearing on the quality of consciousness for the animals in question.

The mediation of consciousness between members of a species is also relevant. Elephants communicate in a number of ways, tactily via their trunks and infrasonically through low-frequency booming sounds. They engage in this communication the whole time they are together in groups. A high level of sociality would seem to correlate with high levels of consciousness. Chimpanzees, bonobos, monkeys, gorillas, for instance, are very social animals, live in communities, produce sounds with basic meanings and engage in reciprocal grooming.

Despite the great advances in ethology (the study of animal behaviour) in the past century or so, no one yet knows what it feels like to be a dolphin, a monkey or a dog, still less a crow or an octopus. At least if these animals had versatile language they could explain their sensations of consciousness to us humans. But that is out of the question, of course. Can we draw a conclusion from this fact? If chimpanzees or dolphins, for instance, do not have a language in any way comparable to ours, is it fair to say that they do not have a language of thought either?

18.6 Where Is Consciousness?

There is no centre for consciousness. It is a distributed feature of the brain. But it is not physical so the image of being spread, like icing on top and around the edges of a cake, is not appropriate. By maintaining that consciousness is a distributed feature is meant that there is no one point in the brain where it is generated. There are important consequences from this fact. The American–Israeli neuroscientist Eran Zaidel has shown that patients with intractable epilepsy and who have had their corpus callosum (the bridge of nerve fibres between the two halves of the brain) severed do not immediately show behavioural abnormalities. Importantly, they still have a single experience of consciousness, but with certain peculiarities. On occasions, the consciousness generated in both hemispheres of the brain can be antagonistic, as when patients remark, for example, that their hands want to do different things, like touch someone and not touch them, grasp something and not grasp it (called the 'intermanual effect'). In all other individuals, consciousness from both hemispheres is coordinated by communication across the corpus callosum and so any antagonism would be resolved before motoric commands are issued to the limbs of the body.[4]

18.7 Consciousness and Attention

Recall the example given above of setting the breakfast table while talking to someone at the same time. Normally you would concentrate on the talking, unless you had to make a conscious decision like choosing which cup to take for your coffee or what type of jam for your bread. You may be forced to change your attention, for instance, if you spilt some milk on the table or dropped sugar from your spoon. The point of this anecdote is that consciousness is about attention, about focusing on some aspect of your behaviour in the moment. Working memory is important for consciousness and is connected to the focus of our attention ('Did I put sugar in my coffee already?'). The brain is like a searchlight which can shine on different things, when called upon. One way in which this can happen is by brief networks of neurons, assemblies of millions of neurons, firing together briefly. These are mid-level processes, located between the firing of individual neurons and the consciousness we experience. The neuronal assemblies are highly transient: they trigger together for a brief period, a fraction of a second, and account for a degree of consciousness.

Continuing the analogy of a searchlight: the light is always on, but its beam is continuously being pointed in different directions. This fact is reflected in language as well: we have various expressions to demand of others that they focus their attention in the direction we want. Consider phrases such as 'Be careful,' 'Pay attention,' 'Watch out,' 'Concentrate on what you are doing,' which are all requests/demands that one's interlocutor point their 'attention searchlight' in a specific direction.

The metaphor of the moving searchlight helps us to recognise a key development in our evolution: we had to continuously scan our immediate

4 In fact, the (potential) antagonism of different consciousness-generating parts of the brain might be reflected in established language use; for example, what does it mean to claim, say of a novel you read, 'Part of me liked it and another part did not'?

surroundings for danger when we were out and about looking for food. We can also observe this behaviour in animals today. Imagine you put some bird feed out in your garden, then watch what the birds do. They survey the scene first; fly in close to the feed and fly away again, then come back, then fly down close to the feed, look around several times, maybe fly away again and come back and only then pick a few nuts or seeds, all the time looking around them, keeping an eye out for danger.

There is an important insight from the considerations here: the brain is selective in choosing what parts of the huge array of external stimuli it is continually presented with it should concentrate on. This is clear when one considers human vision. Our field of vision is quite large, about 180 degrees horizontally and 130 degrees vertically, but the part we are aware of, which we register consciously, is always just a fraction of this, for instance the text lines on a page when reading a book or the lines on a computer screen. It is not that everything else is out of focus, but that the brain is only processing information (and possibly organising a reaction) from a small part of our field of vision. Because this happens in a swift and seamless manner we do not notice how our focus is moving continuously around our field of vision. In fact, if all the information from the eyes were processed simultaneously, the brain would not be able to focus its attention on a small part and would lose the ability to react appropriately in a given situation.

The focus of attention can also be directed inwards as when we introspect on something. If we are lucky, we can also get ideas this way. By concentrating on a particular issue, a useful thought, connected to the matter at hand, can simply pop into our consciousness, as discussed above.

18.8 The Conscious and Unconscious Brain: A Division of Labour

Much of the brain's activity is unconscious and takes place in areas below the cortex. Such activity includes involuntary control of heartbeat and breathing, digestion and temperature regulation in the body. This frees up the cortex to make decisions concerning optional activities. The brain relegates actions, which do not require focus or decisions, to the unconscious. The generation of sentences is a good example: when we speak we generally focus on what we are saying, and rarely on how we are saying something, unless we need to pay attention to this, such as when we are writing an important text. So consciousness not only consists of awareness but also of a meta-mechanism which decides what is to form the focus of our awareness at any one point in time.

We also have a sense called proprioception: the feeling for one's own body. That covers large-scale matters such as the position of our limbs, our movement through our environment. But proprioception does not extend down to the level of cells, for instance, nor does it extend to organs which we do not need to be conscious of, such as our digestive system: acid in the stomach and the peristalsis of the intestines look after digestion. But if we have a pain in our abdomen, from eating putrid food, for instance, our focus is directed to parts of our body we are normally not aware of. The same applies to injuries: for instance, you are not normally aware of your ankles but if you injure one then the pain forces you to focus your attention on it.

18.9 The Quantum Brain?

The issue here is whether there is a demonstrable relationship between quantum mechanical processes and consciousness and what this might be like. At present we are nowhere near making any substantiated claims about such a connection, though there is no shortage of speculation about it. Pondering the issue of consciousness has led the American Stuart Hameroff (University of Arizona) and the British Roger Penrose (University of Oxford), to imagine that the answer should lie on the quantum level, with such phenomena as superposition, where a physical system, e.g. of electrons, can be in more than one configuration at one point in time, and a trace of this situation can be captured, as the famous double-slit experiment shows.[5] But the responsibility of the quantum realm for the rise of consciousness in our brains is not a view supported widely in science, chiefly because quantum effects are found only on subatomic levels and are terminated by the so-called collapse of the wave function once we move up to a larger scale, certainly on the atomic, let alone on the molecular scale.[6] Furthermore, the idea that the quantum states in microtubules, taken by Penrose and Hameroff to be causally linked to consciousness, is part of their controversial theory of consciousness, called 'orchestrated objective reduction'.

18.10 Memory

Imagine you are looking for a picture on the hard disk of your computer. You know it is there somewhere. Then finally you find it in a deeply nested folder and load it. It is a picture of you cutting the wedding cake with your new spouse. The picture shows everything which was captured optically (within the resolution of the camera used then). This would include things like the colour of the window curtains behind you, the number of chairs in the picture, the plates and glasses on the table and, of course, the cake. Everything is shown without any prioritising of information in the picture.

Now consider your memory of the event. What do you remember? Certainly not the number of plates and glasses on the table. Maybe the tiers of the cake and the figurine of a wedding couple on top. You might remember the colour of the icing because you deliberately chose a shade of light green which matched your outfit. And you might remember what

5 Quantum entanglement, which leads to what Einstein in a famous phrase called 'spooky effect at a distance' (German: *spukhafte Fernwirkung*), is another counterintuitive phenomenon which has been linked to consciousness, but just how such a link manifests itself has not been specified.

6 The Swedish-American physicist Max Tegmark has shown that the decoherence timescale for the disappearance of quantum effects is of the order $\sim 10^{-13}$–10^{-20} seconds, whereas both neuron firing and polarisation excitations in microtubules are of the range ~0.001–0.1 seconds, which is far above that of the quantum realm.

the cake tasted like or giving two slices to your father who found it particularly delicious but that your mother did not eat any then because she was full from the turkey of the main course.

So how does our memory work? For one thing, it is not photographic. It does not provide us with optically faithful recollections of events. Rather, memory is constructed in real time from disparate items of information and presented to your consciousness. Consider the remarks on the slices of cake in the previous paragraph: that is not optical information but additional information to the scene you are recollecting, things which 'stuck in our memory' as the saying goes.

Now consider the linguistic side of this complex: when committing something which has been said to memory, we don't remember the form (the syntax of sentences) but rather the meaning and this is transferred to long-term memory. This is why we have such difficulty in recalling some speech event verbatim, for instance in court.

But there is yet more to memory. How accurate is it? Not very, really. I remember that for years I had a very clear picture in my mind of the entrance to a university in South Africa which I had visited once. I revisited the institution many years later only to find that my recollection of the entrance – the view of the building from the front – was actually quite wrong. The lesson for me from this is that our memory is not a faithful record of the past. In fact, if one takes two individuals who both experienced the same event (significant for both of them), one often finds that their recollections differ. And it goes further: test persons in scientific studies have been told by the investigator about events which they were supposed to have experienced and, while initially denying that the events took place, later on, after they were offered supposed details, came themselves to believe that the events did take place, even providing apparent details of the fictitious events.

So why are our memories apparently so unreliable? To answer this one should consider different types of memory and try to identify their functions for the brain. Short-term memory, like remembering a telephone number for long enough to dial it, is relatively straightforward. The instances discussed above all refer to long-term memory, which involves information which had been assessed, sorted, encoded and stored for later retrieval. It is known that the hippocampus plays a central role in assessing and sorting information. It 'decides' what information should be passed on to higher cortical regions for storage (though exactly where and how memories are stored is unknown). We also know that regions of neurons involved in memory show dendritic connections, which can be strengthened or weakened, and that this active process affects the quality of memory. The formation of memories is also dependent on age, with a decline in speed with which they are formed and their quality in advanced age, especially if the brain is not trained to form new memories, such as being repeatedly exposed to cognitive challenges like learning a new language.

The instance discussed at the beginning of this section (remembering your wedding cake) is an example of episodic memory; this concerns the where, what and when of an event. The brain is selective: it chooses to remember the essential parts of an episode, which may be spread across several senses. Memories are constructed by the brain in a convergence zone (according to Antonio Damasio), gathering together elements of the memory (visual and auditory, perhaps olfactory) and combining these on the fly to form a conscious memory. This would explain why memories are often faulty as they are assembled from fragments of a situation which you are momentarily remembering.

Collectively, episodic memory constructs our autobiography, it gives a continuous timeline to our lives. It is instrumental in the maintenance of personality and a sense of self. Your past makes you what you are, a unique individual. If this past is no longer accessible because of a loss of memory, it entails a loss of identity and a change in personality for the individual in question.

Our brains are continually trying to make sense of the world and constructing a life narrative for us. The American cognitive neuroscientist Michael S. Gazzaniga has investigated the human ability to construct a past through the store of memories we continually maintain and reflect on constantly during the present. It is this collection of memories which provides the scaffolding for our identity and our sense of self, imperfect and unreliable as these memories are in a literal sense.

But memory has a proactive function as well. It equips us to deal with situations in the future similar to those which formed the memories. This feature is encapsulated in the common saying, 'We learn by our mistakes.'

A memory may convert from conscious short-term memory to long-term unconscious memory by being associated with the strong emotion of a life experience, either negative, as with an accident or bereavement, or positive, as with a happy professional or private event. If the memory has a neutral context, practically the only way of successfully converting it to long-term memory is by repetition while concentrating on the content of the matter.

On the physical level, the formation of long-term memories requires the activation of genes to grow synaptic connections. This leads to long-term potentiation, encapsulated in the somewhat oversimplified phrase 'neurons that fire together wire together' (sometimes referred to as Hebb's law, named after the Canadian neuropsychologist Donald Hebb, 1904–1985). A consequence of the fact that long-term memory causes anatomical changes is that no two brains can ever be the same, not even those of identical twins, because all individuals have different life experiences.

Short-term explicit memories are associated with the medial temporal lobe, specifically the entorhinal cortex, and the hippocampus. However, when a selection of such memories become implicit, procedural memories they are associated with deeper parts of the brain, with the caudate nucleus, cerebellum and reflex pathways in the nervous system. These unconscious procedural memories are typically associated with skills and habits, such as riding a bicycle or tying your shoelaces.

18.11 Neuroplasticity

Generally, you do not grow new neurons during your life. In fact, only the olfactory system (responsible for smell) and the hippocampus can partially generate new neurons. But brains show plasticity by changing the synaptic connections between neurons. Such plasticity can in principle be of three types, (i) additive where new connections are established, (ii) augmentative where connections are strengthened or (iii) subtractive where connections are weakened or removed. For instance, your memory can be trained to improve performance,[7] and this can be done in adulthood with cognitive challenges, leading to physical enhancements. The Irish neuroscientist Eleanor Maguire, from University College London, showed in a long-term project that trainee London taxi drivers go through an enlargement of parts of the hippocampus during their period of training, given the very considerable demands on spatial memory made when memorising the myriad streets of London (in the days before satellite navigation).

Subtractive plasticity may sound negative but it is in fact beneficial in specific contexts. During the first few months of life babies go through a phase of maximum sensitivity to distinctions in sound, pruning back neuronal connections which are not necessary for the language(s) they are acquiring natively. For instance, sensitivity to tonal distinctions is vital to young Chinese or Vietnamese children but not to those acquiring nearly any one of the languages found in Europe. Equally, the ability to distinguish *r* and *l* sounds is mostly lost in the first year with Japanese children while at the other extreme Russian children attain the ability to distinguish two types of *r* and two types of *l* (technically called palatal and non-palatal consonants, respectively). Those children acquiring English, German, Swedish, etc. (all Germanic languages) maintain the ability to distinguish long and short vowels (as in English *bit* versus *beat* or *full* versus *fool*)

18.12 Consciousness: An Attempted Summary

Three main suggestions for a principled explanation of consciousness have been made, both by philosophers on the one hand and neuroscientists on the other.

7 This can involve a number different types, for instance motoric memory. The brain of Albert Einstein was extracted after his death and examined to see if there was any physical correlate to his extraordinary cognitive powers. A particular part at the back of the right-hand half of the brain was larger than expected. This turned out to be a region associated with finger control for the left hand. The reason for this is that Einstein learned to play the violin as a child.

Suggestions for a principled explanation of consciousness

1. *Reductionism.* This assumes that if one could account for all the physical phenomena of the brain then these in some way would explain consciousness (Daniel Dennett). David Chalmers thinks that this is too close to denying that consciousness has a separate ontological status from the physical substrate of the brain, even if the latter is responsible for its arising in the first place.
2. *Fundamentalism.* This maintains that consciousness is fundamental, it is a primitive given in our universe which does not, and cannot, be explained by anything more basic. Everything has consciousness: there is always a non-zero value for this property. It would then be like magnetism, which always has a value, even in, say, a piece of rock, but obviously a much higher value in an industrial magnet, for instance.
3. *Panpsychism.* This claims that consciousness is universal, it is everywhere. It is not so much a property of matter (as in the previous view) as a non-physical level of our universe to which we connect via our brains. The extent of consciousness depends on the amount of information integration in a system. The human brain is just such a system: an organisational structure which is clearly delimited from its surroundings and consists of an internal network of connections.

There are issues with these views. The first (reductionism) is common among neuroscientists who work on the basis of correlations between the phenomenology of consciousness and physical activity of the brain. These correlations show an ever-increasing degree of sophistication but do not account for why there is consciousness, indeed many neuroscientists would say that is not the goal of their science. The second and third views (fundamentalism and panpsychism, respectively) both suffer from the same weakness: they offer no promise of establishing a causal connection between the physical substrate of the brain and the subjective experience of consciousness. But that must exist. After all, we know from traumatology that damage to specific parts of the brain will lead to certain impairments in our consciousness, so definite connections between the substance of the brain and the mind, which the brain generates, do exist (it is also known that narcotic drugs trigger predictable effects in the brain).

Lastly, one can remark on how the brain can keep generating consciousness despite severe impairment. Take neurodegenerative diseases like Alzheimer's disease, for instance. It arises due the formation of plaques and neurofibrillary tangles (of tau protein), which leads to the atrophy of considerable amounts of brain tissue. The higher cognitive abilities of individuals suffering from this disease are greatly impaired, so their minds are negatively impacted, especially their ability to remember things and to concentrate on even simple tasks. However, their consciousness, as evidenced by their interaction with their environment, seems nonetheless to function as does sensory input (sight, hearing) and the coordination of muscular movements, to a large extent in all but advanced stages of the disease.

18.13 A Final Remark

There is a delicate irony in the fact that the ultimate frontier of science – accounting for consciousness – involves the very organ which enables science in the first place, namely the human brain.

CONSCIOUSNESS WITH EXOBEINGS

If by 'exobeings' we mean forms of intelligent life capable in principle of communicating with us, such beings will be conscious, at least to the extent that we are. Just how this consciousness would arise and manifest itself for exobeings is one of the most intriguing issues scientists could address on discovering intelligent life on another planet. Given the assumption that exobeings could only arise through Darwinian evolution from original single-celled organisms, any consciousness they would show would be based on a physical substrate in their bodies, that is, by their functional equivalent to the human brain. Any other assumption would incur a very challenging burden of proof: it would have to demonstrate how consciousness could otherwise arise.

19

Artificial Intelligence

The term 'artificial intelligence' or just 'AI' is a buzz word tossed around at liberty in many publications and on the internet today. It is often used to refer to technologies for very specific tasks where human labour would be expensive, or subject to error due to endless repetition. Such technology has considerable applications in many fields of present-day engineering, in digitally based manufacturing and in important scientific domains such as medical research, diagnosis and treatment. Where the technology is used to replace human operators, as on assembly lines, it is more accurately known as robotics. The basis for such technology lies in high-performance computers,[1] which have been programmed to perform precise complex tasks. The programming behind such computers

1 Computer power has increased exponentially since the 1950s, when room-filling devices were used to perform fairly simple mathematical operations. Two milestones in hardware development can be mentioned: (i) the rise of integrated circuits, which could then accommodate hundreds, then thousands, later millions of transistors (semiconductor devices) in a single-wafer chip and (ii) the rise of photolithography, which uses light to transfer a circuit pattern onto a silicon substrate coated with a photosensitive emulsion (repeatedly over several different layers). This results in extreme miniaturisation of circuitry (down to scales of five nanometres in 2020), which permits the accommodation of billions of transistors and very rapid processing of operations. This technology is also availed of to produce micro-electrical mechanical systems (known as MEMS), which promise to have powerful applications in areas like medicine.

is generally declarative, that is, the computers are given precise instructions about what they are to do. Computers are also used for procedural programming, where the programs they run take questions they are given and attempt to provide answers. Navigation software is a good example of this: it is presented with a destination and works out the best way of getting there (in terms of distance, speed, etc. depending on input parameters).

In recent years, different approaches have been developed to analyse large amounts of data, providing classifications, recognising patterns and making predictions, for example in forecasts of the weather or of economic developments. Powerful computers are also used for simulations of galaxy genesis and future developments, and for modelling systems of stars and planets.

A stricter use of the term 'artificial intelligence', labelled 'artificial general intelligence', refers to machine intelligence which could successfully emulate human intelligence. This includes various types of reasoning, the integration of information to form knowledge, the construction of memory to guide future actions or the ability to make independent decisions not provided for deterministically in the program code running on a computer.

There is much discussion, not to say hype, surrounding the field of artificial intelligence and predictions are regularly made about when computers will overtake humans in their cognitive power, become 'superintelligent' (Bostrom 2014). This digital superintelligence is sometimes referred to as a species-level risk, as opposed to goal-restricted artificial intelligence, for example in weapon systems,[2] which requires humans to determine its overall behaviour.

Would the development of digital superintelligence be a matter of principle or just of degree? If the latter, all we need are more powerful

2 It is a sad and sobering fact that every technological development in human history has been applied to the military sphere very quickly after its appearance. From the invention of the wheel to the splitting of the atom, the story has always been the same: the potential of the new technology for wreaking death and destruction on others was exploited almost immediately.

computers.[3] If the former, then some fundamental advance will be needed to achieve it.

The worst predictions say that an artificial intelligence explosion will lead to an uncontrolled reaction increasing exponentially and impacting us negatively, ending humanity-centred life as we currently know it, and ushering in a 'post-human' era. The time at which this situation will arise, especially when conceived of as beyond our control, is dubbed the 'singularity' (an unfortunate use of a well-established term in astronomy).

In this context, a remark on just what is meant by intelligence is called for. There are many different types. The ability to perform complex calculations and operations is only one type, there is also emotional intelligence, which is closely linked to wisdom, the ability to make decisions which are maximally beneficial to the largest number of parties involved in a particular situation. Social intelligence is the ability to interact optimally with members of one's community. How one might judge decisions made by 'superintelligent' computers along these lines is difficult to say.

19.1 The Singularity: A Modern Frankenstein?

A lot of reporting on artificial intelligence has been sensationalist, painting dire scenarios of machines taking over the world. The prophets of doom are forecasting that artificial intelligence will come to dominate us and be beyond our control. Such depictions enamour sensationalists with the popular media and ensure that they get prime coverage for their not-well-thought-through ideas about the future of artificial intelligence. Talk of an artificial intelligence which will surpass human intelligence is vague and fuzzy. Just how would one define 'human intelligence' to then judge whether an artificial intelligence surpasses it? In reliability (no distraction, no lack of attention or no boredom) and in the number

3 A well-known early critic of unbridled artificial intelligence, as a means to machine consciousness, the American philosopher Hubert Dreyfus (1929–2017) pointed out repeatedly that an increase in processing power will not lead to computers becoming conscious.

of possible calculations, computers are already way beyond humans. Computers are specialists – they perform calculations blindingly fast – but humans are 'generalists' – we can simultaneously process multiple input channels of sensory information, filter this and process it in a sensible way, all in real time. So what we see, hear, smell, taste, touch, feel provides us with information which, in terms of our evolution, was important for our survival – performing huge super-quick calculations never was, nor is it ever likely to be.

The term singularity, in the technological sense,[4] refers to a putative point in time when computing power becomes so great that it would irreversibly go beyond human control. How could such a situation arise? Probably by supercomputers engaging in powerful deep learning via neuronal networks with back propagation, which would increase their power exponentially and put them beyond the control of the humans who constructed these devices in the first place. Opinions about just when the singularity is deemed to occur vary greatly. There are optimistic predictors, like the American scientist Ray Kurzweil,[5] who has put it at 2045. Some predictions have been sooner, and some later, but the second half of the present century is a commonly mentioned time frame. Those who see it as happening sooner rather than later point to the exponential growth of digital devices throughout the whole twentieth century and into the twenty-first. It is true that technical devices can reach physical limitations; vacuum valves could not be physically reduced to beyond a certain size and even integrated circuit transistors have limits due to material properties and heat considerations at nanometre scales. Again, proponents of the singularity hypothesis point out that there have been paradigm shifts throughout the past century which overcame technical limitations, as with the major shift from valves to transistors in the 1960s.

There is a linguistic aspect to this issue. The use of the word 'singularity' seems to imply that we are talking about something which will soon exist

4 The term 'singularity' was introduced to a general audience in 1993 by the American science writer Vernor Vinge (1944–) in a study entitled *The Coming Technological Singularity*. For a book-length treatment of this topic, see Kurzweil (2005).

5 He has actually founded, together with Peter Diamandis, an institution called Singularity University in Santa Clara, which has been supported by NASA and Google.

and that there will be a certain event beyond which our lives will have changed irreversibly. Such projections are beloved of the Cassandras[6] of our modern world (see Barrat (2013) and Harari (2018) as examples). However, the singularity, if it ever occurs, will certainly not be a single event triggering a runaway situation like a nuclear core reaction which cannot be stopped.

Many of the scenarios envisaged by futurists concerning uncontrollable computers, on whatever physical substrate, are close to science fiction and many predictions, such as that for three-dimensional molecular circuitry in computers the size of a button, are nowhere near realisation now (mid-2022). The statements are suitably vague and often in fact betray a lack of understanding of the actual technology which is supposed to spawn superintelligence and trigger the singularity. Just consider, for a moment, what types of supercomputers are going to start controlling us. Surely not racks of electricity-hungry computer elements in long rows in specially constructed, air-conditioned rooms, as is the case with present-day super-computers. Any supercomputer which is to supersede humans on a broad front, should be able to engage in locomotion; that is, they would have to be robots which could move around freely, make independent decisions and perhaps, in turn, prevent us from undertaking measures to curtail their actions, a sort of latter-day Frankenstein.[7] At present, this type of scenario is pure science fiction.

Nonetheless, it is important to think about possible future developments now and not wait until these become more difficult to contain and control.

6 Cassandra was a figure in Greek mythology, a priestess who was given the gift of seeing the future by the god Apollo, who took a fancy to her. She accepted the gift, but then rejected his approaches for which Apollo added the essential element to her ability to predict the future, namely that no one would believe her predictions. Today, the label is used for a prophet of doom.

7 *Frankenstein* is the title of a novel published in 1818 by the English author Mary Shelley (1797–1851), which tells the story of a scientist, Viktor Frankenstein, who creates a hideous creature in a laboratory, known simply as 'the Fiend' or 'the Creature', which gets out of control and actually kills his creator's new wife. Frankenstein dies in the attempt to find and kill the Creature (in the Arctic), which itself dies after repentance for its misdeeds. The word 'Frankenstein' has come to be used for an uncontrollable creation by humans.

This is basically the aim of 'human-aligned artificial intelligence', which seeks to ensure that present and future technological developments serve to improve situations for humanity and not spiral out of control.[8]

There are basically two types of danger which could result from ever more powerful artificial intelligence. The first is that a computer will be constructed which will then develop an intelligence greater than ours and begin to behave on its own and contrary to our wishes. As just discussed, that is extremely unlikely. The second danger is much more real. This is where powerful artificial intelligence facilities, especially those designed for military purposes,[9] could be manipulated by evil-intending individuals (dictators, terrorists) and used to wreak havoc on our planet. It will be essential to ensure that the control of artificial intelligence systems is in the hands of responsible individuals.

19.2 A Conscious Computer?

The enormous increase in processing power and information management promised, for instance, by proposed quantum computers allows us now to legitimately ask ourselves whether a computer could ever be conscious.[10] In a way, that depends on how one defines 'conscious' (see the discussions in the relevant sections above). Does this just mean making decisions and negotiating one's environment successfully? But surely consciousness encompasses a degree of self-awareness and reflection. And how would we know that a computer possesses these faculties? The computer would have to tell us, it would have to speak. It would have to generate language independently and not just words or phrases which some human had

8 This approach is taken by many academic institutions concerned with these issues, such as the Future of Humanity Institute at the University of Oxford.

9 The military are always interested in scientific developments. This was the case with Galileo's telescope in the early seventeenth century. Some scientists devised military applications of their ideas, like Leonardo da Vinci, who did a sketch for a tank-like vehicle.

10 Some scientists, like the American neuroscientist Michael Graziano, believe that in the mid-term future, 100 to 150 years from now, it will be possible to have consciousness on a non-biological substrate, e.g. a silicon-based computer.

put at its disposal. That is, the computer would have to communicate its sensation of consciousness to us. Recall that all attempts to determine if an animal is conscious, and to what degree, are indirect: the size of their brains, the manner in which they behave and react to one another, recognise themselves as members of their species, etc. (it would be much better if you could just ask your dog how conscious it is and it would simply tell you). Having an inner world of thought which one can express via language would be an essential quality we would expect of any organism or technical object before conceding that it has consciousness comparable to ours. Computers do not have an inner world of thought as we humans do nor do they have emotions (see the discussion in the next section).

Not only that, a computer would need a whole series of sensory inputs to be remotely equivalent to a human. It would at least need a few of the senses which we humans have: sight, hearing, touch, smell, taste and proprioception. And it would have to overcome the hardware–software issue and be a truly integrated system. Furthermore, computers are static; they do not have personalities; they would not grow up and mature. No computer is ever going to achieve wisdom like many humans do in their later years. How old can a computer get? 70 years is a good average for humans nowadays. When computers get old, in 5–10 years, they just become obsolete and begin to break down and are ultimately abandoned.

Nonetheless, the obvious differences between computers and humans have not stopped scientists from considering the question. A well-known early example of this is the Turing test from 1950, named after the English mathematician Alan Turing (1912–1954). He imagined a situation in which a human evaluator would assess a natural language conversation between a human and a computer. If the evaluator could not tell the text produced by both the human and the computer apart, the latter would be deemed to have passed the test and showed intelligence equivalent to that of the human.

A counterargument to the Turing test was produced by the philosopher John Searle in what has come to be known as the Chinese Room argument. It runs as follows: the philosopher is alone in a room and uses a computer program to respond to messages in Chinese slipped under the door to him. The philosopher does not understand Chinese but by using the program he can generate approximate strings of characters, which he prints out and slips back out under the door. There are people outside who assume that the person in the room can understand and produce Chinese. The moral of this story is that the Turing test could achieve results which might lead one to think the computer can understand human language; but all it does is manipulate predetermined strings, it does not use the language creatively as humans would – the computer does not know the meaning of language, it has no semantics. Computers, according to Searle, mimic human minds, but do not have any understanding of them.

There remains the fundamental question: could a silicon-based computer achieve in principle a type of consciousness which is comparable to that of the carbon-based biology of our brains? If a computer, or rather a battery of linked supercomputers, could match the neurons and connections in our brains with transistors and their connections, would that then trigger consciousness in the machine? My hunch is that it wouldn't, but only time will tell. Powerful though computers may be, they are just machines. They have always been good at doing calculations and in recent decades they have got immeasurably better at this. Computers manipulate binary bits in their central processing units; that was the case in 1980 when the first personal computers became widely available and it is still the case now, over 40 years later. True, with vastly increased processing power, computers can be taught to do novel things, and to develop a degree of independence, assuming their software and hardware allows them to do this.

There are many additional issues surrounding the question of consciousness and computers, one of which concerns how a computer would distinguish between conscious and unconscious actions. If a computer were conscious, how would routine tasks become unconscious for a computer, comparable, say, to learning the piano or to touch-type on a keyboard? In fact, given that every action would be a routine for a computer – it would not probably have to go through a learning curve – all actions should be unconscious. And how would a supposedly unconscious computer acquire its consciousness? Would a computer have a critical period for sensitivity towards learning certain things, like language? Could a 6-month-old computer learn better and more quickly than a 10-year-old computer? And would the computer lose its consciousness if someone turned it off? If so, would that then be equivalent to killing the machine? We would then need a society for the prevention of cruelty to computers and computer rights groups would likely arise. These considerations show how absurd the whole issue of computers and consciousness is.

19.3 *Sentio ergo sum* Again

The most famous quote from the seventeenth-century French philosopher René Descartes is *cogito ergo sum* 'I think therefore I am'. But surely that should be complemented by the dictum ***sentio** ergo sum* 'I **feel** therefore I am'. Feeling accompanies our every living moment. We can be sad, happy, bored, fascinated, afraid or at our ease. We can be uninterested or we can be curious. This last quality of human beings is, in a way, the most important. Without curiosity nothing would happen, certainly science and the progress it has engendered would simply not have occurred. All the scientific achievements of humanity were reached because individuals wanted to find out how something worked or to gain a solution to a problem. If humans did not feel the desire to do things, nothing would ever occur; we would not even exist.

Do computers feel? Can computers suffer pain or enjoy happiness? No, that is plainly ridiculous. So they do not experience any motivation to do things. A computer can only do something if it is made to do so by passing it a series of precise instructions. It can never wish to find something out, it can never desire to do something. Computers do not experience curiosity. This is actually good news for those afraid of future artificial intelligence. It would never have a desire to dominate (on its own), this could only arise indirectly, as a side effect of some other aspect of the artificial intelligence, for example, by being programmed to duplicate subsystems exponentially through unlimited cycles of recursion. But that level would be, at least initially, controllable by humans so that the dangers of artificial intelligence are dangers which emanate from humans. Yes, there might be some uncontrollable proliferation of computer systems, but that is an issue for humans to solve, the computers would not take decisions on their own, let alone wish to harm us (though this might happen inadvertently). This issue is sometimes discussed under the heading of the 'value alignment problem', that of ensuring that any artificial general intelligence does not act in a way detrimental to us humans. This discussion seems to imply agency to computers, which is clearly false; it is the makers of the computers who act and who may need to be supervised and controlled to a greater extent in future.

19.4 The Mental Lives of Exobeings

Given that exobeings would most likely not be in sync with us in terms of science and technology, they may well, depending on their head start on us, have developed to stages which we can now only barely conceive of. Could there be machine civilisations which have managed the complete and successful transition from biological to non-biological life? This is complete speculation and one should not get carried away considering options in this sphere (see the previous section). We do not know what role artificial intelligence would play in an exosociety. Would exobeings already exhibit transhumanism or would an exosociety, assuming it was millennia more advanced technologically than we are now, be entirely digital, having done away with unreliable, vulnerable organic material and moved itself to a putatively higher plane of existence? When considering this matter, it is salutary to bear in mind that no technological achievement on Earth has ever come close to the agility and flexibility in terms of locomotion, self-awareness and processing of sensory input as forms of animal life, so it is currently difficult to conceive of robots attaining a similar level of performance across these parameters.

For the moment, I would limit the discussion to that of exobeings which have a biology that would be in principle similar to ours on Earth. Assuming such exobeings would have structures in their functionality similar to human brains, would their 'brains' also be able to produce advanced consciousness? Would their brains also have tens of billions of neurons, trillions of connections, an electrochemical system of signal conduction between neurons and a division into task-specific areas, as well as a grounding in a body continually experiencing physical sensations? Would consciousness be scalar for the animals on an exoplanet, with low consciousness with simple forms of life and greater consciousness, comparable to ours, for the exobeings themselves?

The tantalising question concerning exobeings is what their consciousness would be like. If such beings, capable of advanced science and technology exist, they will have consciousness, in principle comparable to us humans. If beings on other planets only live very largely in the here and now, and only react to external stimuli, they would fall broadly into the category of non-human animals on our planet.

Nonetheless, given that any exobeings could well be much further into their digital age than we are, who have just started, the question of how they might use artificial intelligence is central to current thinking about life beyond Earth. Enhancements of biological life through artificial intelligence could improve survival chances on an exoplanet and thus prolong their presence in our cosmic neighbourhood, making detection that bit more probable.

Part V

Language, Our Greatest Gift

For humans, language is as natural a faculty as the ability to walk or use our hands. On closer inspection, all the languages across the globe are seen to have clear structures with recognisable building blocks and rules for combining them into meaningful utterances. Furthermore, there is widespread, albeit tacit agreement within communities about how to use language and what role it plays in maintaining social ties and forging group identity.

20

Looking at Language

> [Language] differs, however, widely from all ordinary arts, for man has an instinctive tendency to speak, as we see in the babble of our young children; while no child has an instinctive tendency to brew, bake or write.
>
> Charles Darwin, *The Descent of Man* (1871)

> Language is so tightly woven into human experience that it is scarcely possible to imagine life without it. Chances are that if you find two or more people together anywhere on earth, they will soon be exchanging words.
>
> Steven Pinker, *The Language Instinct* (1994)

In order to speculate sensibly about what an exolanguage might be like, it is necessary to take a closer look at language on Earth in a number of separate but related steps.

Five steps to view and analyse language

1. How language is structured and how it can be analysed by linguists
2. How language is related to our brains and cognitive abilities
3. How we produce language given our anatomy
4. How we acquire language in our childhood
5. How language probably evolved on Earth

Throughout this part of the book, references will be made to how some features of human language might be manifested in a possible exolanguage. In this way it is hoped that the speculation about exolanguages will be seen as grounded in established knowledge about language(s) on Earth. The question is not only how did language evolve on Earth but also how humans arose with the language faculty in the first place, so as to then possibly extrapolate from our evolution on Earth to that which might be found on exoplanets. In this context it will be worthwhile considering communication systems among non-human animals to relativise statements about human language. The goal is to establish structural and functional principles which hold for terrestrial communication systems and then consider how these might be realised on an exoplanet with intelligent beings.

20.1 What Is Language?

To begin with, one should distinguish carefully between languages, found throughout the world, and the human language faculty.[1] A singular use without an article – 'language' – refers to the unique faculty humans possess, as part of their biological endowment, this enabling us to acquire a specific language, to understand and to use it (speak and/or sign it). Existing languages share a core of structure common to all, as dictated by the language faculty of humans (see Chapter 23 below on universal grammar).

1 If you are a linguist, or are au fait with linguistics, you will probably not need to read the following sections.

Particular languages – English, Russian, Swahili, Hindi, Japanese, etc. – are products of the language faculty, when it has been active in the members of a social community over many centuries. The English language (in its many varieties), as we hear others speak it and as we see it written, is an external manifestation of speakers' internalised language, which all native speakers possess in their brains and which they use for organising and manipulating their thoughts. In the technical parlance of the linguist Noam Chomsky, the 'language of thought' is an *i*-language ('i' stands for 'internal') and the language manifest in speech and/or signs is an *e*-language ('e' stands for 'external').

There is an order to the sequence of language faculty and languages. The language faculty is what allows actual languages to arise and develop. Languages are the products of the language faculty operating. Without this faculty there would be no languages. So, when considering humans from an evolutionary point of view, the first question concerning language is: 'How did the faculty for language arise?' Scholars are divided on this and there are two main camps, those who adhere to the *continuity view*, put simply: the language faculty arose slowly from a simple beginning and gradually grew more complex; and those who hold the *discontinuity view*, again put simply: the ability to speak rests on a sudden mutation in one single individual, which led to that person acquiring the ability to have language as we know it, with an internal hierarchical and non-linear structure (see Section 32.7 below for further discussion). The latter is a minority view, but one which is held by the most influential linguist of our time, Noam Chomsky, which in itself means that this view is considered seriously in all works on the origin of language.

20.2 The Purpose of Language

This might seem to be a straightforward matter: language is used to convey information or to express emotions; language is a system of communication used by humans. After all, we have the urge to communicate with others, expressed neatly in the German word *Mitteilungsbedürfnis* 'wish or need to communicate' (a term commonly used by the American linguist W. Tecumseh Fitch). This is true, but by no means the whole story. Communication is an external function of language, manifest in exchanges between humans. But language is also a vehicle for thought, an ability which we carry in our brains to organise and direct our cognition. In this respect the organisation of language is probably not primarily linear with a temporal ordering of elements. (I will return to this issue later, see Section 32.7.)

But there is a strong social component to language as well. It is used to maintain social relationships and to identify with a certain section of society. This means that all human languages have two sides: an internal structure, concerned with the organisation of linguistic information necessary for thought; and an external aspect for communication, where the manner in which language is expressed carries social significance. When one considers the first aspect, the internal organisation of language, one can see that in the course of human evolution our ability to speak would appear to have become self-contained and independent of our environment. Not only that, the levels within language, those of sounds, words and sentences, would also seem to have become largely independent but with connections linking them. This modularisation is a distinct advantage to the organisation and maintenance of language and is the reason for treating the levels separately in books on language.

INTERNAL AND EXTERNAL ASPECTS OF AN EXOLANGUAGE

Given that exobeings would arise through Darwinian evolution, there are grounds for assuming that their language(s) would have arisen through intraspecies interaction and then become internalised and passed on to following generations. In addition, there would probably be, at any time, external aspects to language(s), determined by the social significance which linguistic forms have in the communities using these languages.

20.3 Definitions of Language

Normally there is tacit agreement among linguists as to what constitutes language. However, when pressed on the matter, they find it difficult to come up with a single definition which satisfies everyone. By their very nature, definitions of language try to compress into a single sentence the essential elements that characterise it.

The American anthropologist and linguist Edward Sapir (1884–1939) defined language as 'a purely human and non-instinctive method of communicating ideas, emotions and desires by means of voluntarily produced symbols' (Sapir 2004 [1921]: 3). This definition highlights the fact that language is used for communication by the use of symbols, usually auditory, but also possibly visual, as in sign language. Whether language is non-instinctive, as Sapir maintains, is a contested issue.

There are many other definitions of language and that by Noam Chomsky (1928–) stands out by emphasising the centrality of grammar and its ability to produce a potentially limitless number of sentences: 'I will consider a language to be a set (finite or infinite) of sentences, each finite in length and constructed out of a finite set of elements'. The common ground shared by the very different definitions just given can be usefully summarised.

Characteristics of human language

1. Language is a system of communication.
2. It involves symbols (usually sounds) with arbitrary values.
3. It is a rule-governed system, which is open-ended.
4. It is used by humans.

20.4 Design Features of Language

Definitions of language are by their very nature single-sentence renderings of what scholars see as the essence of the subject matter. But some scholars have gone beyond that to list and discuss features which they see as true of all human languages. Such features would have arisen before the last dispersal of *Homo sapiens* from the continent of Africa about 70,000 years ago, as they are found to hold everywhere, irrespective of the observable differences between the world's languages in pronunciation, grammar and vocabulary.

A well-known classification of the design features of language was offered by the American linguist Charles Hockett (1916–2000), see Hockett (1960), and represents an effort to comprehensively list the essential features of human language. The following list is based on the original one by Hockett but has been expanded to include further relevant information.

General Features

Relationship of words to concepts/objects is arbitrary: 'Arbitrary' in linguistics denotes a relationship between linguistic signs (words) and what they stand for (concepts, which typically refer to objects in the outside world). This relationship is not fixed or determined by the nature of the objects. In any given language, the relationship between word and concept/object is set by convention, for instance there is no reason why a female bovine animal should be referred to as *cow* [kau] in English, *bó* [boː] in Irish, *vache* [vaʃ] in French or *korova* [kʌˈrovə] in Russian. Nonetheless, to speakers of these languages, their word seems to be entirely appropriate for this animal.

Stimulus-free: As opposed to most animal communication systems, human language does not need a trigger such as danger or the search for food or the desire for procreation. In essence, we can speak without any external motivation.

Structure-dependent: Language does not consist of a string of random elements. The sounds and words of language are arranged in a certain meaningful order determined by the rules of the language.

Duality of patterning: A major organisational principle of language is that it involves two levels of structure, one of units and one of elements used to build these units (Figure 20.1). Words consist of sounds which in

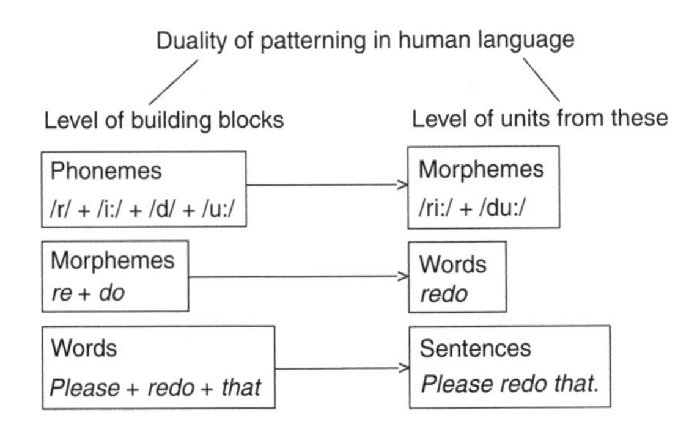

Figure 20.1 Duality of patterning

themselves have no meaning. For instance, one cannot say that the sounds /p/ *p*, /ɛ/ *e* or /n/ *n* have a meaning on their own but the combination /pɛn/ *pen* does. The same applies on a higher level: words combine to form sentences which they do not form on their own.

Discreteness: This requires that one has an exact realisation for each sound in a language. It represents the essential difference between noise and the sounds of human languages. Noise can vary at random but the sounds of a language must be kept apart clearly, that is, they are 'discrete'[2] in the technical sense. In English, one cannot use a sound which is intermediary between /p/ and /b/ as this would not be sufficiently separate from both of these. This applies equally to vowels. Again, in English one must distinguish clearly between vowels as, for example, in *bid, bed, bad, bud, booed, bard, bide, bowed.*

Productivity: The number of utterances one can make in a language is not limited: new sentences are produced by taking one of a limited set of sentence structures and filling it with words from one's vocabulary. By these means, one can produce a theoretically unlimited set of sentences (see the discussion of digital infinity in Section 32.7 below).

In word formation, one can also see this principle at work. Take the example of the ending *-wise*, which is used to make adjectives from nouns. This can

2 The word *discrete* (spelt in this way) means, in linguistics and other sciences, 'separated, detached, without any perceptible transition.' *Discreet* (with double *ee*) has a different meaning, namely 'prudent, careful, unobtrusive.'

be applied productively to virtually any noun, irrespective of whether the new word already exists or not, such as *Computerwise the department is well equipped*. What enables productivity is the application of rules to any input element to gain a new structure, be it from grammar or vocabulary.

Gaps and prohibitions: A gap is a permissible form in a given language, which happens not to be attested. Prohibition refers to the exclusion on principle of some forms from a language. Consider the form *blick*, which is a possible (but unattested) word in English and *bnick*, which is not permissible because no English word can begin with a stop and a nasal.

Physical Features

Vocal-auditory channel: This refers to those parts of the human anatomy which are used for the production and perception of speech. Note that the organs of speech are not primarily designed for language production (they all have some other more basic function, like chewing for one's teeth). Our system of hearing was not originally developed to perceive speech (but just the sounds of our environment).

Rapid fading: Spoken language dies away quickly, thus freeing the channel for the next message. This increases the quantity of signals per unit of time. Contrast this with animal communication devices such as pheromones (chemicals secreted to evoke a response in other members of a species who smell them), which are slow and liable to mixing so that they can only encode a limited amount of information. The same is true of other potential means of communication such as electrical fields (both for electroreception and electrogenesis) found with certain fish, including some eels, which basically use this mechanism to locate and possibly stun prey.

Broadcast transmission:This refers to the fact that the transmission of sound is omnidirectional, which means one does not have to face one's partner in conversation and, furthermore, information can be conveyed in darkness. This does not of course apply to sign languages.

Directional reception: This permits the location of the speaker by the hearer.

Features Involving Meaning

Semanticity: This refers to the existence of meaningful ties between elements in a language (words, phrases, sentences) and concepts/objects in the real world.

Reflexiveness: The capacity to use language to talk about language itself. This is part of the general ability of humans to reflect on themselves.

Prevarication: The possibility of deliberately telling untruths and talking deceptively; for instance, language can be used in ways which are at variance with its communicative function.

Displacement: The ability to refer to things or events remote in time or space.

Learnability: The ability to acquire another language (or more than one simultaneously). The language faculty does not per se put an upper limit on either the nature or the number of languages which can be acquired.

Interchangeability: This refers to the ability of any human to act as transmitter and receiver alternatively. This does not necessarily apply to animals; bees, for example, convey information during their dance, but they do not receive feedback from others.

Complete feedback: Speakers can perceive their own signal totally (and thus monitor it carefully) – we hear all the sounds of language which we produce.[3] This is obviously not the case where facial gestures are involved, e.g. with chimpanzees.

Iconicity: The use of symbols which bear a recognisable resemblance to the object they denote, see the discussion below.

In general, human language can be characterised as displaying *efficiency* and *distance*. Efficiency is achieved by rapid fading, discreteness, broadcast

3 That this monitoring of one's speech is important can be seen with individuals who become deaf: their speech gradually becomes indistinct because they have no acoustic feedback from their own talking.

transmission, etc. Distance tends to remove language from the physical environment in which it is produced. Here, the essential aspects are displacement, reflexiveness, prevarication, interchangeability and arbitrariness.

DESIGN FEATURES OF AN EXOLANGUAGE

For an exocivilisation to arise, one or more exolanguages would be required. Furthermore, the general features just outlined might well apply, above all productivity, which would allow an exolanguage to be used flexibly in all kinds of situations on an exoplanet. It is less certain whether the physical features would apply, as we do not know what the anatomy and physiology of exobeings might be like.

20.5 Structural Notions in Linguistics

In the history of linguistics, various scholars have made a number of distinctions helpful in better understanding the manner in which language is organised and operates. The Swiss–French linguist Ferdinand de Saussure (1857–1913), who taught in Geneva at the end of the nineteenth and the beginning of the twentieth century, is particularly important in this respect. He did not write any linguistic books but after his death his pupils put together notes taken during his lectures into a book and published it posthumously as *Cours de linguistique générale* 'A course in general linguistics' in 1916. In the *Cours*, Saussure introduces a number of dichotomies which apply to all languages. Four of the most important are briefly described in the following.

Synchrony and diachrony: Synchrony is the investigation of language at one particular point in time (which may, but does not necessarily have to be, the present). Diachrony is the investigation of language over time and is really the investigation of several synchronic 'slices', ordered one after the other (Figure 20.2). Changes over time are hence due to factors operating in each successive synchronic stage, each 'slice' of a language.

Diachrony (historical viewpoint)

———————————————————————————> time axis

| | | | | | | | | | | | | | | | | | | | | | | | | | | | | | | | | | | | | | |

Synchronic ‘slices’ (points in time)

Figure 20.2 Diachronic and synchronic views of language

Langue and parole: Saussure’s second major dichotomy is that between langue and parole. By *langue* is meant the system of the language; by *parole* is meant actually speaking a language. These two notions are close to those used by Noam Chomsky in the early 1960s, namely *competence* and *performance*. There is a difference between langue and competence, however. The former stresses the system of a language as the common core of linguistic knowledge in a community of speakers (here Saussure was under the influence of leading nineteenth-century sociologists like Émile Durkheim) while the latter refers to the abstract ability of speakers to produce and recognise correct sentences in their own native language on the basis of structures which they have abstracted during the period of language acquisition in their childhood.

Signifiant and signifié: For Saussure, a language is a sign system, which consists of two essential parts. The first is the *signifiant* (that which signifies: the sound shape of a word) and the second is the *signifié* (that which is signified: the concept in the mind). The concepts are usually linked to objects in the outside world, but that is not necessarily the case (Figure 20.3). For instance, the word *unicorn* is a *signifiant* which points to the concept, the *signifié*. But that is where it stops because there are no unicorns in reality.

From this account it is clear that Saussure sees the concept as something mental, which is independent of language and of the external world. Take

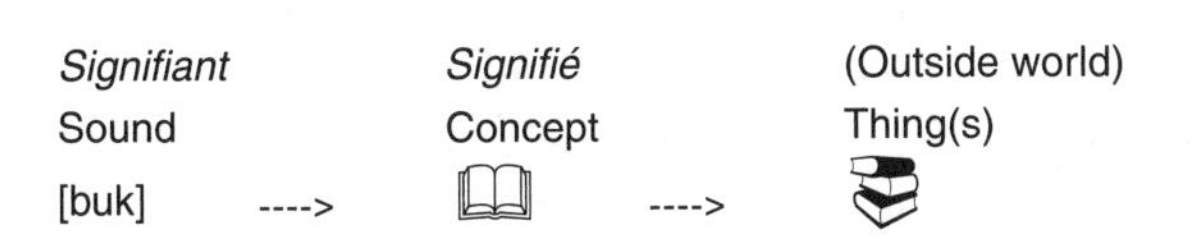

Figure 20.3 Signifiant and signifié after Saussure

as an example the concept of 'book'. For Saussure, we have a notion of 'book', which points to our mental concept of 'book', which then in a further step denotes an example of the physical object 'book'. This last step can be present, as in the sentence *The book I read last week* (a particular book in the outside world); or not, as the case may be, such as *Books are a key source of reliable information* (a generic statement, no specific object reference).

The relationship of the word to the concept is arbitrary by which is meant there is no necessary connection between the sounds used and the concept being referred to. This type of relationship is what gives human language its specific quality as a communication system. Once language no longer reflected a word's contents in its form (as with onomatopoeia, see the example of *cuckoo* in Section 21.1 below) it was free to develop new forms which needed only to be accepted (conventionalised) by the speech community. Each language consists of thousands of conventions which determine the sound shapes used for concepts (of objects or ideas). The arbitrary nature of these conventions can be seen in the variety of sound shapes used for one and the same concept, such as 'tree' which is *arbre* in French, *Baum* in German, *derevo* in Russian, *dentro* in Greek, *crann* in Irish, *zuhaitz* in Basque, etc. (see the comments in *Relationship of words to concepts* above).

Paradigm and syntagm: A syntagm is a series of linguistic units arranged horizontally (e.g. a phrase or sentence). Each syntagm consists of a number of 'slots' into which various elements can be placed. A paradigm is then the set of elements which can occupy a single slot in a syntagm (Figure 20.4).

Apart from the four Saussurean dichotomies just discussed there are others, which are part of the bedrock of linguistic analysis.

syntagm ———————————————————————— >

| Fiona	| bought	| a	| new	| car
| Fergal	| made himself	| a	| cheese	| sandwich
| That man	| taught himself	| some	| difficult	| language
paradigm	*paradigm*	*paradigm*	*paradigm*	*paradigm*

Figure 20.4 Syntagm and paradigm

Open and Closed Classes

Closed class: This refers to those elements in a language which are limited in number. For instance, the distinctive sounds of a language are limited, a figure of not much more than 40 such sounds is a typical average. The group of prepositions or the number of verb forms in a language also constitute closed classes. These are acquired in early childhood, retained fully throughout the rest of one's life and are virtually unalterable (though language change in this area can lead to slight shifts).

Closed classes tend to contain polyfunctional elements. The motto here would seem to be 'make maximum use of fewest elements'. For example, the set of grammatical endings is limited in a language, but its elements often have several functions. In English the inflectional /-s/ signifies (i) genitive singular, as in *Fiona's hairslide* (note that the apostrophe has no phonetic value) (ii) plural of nouns, as in *hundreds and thousands* and (iii) third person singular in the present tense of verbs, *Fiona likes baking*.

Open class: This is a group of elements which can change in size, by adding new elements and of course by losing others. Vocabulary is an open class as words come and go in a language.

A speaker's awareness of open classes is much higher than that of closed classes. This stands to reason: if one can add or remove items from a class, then one is aware of this. Furthermore, this means that when non-linguists reflect on change in their language they invariably mention words which have been added or have been lost. Speakers do not generally note ongoing structural changes to their languages.

Characteristics of closed classes

1. Small number of units
2. Polyfunctional
3. Acquired in early childhood
4. Low or non-existent awareness for non-linguists

Rules and Processes

What is a rule? In linguistics, the term 'rule' is an explicit statement of a process in a language which attempts to capture both its regularity and its obligatory nature. In this sense, rules are artefacts of linguistic description but, if they are accurate, they should capture some generalisation about how a particular part of a language works. So, ultimately, rules should derive from the structure of a language and hence have an existence independent of the linguist describing them.

Many different kinds of rule exist, such as rules governing the production of sentences, the formation and pronunciation of words. Given a natural human desire for order in any field of study, speakers often expect rules in language to be watertight. However, language does not work like that. There are rules of course, but most of them allow for various exceptions. The fact is that speakers can deal with a great deal of exceptions in language. And because languages develop over long periods of time and they are never overhauled and pruned back, so to speak, they tend to collect exceptions to rules as they go along.

Rules are generally invisible to native speakers: they are not conscious of them although they will abide by them in their speech. Consider the rules governing the formation of acceptable sentences. Few if any speakers are in a position to verbalise the rules of sentence structure and yet we all adhere to them in our native language. For example, verbs which take two objects frequently allow the direct object to be preceded by the indirect one, rather than it coming afterwards with the preposition *to*: *Fiona told Fergal the good news* (for *Fiona told the good news to Fergal*). One might expect that a verb like *explain*, which also involves telling, would allow the word order 'indirect + direct object', but this is not the case: **Fiona explained Fergal the computer program* (a preceding asterisk indicates that a sentence or word is not well formed in a language) is not a possible alternative to *Fiona explained the computer program to Fergal.*

21

Talking about Language

The great German poet Johann von Goethe (1749–1832) once said that because people can speak they think they are entitled to speak about language.[1] The point he was making is that because we have language we think we have the necessary knowledge to make pronouncements on language. However, this is not true. In order to make objective statements about the structure and use of language one needs training as a linguist. So why does one need technical vocabulary when talking about language? The reason is that, although we all speak our native language effortlessly, there is a lot of internal structure involved in this process and we are normally unaware of this. To describe all facets of language one needs an array of technical terms. Bear in mind that most of our knowledge of language is unconscious, like an iceberg where nine-tenths are hidden below the water's surface. We internalised this knowledge in our early childhood without being aware of it during the language acquisition process. Now if you wish to discuss all the conscious and unconscious aspects of language, especially the latter, you need at least some technical vocabulary to

1 The German original reads: 'Ein jeder, weil er spricht, glaubt, auch über die Sprache sprechen zu können' and comes from Goethe's maxims and reflections.

do so. To help readers along, I will define terms as they are required and do my best to find everyday analogies to explain them.

21.1 How Words Represent Meaning

Onomatopoeia and Sound Symbolism

The number of words in a language which more or less directly represent the object or beings they refer to – this is known as *onomatopoeia* – is very small indeed. These are in the main restricted to the sounds which animals are supposed to make such as *moo* (for cows) or *meeow* (for cats) or to sounds without specific meanings but agreement on the context in which they can be used, for example. *shhhh!* for 'be quiet', *gosh* to express surprise or *oops* when doing something accidentally. The notion of onomatopoeia runs counter to the principle of arbitrariness (see Section 20.5) which demands that there be no necessary connection between the sounds of a word and what it represents. Having said this, one can see that languages sometimes have an indirect representation of phenomena of the world around us, for instance in English an initial /fl-/ often indicates the movement of liquids or of objects through air as in *flush, flow, flux; flip, flick, fly, fling* etc. Other examples of similar meanings and phonetic form are /sl-/ in *slip, slither, slide, sling* (and perhaps *slope*); /-ʌmp/ (the inverted 'v' represents the vowel sound in these words) in *bump, thump* (dull strike), *clump, slump, stump* (low, undefined shape); word-final /ʃ/ (this phonetic symbol corresponds to the letters <sh> in writing) in *splash, swish, gush, slush, mush,* which again tends to refer to liquids or quasi-liquids, including figurative uses; initial /z-/ in words like *zip, zap, zigzag, zing, zizz,* which seem to imply uncontrolled irregular movement.[2] Further examples involve whole words, such as *boom* for a low-frequency sound,

2 The adjective *zany* might appear to fit into this category with its implication of a degree of craziness. It is actually a loan-word from the Venetian Italian pronunciation of *Gianni,* short form of *Giovanni,* and referred to stock characters in comedies.

bang for an explosion, *thud* for a heavy object falling to the ground. This phenomenon is called sound symbolism or phonaesthesia.

Both onomatopoeia and sound symbolism are probably remnants of very early stages of language before the relationship of sound and meaning became arbitrary and vocabulary expanded. There may be a very few deeply conserved words where the sound somehow suggests the meaning, consider English *cuckoo*, German *Kuckuck*, Italian *cuculo*, Russian *kukushka* /kukuʃka/, Irish *cuach* /kuax/, all of which share the initial sound /ku-/.[3] However, the number of words where this is the case is very small indeed and not of any relevance to the structure and semantics of modern languages.

Iconicity

Although onomatopoeia is not a dominant principle in language the question as to what extent linguistic structures reflect the organisation of the world is a valid one. This more indirect relationship between language and external reality is termed *iconicity*, from icon meaning symbol, which is something that stands for something else.

A possible instance of this would be the Irish word *bog* /bʌg/ 'soft', which only contains voiced or 'soft' consonants; the Irish word for 'hard' begins with a voiceless or 'hard' consonant: *crua* /krʊə/. In a similar vein one could quote words for 'large' such as English *huge*, *enormous*, German *groß* /groːs/, Irish *mór* /moːr/, Japanese 大きい ōkī, Italian *grande* /grandə/, which all contain back or low vowels, usually long (English *big* is a notable exception). Contrariwise, words for 'small' often have a high front vowel: English *little*, colloquial *teeny* /tiːni/, Scottish English *wee*, German *winzig* 'very small', Irish *bídeach* 'very small', Italian *piccolo*.

In sentence structure there are examples of iconicity. A correlation between the linear order of sentence elements and the natural sequence of events

3 The languages listed here belong to the Indo-European family. Those from other families may vary here, e.g. *cuckoo* in Finnish is *käki* and in Japanese is カッコウ *kakkō* (where the first consonant but not the vowel is shared with the other languages).

is often found, such as *After Fiona got up, she made herself some coffee,* which means, first she got up then she made the coffee; *Paddy hit Brian,* in other words, Paddy initiated an action which resulted in Brian being hit. Statistically, there are more languages with the word order subject + object than with object + subject and the former usually corresponds roughly to the sequence of actor and the person/thing affected by the action in a sentence. However, a correspondence between the occurrence of elements in a sentence and the sequence of events in the narrative is not mandatory. Frequently, speakers wish to emphasise a certain element of a narrative and put it at the front to highlight it: *Fiona made herself some coffee after she got up* (she normally drinks tea first thing in the morning); see Section 32.7 for further discussion.

21.2 Linguistic Relativity

By and large it is true to say that languages have words for the objects of the world, the thoughts and feelings which its speakers have and experience. And to a certain extent it is the case that separate words for objects tend to reflect their relative importance for speakers. For instance, English has a special word for *thumb*, the finger on the inside of the hand at an angle to the others. But the equivalent on our feet, the big toe, does not have a special word.[4] One could say that one uses one's thumb more and one sees it more often and so there is a separate word for it. But not all languages work like that. Indeed, some do not even have a separate word for 'toes'; for example Irish uses a form *méara coise* 'fingers of the foot' (as does Turkish). In this case the pitfall is to imagine that the Irish (or the Turks) pay less attention to their toes because they do not have a separate word for them.

Here is a parallel example from the realm of thoughts and feelings. English has a word for 'remember' but there is no special word to express what

4 The general way of distinguishing toes is to use an adjective for size; for instance, German *grosser/kleiner Zeh* for 'big/small toe' but French has both *doigt de pied* (finger of foot) and *orteil* (toe).

happens when one suddenly has a feeling again which one had in one's past, a type of emotional remembering. What speakers of course do is to use a phrase when a single word is not at their disposal, such as *I remember the feeling I had on the morning of my wedding day*. The danger here is imagining that the existence – or lack – of a special word for a particular matter somehow reflects its importance for the speakers of the language in question.[5]

During the first half of the twentieth century many linguists in America were concerned with describing the remaining native American languages before they died out. There are several hundred of these languages belonging to a couple of large families and they are unrelated to languages in Europe and structurally very different from these. This fact led linguists to reflect on the possible influence which the structure of language has on thought. Two linguists in particular are associated with this idea: Edward Sapir (see Section 20.3 above) and his student, the anthropologist Benjamin Lee Whorf (1897–1941). The view that language substantially influences thought is known as the *linguistic relativity hypothesis* (formerly the *Sapir–Whorf hypothesis*).

This hypothesis has a strong version, namely that language determines thought and it is this which Whorf apparently adhered to. He assumed that the structure of Western languages is quite similar and termed this Standard Average European. The structure of the native American languages which he investigated, above all Hopi (a native American language of the Uto-Aztecan family, mainly spoken in north Arizona), is radically different from European languages. The proponents of the linguistic relativity hypothesis maintain that such structures in language influences the thinking of native speakers. Critics of such an extreme view state that distinctions which are not formally encoded in a language can always be expressed via paraphrase. A closer look at the native American languages,

5 German has a special word used for animals eating, *fressen* (etymologically related to English *to fret*), which is distinguished from the word *essen* used for humans. But it would be wrong to imagine that the act of eating by animals has some special significance in German culture.

LINGUISTIC RELATIVITY AND EXOLANGUAGES

Would the languages on an exoplanet determine the way exobeings think? If the languages have arisen in a manner similar to that on Earth, there would be a connection between language and thought, but the former would not necessarily determine the latter. What would be of interest to linguists from Earth would be the elements of their lives and world which became incorporated into their languages and were reflected in their structure.

which Whorf originally considered, shows that they also have the option of paraphrase or have availed of other methods to overcome what they seemingly lacked. Hopi, for instance, has borrowed words from Spanish, e.g. *kawáyo* 'horse' from *caballo*; *karéeta* 'waggon' from *carreta*.

There is probably some validity to a weaker form of the linguistic relativity hypothesis. The structure of a language compels us to focus on certain aspects of what languages are used for and it can also highlight certain features of extra-linguistic reality. Languages compartmentalise our experience of the world in different ways and segment the continua of our lives. They play an essential role in organising the world we experience into categories which we can handle with cognitive ease.

21.3 Language as a Reflection of Speakers' World

Exobeings would live in the same three-dimensional universe as we do. They would be subject to the same arrow of time which we perceive on our Earth (you cannot be 'unborn,' a broken glass cannot be 'unbroken'). They would furthermore have a conception of actions in time. How could they not have such notions, such as to run, turn, jump, slip, etc.? They would have notions of objects, their sun, their planets, possibly a moon or moons. They would have a notion of subject and object, that to eat something or to be eaten by it are two very different things. Some words (our nouns) could be animate, like themselves or other animals, and some inanimate, like a rock or a stick. They would perceive differences in size, such as big rocks and small rocks; angry animals and peaceful animals. And they would recognise different spatial relationships in their world, the animal is under the rock, in the tree, beside the river, on the strand, etc. Furthermore, they would recognise differences in the manner in which actions are performed, to run quickly (escaping a predator), to walk slowly (creeping up on prey). And exobeings would make decisions by weighing up the factors in a given situation, for instance when considering 'fight or flight' on being confronted by aggressive animals.

All these cognitive distinctions would result from the environment in which they live just as they have in our environment on Earth. It is reasonable to assume that an exolanguage, with a similar level of differentiation to ours, would have means to express these distinctions. So how would this work? Consider how it is done in human languages (Table 21.1).

Table 21.1 Cognitive categories and their possible expression in language

Cognition	Language
Objects	Nouns
Qualifiers	Adjectives
Actions	Verbs
Qualifiers	Adverbs
Spatial relations	Prepositions (or similar elements)
(In)definiteness	Articles (not present in all languages)

With this very simple arrangement we already have five major word classes. Another extension would be to use spatial elements for temporal relations; for example, consider the following two sentences, where the first meaning of *near* is spatial and the second temporal: (i) *The house is near the sea*; (ii) *The election was near Easter*. Another extension is where literal meanings become figurative – a strong tendency among languages – consider (i) *The forest is beyond the river* and (ii) *This is beyond our capacities at the moment.*

WOULD AN EXOLANGUAGE REFLECT LIFE ON AN EXOPLANET?

The simple answer is 'Yes.' Bearing in mind the close relationship between the structure of language and our cognition and the manner in which we use the latter to encode our knowledge of the world, it is fair to assume that exobeings would act similarly. Of course, there would be areas in which an exoplanet would differ considerably from our Earth, for instance in the occurrence of light and colour, the extent of heat and gravity as well as the movement of the planet. It is legitimate to expect that these aspects of an exoplanet would be encoded in the language or languages which occur there.

Apart from the categories listed in Table 21.1, languages generally have conjunctions which serve to link parts of sentences which often stand in a certain kind of relationship to each other, such as: *Fiona likes astronomy and physics* (*and* links two elements), *Fiona is studying astronomy although she prefers physics* (*although* is a concessive conjunction). Furthermore, there are languages where relations, such as those involving space or time, are expressed via endings on words; this depends on the 'type' of a language, a term referring to the overall structural organisation of a language. In Finnish, for instance, the relations shown by prepositions in English are normally expressed by endings; for example, *He tapasivat talossa* 'They met in the house' where *-ssa* 'in, inside' is an ending added to the word *talo* 'house.'

21.4 Names and Language

All languages distinguish classes of objects and individual instances of such classes. For instance, there are dogs and cats - classes of animals - and there are individual animals - single dogs and cats. People use names for individuals to which they have a specific connection, such as Fido for your pet dog or Sandy for your pet cat. Names are used to reference persons, so much so that all human beings have names and these are a key part of personal identity. Complex multi-part names are often used to distinguish people from each other, especially because it is common in cultures for names to be selected from a small set. Just think of how many males are called John or Michael and how many females are called Mary or Anne (in countries with a Christian tradition).

All languages have names too. The name used for a language may derive from its community or be that used by outsiders. A name may refer to a region of origin, as with 'English,' which means settlers from Angeln, an area in the far north of Germany, close to the Danish border. A name may simply derive from something like 'we, of the people' as with German *deutsch* and the related Dutch word *duits* or Swedish *tysk*. The etymology (antecedents of a word in history and its meanings) can be traced back to Proto-Indo-European **teuta* 'tribe' (note that the asterisk in this case indicates a very early form, which is postulated but not actually attested).

Objects tend to get names when people devote increased attention to them. All the planets in our Solar System have names. Those outside our system are referenced using a numbering system, often with a leading name tag such as the Gliese set of exoplanets, named after the German astronomer Wilhelm Gliese. There is also a group of exoplanets named after the Kepler Space Telescope, which was instrumental in finding them. For stars and constellations, mythological figures from classical Greece and Rome have served as a good source of names, usually with a certain semantic motivation, so that Jupiter, the largest planet, is called after the chief Roman god; Mars, the red planet, after the god of war; Pluto, the furthest planet (now demoted from this status), after the god of the underworld.

NAMES IN AN EXOLANGUAGE

Naming is a practice characteristic of all human cultures in order to reference individual members of a group. A similar practice would probably apply on exoplanets as the distinction between classes and instances of classes would hold. The sounds of an exolanguage would be used to yield individual sound shapes for each name. An exoplanet would likely have a name for itself and for celestial objects in its immediate surroundings, for stars, planets and moons. Whether such names would have specific meanings would be worth investigating if broad-based contact were ever established with an exoplanet.

21.5 Language, Environment and Culture

Languages reflect the environments and cultures in which they are embedded. The more general of these two situations concerns environment as can be seen when one considers the terms used for basic colours across the world. This phenomenon was investigated in a well-known study by two American linguists, Brent Berlin and Paul Kay, in a book published in 1969 called *Basic Color Terms: Their Universality and Evolution.* They suggested that languages can exhibit seven levels of colour distinction, starting with dark-cool (approximately 'black') and light-warm (approximately 'white') proceeding through red, followed by either green or yellow, then blue, brown and, finally, other minor colours such as pink, purple, grey, orange. The first universal level, with 'black' and 'white' archetypes, rests on the division of night and day; the next level 'red' most likely has to do with the colour of blood and the general association of red with vitality. The colours green and yellow reflect common colours in nature, etc. From their investigation Berlin and Kay concluded that there are 11 basic colours, which occur in a specific implicational order, for instance the authors claimed that if a language had a word for 'green' then it had a word for 'red', if 'brown' then 'blue', etc. The ordering for their eleven colours is as shown in Table 21.2.

The basic colours found on an exoplanet would presumably have a similar motivation: the environment in which exobeings live would determine the colour distinctions their languages would make.

Table 21.2 Implicational colour hierarchy (items on the right imply those on the left)

	←	←	←	←	←
White	Red	Green	Blue	Brown	Purple
Black		Yellow			Pink
					Orange
					Grey

COLOUR TERMS ON AN EXOPLANET

If an exoplanet was orbiting a red dwarf star, colours in its languages would probably have a different significance. Red would be primary, as in human languages, but perhaps with many basic terms for shades of red, going from bright to dark. The colours green and blue would be less important, indeed basic terms for these colours might be missing entirely if the exoplanet had no green plants and no blue sky.

Kinship Terms in Languages

Kinship terms rest on the value which humans in different cultures place on membership of the family on different levels. Languages have words for mother and father (parents), son and daughter (children), brother and sister (siblings) and usually for aunts and uncles (siblings of parents). The kinship system found in a given language will first have lexicalised terms for the members of the nuclear family. Relations beyond this unit, either vertically backwards (i.e. grandparents and great-grandparents), or vertically forwards (i.e. grandchildren or great-grandchildren), may have so-called lexicalised terms, single, everyday words. But these can also be constructed from elements of the nuclear family, for instance something like 'mother-mother' for grandmother. Horizontal distinctions are a further plane on which relationships can be distinguished: one has uncles and aunts on the level of parents, nephews and nieces on that of one's children. From the children's point of view, within a generation, these individuals are cousins. Where languages have cover terms, such as cousins for all children of parents' siblings, gender distinctions may or not be made, in English they are not but in French, with *cousins* 'cousins' (masculine) and *cousines* 'cousins' (feminine), they are, but this is due to the fact that French shows gender distinctions across a host of grammatical categories.

A further axis of distinctions in kinship systems is the side on which relations are located. Relations can be identified as either paternal or maternal. Languages have ways of indicating this: English uses a special Latinate

KINSHIP ON AN EXOPLANET

For kinship and kinship terms to exist on an exoplanet, a system of sexual reproduction would have to hold there. We have no way of determining this now, but if two sexes did exist and if they were sufficiently differentiated on an exoplanet, this would most likely be reflected in language. Assuming two sexes, there would be equivalents to mother/father, son/daughter, brother/sister. A more general, but hard-to-answer question is what significance these distinctions would be accorded by the cultures of an exoplanet.

adjective before the kinship term in question. But some languages have lexicalised terms for paternal and maternal relations, respectively. Swedish, for instance, uses single-word combinations of 'mother' and 'father' to indicate the side on which grandparents are located; for example, *morfar* 'mother-father' is the maternal grandfather, *farfar* 'father-father' is the paternal grandfather.

Kinship terms may involve age distinctions within a single level, such as older brother, elder sister, leading to very complex matrices of terms, as in Chinese or Thai. They may also involve terms for marital rather than blood relationships like mother-in-law, brother-in-law. Finally, languages generally have colloquial terms for relatives closest to an individual, for example *mum, mam, mummy, ma(ma)* for 'mother', *dad, da, daddy, pa(pa)* for 'father'; some of these can extend two generations back, as with German *Opa* and *Oma* for 'grandfather' and 'grandmother', respectively.

21.6 What Do Speakers Know about Language?

When non-linguists think of knowledge they think of conscious knowledge, that is, of something which they can express and reflect on. For instance, if someone asks you whether you know how to play chess, you would answer 'Yes' if you felt able to list the rules of the game. Knowledge of chess would be seen as equivalent to the ability to use the pieces and explain the rules governing their movements to others. This is a typical instance of conscious knowledge as players are aware of the rules and reflect consciously on them during the game.

Language is organised quite differently. The distinction between unconscious and conscious language not only refers to what words and structures are well formed but also to what is not possible in a language. Imagine you had to invent a name (in English) for a new brand of dog food and someone suggested the word *fnoppy* to you. It is unlikely that you would accept it because you know (unconsciously) that in English /fn-/ is not a permissible beginning for a word (technically a syllable onset), although it existed in Old English (450–1066 CE), as in *fneosan* 'to sneeze', and is allowed in other languages like Russian (as *vn-*, e.g. *vnimaniye* 'attention'). In fact, you know (again unconsciously) that only a fricative (a sound produced without interrupting the air flow fully) produced at the same point in the mouth can occupy the slot before a nasal sound here: *snoppy* would be a permissible sequence but *thnoppy* or *shnoppy* would not, as neither *th-* nor *sh-* is produced at the same point as *n*. Now the fact that English speakers cannot formulate the restriction on word beginnings (syllable onsets) simply means that this knowledge is unconscious. It cannot be verbalised by non-linguists. But that does not make it any less valid or true.

Take another example, this time from grammar (syntax). The following sentence is unlikely to be accepted by speakers of English: **Saw I down on the strand her*. The reason is simply that they know that in statements the subject precedes the verb, so we have *I saw ...*, and that the direct object always precedes a prepositional phrase, as in ... *her down on the strand*. You would never think of producing such a sentence because when speaking you automatically avoid structures which are ill-formed in English. However, the order of sentence elements varies greatly across languages. For instance, in Irish the order of elements in the ill-formed sentence of English just quoted is quite normal, as can be seen from the Irish equivalent of this sentence: *Chonaic mé shíos ar an trá í*, literally 'saw I down on the strand her'. Such examples show that there is variation in word order across languages, that speakers recognise this and that they unconsciously know (from language acquisition in early childhood) what the order is for their native language, that is what order will result in well-formed sentences acceptable to other native speakers. So people do not think of the rules of syntax when they speak. It is valid to assume that there are many rules governing well-formedness in sentence structure; however, these rules cannot normally be listed by native speakers (unlike the rules of chess) unless the speakers in question have received specialist training in linguistics.

Unconscious knowledge must exist, otherwise our speech would be incomprehensible. Here is a comparison to illustrate what is meant: if you think about how you walk, then you probably imagine yourself putting one foot forward, shifting your body weight onto this and then moving the other foot forward, then shifting to this other foot and so on. This would be the equivalent of conscious knowledge of language. But walking requires a very intricate interplay of nerves, muscles and tendons as well as feedback to the organ in the head responsible for balance to ensure that one remains upright during the action. The neurological and muscular activity involved is not 'visible' to the person walking and is similar to the unconscious knowledge of language which is active 'in the background' when we speak. It is right for the human organism to keep this knowledge hidden from speakers as too much consciousness would render speech too difficult – the main thing is that the unconscious knowledge works properly, which it does nearly all the time. The only exceptions being unusual situations, temporary ones such as nervousness, tiredness or inebriation, when one's speech is somewhat uncoordinated. There are also more lasting disturbances, known from language pathology, which can arise after an accident or brain disease such as a tumour or a stroke.

Our unconscious knowledge about language is in part innate – the universals of language which we inherit in the genetic code of our DNA – and in part it is acquired in the early years of childhood. Such unconscious knowledge works very well, which is why a language that is acquired early is entrenched for life. After puberty the ability to acquire knowledge with comparable competence declines rapidly, which is why adults have difficulty in learning a foreign language.

In the evolution of our language faculty various elements of language were relegated to the unconscious. Several advantages accrued from this development. One is speed: unconscious actions are always faster than conscious ones, just think of how quickly you can tie your shoelaces if you do not think about it. Another advantage is that it frees up conscious resources for more important aspects of language. Consider that you do not make a conscious decision for each item in language. Imagine deciding consciously to pronounce /f/ then /u:/ then /d/ when saying the word *food*. And, when you say a sentence, you do not consciously choose a syntactic structure and decide what precise elements to put into it. When saying something like *What is the food like in that hotel?* you think of the meaning you wish to convey and the well-formed sentence is generated unconsciously. On reflection you realise that, if you are a native speaker of English, you certainly will not say something like *What that in hotel like food is the?* This brings us to the issue of intuitions about one's native language.

21.7 What Are Speaker Intuitions?

If linguists are not sure what structures are valid in a language, they may choose to interview a representative selection of native speakers on this issue. For instance, some verbs in English take an infinitive and others take a participle (technically called a complement). To determine what options are valid for a particular verb one could elicit responses from test persons by presenting them with templates like the following and asking them to fill in the empty slot.

Fiona considered ____ home   Fiona wanted ____ home

The answers would probably be as indicated below (a tick shows the acceptable and an asterisk the unacceptable sentences to speakers of standard present-day English; a question mark indicates that speakers might be a bit doubtful).

✓*Fiona considered going home*	✓*Fiona wanted to go home*
**Fiona considered to go home*	*? Fiona wanted going home*

There are a number of important generalisations one can make from the preceding sections. The first is that in the evolution of the human language faculty certain elements migrated from conscious to unconscious knowledge, which freed up capacity for conscious reflection on meaning in language production. The second is that the knowledge of our native language, which is stored unconsciously during the first language acquisition process in early childhood, has the effect of yielding a large amount of agreement among speakers about what structures are well formed in their language and what ones are not. This in turn has made languages more efficient as communication systems in speech communities, a quality which they retain to the present day.

WHAT WOULD EXOBEINGS KNOW ABOUT THEIR LANGUAGE?

Assuming that an exolanguage would have evolved on an exoplanet in a manner in principle similar to that on Earth, one could expect that the migration of elements of language, not relevant to the immediate goal of thought expression, to the unconscious would have taken place in a similar manner. A low consciousness of grammar would free up cognitive space for meaning construction and hence lead to a more efficient conversion of thought into language.

22

The View from Linguistics

Linguistics is the study of the human language faculty and the languages it engenders. It enjoys interfaces with other sciences such as cognitive psychology, neurobiology and physiology. The borders between linguistics and other sciences are fluid and have shifted with increasing research in the field. Where we draw the lines is a matter of debate among scientists, but we do draw them because science compartmentalises reality for the purpose of inquiry and analysis.[1]

For some two centuries linguistics has been an established branch of science, and considerable strides have been made in researching how humans speak, how they acquire their native language and in documenting the known languages of the world. This knowledge has been arrived at by devising means for analysing human language and by adopting approaches based on notions which are likely to be quite different from

1 This issue affects virtually all sciences, just consider the problem of distinguishing physics and chemistry or psychology and psychiatry. Of course, guidelines and yardsticks can be offered, for example psychiatry is about pathological conditions. But when is something pathological? Who has the right to determine this? Physics is about atoms, particles and fundamental forces; chemistry is about molecules, their interactions and substances on a human scale. But electrons are particles which are central to both branches of science and their behaviour falls into both domains.

the views which non-linguists will have about the subject. Thus, one of the difficulties for linguists in their attempts to explain the nature of language is that non-linguists will always have ideas about language already. These derive from reflections on the more conscious parts of language such as vocabulary; on the widely held belief that language change is language decay; on ideas about what constitutes good and bad language, in particular in connection with style and social class, and on an inordinate reliance on the written word; and, finally, on a general confusion of language itself with the people who use language.

In all these areas linguists will find that the ideas of non-linguists need to be adjusted and set in a new direction. For example, linguistics is often more concerned with the less conscious parts of language, such as the levels of sound systems and sentence structure, and it views language change as an inherent and necessary aspect of language, seeing as how it has its ultimate roots in the continual choices made by humans when speaking. The goal of linguistics is not only to describe languages but language as a whole and to reach conclusions about our unique ability to speak, that is, the human language faculty. Linguistics is certainly not concerned with dictating usage to others, called prescriptivism, and it is primarily oriented towards the spoken word, the spelling system of languages being a secondary phenomenon arising much later in history due to needs in societies.

22.1 The Complexity Envelope of Language

The complexity envelope of human languages is an issue with two different but related aspects, a physical and a cognitive one. The physical envelope is defined by the sounds which humans can produce with their organs of speech (see Section 22.4 below). The cognitive envelope is determined by our mental capabilities regarding language. There is no easy metric for measuring these, but we know that the structural similarities across human languages are ultimately determined by our cognition. Looking at the documented languages of the world one sees that there are great differences in relative complexity of their subsystems.[2] Take sound structure, for instance. All languages have a set of key sounds used to distinguish meanings when forming words, thus *p* and *b* are used to distinguish word pairs like *pat* and *bat*, *pear* and *bear* (technically such sounds are called phonemes). The discussion here concerns sounds, not letters, so a word pair like *cease* and *seize*, while written very differently, only have one sound difference, in the 'hard' or 'soft' *s* at the end.

The numbers of these key sounds (phonemes) varies greatly. English, with 40+ (the exact number depends on certain interpretations and on the variety of English one is considering), is somewhere in the middle. But there are languages with fewer phonemes, such as Hawaiian with only 13 (8 consonants and 5 vowels) while other languages, like those found in the Caucasus (the mountainous region between the Black Sea and the Caspian Sea) can have as many as 70–80. The lower limit is probably determined by the number necessary to provide sound shapes for the words of a language. It would be very difficult to attain different sound shapes for thousands of words if a language only had three phonemes, for instance. The upper limit for phonemes is probably determined (i) by processing difficulties with huge numbers of phonemes and (ii) by the limits in sound variation which a human can realise. Furthermore, in languages with very high numbers of phonemes, usually only some occur in certain contexts, at the beginnings or ends of words, only in the middle, only in certain clusters, etc. This again facilitates the processing of sounds by speakers and reduces the cognitive burden of handling a very large number of phonemes.

Languages also vary considerably in their grammar, specifically in the part of grammar concerned with cases (nominative, accusative, genitive, etc.), number (singular, plural), gender (masculine, feminine, neuter), tense marking (present, past and future) and the formation of words. It is well

2 Complexity is not the same as 'sophisticated' and less complexity does not imply that a language is in any way 'primitive'. Consider that English is not grammatically complex and the languages of the highlands in Papua New Guinea are languages spoken by non-industrialised, non-technical communities.

THE COMPLEXITY ENVELOPE OF AN EXOLANGUAGE

It would be entire speculation to venture an opinion on how complex or simple the grammatical or sound structure of an exolanguage might be. The first unknown factor would be the cognitive power of exobeings. If they were cognitively more advanced than we are, they could well handle more complex languages. Furthermore, factors which favour the rise of linguistic complexity on Earth, relative isolation of ethnically homogenous communities, might apply differently on an exoplanet or not at all.

known that German and Russian have complicated grammars with three genders (masculine, feminine and neuter) and lots of endings. Finnish and Hungarian have very complicated case systems but no grammatical gender.

There are many views on why some languages should show greater complexity in grammar than others. Those which have little, such as English or East Asian languages like Chinese languages or Vietnamese, get on perfectly well with few endings and may have other mechanisms for making key distinctions in language, for example by using different tones in words, as in the different languages in China. But why should a language like English be grammatically less complicated than German, to which it is related, or Bulgarian compared to Russian, to which it is also related? One reason might be that contact with other languages leads in time to a simplification of the grammar because there are many second language speakers in the community and they use simplified grammar. The converse would seem to be true: languages which have long existed in isolation tend to have complex grammars, like those isolated in the Caucasian mountains for centuries on end.

22.2 Levels of Language: Modular Organisation

Language is a phenomenon which evolved over tens of thousands of years and can be compared with parts of the human organism more than with man-made structures. In order to maximise efficiency and minimise the effect of damage to some part, language evolved into a system which is modular in its organisation (for more information see Section 32.7 below). In this respect it can be compared to a part of human physical make-up, like the immune system (see Section 32.8 for further discussion). Each module of language is self-contained with its own rules and representations – for instance, phonology (sound structure) and syntax (sentence structure) have quite different internal structures. There are, of course, interfaces between each module so that they can interact together and appear as a whole in actual speech. What unifies all modules is the purpose of the system. Just as the immune system has the superordinate goal of protecting the body from infection, language has the function of organising and expressing thoughts, thus enabling communication: ultimately, all elements of language work towards this two-pronged end. However, because language is a cognitive ability of humans it also has secondary functions. For instance, it has a frequent role as a carrier of social attitudes, something which has nothing to do with the simple communication of messages but which has been superimposed on the system. This is because speakers have always forms of language associated with those who use it and are aware that the system can be used to differentiate speakers socially without a loss of the primary function.

It is common procedure to treat the various levels of language (Figure 22.1) separately, as in textbooks on linguistics. This has the tuitional advantage that one can deal with them concisely and neatly in separate sittings of a

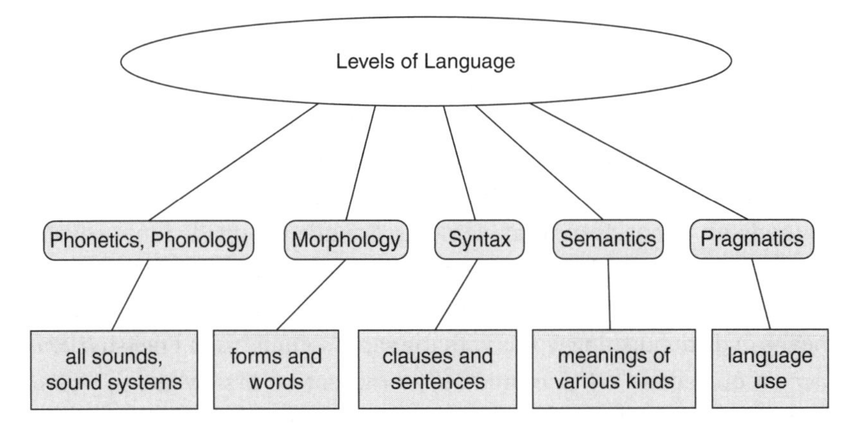

Figure 22.1 Levels of language

Table 22.1 Bottom-up approach to language organisation

Object of study	Name of field	Size of unit
Language use	Pragmatics	–
Meaning	Semantics	–
Sentences, clauses	Syntax	Largest
Words, forms	Morphology	↑
Classified sounds	Phonology	
All human sounds	Phonetics	Smallest

course. However, one should emphasise that the division is not something people are usually aware of when speaking. Because of this they do not always expect language in linguistics to be divided into levels. Arguments in favour of the psychological reality of the different levels can be put forward, for example by showing that the rules of phonology are quite separate from those of syntax despite the interface which exists between the two levels. In the following, the different levels are dealt with from the bottom up, as shown in Table 22.1.

Phonetics, Phonology

This is the level of sounds. The set of possible human sounds constitutes the area of *phonetics*, while the set of system sounds used in a given language constitutes the area of *phonology*. Phonology is concerned with classifying the sounds of language and saying how the subset in a particular language is utilised, for instance what distinctions in meaning can be made on the basis of what elements. Sounds are divided into vowels and consonants, which are arranged into combinations to yield the sound shape of words. Many languages also avail of accent and pitch as possible means of distinguishing words with the same sounds, as in English *below* (accent on second syllable) and *billow* (accent on first syllable); *record* (accent on first syllable) and to *record* (accent on second syllable). See Section 22.4 below.

Morphology

This is concerned with (i) grammatical endings and (ii) the internal structure of words. It is what one normally understands by grammar (along with syntax). In morphology, the minimal forms in language are analysed. These are in turn comprised of sounds which are used to construct words that have either a grammatical or a lexical function (see lexis).

Syntax

This is the level of sentences. It is concerned with the meaning of word combinations. In particular it involves analysing the internal structure of sentences by looking at the relations between functional elements like subjects, objects and verbs in various combinations. Syntax is further concerned with word order and deals with the relatedness of different sentence types and with the analysis of ambiguous sentences.

Vocabulary

The study of vocabulary has two main aspects. The first is that of meaning, which is related to semantics, and the second is that of form, which is part of morphology, more precisely, word formation. The vocabulary of a language is an open class, that is it can take in new elements all the time, and has done so throughout the history of each language, usually by borrowing from other languages. The items in vocabulary are primarily words for things - nouns - and words for actions - verbs. There are also additional word classes, such as adjectives, adverbs, prepositions, etc., along with subdivisions of these word classes. Languages have concrete vocabulary, for objects of their environment, and more abstract vocabulary for expressing ideas and concepts.

Semantics

This is the area of meaning and extends across the other levels of language, specifically morphology and syntax, leading to different kinds of meaning.

For non-linguists, the most obvious type is probably lexical meaning – that of single words – which answers the question, 'What does this word mean?'; for instance, *What does 'geodesic' mean?* However, there are other types of meaning: (i) grammatical meaning, carried by the endings of words, for example, the *-s* in *cats,* which indicates plurality; (ii) sentence meaning, which results from the order of elements in sentences, compare *The dog chased the cat* and *The cat chased the dog,* which have opposite meanings; and (iii) utterance meaning, a part of pragmatics which rests on the context in which an utterance occurs, see the following section.

Pragmatics

This area is concerned with the use of language in specific situations. The literal meaning of sentences need not correspond to the intention of the speaker in a given context; for instance, *It's cold in here* is a statement of fact, but when said entering a room with the window open, for instance, it would normally be interpreted as a request to have the window closed. In the latter case one speaks of utterance meaning, that is, meaning in a specific speech context. The area of pragmatics relies strongly for its analyses on the notion of a speech act which is concerned with the actual performance of language. This involves the notion of proposition – roughly the content of a sentence – and the intent and effect of an utterance.

INTERNAL ORGANISATION OF AN EXOLANGUAGE

Assuming that the division of systems into interconnected sets of subsystems is a valid principle of evolutionary biology, one can take it that an exolanguage would consist of just such a set of subsystems. If an exolanguage used sound to communicate between individuals, there would be a level of sounds. Meaning would also be a level for an exolanguage because any utterance would express the intention of the speaker to convey something to another individual via meanings contained in words. Given that words are expressed in real time, there would be a temporal aspect to the ordering of words (meaning units) and hence a level comparable to our syntax would also exist in principle.

22.3 Language Typology

Language typology attempts to classify languages according to higher-order principles of grammar (morphology and syntax) and to reach generalisations across different languages, irrespective of their genetic affiliations, that is, what language family they belong to. It is also concerned with making statements that apply to all or nearly all languages, which are called universals. The number of universals is actually quite small (but essential to all languages), so language typology is often concerned with determining what the most common value for a parameter is across languages. Take word order, for instance. All languages have what is called a default value for word order, that used in simple declarative sentences. In English the default value for word order is subject–verb–object (S–V–O) though this may vary, for example in interrogative sentences, as in *Can we leave now?* Some speakers will accept as well formed sentences which have the order O–S–V, as in *This girl I don't know*, though such word orders are not really part of standard English.

Some default word orders are frequent across the world's languages. S–O–V, with the verb at the end, is found in languages like Turkish, Basque and Japanese, though S–V–O is the most common word order when languages do not have many grammatical endings to distinguish subjects from objects. Other word orders, like V–S–O (verb-initial languages) are much less common, Irish and Welsh being two notable examples in the European arena, Classical Arabic and Tagalog (a native language of the Philippines) are other examples.

The demographic situation of a language can play a role in what categories are present. For instance, languages which have been in geographical isolation for many centuries tend to develop certain categories, as opposed to languages in scenarios of high contact with other languages. For instance, verb forms which denote evidence for a statement tend to occur only in small, close-knit communities, where indicating responsibility – or lack of it – is expected of speakers who are known to each other as in Eastern Pomo (a Native American language in Northern California).

WOULD EXOLANGUAGES SHOW DIFFERENT TYPES?

Given that languages diversify over space and time and assuming that these parameters would also apply on an exoplanet, one can assume that the exolanguages would differ in their structure and organisation. What the details of this variation would be like is something which we cannot predict with confidence from Earth, but positing the principle of language variation on an exoplanet would seem to be valid.

22.4 Language Production

Humans have very controlled breathing (when at rest or just walking), with about 10 per cent of the time devoted to inhaling air and the remaining 90 per cent to exhaling. It is during the period of exhalation that we produce speech. This is an unconscious process and we do not perceive of breathing in as gulping air.

By comparison, when we run, we usually coordinate our breathing with the rhythm of our running and so cannot devote 90 per cent of the time to exhaling and speaking. The intake and release of air takes about the same amount of time, which makes it difficult to run and speak at the same time.[3]

Bipedalism probably assisted greater breath control among early *Homo* species, as the movement of the legs and the rhythm of breathing were no longer as closely connected as with quadrupedal animals. With bipedalism, the forelimbs were not primarily involved in movement in our surroundings and the further evolution of our forelimbs to arms with hands having fingers, with an opposable thumb, provided us with great dexterity for additional tasks apart from basic locomotion. These developments further disconnected breathing and the use of our limbs.

3 Ingressive speech does occur, and is often a characteristic of a certain individual, and not typical of the entire community that person belongs to. Examples of this would be saying 'Yeah, yeah' while inhaling air. One can only do this for a very short time as the lungs quickly fill up with air. Equally, we cannot laugh easily while running, though we can smile.

WOULD EXOBEINGS BREATHE LIKE US?

If one assumes that exobeings have a mechanism for the intake of oxygen from their surroundings, one could ask the question whether they would breathe like us through an opening in the face, down a pipe to lung-like organs, with vocal cords to produce sound during exhalation, the latter being modulated to realise the phonetics of their language(s). It is not too presumptuous to suppose that they use oxygen for various biochemical processes, but to what extent their physiology would be similar to ours, at the level of detail of our breathing apparatus and organs of speech, is entirely unclear. But if their languages were transmitted via sound then exobeings would have to have a method of producing this sound. As sound consists of waves generated by the vibration of air, they would require some kind of air flow with a modulation mechanism to realise various sounds.

22.5 The Human Tongue and Throat

Babies when born have their larynx, the 'voicebox', where the vocal folds are located, quite high in the throat, in fact so much so that they can breathe and swallow food at the same time, as can many animals. But after about three months the larynx descends into the throat, producing a narrow, inverted fork-like configuration which allows us to produce sounds with our vocal folds quite far down in our throats, and we have a resonance chamber above it to amplify the sound we are producing. This dual-purpose configuration of food plus windpipe makes us prone to choking, as food can easily enter the trachea (windpipe) if it is not closed off by the epiglottis (Figure 22.2).

But another advantage of the lowering of the larynx, for us as a species and for each of us individually, is that it allows a lowering of the tongue root, which in turn allows the tongue to move easily along a front-to-back (horizontal) axis, not just an up and down (vertical) axis in the mouth. This means that the human tongue after early childhood can adopt any of the positions for vowels which might be necessary for any language.

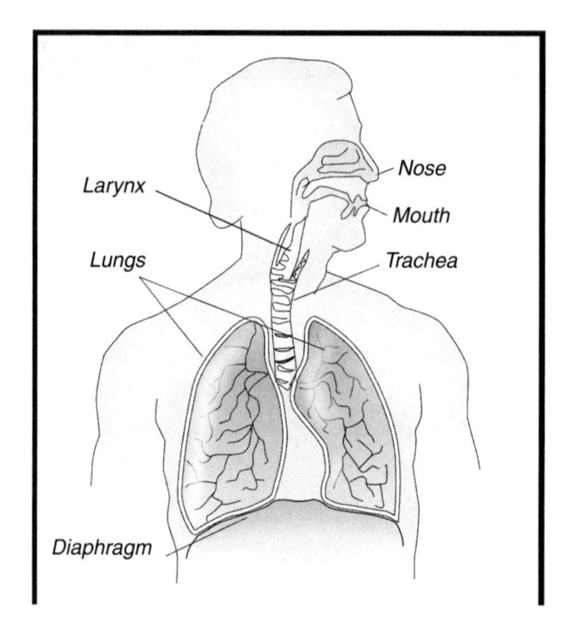

Figure 22.2 Human breathing apparatus

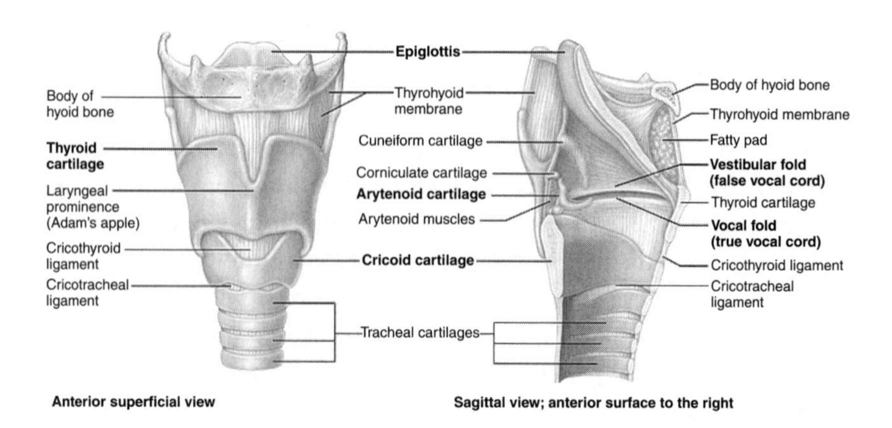

Figure 22.3 Human larynx (anterior and sagittal view)

Much has been made of the lowered larynx in early humans as possible evidence of the ability to produce speech (Figure 22.3). But research by various scholars, such as W. Tecumseh Fitch (Fitch and Reby 2001), has pointed out that a lowered larynx is regular with other mammals, like dogs

ANATOMICAL/PHYSIOLOGICAL AND COGNITIVE EVOLUTION

A suitable anatomy and physiology (structure and relevant functions) for the production of speech is a necessary condition for speech. But is it sufficient? Does the anatomy, largely evident in fossil records, necessarily imply that a species had speech? For this to have been possible at an earlier evolutionary stage of *Homo sapiens*, cognitive abilities would also have been required and it is not known to what extent the cognitive and anatomical/physiological developments proceeded in tandem or if, indeed, the former preceded the latter.

when barking or stags when roaring. But these animals appear to lower their larynx only when producing their characteristic sounds whereas humans have the larynx permanently lowered. Nonetheless, the rapid development of the primate larynx is disproportionate to the development of body size with these animals, not just humans (Bowling et al. 2020).

The crucial distinction between mammals and humans lies not so much in anatomical differences but the neuronal circuitry controlling the vocal organs. This can be seen, for instance, in the size of the hypoglossal nerve canal, which in turn indicates a sophisticated nerve bundle, innervating a flexible and agile tongue necessary for speech (Hurford 2014: 85). The same is true of the nerve canals which innervate the chest muscles and the region of the throat.

22.6 What We Hear

Would exobeings be able to hear? Clearly, they would need to receive acoustic input from their surroundings and from others. To process spoken language, one must be able to hear. For the latter, one needs ears or some functional equivalent. Higher mammals have ears and can generally hear as well as we can, or even better, in many cases. So they have one of the preconditions for successful (spoken) language. Our hearing is especially attuned to sounds between about 20 hertz and 15,000 hertz (less at the high end as we get older), which is the range of human speech. Non-linear sounds, big jumps in frequency, very high-pitched tones, very low booming sounds, trigger the startle reflex, before your conscious brain knows it. This is a primitive reflex which prevents us from being attacked and possibly eaten by some nearby predator.[4]

The range of sounds for human speech is also that of noises in nature, such as rushing air, moving water, leaves brushing up against each other or of some animal walking in the undergrowth, all noises the perception of which were essential to species survival during our long evolution. These sounds would be found in a similar wavelength range on an exoplanet so it would make sense to have hearing which is attuned to such sounds and to have the sounds of an exolanguage located within that range.

Our hearing is especially sensitive to the sounds of language in that we can recognise language in any set of noises we happen to be picking up at any given time. This ability would seem to be due to the processing of incoming audio signals in the primary auditory cortex of our brains. It is here that we can allocate sounds to the categories in the system of the language we are listening to. This is no mean feat and involves normalising the input signal, which means taking into account the relative frequency of a voice (male or female, child or adult) and the tone of voice of the speaker. And this is done in real time while also deciphering the grammar and meaning of what one is hearing.

22.7 Vowels and Consonants

All languages have a distinction between vowels and consonants. The vowels are produced (i) by allowing the vocal folds in the larynx to vibrate in the escaping air from the lungs and (ii) by adopting a certain configuration of the oral cavity (the space in the mouth) by moving the tongue

4 We have also added some more recent sounds to our catalogue of those startling us: screeching brakes, crashing cars, shattering glass would be examples.

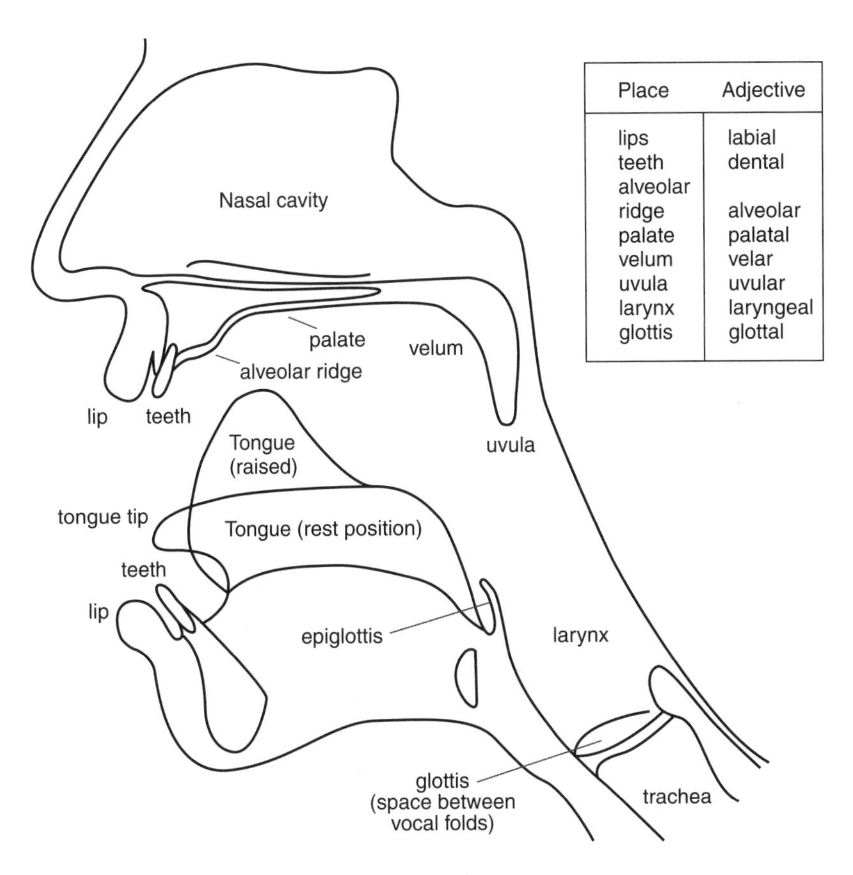

Place	Adjective
lips	labial
teeth	dental
alveolar ridge	alveolar
palate	palatal
velum	velar
uvula	uvular
larynx	laryngeal
glottis	glottal

Figure 22.4 Vocal tract with points of articulation

into a certain position (Figure 22.4). Consonants on the other hand are all produced by obstructing the escaping air in some way or other. There are consonants which are vowel-like, involving little obstruction, such as <r> and <l> in *rip* and *lip*, respectively, and others which not only involve a more radical obstruction but also do not require any vibration of the vocal cords, such as <t> or <s> in *tip* and *sip*, respectively, or the <(t)ch> sound in a word like *chip* or *catch*.

The rise of consonants in early human language is not surprising given that the tongue, lips and velum are part of the oral tract, which are moved about repeatedly when we are biting, chewing and swallowing food. The human tongue was agile enough at the beginning of the human lineage to produce primitive sounds, which became increasingly sophisticated in their articulation with the continued evolution of humans and their speech.

The articulation of vowels and consonants

Vowels

1. Articulation: Use the source of sound in the larynx (vibrating vocal folds)
2. Variation: Use different configurations of the tongue in the mouth; air flow through the mouth and/or nose

Consonants

1. Stops: Close off the air stream in the mouth or throat (as in <t>)
2. Fricative: Constrict the flow of air out through the mouth (as in <s>)
3. Affricates: Combine stops and fricatives to a single unit (as in <ch>)
4. Nasals: Let the air flow escape through the nose (as in <n> or <m>)
5. Laterals: Let the air flow escape through the sides of the tongue (as in <l>)
6. Place of articulation: Use different points in the mouth (from the lips back down to the glottis (the gap between the vocal folds)

How Are Vowels Produced?

Imagine you go to a concert. Along with all the others you take your seat, the members of the orchestra come onto the podium and sit down. When everyone has settled, the chief violinist stands up and gives a sign to the oboe player, who then plays a single long note, an A, and the rest of the orchestra chime in by playing the same note, aligning the pitch of their instruments to that of the oboe. An A is about 440 hertz, though some orchestras have a slightly higher value (a few hertz more) to create a more brilliant sound. But the question is: if all the instruments are playing the same note why do not they all sound the same? The answer lies in the harmonics produced by each instrument. The sound of an instrument consists of a note at a certain pitch but also of various frequencies above that note which have stronger or weaker values, depending on the type of instrument (wind or string, for example) and its shape (trumpet, horn or tuba, for instance). So the vibrations of the air when an instrument is being played consists of the basic pitch of notes, called the fundamental frequency, and a highly complex pattern of additional vibrations above that pitch, called overtones, which results in the typical sound of an instrument. So the pattern of overtones produced by an oboe, a clarinet, a flute or a horn are all very different.[5]

So what does this have to do with language? It explains why the vowels of a language all sound different. Imagine you produce the vowel of the (English) letters 'AH' (as in the word *shah*), long and clear. Now keep the pitch you are using for that vowel in your throat (produced with your vocal folds) and produce the sounds for the (English) letter 'E' (as in the word *see*) and then the letters 'OO' (as in the word *soon*). So now one could ask: if the pitch is the same for all three vowels, why do the AH, E, OO[6] vowels all sound different? The answer lies in the variable amplitude of the overtones, which in phonetics are called formants. Formants are derived from the different configurations of the tongue (Figure 22.5) and hence the difference in the shape of the oral cavity through which the sound generated by the vocal folds passes, on its way out through the mouth.

MAKING USE OF WHAT NATURE GIVES US

The air from the vocal folds normally escapes through the mouth when speaking. But it is also possible to open the nasal cavity at the same time by lowering the velum at the back of the mouth. Some individuals lower their velum slightly

5 This is a simplification of the actual situation in an orchestra. For instance, the various string instruments (violin, viola, cello, double bass) are similar but play their notes at different pitch levels called octaves, which are eight notes apart (either up or down).

6 The letters used in English to represent sounds are different than in most European or other languages. So linguists use a phonetic script, the symbols of which are placed in square brackets. In this script a long vowel is indicated by a colon. This means that 'AH,' 'E,' 'OO' would be [a:], [i:], [u:] in phonetic script.

MAKING USE OF WHAT NATURE GIVES US (CONT.)

during vowels in English, this is then just an idiosyncrasy of their speech. But if a whole community does this consistently, the language may develop nasalised vowels as happened in French, cf. *quand* [kã] 'when' (historically, the nasal quality of the <n> spread to the left into the preceding vowel and the [n] sound was lost; Polish also developed nasalised vowels in a similar fashion). Such extensions to the palette of possible sounds could be expected in exolanguages as well: anatomical features which would increase the differentiation among sounds would likely be co-opted into the sound systems of at least some of the exolanguages of a given exoplanet.

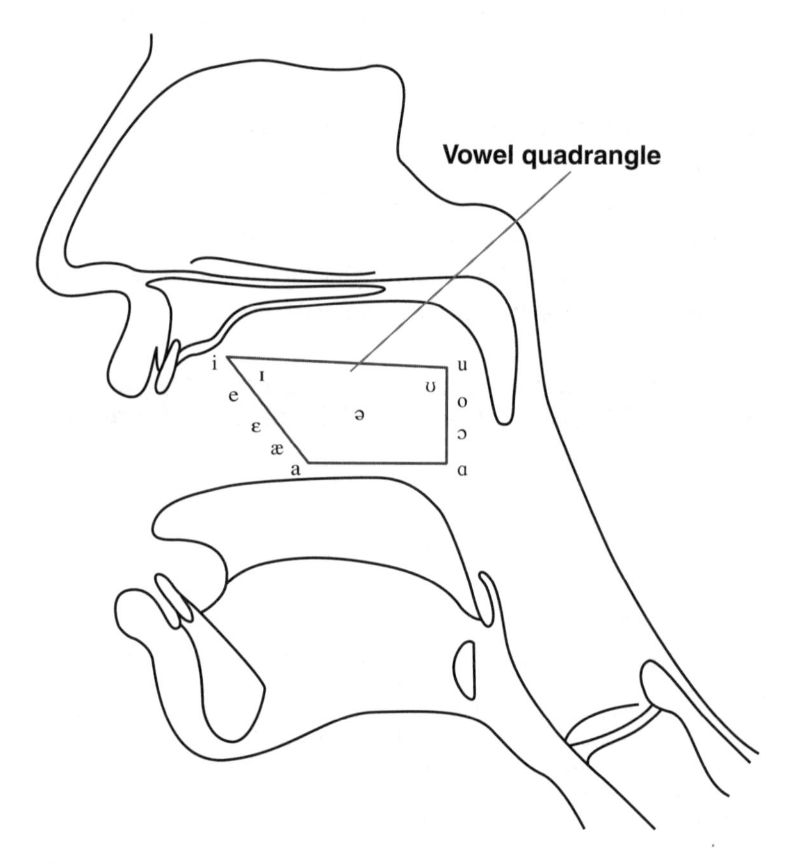

Figure 22.5 Vowel quadrangle

22.8 Convergent Evolution and Language Production

For scientists there is a problem with something which only occurs once. Is it a fluke occurrence or a common development, given the right conditions? Humans have a set of folds in their throat which vibrate to produce sound. Was this development unique or are there independent parallels elsewhere in the animal world, which would support the idea that this anatomical feature was not just a freak of nature? If one looks at some other mammals, in this case toothed whales and dolphins (odontocetes), one sees that they have phonic lips in the part of their anatomy which corresponds to the nasal cavity in humans. When the phonic lips are brought together during exhalation by the animal, the ambient tissue vibrates producing sound analogous to that generated by humans with their vocal folds (in the larynx).

This is a clear example of a convergent development, which shows that the production of sound for echo location in odontocetes and that for voice with humans avails of similar anatomical structures, supporting the view that these are naturally evolved mechanisms for sound generation.

23

The Language Faculty and Languages

> What many linguists call 'universal grammar' may be regarded as a theory of innate mechanisms, an underlying biological matrix that provides a framework within which the growth of language proceeds.
>
> Noam Chomsky, Rules and Representations (1980:187)

23.1 The Nature of Language Acquisition

The language faculty, the ability to understand and acquire human language, is a feature of our neurobiological make-up which has been passed down through the generations in every human being as part of our genetic endowment.[1] It is the language faculty which allows us to acquire any

1 Chomsky has spoken of the language faculty as an organ. However, it is not physically delimited like the heart, liver or kidneys. But functionally, the language faculty has in common that its operation is distinct from that of other human faculties, just as the heart is the organ for pumping blood, the liver for detoxing the body, the kidneys for managing liquid waste, etc. Given that the brain is anatomically and physiologically recognised as an organ, it might be best to treat the language faculty as a self-contained component of the brain, highly connected with other components.

language as long as we are exposed to it in our early childhood. Although it cannot be directly observed, the language faculty imposes structural conditions which must be met by all the languages of the world, that is, it provides limits to what can occur in a human language by containing a framework within which language variation can arise.

The language faculty is the same for everybody no matter where they might come from. This can be shown quite easily by the fact that a child from one part of the world, if taken to another, at an early enough age and exposed to a language there, will acquire that language as a native speaker, indistinguishable from all the others whose parents are from that part of the world. For instance, if a British couple moved to Vietnam with a newborn baby and stayed there for some years (to allow language acquisition to take its course through interaction with other Vietnamese), then the child would learn Vietnamese[2] like any child of Vietnamese parents.

It is clear from the above remarks and from basic reflection that we have the ability to acquire and understand any language, but that we are also native speakers of one or maybe two, rarely three, of the over 6,000 languages spoken on Earth. These are all external languages, by which is meant the expression of the internal language faculty, manifest in pronunciation, grammar and vocabulary when we speak.

The framework of constraints and conditions which the language faculty puts on externalised languages has been labelled 'universal grammar' (see above) and is manifest in the similarities in underlying structure of all languages. Knowledge of this structure is passed on genetically to each generation and in our long childhoods we take this abstract structural knowledge and combine it with the specifics of the language or languages we are exposed to during the process of language acquisition. Just what features constitute this universal grammar is contested among linguists but the existence of the framework, as evidenced by the deep structural similarities across all languages, is far less disputed.

2 Vietnamese is a language of the Austroasiatic family in South-East Asia. In pronunciation (it has tones which are significant for meaning distinctions), in grammar and in vocabulary it is totally different from English, or rather it appears so. But on a deeper structural level it is organised like any other human language.

THE LANGUAGE FACULTY WITH EXOBEINGS

The evolution of a language faculty shows how, with time, systems in our bodies gravitate towards ever greater abstractness and efficiency. Passing on this faculty genetically, and working out the details of specific languages during childhood exposure, renders the language acquisition process efficient not just with regard to the languages which exist on Earth but also regarding any future languages which might arise. Assuming a similarly long period of evolution for exobeings, one would expect that their cognitive systems, like others in the body, would become increasingly efficient with time. Thus a system similar to our language faculty would be expected in exobeings who had, in principle, similar evolutionary paths to ours.

The extended childhood of humans allows them to develop their faculties, such as language, more fully than if they reached adulthood more quickly.[3] There is also a sense in which humans are born too early (from a purely physical point of view). A possible reason for this could be to allow the child to pass through the mother's birth canal with its large head. A consequence of this is that the child enters the outside world quite early and can receive and process language and other stimuli from this source at an early point in its physical development.

23.2 The Question of Modality: Sound or Gestures?

Most animals use sounds in the audible range to communicate with each other. It is true there are bats which use ultrasound (above 20,000 hertz) to locate their prey. One might think that exobeings could also use ultrasound for their communication and so their language would be inaudible to us.

3 The childhoods of *Homo* species were continuously extended after their development from earlier *Australopithecus,* with the age of puberty (sexual maturity) extending into the second decade of life.

But how likely is this? In our animal world, ultrasound is the exception rather than the rule, although many more animals have been discovered to use ultrasound, such as tarsiers, and that might be true for exobeings as well. But consider that ultrasound (used in echo location) is mostly typical of nocturnal animals, especially those which are insectivorous. Of course, if you had beings on the dark side (or twilight zone) of a tidally locked exoplanet, they might avail of ultrasound.

Compare this situation to temperature range: most animals live in a range which does not exceed 40 °C. But there are extremophiles which can live around or indeed above boiling point. However, that is very much the exception in our world. So maybe sounds of the audible range we use, like the normal temperature range for carbon-based biology, are the most typical in the universe and so the most likely to be used by exobeings.

However, the fact that sound is most likely as a modality does not mean that it is the only one (see following table). Recent research has also shown that the modality for expressing internal language can be either by vocal-auditory means (speaking and hearing) or by signing (using one's fingers and hands, possibly with facial expression and mouthing, in precisely structured movements for others to see and interpret).

Modality	
Production	Speech, signing, writing
Reception	Sound, sign interpretation, reading

Goldin-Meadow (2012) found out that young children have two-part communication strategies to begin with: they point to something and utter a word and later, when the two-word stage develops, they drop the gesture. If this is a case of ontogeny recapitulating phylogeny (the growing individual going through similar phases as in evolution), it might be evidence that gestures preceded sound for language expression in the history of our species.

23.3 Sign Language

There are advantages to sound: it can be heard over several metres, one does not need to be facing the source to hear it and it can be heard in the dark. It fades rapidly so that a reasonable amount of information can be packaged in a given stretch of time. Sound can be varied in various ways, for example by loudness, pitch and frequency, so that language can utilise these features, either for the basic distinction of words and for carrying additional information relevant in a discourse. This means that in evolutionary terms sound may well have superseded gestures, given its advantages. However, in situations where sound it not available, for instance with congenitally deaf people, gestures can and have been used. Gestures and language are closely linked so that it is no surprise that sign language arose as an alternative modality. Over the past few centuries flexible and sophisticated sign languages have been constructed, which act as a viable alternative to language in sound.

Sign languages have been developed for most of the large countries of the Western world, such as British Sign Language and American Sign Language. These languages are not necessarily related to each other (though American Sign Language and French Sign Language are), nor do they derive from the spoken languages of the countries where they are found. In the English-speaking world alone, there is a plethora of sign languages which have arisen in the past few centuries or so: South African Sign Language, Australian Sign Language, New Zealand Sign Language, Irish Sign Language, etc. (these languages share features with British Sign Language, less so with American Sign Language).

Of particular interest to linguists have been the cases of new sign languages arising spontaneously in communities which have been relatively isolated from the speech communities around them and/or which have a high incidence of inherited deafness. For linguists, the essential question is how structure and conventions arise in such recent sign languages. There are four well-known instances.

Sign languages in isolated communities

1. *Nicaragua Sign Language.* This was developed spontaneously by deaf children who were left to their own devices in schools in Nicaragua during the turbulent 1980s in that country.
2. *Central Taurus Sign Language* In the central south of the Taurus mountains in Turkey a sign language arose in a group of three villages for deaf individuals who communicated to each other via signs. It is independent of the more general Turkish Sign Language.
3. *Al-Sayyid Bedouin Sign Language.* This is another system of signing which arose among about 150 deaf members of the Al-Sayyid Bedouin tribe living in the Negev desert in the south of Israel.
4. *Kata Kolok / Bengkala Sign Language* A further case of sign language arising due to the presence of a gene mutation in a closed population, this time in the north of the Indonesian island of Bali. Given that deafness is widespread in the central village of Bengkala, this sign language came to be widely used (the relationship is about 40 deaf to 1,200 non-deaf signers).

In these cases it has been observed that the languages moved rapidly from an initial uncoordinated stage of idiosyncratic signs to a more conventionalised set of signs exhibiting an internal structure in the composition of multi-word phrases (corresponding to syntax in spoken language). The British linguist Simon Kirby has shown in experiments that this can happen quite quickly and that users abide by the conventions established in the group, thus enabling a transmission chain across future generations (see 'Structure and Iterated Learning' in Section 32.5).

The generalisation to be made from such observations is that human language may well have evolved in this manner itself at its earliest stages, with the conventionalisation of signs and the fixing of structures providing a consensus framework in a community for the use and transmission of structured language.

23.4 Communication by Touch?

Imagine you wished to communicate with someone you knew who was deaf and you did not wish to use sign language for some reason. You could devise a system whereby you tapped various rhythms with different fingers on the forearm of that person and vice versa. You would need to come to an agreement about what sequences meant and you could slowly build up a small vocabulary and some basic syntax, with simple sentences like actor–action–goal – as in *I made some coffee* – or some simple questions like *Would you like some coffee?*, all worked out via a system of taps on the forearm.

Now imagine a situation where some infants were born both aurally and visually challenged and they were encouraged by their caretakers to tap each other on the forearm to communicate with each other. The question is whether these individuals would develop a fully functional language consisting of taps on the forearm? The proponents of the view that internal language can be externalised (expressed) through a variety of modalities would say that this is in principle possible. It would not be very efficient, the tapping would be relatively slow, would not allow for the amount of structural variation of human speech, or signing for that matter, and would be cumbersome, as you would need to be close to your interlocutors and their forearms would have to be bare (not very practical in cold climates). But in principle it should be possible, it would be like playing the piano with one hand on someone's forearm (note that there are instances where signers use touch to communicate with others, as in situations of darkness).

23.5 Receptive Modality

We normally listen to sounds and decode the semantics of the message in real time without any effort. But we can also scan written language, typically on a page, effortlessly composing the meaning the author of the text intended, again in real time. And the medium can vary: written text can be handwritten, printed in various fonts and size, upright or slanting, on paper of different colours, on metal, stone, wood, etc. or on a computer, tablet or smartphone.

When reading we scan the beginning of each word and, by means of an internal predictive algorithm, recognise words immediately and so can skim from word to word very quickly. We also switch from the end of one line to the beginning of the next without noticing it. Another feature of this visual recognition of speech in writing is that the image we produce in our visual cortex, from the input the retinas of our eyes pass back along the optic nerve, is stable whereas it should in fact show drag as our eyes move along each line of writing and then jump back and down to the beginning for each new line. Why do not we see the drag of eyes moving as we scan the lines on a page? The answer is that the image we see is one which is *constructed* in our visual cortex from retinal input. And this input is also scanned linguistically to allow us to understand the language which is printed in front of us. It is the processed image of the page and the linguistic interpretation which is presented to our consciousness. All of this happens effortlessly and seamlessly as we read text on a page. So where, you might ask, is this *processed* image in the brain? It is not really at any one point. We know that, when we are reading, both the visual cortex and Wernicke's area (responsible largely for understanding language) are active. But information is also passed to and fro between these areas and between the visual cortex at the back of the head and the prefrontal cortex, which is responsible for most cognitive processing in our brain.

The net result of all this brain activity is that we perceive the page we are looking at as a three-dimensional image, projected outwards into space in front of us. This 'illusion' is created by the brain on the basis of retinal input passed to it through the optic nerve. Experience has shown us that the images we process from this input seem to correspond well to reality. After all, you might say, we can reach out to the paper with the print on it, we can measure the printed lines, count the number of words on a page, etc. But all this information is gained through a similar complex interaction of sensory input to the brain – from our eyes and our limbs (probably hands/fingers in this case).[4]

If the image we think we see is really constructed in our brain, what do the eyes actually see? Well, light does enter the eyes and strike the retinas, leading to electrical signals being generated which are then bundled and passed along the optic nerve to the visual cortex. But the light passing through the lens of the eyes is flipped so that it is projected upside-down on the retina. Furthermore, although the image you see when you look out at the world seems to be in focus, your eyes actually only focus on a small part in the centre of your field of vision. Try the following little experiment to demonstrate this. Pick any line of text on the page you are now reading. Focus intensely on a single word in the middle of this line. And without moving your eyes see if you can recognise the words to the left and right of the one you are focusing on. You probably can only recognise one word to the left and right, two if you are lucky, but that's all. Of course, under normal circumstances, if you wish to read words elsewhere on the line or page, you just move your focus to those words, like when you reread a line or skip ahead. But the point is clear: all you can see, in the sense of focus on, is a tiny patch in the centre of your field of vision. Your visual faculty is not like a camera where everything which is not in focus is blurred. Instead, it creates the illusion that the entire field of vision is in focus. But what in fact you are doing is endlessly and seamlessly shifting the centre of focus around your field of vision to take in information from what is in front of you. And you do this all day long, not just when you are reading as you are now.

From this simple example of how we read text on a page you can recognise that the brain is a highly complex, indeed manipulative organ, which constructs consciousness inside your head allowing you to manoeuvre through the outside world and to take in and utilise information as you do so. It is the seamless manner in which the brain processes the sensory input it receives that gives us the feeling that what we perceive is exactly

4 When we move our limbs, the brain is given information about their position in three-dimensional space through a process known as proprioception (information about the position of one's body in space passed back to the brain).

SIGN LANGUAGES ON AN EXOPLANET

Sign languages presuppose hand-like limbs with flexible fingers. These would be a precondition among exobeings who are capable of constructing artefacts. This means that there is no intrinsic reason why a modality, functionally equivalent to terrestrial sign languages, could not develop on an exoplanet.

what is outside in the real world. In our everyday lives,[5] there is no time when the brain flashes a message, 'Processing, please wait', with a little egg timer across our eyes, so we do not notice what it is going on below the level of our consciousness.

23.6 Language and Writing

We are not pre-wired to write languages, but in those countries with written forms of language (most) children learn to write from an early age. The key here is the specification 'from an early age'. If children learn to write before the age of four or five, they master the writing system of the language in question. There are a number of hurdles in achieving competence in writing. The children have to master the principles of the writing system.

Principles of writing systems

1. Direction of writing: left to right (like English, Russian, Greek) or right to left (like Arabic, Hebrew, Persian, Urdu). While some languages can be written from top to bottom, such as Japanese, and English on various types of signage, this is not the default direction of writing.
2. Relationship of letters to sounds, such as one letter to one sound, the so-called phonetic principle found in English, French, Italian, or one symbol

5 There are, of course, special situations where people suffer from hallucinations, for instance when under the influence of drugs or if they suffer from a psychiatric disorder. However, such situations are exceptional, interesting as they may be for the insights they provide into the way we perceive the world.

per word or at least per syllable, as in Chinese, Japanese and Korean. Some languages have mixed systems, like Japanese which used *hirigana*, a system of one symbol per mora (short syllable) for native words and *katakana*, which is a slightly different system found with newer words and borrowings from other languages.

3. The degree of consistency in (2). This is a serious matter: for languages like English children must master a high degree of irregular correspondences; for instance *gauge* is pronounced like *g* + *age*, *gaol* like *j* + *ale*. They must learn that a sequence of letters such as *-ough* can be pronounced in a variety of different ways, consider the words *plough, although, through, cough, rough*. Speakers of other languages, such as Spanish, Dutch or Finnish, have a much easier time as there is an almost perfect correspondence between letters and sounds.

Written Documents of Language

Writing is not the primary language activity, speaking is. This becomes abundantly clear when you consider the facts of writing as a secondary activity.

Writing as a secondary activity

1. Written language arose tens, if not hundreds of thousands of years after spoken language.
2. For each individual, spoken language comes first; learning to write, a conscious exercise, comes later with schooling.
3. You need an instrument to write; we are not genetically endowed with the means to write, but we are for speech.
4. You do not need to write, most of the world's 6,000+ languages are not available in written form.

MEDIA FOR CONSERVING AN EXOLANGUAGE

This question would be central to any civilisation wishing to preserve knowledge across generations and thus avoid the necessity to discover/learn everything afresh. Exobeings could be expected to have a medium for the conservation of language. But whether they would know paper would depend on several factors, not least of which is whether they would have trees or equivalent sources in their environment from which to gain cellulose fibre for paper production. If theirs were a digital civilisation, electronic means of storing data could be expected, corresponding functionally to data storage on computers on Earth.

Despite these obvious facts, writing has an inordinately high status in Western societies. The main reason is because the written form of language is more permanent. It is used to pass on knowledge across the generations, it is used to draw up legal documents, charters, contracts, wills and the like. The legal value of the written word extends to handwriting: a signature has greater legal status that a verbal agreement. This is obvious given that one can produce a signature for others to view (e.g. lawyer or judge) if an agreement is contested whereas claiming that someone entered a verbal commitment is just one person's word against another's.

Could a Civilisation Have Language but No Writing?

This is a crucial question for two main reasons: (i) Writing is not biologically given as opposed to speech. So the exocivilisation would have had to develop a system for writing. (ii) A writing system is indispensable for the development of a civilisation as without one there is no obvious means of transmitting knowledge across the generations. If this is not the case, such knowledge must be either (a) genetically encoded or (b) acquired afresh in each generation.

There are a few occasions where a civilisation did not have writing. That which built Machu Picchu in the Andes did not have a writing system. It was built around the mid-fifteenth century (the site was later abandoned in 1572 after being inhabited for only about 100 years) in the Cuzco region of present-day Peru by Inca people who spoke a Quechuan language.[6] These languages did not come to be written down until the Spanish introduced a system based on the Roman alphabet (which has since been modified in modern Peru). For the Incas, the knowledge of architecture was passed down orally through the generations. However, the great civilisations of the West, in Ancient Greece and Classical Rome, left behind a huge body of written literature, which tells their stories in considerable detail.

Language and Symbolic Art

Could we have art without language? If abstraction and symbolic representation are an essential part of language, any group of hominins, such as the Neanderthals, who had engaged in their later stages in symbolic art, most probably had language close to our modern understanding of the term.

23.7 Linguistic Diversity on Earth and Beyond

The human language faculty finds its expression in the large number of languages across our world – the estimates vary between 6,000 and 7,000 depending on how you define a language.[7] At first sight, that might seem a lot of languages, but the peak of linguistic diversity was probably in the

6 The term 'Quechua' refers to a family of indigenous languages spoken by about 8–10 million people in the Andes mountains and highlands of Bolivia, Peru and Ecuador in South America. These three countries have recently recognised Quechua as official languages within their borders.

7 Two varieties are generally regarded as separate languages when they are no longer mutually comprehensible. This can happen over time, as with Dutch and German, and is often connected to questions of national sovereignty, and oftentimes political issues hold sway over purely linguistic considerations. Thus, Norwegian and Swedish or Romanian and Moldavian are regarded as separate languages although they are historically closely related and were/still are mutually comprehensible. The opposite is also true: there are a large number of historically related languages spoken in China but there is a political reluctance to recognise all these as separate.

fifteenth/sixteenth century, with perhaps as many as two to three times the present number (a rough estimate, given the lack of data). Today, 23 languages account for those spoken by over half the population of the world (close on four billion) and about 40 per cent of present-day languages have a status of 'endangered language' (www.ethnologue.com/guides/how-many-languages).

No matter how many speakers a language has, each one is an external manifestation of the internal language faculty which is common to all humans. All humans have the same type of vocal tract and organs of speech. This means that the physical substrate for all the 6,000+ languages on Earth is the same. Not only that, the design features (see Section 20.4 above) are the same for all our languages, again representing a common core of organisation and structure across the Earth.

Geography and Language

The exoplanets, where one can expect exobeings to have developed, would likely be rocky water planets, not dry rock planets like Mercury, Venus and Mars on the one hand and not gas giants like Jupiter, Saturn or ice giants like Neptune and Uranus on the other. Consider the geology of such an exoplanet: it would probably have a rocky surface interspersed by oceans and lakes containing water (the ocean might be saline like ours, but we do not know). If there were oceans of water, these would divide sections of land from each other, and such an exoplanet would then have continents. In that case it would likely experience plate tectonics like the continents of Earth, given that the interior of an exoplanet would probably be molten (relatively near the surface). Now if there are continents, the populations on these continents would be separated from each other, if not by oceans or huge lakes then at least by mountain ranges and stretches of more or less narrow land.

If an exoplanet had a historical background of geographical spread across their planet, we might assume that they have entities roughly similar to the countries on Earth and that these could be associated with certain languages. Given time, the separate groups would develop differently, both culturally and linguistically. So it might be reasonable to assume that exobeings would speak many different exolanguages distributed across the land surface of their planet.[8] After all, this is the situation on Earth.

It could possibly be that exobeings would have reduced the number of languages on their planet, either through linguistic dominance of one language, like English currently on Earth, or by actively eliminating languages. Whether the latter could happen is somewhat doubtful because language is part of one's identity and, if Earth is anything to go by, people resist having their identity taken from them by other groups.

Linguistic dominance is responsible for the inexorable spread of languages like English on Earth. Consider that more than 80 per cent of the world's languages are spoken by less than one per cent of the Earth's population. There are regions/countries which have an inordinately high density of languages, such as the Caucasus between the Black Sea and the Caspian Sea, the island of Vanuatu in the South Pacific or the Melanesian island of Papua New Guinea, north of Australia. The latter country has a difficult terrain in its mountainous interior, which historically led to the isolation of small communities whose languages diverged to such an extent that their relationships are not always easy to determine, even for linguists. In the region of 800–1,000 languages are spoken in Papua New Guinea accounting for about 12 per cent of the languages on Earth. Such regions might well exist on exoplanets, perhaps with isolated peoples living there, like the inhabitants of North Sentinel Island in the Andaman Islands chain in the Bay of Bengal. Such inaccessible regions could well harbour old languages unaffected by contact with outsiders.

The above remarks refer to the range of possible linguistic situations on an exoplanet. This would be embedded in the wider geopolitical layout of the planet, which would quite likely show a similar cline of power and dominance between states as can be seen only too clearly on Earth. And a rivalry between superpowers, as we see on our planet, could equally well hold for an exoplanet. This would mean that the language of space exploration for

8 Most of the Earth-analogue exoplanets thus far discovered are larger than our Earth and so could easily harbour over, say, 10,000 different languages, given a reasonable population size of several billion and the conditions of evolution and relative isolation (many mountains, lots of islands) which would promote language diversity.

exobeings is likely to be that of the dominant group, or a small number of such groups. This might well be detrimental to linguistic diversity on an exoplanet but would render communication with the inhabitants of other planets that bit easier as we would not be required to decipher a series of codes.

23.8 Was There One Original Language?

Can one reverse engineer language? Unfortunately not, because even the oldest remnants of language – written documents about 3,000–4,000 years old – already show human language with all the features typical of language today. Older languages are not simpler versions of modern languages.

The idea that all humans once spoke just one language has been around for a long time. Before the dawn of modern linguistics in the early nineteenth century, people were bound to a theological view of the origin of language and thought that Classical Hebrew, the language of the Old Testament of the Bible, was the original language. In modern times, the idea of a single language was resurrected and discussed by the Italian scholar Alfredo Trombetti (1866–1929), who published a book on the subject in 1905.

But was there one original language from which all others derive? That depends on how the evolution of language is viewed. If it is seen as having arisen slowly over a considerable period of time with ever-increasing complexity, there was no one original language, in our modern sense, but rather a series of stages through which humans (*Homo sapiens* and possibly Neanderthals as well) and their forms of language developed. If, however, you believe the modern language faculty sprang into existence in a super-saturated, language-ready brain (see 'Grammar: How Did Compositionality Arise?' in Section 32.5 for more on this view), whatever resulted from this sudden event, in terms of an external manifestation of the language faculty, would have been the original language.

The First Language: Proto-World

Was there such a thing as Proto-World? The comparative method, used in historical linguistics, does not help us as it can only use written records going back about 3,000–4000 years, with speculation about what went before that. Undoubtedly, the various language families across the world are related to each other far back in prehistoric time. The difficulty for linguists has been determining what these relationships are on the basis of the earliest attestations for the language families we find on Earth today. All one can do is compare words and perhaps short phrases (usually inscriptions) and try and recognise possible relationships in the fragments of the earliest documented languages. Linguists nowadays tend to keep to more recent stages of language development but there have been attempts to reach greater time depth, notably with the Nostratic hypothesis, first proposed by the Danish linguist Holger Pedersen (1867–1953). This sees language families as stemming from a common source, families such as Indo-European,[9] Afro-Asiatic (which includes Arabic and Hebrew) as well as a number of Caucasian languages and some from central and eastern Asia along with Dravidian from present-day southern India. The basic technique for reconstruction, used by linguists, such as the Russian–Israeli Ahron Dolgopolsky (1930–2012), involves the comparison of basic vocabulary across many languages. But the gradual loss of this vocabulary imposes a limit of about 10,000–12,000 years on its use for linguistic research. This means that the validity of the macro-language family Nostratic is contested among linguists, not least because of the additional effects of persistent language contact which cloud the issue considerably. This is true for other superfamilies proposed for other parts of the world. The American linguist Joseph Greenberg (1915–2001) postulated that there were three waves of migration into the Americas and that this is the basis for his tripartite division of the native languages there into (i) Amerind, (ii) Na-Dene and (iii) Eskimo-Aleut. There has been support for this view from genetic investigations of various native American groups done by Luigi Cavalli-Sforza and his team at Stanford University, but many linguists remain sceptical about trying to reconstruct languages at such distance in time as there is no direct evidence available and strongly contest Greenberg's division. But even if allowed, such attempts would represent a time-depth of not more than

9 The development of the Indo-European languages from a postulated initial stage – called Proto-Indo-European – spoken about 4,000 BCE probably in the steppe region north of the Caucasus, is the most well-documented case of language dispersal and change.

20,000 years, probably only a quarter or fifth of the way back to the earliest human languages, subsumed under the title Proto-World.

Although such reconstruction is not possible, it is still interesting to pose the question of whether language started with just one individual or at least one small community? That would greatly depend on what type of primitive communication system one would regard as a language. Or did language arise at various places at different times? This view is connected to the divergent views on where *Homo sapiens* began, in one location or several locations independently. The latter view is definitely a minority view in present-day palaeoanthropology, but it has had its proponents, notably the German Franz Weidenreich, who worked in the early twentieth century at the fossil excavation site in Zhoukadian near Beijing. While multiple locations for the origin of modern humans outside Africa are not supported nowadays (haplogroup evidence speaks strongly against this), there is still the open question of whether *Homo sapiens* arose at one location in Africa or several. The existence of universal organisational principles in all modern languages would favour a single-location view, or a small cluster of locations. Depending on how sections of the original *Homo sapiens* group dispersed within and out of Africa, various serial founder effects can be observed: subsets of linguistic features cluster in areas outside Africa depending on the original *Homo* populations which moved there, a view supported by the New Zealand linguist Quentin Atkinson.

ONE ORIGINAL EXOLANGUAGE PER EXOPLANET?

The situation on a planet with beings which have evolved in manners essentially similar to ours would probably be not very different, again in principle. There would have to be some beginning for an exolanguage. Whether that would be known to the inhabitants of an exoplanet is anyone's guess. What would be interesting to know is whether exolanguages would have arisen independently at different locations on an exoplanet. This would give rise to the further question of whether the descendents of such early exolanguages would show structural and organisational similarities.

23.9 Language Change

As a research field, language change looms large in linguistics. It would be impossible to do anything near justice to the nuanced insights in this field within a few paragraphs. Here it must suffice to mention a few general principles of change which would likely hold for exolanguages as they move through time.

If we contacted exobeings we would find them at one point in their development. They would have a past behind them, just as we do. The same is true for an exolanguage: it would be a snapshot of its development through time. Assuming the existence of societies in principle comparable to those on Earth, the contact between speech communities from different societies would have led to language mixture over time. This would represent an instance of externally motivated language change, as would that triggered by the behaviour of speakers within a community in their desire to make their speech more like some subgroup, or less so, as the case may be.

There is also internally motivated language change, which is typically found in the transgenerational transmission of language. In this scenario speakers, usually in their early childhood when acquiring their native language, make slight changes to the system on the basis of internal considerations, often to render some set of forms more regular, historically verbs like *help* : *halp* > *help* : *helped* or noun plurals as in *fish* : *fish* > *fish* : *fishes*.

The internal and external forces in language change would lead to constant shifts in the form of externalised language across an exoplanet. However, if they had an internal language faculty, as humans have on Earth, this would remain stable and provide the framework within which variation would constantly occur.

On Earth, language diversification probably happened as follows: a language would have arisen in a group by its members using it for communication purposes. Within a community of speakers using this language tiny variations would have appeared over time. These variations would have been passed on to future generations and, given enough time, the variations would have led to major changes, resulting in later generations not necessarily understanding the language of many generations back. The rate of change in natural languages on Earth varies, depending on a variety of factors, such as contact with other groups. However, variation and change is found in all languages. If the variation is minor, we talk about dialects, different but mutually comprehensible forms of one language. Such a dialect becomes a separate language when this mutual comprehensibility no longer holds. An instance of that would be the development of modern Dutch from earlier dialects of German,[10] or the diversification of forms of Latin into the Romance languages we know today.

There are other reasons for language change. One very important one is the manner in which children in any given generation construct their language system from what they hear their parents/carers saying around them. If their system is not exactly the same as that which the previous generations internalised in their childhood, change has taken place.

So the genesis of languages is a bit like biological speciation: variation arises and at some later stage interbreeding is no longer possible and an independent species then exists. Similarly, variation in language, for the reasons just outlined, can lead to mutual incomprehensibility and then a new language is taken to exist. Like species, languages can be organised into groups and then larger groups.

The upshot of these considerations is that exoplanets, assuming they have a sizeable population, say hundreds of millions, if not billions, will have

10 Given the political independence of Holland and surrounding provinces, later the Netherlands, and the rise of a separate spelling system for the standard form of Dutch, the language became even more different from bordering German dialects.

many exolanguages as well. Some of these will be spoken by large numbers of speakers and some by smaller groups, just as on Earth, for reasons which have to do with the history of certain countries and the dominance of some groups over others. Exoplanets might also be subject to globalisation which leads to a reduction in variation across societies on a planet. This would in turn lead to most smaller languages disappearing, as is happening rapidly on Earth where the number of languages will be drastically reduced during the course of the twenty-first century.

24

Language and the Brain

The ability to speak a language rests on physical aspects of our brains. We can identify areas which are especially important for language, and we can examine individuals with language impairments to gain some insights into the manner in which knowledge of language is stored in the brain. This study of language in relation to the brain is called *neurolinguistics*. It is a special field which is becoming increasingly a focus of interest for linguists. It is true that it is not possible to pinpoint linguistic activity in the brain, to put the transmission of minute electrical currents between nerve cells in correlation with the production of language. Nor can linguistic structures be assigned to the information stored in these cells. Although the ultimate goal of linking biochemical and electrical processes in the brain with the production and reception of language is still a very distant one, the field is one which has produced significant research results, for instance in the area of aphasia – disturbances in the normal functioning of language, for whatever reason.

24.1 Language Areas in the Brain

The *cerebrum* is the part of the brain under the skull and is divided into two sections of equal size, called hemispheres (see Section 15.3 above for more details). These are joined by a 'bridge' of thick fibres called the *corpus callosum*. Each hemisphere of the brain controls the opposite half of the body: the left hemisphere is responsible for controlling the right half of the body and vice versa (technically known as contralaterality). The outer layer of both halves is called the cortex (and is greyish in colour when the brain is prepared after death, hence the colloquial term 'grey matter' for the brain). The entire brain contains a vast number of neurons (nerve cells) – something in the region of 85–90 billion) – all of which are interlinked by nervous fibres, which allow communication between the nerve cells. The communication away from a nerve cell takes place along axons, which are coated by a sheath of myelin. The pathways to the nerve cell are along dendrites. The central area around the nucleus of the cell is called the soma.

Signals between nerve cells must cross synapses, gaps in the membranes which are found between neurons and axons (Figure 24.1). These synapses are junctions across which information can flow or be impeded, depending on the concentration of chemical substances known as neurotransmitters, which regulate the flow of information through the brain via neurons. These are released when an action potential (an electrical charge) builds up on one side. They cross the synaptic gap and dock into receptors, leading to the action potential being generated on the receiving side. After this there is a re-uptake of neurotransmitters into so-called vesicles, which hold them until required for a future signal. The same principle is used not just in the entire brain, including parts responsible for language, but between nerve cells and muscle cells in various parts of the body.

If neurons are fired for a specific task, such as speech, can one ask if anything is happening in the brain when one is not performing any specific task. The answer would appear to be 'Yes'. There is a state labelled the *default mode network*, which basically refers to the state when you are lying there in a wakeful state, not engaged in a specific cognitive task. This state shows blood-oxygen flow in at least the medial prefrontal cortex, the posterior cingulate cortex and the angular gyrus, which is suspended when other regions become active for specific tasks, such as speaking.

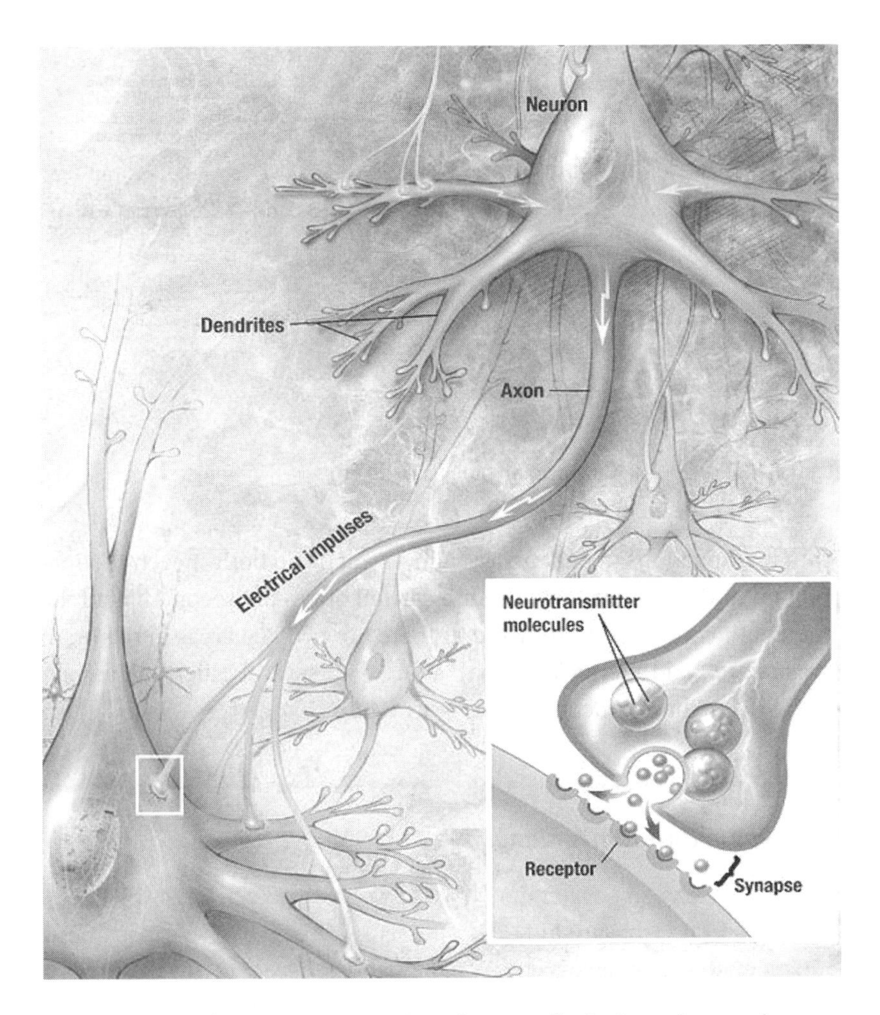

Figure 24.1 A schematic representation of nerve cells, the inset shows a close-up of a synaptic gap

Many functions of the brain are associated with one of the hemispheres. The final assignment of these functions to a certain hemisphere is called lateralisation, completed before puberty. This means that a dysfunction of an area of the brain after this watershed cannot normally be compensated for by the transfer of the associated functions to the opposite hemisphere. Puberty is also the cut-off age for acquiring a language with native-like competence.

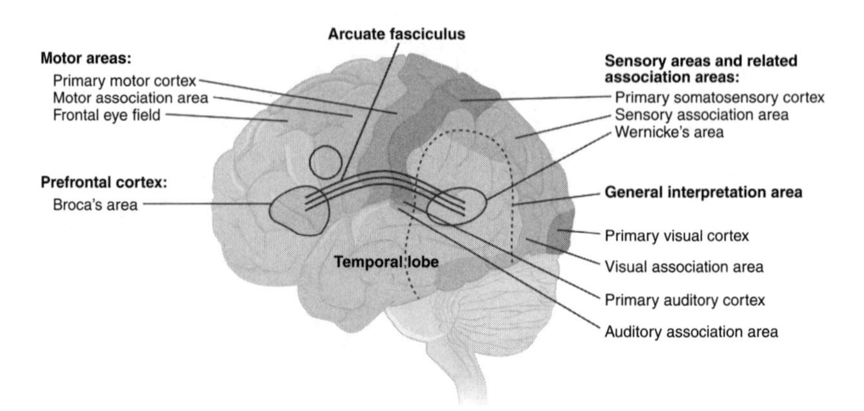

Figure 24.2 Regions of the human brain (simplified)

From investigations of patients with impairments to both speech production and understanding, which were carried out in the second half of the nineteenth century, it is known that there are two key areas in the brain (Broca's area and Wernicke's area), along with several others, all associated with language functions (see Figure 24.2).

Key areas in the brain for language

1. *Broca's area* is named after the French surgeon Paul Broca (1824–1880), who described the area in 1864 on the basis of information gleaned from patients with speech impairments. This general area is approximately an inch in size, located towards the temple in the left hemisphere of the brain and is associated with language production, covering a set of subareas. It relays instructions about pronunciation to the motor cortex so that phonation (sound production) can be initiated. Broca's area is responsible for syntactic processing as well. This area is about seven times larger in humans than in chimpanzees. In recent research it is more often referred to as the inferior frontal gyrus and includes the area indicated in Figure 24.2 above Broca's area.

2. *Wernicke's area* is named after the German scientist Carl Wernicke (1848–1905) who described it, in 1874. The general area is found behind the left ear and is responsible for the reception and comprehension of language.
3. The *primary auditory cortex* is the name of an area in the superior temporal gyrus, next to Wernicke's area towards the front. Initial sound processing is performed in this region, with pre-processed nerve signals passed onto the parietal and frontal lobes of the cerebral cortex, after which the sounds are perceived consciously. The area involved here can be larger than that indicated in Figure 24.2, spreading diagonally downwards into the temporal lobe.
4. The *arcuate fasciculus* is a thick band of nerve fibres joining the receptive areas on the left-hand side of the head with the (pre-) motor areas of the brain towards the front. It is responsible for feedback in language. Damage to the arcuate fasciculus leads to conduction aphasia where patients fail to connect what they hear to what they say.
5. The *primary motor cortex* is a strip which runs over the top of the brain with the part to the rear forming the somatosensory cortex. In front of the motor cortex is a region called the *supplementary motor area*, involved in muscular movements and coordination during speech production.
6. The *primary visual cortex* (at the back of the head) is not involved in the direct generation of speech but is responsible for the recognition of writing, both for scanning written text and when reading out loud.

24.2 The Binding Problem in Language

Just as there is an issue in binding the various sensory inputs to the brain into a seamless experience of consciousness there a similar problem in both language production and perception in accounting for how pronunciation, grammar and meaning blend in seamlessly when we produce and listen to language. Evidence from the structure of human languages and from language pathology support the ontological status of the different levels of language. They are organised differently but interact closely in the production and reception of language. Just how this happens on the levels of neuronal activity has yet to be demonstrated although it is clear that there are complex circuits of short connections in different cortical regions of the brain and their firing can be tracked in fMRI scans (by tracking blood-oxygen flow). For sounds, the pathways are evident as Broca's area communicates with the motor cortex/supplementary motor area to initiate pronunciation. But just how meaning and grammar interact and generate sentences on the neuronal level at stages prior to speech initiation is not known (see Section 15.5 above for a more general discussion of this issue).

Imaging studies have revealed that the inferior parietal lobule, a part of the parietal lobe, also called Geschwind's territory after the American neurologist Norman Geschwind (1926–1984), is strongly connected by nerve-fibre bundles with both the Broca and Wernicke areas. It may thus represent a second, alternative language pathway, parallel to the general pathway between these two areas via the arcuate fasciculus.

24.3 Evidence from Language Impairments

Language pathology is an area of linguistics concerned with language impairments, which can provide insights into the nature and organisation of language in our brains. Pathological aspects of language production and reception are subsumed under the general label *aphasia*, lit. 'lack of speech'. There are many subtypes which manifest themselves in different kinds of speech and/or hearing impairment. Insights from aphasia can be used for a number of purposes, for instance, in remedial linguistics where language specialists try to help patients, such as those who have suffered a stroke, to regain as much language ability as possible. But aphasia is also relevant to general linguistics, as the breakdown of language, and how this proceeds, can help us better understand how language in general is structured.

Dementia refers to a decline in cognitive ability due to brain damage or as a result of old age (senile dementia). This usually entails a gradual decline in language performance. *Dyslexia* is a condition in which individuals often fail to connect the written and the spoken word. Such persons have difficulty with reading and spelling, irrespective of their level of education. The diagnosis of dyslexia is problematic as it occurs to varying degrees and it is notoriously difficult to determine when it is pathological. Dyslexia may be an acquired condition with individuals who were previously literate.

24.4 Types of Aphasia

Aphasiac disorders typically involve damage to either the speech production (Broca's) or speech reception (Wernicke's) areas in the brain or perhaps the arcuate fasciculus or the supplementary motor area (see above). When discussing aphasia, scholars use the label *fluent aphasia* for any kind of language impairment in which the motoric aspect of speech production is not affected (Table 24.1). Where this is the case, the term *non-fluent aphasia* is used.

Broca's aphasiacs have typical disturbances in their speech. It is slow and difficult, such speakers seem not to manage grammatical rules, though their vocabulary is normally intact. The type of aphasia where speakers show a lack or confusion of grammatical words by, for instance, omitting small, grammatical words and often inflectional endings is termed *agrammaticism.*

Speakers with disturbances in Wernicke's area can speak normally, though what they say often makes little sense. Their sentences are semantically incongruous.

Table 24.1 Features of four major aphasias

Type	Fluency	Comprehension	Repetition	Naming
Broca's	–	+	–	–
Wernicke's	+	–	–	–
Conduction	+	+	–	–
Anomia	+	+	+	–

Interestingly, those with Broca's aphasia are aware that their speech is deficient and tend to respond to remedial treatment whereas Wernicke's aphasia patients do not, to anything like the same extent, as they do not perceive their speech as deficient. Both types of aphasiacs suffer from the inability to find words, technically known as *anomia*. Broca's area lies quite close to that which controls motor movements whereas Wernicke's area is close to the auditory area. The set of connective fibres between these two areas, the arcuate fasciculus, can also experience damage, causing *conduction aphasia*, whereby speakers can speak to some extent, retain good comprehension but are incapable of repeating what is said to them.

There is also a rare symptom called *isolation aphasia* in which the production and comprehension areas become severed from the rest of the brain. Speakers with this impairment can neither produce nor comprehend sentences and only repeat what was said to them or set phrases which were learned in childhood.

It is interesting to note that aphasia affects people's ability to use sign language just as it does with speech. Different types of aphasia can give us information about the structure of normal language. For instance, people with left-hemisphere strokes may sometimes use nouns but not verbs. This is also true when the nouns and verbs have the same form, so that speakers could understand and produce a sentence like *She prefers butter to margarine* but not *She buttered the toast while it was still hot*. This would imply that nouns and verbs are stored differently.

Aphasia can also help answer a key question in current research on language evolution: 'Is thought possible without language?' Investigations of persons with global aphasia, such as those reported on by Fedorenko and Varley (2016), with little or no ability to either produce or understand language, can be helpful here. These individuals, despite their near total loss of language, can perform basic arithmetical operations (addition and subtraction), solve problems involving logic, listen to music, engage with the thoughts of others and move around their environments with relative ease. In addition, the investigation showed that healthy individuals when engaging in similar activities do not show firing of neurons in language

areas of the brain. This comparison of aphasic with non-aphasic test persons indicates that, for some cognitive activities at least, the language areas of the brain are not active and that thought and language are indeed largely separate. There is a further conclusion to be drawn, this time from the examination of inherited mental disabilities, namely that they do not affect the language competence of the individuals in question to the extent that one might expect. This is additional, if only partial evidence for the separation of cognitive and linguistic abilities.

Specific Language Impairment

Very small variations in genes can have major consequences for the organism which is developing during pregnancy while other variations may be of no consequence. The *FOXP2* gene, which is crucial for our ability to produce vocal signals, can show a very slight variation that can have devastating consequences for the affected individual. This became clear in the case of the KE family investigated in Britain in the 1990s. Several members showed a slight mutation of the *FOXP2* gene which, in its expression, meant that the individuals suffered from facial rigidity, the inability to pronounce words like other native speakers, severe difficulties in the acquisition and use of vocabulary along with anatomical irregularities and cognitive deficiencies, resulting in the pathological condition known as *specific language impairment* (abbreviated *SLI*). Children with this condition would seem to be unaware of the existence of grammatical rules. The Canadian linguist Myrna Gopnik showed in her study of the KE family that some 16 of 30 members suffered from the defect over three generations. While media reports of a 'grammar gene' were somewhat sensational, it was shown by later research in the University of Oxford that *FOXP2* (located on the long arm of chromosome 7 in a region of about 70 genes) was the main gene involved in linguistic abnormalities, a result confirmed by the investigation of a further young male from a different family with the same defect. This would seem to imply that the linguistic abnormalities are genetically transferred. The implication of this is that the ability to grasp the rules of grammar in first language acquisition and to achieve native-like adult vocalisations is genetically encoded.

Williams Syndrome

There is a medical condition, called *Williams syndrome,* after the New Zealand cardiologist John C. P. Williams, who first described it in 1961. Patients with this condition can have quite severe disabilities with distinctive abnormalities in their appearance, both as children and adults. They have difficulties counting properly or carrying out simple tasks like tying their shoelaces. The condition results from the deletion of some 25–27 genes from the left arm of chromosome 7, which accounts for something over 5 per cent of our DNA (about 159 million base pairs).

However, the affected individuals are fairly normal speakers of their native language and just show a slight tendency to overgeneralise (they might say *speaked* for *spoke*, overgeneralising the weak verb type in *-ed*). They have a good command of grammatical rules, which shows that their language faculty is intact. Again, the implication of this is that our ability to speak language is largely separate from other cognitive abilities.

The Special Case of Reading

We are not born to read. In contrast to language, reading is not hard-wired into our brains. It is a learned ability and one which only arose recently in humans with the invention of writing (see Section 23.5 above). If we are not taught to read, we cannot perform this task. So how is it that people can read so effortlessly if they learn in early childhood? After all, school-educated

LANGUAGE IMPAIRMENT ON AN EXOPLANET

This question is embedded in the larger one concerning health and disease among exobeings. As has been emphasised repeatedly in this book, exobeings could only arise through a process of biological evolution similar in principle to ours on Earth. In such a process, variations would occur among organisms across generations during the genetic development of species. Some of this genetic variation would be deleterious and hence lead to impairment on various levels, including the cognitive/linguistic area.

individuals can read text shown in different fonts and sizes, with attributes like italics or bold (when printed), and can generally read handwritten text when letters are sufficiently separated from each other.[1]

The French cognitive neuroscientist Stanislas Dehaene (1965–) has maintained, on the basis of his research, that individuals acquire the ability to read by co-opting circuits in a certain region of the brain, the ventral visual cortex, dedicated to recognising shapes, for the task of assigning letter shapes to the sounds of their language. This region is also implicated in long-term memory, which facilitates the acquisition of the ability to read; it extends from the visual cortex, at the back of the head, under the brain towards the inferior temporal gyrus, where information from the visual cortex, including shapes of letters, are streamed for processing and then passed to circuits responsible for understanding spoken language. Learning how to read involves connecting letters to sounds, which in neuronal terms means establishing long-term connections between circuits for recognising shapes and those responsible for processing spoken language. After children have mastered the letter-sound correspondences of their native language they can then process these in whole groups, as words. In fact, with increasing skillfulness in reading, children (and later adults) only need to recognise the initial letters of words to recognise them and determine their meaning. This happens with such ease in real time that we do not even register the task of letter–sound matching but concentrate solely on the meaning of any text being read.

READING AN EXOLANGUAGE

Could exobeings read? To answer that question, one needs to consider how they might render their language in more or less permanent form. If this were by means of visual symbols, they would surely be able to scan these and recognise them as a means of visually representing spoken language. Just what such symbols might look like is, at present, anybody's guess.

1 There are, of course, individuals suffering from dyslexia, who experience difficulties relating the sounds of language to the letters used to represent these. This disorder can occur to varying degrees and is not incapacitating for the people afflicted by it. Dyslexia is different from general difficulties in reading, which speakers of languages like English or French can have where words are spelled in unexpected ways, as in English *gauge* [gɛɪdʒ] or French *faisaient* [fəze] '(they) did/were doing'. See remarks in Section 24.3.

25

Acquiring Language

The limits of my language mean the limits of my world.

Ludwig Wittgenstein (1889–1951)

Language learning is doubtless the greatest intellectual feat any of us is ever required to perform.

Leonard Bloomfield, *Language* (1933)

You have already performed the greatest feat of your life, although you most likely are unaware of it. This is the acquisition of your native language. Within the first few years of your life, you went from nothing to a fully competent speaker of the language(s) you were exposed to. That happens unconsciously, without any instruction,[1] in a very short time, with native-speaker competence as the result.

1 In the early days of generative grammar (late 1950s and 1960s), Chomsky talked of a language acquisition device (abbreviated to 'LAD'), a type of black box (in our brains) containing everyone's ability to acquire a native language in early childhood.

Would the same hold for exobeings? Indeed, would it be valid to assume that exobeings have a division of their lifespans into childhood and adulthood as with humans? Recall that for Darwinian evolution to occur there must be some way for an organism to reproduce and be gradually subject to natural selection. If sexual reproduction applied in the animal world of an exoplanet, reaching sexual maturity would be a feature of animal life. The period before this maturity would be the equivalent to childhood with humans, and the period after maturity is reached would correspond to our adulthood. It is probably fair to assume that the maximum degree of cognitive flexibility would apply before sexual maturity, as this is a developmental learning stage, and hence the acquisition of language would be likely to take place during this key period of an exobeing's life.

25.1 Are We Predestined for Speech?

It would seem so. Young babies shortly after birth show a tendency to pay attention to speech more than other sounds in their surroundings (this can be shown by tracking eye gaze with infants). It is as if they were predestined to focus on human speech and begin straight away with the job of listening to and dissecting speech as a preliminary to language acquisition.

The acquisition of one's native language is part of one's general cognitive maturation. Humans have a relatively long childhood, compared to higher primates, and it has often been speculated that in the evolution of the human species childhood was extended to accommodate the acquisition of a complex language system.

Despite the long human childhood, the acquisition of language is remarkably fast. Childhood lasts until puberty – somewhere between 11 and 13 years of age – but structural knowledge about language is actually acquired in less than half this time. By the age of five or six children have acquired mastery over all the closed classes of their native language, the pronunciation and grammar system. There would appear to be a critical phase in the first year of the child's life when it becomes attuned to the sounds of the language it is exposed to, with grammar following somewhat later and perhaps extending for the most intricate structures to later years of the first decade of life.

It is extraordinary that language acquisition is a successful process given the poverty of linguistic input and the fact that children receive little or no guidance in language acquisition from those around them. By 'poverty of linguistic input' is meant the unstructured and fragmentary language which children are exposed to. Even those children for whom this is true to an extreme extent, like children in homes or orphanages or with absent parents/carers, still acquire their native language as satisfactorily as those who grow up in an intact nuclear family. In contradistinction to second language learning, children are not first presented with simple sentences, progressing gradually to more difficult ones.

Evidence for the innate predisposition to language, the genetic headstart so to speak, can be provided by demonstrating unequivocably that, after acquisition of their first language, individuals have grammatical structures at their command and are aware of mandatory interpretations for which they received no evidence in the linguistic input from their environment during the acquisition process. Consider the following sentences: *Fiona believes she is intelligent, Fiona believes her to be intelligent*. In the first sentence, *she* can refer to *Fiona* or someone else, but in the second it must refer to someone else. Although mothers do not explain such differences to their children, the latter grasp and observe the distinction, given their understanding of underlying grammatical structure.

This brings us to what is called the 'bootstrapping problem': all that children are exposed to from birth onwards is a continuous stream of sounds. There are no pauses in this stream, no audible 'spaces', to help children segment the phonetic stream. So how do children know what to do? Nobody tells them to organise the sounds they hear around them into words and store these as units of language which they can retrieve later when producing language themselves. And yet this is precisely what children around the world do constantly, no matter what the language of their environment is. The only conclusion one can draw from this is that children have an a priori notion of word; that is, before they hear language, they know what to do with it, how to segment and acquire it. This ability is innate, it is part of the language faculty.

What such observations would also seem to imply is that language acquisition proceeds on the basis of predefined steps, which are determined by innate knowledge about language, this knowledge informing the children about how to manage and structure the information which they pick up about the language in their immediate environment. Another important observation supports this thesis: children correct themselves, given time. For instance, children often start off by treating all verbs as weak and say things like *singed, taked, goed,* or overapply the *s*-plural to all nouns, such as *foots/feets* for *feet.* Nonetheless, given time, they arrive at the adult forms of their own accord. Such features are technically called 'errors' rather than 'mistakes', which are irregular and more characteristic of second language acquisition. Errors are non-adult, overgeneralised features which occur because of the stage at which the child is at a given time (acquisition in as yet incomplete). They right themselves with time, when the child appreciates that many word classes contain a degree of irregularity.

The role of caretakers, above all the mother, has led some researchers to claim that their speech – often termed *motherese,* a deliberately simplified form of the adult language – is important for the child's acquisition of language. Opinions are still very much divided on this point and the evidence of neglected children who acquire language normally would point away from any importance of motherese, despite its obvious value as evidence of the mother's affection and concern for the child.

WHAT IS YOUR NATIVE LANGUAGE?

Your native language is the one which you acquired in early childhood, roughly up to the age of five or six. It is the language which children can speak before they go to school. Importantly, knowledge of this language has a depth and quality which cannot be attained later. Furthermore, people have a strong emotional identification with the language acquired in early childhood. A language learned later can never match the first language. Some individuals, so-called 'balanced bilinguals', may have two native languages, assuming that they have had roughly equal exposure to both in the early years of life.

25.2 The Absence of Exposure to Language

If children are not exposed to language during the first decade of childhood, they will not acquire it later. There have been some instances where this has happened, most of them also involving extreme traumatisation of the children. A well-known case from the mid-twentieth century is that of a girl, just called Genie to protect her true identity. She was kept confined to a single room, physically abused by her psychotic father and exposed to no language whatever until the age of 13, when she was seen by chance when her mother went with her to a local council building, stumbled into the wrong room and some staff immediately noticed the girl's extremely neglected and retarded state. After she was liberated from her domestic situation Genie was accompanied by psychiatrists and linguists in an attempt to provide a new world and a semblance of normality for her. She made some initial progress in general behaviour but none of any substance in language acquisition. Whatever the role her decade-long abuse played in the stunting of her personality, it became obvious that Genie would not acquire language to any degree comparable with other children. While she did manage to learn some words, she had no command of grammar.

An older case is that of Victor of Averyon (*c.* 1788–1828), a young boy who was found in 1800 in woods in southern France and assumed to be a feral or 'wild' child, who spent his life hitherto surviving on his own in the countryside without any human contact. Victor's case was studied in some detail by the French physician Jean Marc Gaspard Itard (1774–1838), who had taken Victor into his home and accompanied him for several years. Again, the progress made in language acquisition was truly minimal. Victor could only speak a few words and could not create any sentences or use French grammar in any way. There has been much discussion of this and other supposed cases of feral children. Nowadays many scientists doubt if such children survived in the wild since infancy, as that would have been well-nigh impossible. In the case of Victor, it is assumed that he was abused by his parents before running away, at which time he must have been at least six or seven, otherwise he could not have survived on his own in the wild.

The details and discussions surroundings such cases, and others like them, are not relevant to the present treatment of language. However, what they do show is that if children do not have exposure to language in the first decade of their lives then the window of opportunity closes and after that they cannot master grammar but can learn some words. This supports the Critical Period Hypothesis in linguistics (actually a hypothesis about a group of sensitivity periods), which maintains that language acquisition must take place and be completed before puberty if normal levels of language competence are to be attained.

The American linguist Susan Curtiss, who accompanied Genie for several years, is convinced that she had cognitive abilities on a par with normal intelligence but without a mastery of language. This ascertainment is of relevance to the ongoing discussion in linguistics of the relationship of language to thought. If children who have been deprived of the opportunity of normal acquisition in childhood nonetheless show approximately normal cognitive competence, this would support a weak connection between language and thought, if not indeed go a good way to confirming that the latter is possible without the former.[2]

25.3 Characteristics of Language Acquisition

The ability to learn one's native language is part of one's genetic endowment, ultimately encoded in DNA.[3] We have an instinct to acquire language, which is activated by birth and unfolds in infancy just as does the instinct to walk upright. Just like walking, it takes a while to get going but it will continue automatically, circumstances allowing. Thus, language

2 In a more recent case study of a woman known as Chelsea, even more compelling evidence for the modular separation of language and thought, including arithmetical ability, was found by Curtiss.

3 Mothers and care givers sometimes correct children but this is not necessary – children self-correct with time. Furthermore, there is no linguistic advantage in adults using supposedly simplified language – baby talk – as no child ends up speaking like that. Needless to say, showing attention to children is important for their normal psychological development.

acquisition can be compared to other instincts such as that to use one's hands or to develop telescopic vision or binaural hearing.

Intelligence has no direct bearing on acquisition, and children of different degrees of intelligence all go through the same process of acquiring their native language, although individuals can and do differ in their ability to manipulate open classes, such as vocabulary, and in the style of written language they master consciously.

25.4 Stages of Language Acquisition

Language acquisition consists of a sensorimotor component, responsible for pronunciation (phonetics/phonology), and a cognitive component, responsible for the non-physical aspects of language, the grammar and vocabulary. Children pass through clear stages of acquisition in the first five or six years of their lives. Within each of these stages there are recognisable characteristics. For instance, up to the two-word stage, nouns and/or verbs mainly occur. No children begin by using conjunctions or prepositions, although they will have heard these word classes in their environment. Another characteristic is *overextension*. Children always begin overextending, for example in the realm of semantics, by using the

---

Summary of main facts of language acquisition

1. Children automatically acquire the language spoken around them (or two if this is the case). No child refuses to acquire language in early childhood.
2. Children acquire their native language with ease and without instruction.
3. Children acquire their native language in a relatively short time (by five or six they can speak fluently like adults, though obviously not about adult themes and not with the vocabulary of an adult).
4. Children acquire their native language fully. They do not falter or get stuck for words as adults do in a language which they have learned after puberty.

---

word *dog* for all animals if the first animal they are confronted with is a dog. Or by calling all males *daddy* or by using *spoon* for all items of cutlery. The conclusion one can draw from this behaviour is that children move from the general to the particular. To begin with their language production is undifferentiated on all linguistic levels (Table 25.1). With time they introduce more and more distinctions as they are repeatedly confronted with these from their surroundings. Increasing distinctions in language may well be linked to increasing cognitive development: the more discriminating the children's perception and understanding of the world is, the more they reflect this in language.

These divisions of the early period of first language acquisition are approximate and vary among individuals. It has been noted that bilingual children start speaking slightly later than their monolingual counterparts, most likely because they have to absorb information from two languages before beginning to reproduce these.

Table 25.1 Stages of acquisition

Stage	Approximate age (in years and months)	Articulation/language
0	0.0–0.3	Organic sounds, crying, cooing.
1	0.4–0.5	Beginning of the babbling phase (random sounds are produced, often those not in the target native language).
2	0.10–1	The first comprehensible words. After this stage, follow one-word, two-word and many-word sentences. The one-word stage is known as the holophrastic stage. Telegraphic speech refers to a type of speech with only nouns and verbs.
3	2.6	Grammatical endings appear, negation, interrogative and imperative sentences are used.
4	3.0	A vocabulary of about 1,000 words has been attained.
5	5 / 6	The main syntactic rules have been acquired. Children are fluent speakers of their native language.

25.5 Abduction and Ambiguity in Language

In progressing through the stages of language acquisition children use a form of imprecise logic called abduction in which they choose the easiest and most likely explanation. Consider the following instance of abduction, consisting of two premises, but a conclusion which may well be false: (i) tigers have stripes, (ii) this animal has stripes, therefore: (iii) this animal is a tiger. Not necessarily, there are animals with stripes that are not tigers, such as zebras. Now consider this example from language: (i) adverbs end in *-ly*, (ii) this word ends in *-ly*, therefore (iii) this word is an adverb. Not necessarily, there are adjectives which end in *-ly*, such as *friendly, likely*, as in *a friendly letter* or *a likely story*. While children use abductive logic as a working hypothesis during acquisition, they revise their assumptions as they are exposed to evidence which contradicts these.

Languages contain ambiguity, some quite a lot if they do not formally mark grammatical categories or agreement patterns. However, speakers can handle this ambiguity because context and expectations regarding likely interpretations assist them. Research by the American linguist Barbara Lust (Cornell University) has shown that children under the age of three can already deal with a grammatical form which has more than one semantic interpretation, such as *Ernie took a bite of his banana and Bert did too*, where the preferred assumption by the test children is that each took a bite of their own banana. This interpretation is probably prompted by experience of the world: usually people eat their own and not each other's food. Our cumulative experience during childhood results in a store of encyclopedic knowledge about the world, which later guides us in making correct interpretations of vague or ambiguous language. Significantly, artificial intelligence does not have a comparable store of knowledge or direct experience of the world; this is the main obstacle to constructing software which can provide semantically acceptable interpretations of linguistic input.

25.6 Localisation of Language and Early Childhood

At the very beginning of language acquisition, in the first year and a half, neuronal activity when producing words and reacting to speech tends to be spread across both hemispheres of the brain. With the increasing knowledge of language, the left hemisphere becomes dominant, with the right hemisphere finally being removed from the process of language perception and production. However, this fixation on the left hemisphere is not unalterable. There are reported cases of children who had their left hemisphere removed or disconnected, usually to put an end to debilitating seizures, with the loss of language. However, within a couple of years such children can show progress is utilising the right hemisphere for the acquisition of language.

25.7 Language Transmission

Language is obviously passed on from parents to their children. But on closer inspection one notices that it is the performance (language in use) of the previous generation which is used as the basis for the competence (abstract knowledge of language) of the next. To put it simply, children do not have access to the linguistic knowledge of their parents. An initial, simplified form of this transmission can be formulated.

Transmission of language across generations (greatly simplified)

1. Linguistic input from parents (performance of previous generation)
2. Abstraction of structures from the sound stream by children
3. Internalisation of these (competence of the present generation)

Language competence is the abstract ability to speak a language, meaning that knowledge of a language is independent of its use. It is constructed during early childhood by combining innate knowledge of language in general (in the language faculty) with actual linguistic input from the child's surroundings. Performance is the actual use of language. Its features do not necessarily reflect the speaker's language competence. For example, when one is nervous or tired one may have difficulties speaking coherently, but one's competence remains.

The three stages in transmission listed above account for why children can later produce sentences which they have never heard before: they commit the sentence structures of their native language to long-term memory and have a store of words as well. When producing new sentences, they use the internalised structures and the vocabulary they have built up for themselves. This process allows children to produce a theoretically unlimited number of sentences in their later lives. However, certain shifts may occur if children make incorrect conclusions about the structure of the language on the basis of what they hear. Then there is a discrepancy between the competence of their parents and that which they construct. This is an important source of language change.

25.8 The Logical Problem of Acquisition

The following problem exists with regard to early language acquisition: it would seem impossible to learn anything about a certain language without first already knowing something about language in general. That is, children must know what to expect in language before they can actually order the data they are presented with in their surroundings, extract structures from sentences and ascribe meanings to the words they encounter.

25.9 The Evidence of Pidgins and Creoles

Children who have very poor input in their surroundings tend to be creative in their use of language. Grammatical categories which they unconsciously regard as necessary but which are not present in the input from their environment are then created by the children. This has happened historically in those colonies of European powers where a generation was cut off from its natural linguistic background and only supplied with very poor unstructured English, Spanish, Dutch, etc. as input in childhood. Such input, known technically as a *pidgin*, was then expanded and restructured grammatically by the children of later generations and is known in linguistics as a *creole*. Here one can see that, if the linguistic input from their environment is deficient, children create the structures which they sense are lacking, going on their own innate knowledge of language.

The implication of the above situation is that children look for language and if they do not find it, they create it somehow, so that they can externalise their internal linguistic knowledge. In this sense language is a true instinct because it starts to unfold of its own accord and does not need to be consciously triggered. During the process of creolisation children add considerable regularity to the largely unstructured pidgin input of previous generations (see 'Iterated Learning' in Section 32.5 below).

25.10 Is There a Gene for Language?

The simple answer to that question is 'No.' Much has been made in the media of the discovery of the *FOXP2* gene (see Section 24.4 above). This is responsible for the acquisition and control of vocalisation and is closely connected to our ability to speak. However, it is not the sole gene implicated in language. In fact, it is only to be expected that a system as complex as human language would not be triggered by the transcription/translation of a single gene.

The *FOXP2* gene was also present in Neanderthals, though not in identical form. It would, however, imply that they had a comparable degree of vocal control, which would have provided them with one of the preconditions for language.

ACQUIRING AN EXOLANGUAGE

To reach an advanced technology, exobeings would have to possess a language comparably powerful to human language. This is turn would require a degree of time to be transmitted across the generations and acquired by children (unless there was some biological process, which we do not know about, which would accelerate this acquisition). If the evolution of an exolanguage proceeded in principle in a manner similar to human language, acquisition would require some considerable time and involve the internalisation of knowledge about actual languages in order for exobeings to then produce language and transmit this to a following generation.

25.11 Constructed Languages

In the late-nineteenth and at the beginning of the twentieth century various scholars turned to the task of devising an artificial language, which could be used as a general means of communication among speakers who do not understand each other's language. This function is fulfilled by English nowadays but before the spread of the latter as a worldwide common language (technically called a *lingua franca*), other practical suggestions were put forward about how to fill the perceived gap.

A proposal which enjoyed brief, if intense, popularity was *Volapük* 'world talk' (in this language), put forward by the German priest Johannes Martin Schleyer in 1879. He devised the language 'out of pure love for troubled and divided humanity.' It was initially very successful, with conferences and public discussions. Even an academy for Volapük was founded in Paris in 1889. However, the language was very cumbersome with too many endings. The community of Volapük supporters found the language too much of an effort to learn and, around 1890, most of them switched allegiances to the following.

Esperanto is another artificial language invented by the Polish scholar Ludwig Zamenhof (1859–1917), again in the late-nineteenth century. It was intended as an easy-to-learn, regular language, which would, like Volapük, further international communication and understanding. The language is based on Romance elements and was intended to be easy for Europeans to learn. The name clearly suggests Spanish *esperanza* 'hope'. Of all the various proposals for an artificial language, Esperanto is the only one which can be said to survive to this day. Although it is not used widely, there is nonetheless a dedicated community of Esperanto users worldwide.

In the course of the twentieth century, various modifications of Esperanto have been put forward, largely to rid it of what were perceived as unnecessary difficulties in the language. Even the great historian of English, the Dane Otto Jespersen (1860–1943), offered a greatly modified version called *Novial*, which, like so many others, did not catch on.

Various suggestions have also come from Romance scholars, such as *Interlingua*, to devise a sort of common-core language which would be comprehensible for all speakers of Romance languages.

In the English-speaking world, attempts have been made to provide non-natives with simplified versions of the language for rudimentary communication. The English literary scholar C. K. Ogden put forward his proposal for Basic English in 1930. This consisted of only 850 words, but interest in the suggestion quickly waned because it was impracticable.

And then there are languages created for science fiction. The best known of these is Klingon, used by the *Star Trek* franchise as the language of a group of particularly belligerent people within the series of films. It was devised by the American linguist Marc Okrand in the 1980s and expanded in later years, leading to a considerable community of enthusiasts for the language. Klingon only uses sounds found in human languages but in combinations which are not indicative of any single language and which contribute to its overall strangeness. It is characterised by many retroflex and uvular sounds giving it a general 'throaty' character and has a syntax which contrasts with English, for example, by using the basic word order object–verb–subject (O–V–S). This is very unusual cross-linguistically, attested for the indigenous South American Carib language Hixkaryana, spoken by only about 500 people around a tributary to the Amazon river. However, it is common as a highlighting word order in many languages, for instance German, as in *Dieses Auto möchte er, nicht das andere,* literally 'this car (O) would like (V) he (S), not the other one'. Marc Okrand has said that he deliberately chose the word order OVS for its unusualness among human languages. Such a bias is rooted in the fact that the constructed language Klingon was tailored to the requirements of the entertainment industry and was not in any way intended by its author as an attempt to represent how he thought actual exobeings on a real exoplanet might speak, should they exist.

COMMON LANGUAGE ON AN EXOPLANET?

It is unlikely that on an exoplanet, with different civilisations and histories, only one language would arise. Furthermore, with an advanced technology encompassing the entire planet, the desire for one language would obviously have arisen. Whether this would be a natural language of that planet, like English on Earth today, or an artificial language like Esperanto, it is not possible to say.

26

Humans and Animals

> I argue that most components of the human capacity to acquire language are shared with other species and open to a comparative approach.
>
> W. Tecumseh Fitch, *The Evolution of Language* (2010)

On our Earth the dividing line between animals and humans is determined by the ability to speak. Assuming that evolution is the only manner by which complex biological structures can arise in our universe, the exobeings of any exoplanet must also have arisen through an essentially similar process of evolution as we humans have.

Language separates us from the rest of the animal world[1] and makes us what we are, beings who live in intricate social networks, with considerable cognitive powers and the ability to convey knowledge from one

1 This intuitive recognition of the greater cognitive powers of humans over animals led in previous centuries to the notion of a *scala naturae*, 'a scale of nature,' in which life forms were arranged on a vertical scale with humans at the top (just below God). See also Chapter III Comparison of the Mental Powers of Man and the Lower Animals – Language in Darwin's *The Descent of Man* (Darwin 1871).

generation to the next and thus build up vast bodies of knowledge which extend far beyond what a single individual could achieve in a lifetime. All this is possible because we can formulate our thoughts in language which others can then understand and engage in exchanges. Without this foundation there would be no science and none of the technological advances which we enjoy in today's world, built on the discoveries and insights of science. And language is used for much more, for the expression of literature, for the recording of history and for the myriad types of communication in which humans are involved in societies across the globe.

The basic question here is whether animal communication, specifically that found with higher primates, and communication among humans is different in kind or just in degree. Is there a continuum between our systems of communication and theirs? The research community is divided on this issue.

26.1 How Intelligent Are Animals?

Intelligence is difficult to define satisfactorily, let alone measure in the animal kingdom. There are different kinds of intelligence, the main two types being general and emotional. General intelligence among animals can be seen in their ability to perform cognitively challenging tasks and/or to influence their environment to their advantage. A well-known example is that of New Caledonian crows, which are able to prepare a stick with a hook-like end to extract grubs from places not easily accessible. They also show this capacity in captivity and display a learning curve in mastering such tasks.

Signs of emotional intelligence can be observed in advanced mammals, two of which are empathy and mourning. Empathy is feeling with other beings, subjectively appreciating their predicaments. Dolphins show empathy across the species barrier and have been known to assist humans who get into difficulty when out of their depth in water. Many dogs also show empathy, especially intelligent ones like Newfoundlanders, who can rescue humans from drowning. It could be objected that such dogs see all humans as potential food providers and hence have an intrinsic motivation to save them, but for dolphins (living in the wild) no such assumption can be made.

Humans mourn their dead. Cultures have rituals they engage in when a member of their group dies. Burial rituals can take on a variety of forms around the world and are usually connected with taking leave of dead individuals before they depart for 'the other world'. Such leave-taking involves mourning, spending time around the body of a deceased person and experiencing sadness at their death. Such behaviour has been observed with dolphins and many advanced mammals, such as elephants and chimpanzees, especially on the death of one of their young. Many non-mammals, such as egg-laying tortoises or spawn-laying frogs, do not stay around for their many young to mature and so would not even know if they were alive or not.

Mourning the dead is connected with aging in populations. If the latter contain older members, death by natural causes is common and mourning, possibly with rituals, will also be widespread. What role age plays in animal groups can help us determine their probable cognitive powers. Some groups of mammals may consist of more than two generations, such as elephants, certain whales and monkeys. In these cases older females ('grandmothers') can provide care for their grandchildren and the group can further benefit from the experiences of the older animals, for instance in procuring food. Transgenerational animal behaviour requires bonding beyond a single animal's own parents and implies higher levels of cognition, an issue which has been studied in detail by Anne Innis Dagg (see Innis Dagg 2009).

Deception: 'It Takes a Thief to Know a Thief'

The ability to deceive others may well be regarded as a quintessentially human trait. But among the more intelligent animals there are instances of this. Consider the behaviour of jays (Emery and Clayton 2001). One jay takes food and stores it, the second jay observes this and raids the cache.

Now the second jay is allowed to store food while being observed by another jay. When the observing jay is gone, the caching jay re-hides the food so that the observing jay will not know where the food is. One might think that this behaviour is evidence of a theory of mind among these birds. But it is more likely a case of the 'evil eye' hypothesis: animals react to being watched and it is eye contact rather than a theory of mind which is key to their behaviour, in this case deceptive behaviour to outwit the observing bird.

Cognitive Tradeoff Hypothesis

Our working memory is poor compared to that of chimps. They live in the here and now and need to count, for example, in an instant, to work out how many enemies they might be facing in an encounter. But we evolved language and the cognitive abilities to plan ahead into the future, something which chimps do not do. So we developed imagination, the ability to conceive of situations which go beyond the here and now. This is also central to sharing in social groups. One interpretation of this development is that we sacrificed some of our working memory potential to invest it in augmented imagination and language capacity, key abilities for surviving as collaborative social groups in dangerous environments in which many animals, stronger than ourselves, saw us as potential prey. This view is known as the cognitive tradeoff hypothesis, proposed by the Japanese chimpanzee researcher Tetsuro Matsuzawa (see Matsuzawa 2010).

Joint Attention in Humans

The comparison of humans with chimpanzees renders the profile of our specific traits all the clearer. Among the chief scientists in this field is the American linguist and ethologist Michael Tomasello, who has investigated and contrasted the early development of hominins. In his empirical work Tomasello showed that very small children from about nine months old onwards track eye contact with others, both children and care givers, and engage with them, behaviour which is not found with chimpanzees. With humans, eye contact is easy to recognise as we have visible whites of the eyes, which allow us establish the precise direction someone is looking in. This task is much harder with other non-human primates whose eyes are largely dark, rendering the direction of gaze difficult for others to determine.

From early childhood we humans also engage in activities which demand shared attention, such as looking at a book or playing a game together. Children also share experiences with others, for instance by showing a caregiver a particular object which currently occupies them.

Shared attention is not a feature of chimpanzee behaviour. They may engage in a task together but that simply serves the purpose of one of them gaining the spoils more easily. The collaborative behaviour of children with others, which is implicit in their shared intentionality, is clearly deliberate and involves commitment. Breaking this commitment, for instance by prematurely ceasing activity with an individual, and going to play with someone else, triggers guilt and is usually accompanied by an excuse for the break being offered to the other person hitherto involved in the shared activity.

Tomasello sees in the genesis of this human behaviour the seed of normative forces, indeed of human morality, which is grounded in commitment to common activities. A further correlate of this sharing among humans is that the gains of any communal behaviour must be shared equally among the participants, something which chimpanzees show no tendency towards.

A further corollary of this kind of human behaviour is our natural curiosity about others. We humans spend an inordinate amount of our time wondering what other people are thinking, not just pondering on what they have done in the past or what they might do in the future, but also what they are thinking about us. We are very sensitive to the opinions of others and if these are negative towards us this can be a major source of stress. From what we can tell (admittedly indirectly by observing behaviour) there is no direct equivalent to this aspect of the human psyche in the animal world.

Communication Across Species

Among animals, communication across the species barrier can occur and usually has a very specific function. For instance, there is a practice among several types of deer, above all gazelles, called stotting (or pronking from Afrikaans in South Africa, common with springboks) in which the deer jump up in the air for no apparent purpose, that is, they are not trying to overcome an obstacle when moving forward. This behaviour has been interpreted by ethologists (scientists investigating animal behaviour) as a signal to potential predators, such as large cats, that the deer is fit and healthy and would probably successfully escape a predator in any attempted chase. Another example would be the bright yellow colouring of the poison dart frog, causing it to be avoided by predators, or the orange colouration on a sponge which implies bitterness of taste thus convincing predators not to attack.

The study of cross-species communication is part of signalling theory and has found only a few types of such communication, mainly warning possible enemies not to attack or devour. Within a species, signalling effects are often associated with sexual selection, as the American evolutionary psychologist Geoffrey Miller has stressed, for example with peacocks and their flamboyant tails, an instance of what is know as Fisherian runaway.[2]

Only Once? Language Analogues in the Animal World

It is always difficult working with a set of one – one Earth, one species with language, etc. We do not know how unusual the single member of the set is. Help might be forthcoming in looking for cases where similar features,

2 This is a selection mechanism, here for the purpose of enticing females to mate in sexually dimorphic species, which seems to have gone far beyond what might be imagined as sufficient. The feature is named after the British biologist Ronald Fisher (1890–1962).

called shared traits, developed independently of each other; they are as close as we can get to a multi-member set. These are cases of analogous or convergent evolution, which are separate parallel developments, as opposed to homologous features that are descended from a common ancestor, however distant in the evolutionary past.

In the case of language there are some instances of partial convergent evolution which can be helpful. For this we have to look beyond our closest evolutionary neighbours. In this context, the American linguist W. Tecumseh Fitch, quoted at the beginning of this chapter, made the following observation:

> Some of these shared traits, like the capacity for vocal learning ... are not shared with primates, but instead have evolved independently in other lineages like birds or whales ... Therefore, in the vertebrate brain in particular, there are many good reasons to expect cross-species comparisons to yield deep insights, and even convergently evolved traits are often based upon shared genetic or developmental mechanisms. (Fitch 2010: 23)

Comparisons of language with animal vocalisation can be insightful, but often not in all aspects. Consider songbirds: it has been observed that young birds do not sing to their parents to communicate with them. Instead they apparently listen and internalise the song patterns they hear and, many months later, when their parents have left them, they try to sing, practising a kind of 'subsong', until they get it right and sing like other adults of their species. But despite the possible parallels between birdsong and human language it is obvious that birds use this song for very different purposes from the transgenerational communication which characterises young humans in the company of their parents.[3]

3 Cases of birds imitating human speech, as with parrots, do not constitute animal language. Nonetheless, there have been impressive instances like the African grey parrot, Alex, which could use about 50 words and combine these sensibly.

RELATIVE INTELLIGENCE AND COMMUNICATION ON EXOPLANETS

Could animals on exoplanets have communication systems of their own comparable to that of exobeings? That depends on whether life on exoplanets would show the same 'cognitive gap' between exobeings and other forms of life. Our Earth may be quite unusual in this respect: exoplanets may show a more gradual increase in intelligence across their range of life forms. Incidentally, this means that exobeings would probably not have the sharp distinction between themselves and other animals. How they might behave towards others similar to themselves but somewhat less intelligent does not bear thinking about. If humanity on Earth and the way it has treated those of equal intelligence throughout history is anything to go by, the prospects for less intelligent exobeings would be bleak indeed.

Part VI

Life and Language, Here and Beyond

Intelligent life on Earth arose due to the confluence of a series of very specific circumstances, which favoured its genesis. In principle, many of these would have had to apply on an exoplanet, for intelligent life to have arisen there as well. Importantly, to be in a position to communicate complex ideas efficiently and to implement many of these successfully in advanced technology would require very particular developments, above all in language.

27

Preconditions for Life

The discussions in this book so far have been about how life evolved on Earth and what paths it took, with the possible situation beyond Earth considered at regular intervals against this background. Some might say this approach is too conservative and that we should think outside the box for a while. After all, life could not just look, but also be very different, in principle, from life on Earth.

Is this really the case? Let's recap on some of the preconditions for life discussed in previous chapters. To reach the level of molecular sophistication, which we observe on our own planet, the biology of an exoplanet would most likely have to be carbon-based because no other element has the same potential to form such a huge array of different molecules. Another point to remember is that high-level functions, like those humans exhibit, with their large brains and intricate physiologies, would require a complex physical substrate. We will not find planets with rocks that talk, mice that write poetry or insects that compose music. But the differences will most likely be in appearance, in the phenotype of exobeings, and not in fundamental principles of organisation.

So while it is impossible to prove that life on an exoplanet will not be completely different from that on Earth, what is much more likely is that we encounter an exoplanet on which life evolved in a manner similar, in principle, to that here but where life has progressed further in time than we have, say a million years into a digital age. In fact, if we encounter an exoplanet with exobeings at approximately the same level of development as ourselves on Earth, at the beginning of their digital era, this would be sheer coincidence (see sections 1.4 and 27.7). Nonetheless, no matter how far ahead a civilisation on an exoplanet might be compared to us, its beginnings would show an evolutionary trajectory in principle similar to that we traversed on Earth.

So what might the situation be like on an exoplanet? Might there be an array of *Homo*-like species, each with a language faculty and languages to differing degrees? Or would there be only a single *Homo*-like species with language? If the latter were the case, then how would that have come about? Would there have been only one *Homo*-like species to begin with or did this species outcompete other previously existing ones as with humans versus the Neanderthals and Denisovans?

If exobeings were cognitively further advanced than we are, would their exolanguage(s) also be more advanced than our language(s)? Human language is adequate for expressing our thoughts and feelings in all the domains of life which we experience. So it would be difficult for us to imagine what a more powerful kind of language would be like – one with more words, more sounds, a more complex grammar? Nonetheless, the difficulty we humans might experience in imagining such a more complex language does not exclude its possible existence.

Assuming that exobeings have a language faculty, they could also be expected to have a 'language of thought' – cognitive linguistic mechanisms by which they organise their thoughts and process input from their senses. Furthermore, an evolved language faculty among exobeings would entail the inheritance of this faculty by all exobeing children and the ability to acquire any exolanguage equally well, irrespective of the language of their parents.

Again, one could object that this is pure speculation. However, if an exosociety with advanced technology exists, there must be a mechanism by which their exolanguage is passed on through the generations. And we have observed on Earth that, given enough time, Darwinian evolution leads to an increasingly generalised and optimised organisation of an organism's subsystems (see the Section 32.8 on the immune system below). This would promote the rise of a language faculty, which would then allow young exobeings to acquire the specific exolanguage they are exposed to.

27.1 What Can the Range of a Search Be?

We should remind ourselves that when talking about life beyond Earth in any practical terms we can only consider its possible occurrence in our immediate neighbourhood of the Milky Way galaxy. Ours is a spiral galaxy of roughly disk-like shape with a central bulge. We cannot see the stars on the other side, and any planet roughly at our position on the other side of the galaxy would be in the region of 40–50 thousand *light* years away from us. Consider that to send a signal to the other side of our galaxy at 300,000 kilometres per second would take 20–30 times the length of time from the Roman Empire to today. And that is just sending a message, the return journey would require that same length of time again.[1]

When we talk about possible life in our galaxy and whether we can detect it, we are actually referring to the possibility of life within a radius of not more than, say, some few hundred light years around us. And this would just be for signal detection. Any practical two-way communication would be restricted to a distance of a few light years from Earth. It is a sobering thought to bear in mind that this is only a tiny patch of our Milky Way galaxy, which itself is only one of hundreds of billions of galaxies in the observable universe, let alone the entire universe, most of which is beyond our field of vision.[2]

1 Of course, one-way communication would be possible and we might indeed some day receive a signal sent from within a distant solar system. But distance would be critical if we wished to engage in two-way communication.

2 The continuing expansion of the universe must also be factored in here: this makes it less likely for light from distant galaxies to reach us, and this situation is exacerbated by the fact that the rate of expansion of the universe is actually increasing.

27.2 The Panspermia Hypothesis

Despite all the advances in evolutionary biology in the last century and a half we must not lose sight of the fact that we still do not know how life got started in the first place. This is the issue of abiogenesis and experience with an exoplanet might well throw light on the matter. Previous notions of a force which could lead to the spontaneous generation of life (a view going back to Aristotle) are not of any help here.

Some scientists concerned with the matter have posited the idea that life did not arise on Earth but was brought here from beyond. How? Via asteroids, meteorites or comets during the early days when the Earth was still being formed. It is true that bacteria could have been contained in the rock of the asteroids/meteorites and the ice of the comets, which then landed on Earth (surviving entry through the atmosphere) and spawned life on our planet. Certainly, asteroids and comets are known to contain organic compounds so it is not inconceivable that they could have carried simple single-celled organisms (prokaryotes) to Earth. Life on an exoplanet may itself stem from some other outside source such as an asteroid or comet impact containing microbial life from elsewhere.

Nonetheless, the so-called panspermia hypothesis does not explain the origin of life, it just says it may have come from elsewhere; it pushes the question back, it does not answer it. A crucial issue is self-maintenance and reduplication of cells. We know that cells can form easily by material being encapsulated within a lipid bilayer membrane, with increasing differentiation within the cells occurring later. But what triggered the duplication and division of cells in the first place? And did cell-like structures arise with only very few in one place, starting to self-maintain and self-replicate, or did these features arise at several locations on Earth more or less simultaneously?

27.3 What Can Be Assumed about Exolife Forms?

Bear in mind that the very basic components of life forms, the elements carbon, nitrogen, oxygen and hydrogen (along with some phosphorus, iron, calcium, sulphur and a few other elements), are abundant across the universe. This is also true for many organic molecules (complex carbon-based substances) and for water, the ideal biosolvent. When considering planets in the habitable zone of their parent stars it might seem reasonable to assume that ever-increasing complexity will arise starting with these basic elements. We know from Earth that atoms and simple molecules have a staggering ability to aggregate and form more complex structures, resulting in ever larger levels of scale with properties and abilities not discernible on lower levels (strong emergence). We also know that life can sustain itself in inhospitable environments, such as fish in the saline pools of Death Valley, California or at the high temperatures of geysers or in the vicinity of black smokers on the ocean floor.

It is true that extremophiles can exist in sulphurous pools at temperatures above boiling point on the one hand and others can survive for long periods in sub-zero temperatures on the other (by storing glycerine in their cells, which lowers the freezing point and prevents the water from turning to ice and bursting the cells). But is it likely that exocivilisations would arise at these extremes and not in the broader mid-range, in temperatures between say 15 °C and 30 °C, where chemical reactions can take place easily in liquid water as a biosolvent?[3]

Life is opportunistic: there is no reason why advanced forms of life should not arise in a seething sulphurous pool. Indeed, they might develop some means of dealing with this situation. However, the question is whether the complex forms of life, which we can expect of an exocivilisation, would be able to evolve in such an environment? And there is a lot of water in the universe: in our Solar System there is our Earth but also places like the Saturnian moon Enceladus and the Jovian moon Europa (see Section 8.10 above); and there are many ice comets whizzing around and such celestial objects probably provided water to planets and moons in the distant past. However, the consideration of extremophiles is important for the likelihood of life elsewhere. The reasoning is that if life can survive in extreme environments, it is all the more likely to arise in more moderate environments on other planets.

3 There have been criticisms of this somewhat conservative view of exolife, some might say myopic, requiring a rocky planet with a temperature range of about 15–30 °C with an abundance of water.

An essential part in the process of increasing structural complexity is that of life itself: the replication of material within a cell and the subsequent division into two and then more and more cells. What is probably necessary here is the presence of liquid water, the right temperature for chemical structures to arise and persist and to partake in reactions. A lack of external destructive forces, that is to say environmental stability, would be necessary over long periods of time, at least some hundreds of millions of years, even better, a few billion years like on Earth. If an exoplanet has these conditions, can we assume that life will arise there because it has done so on Earth? This will depend on whether we are somehow special in a material sense and not typical for our universe. Can we assume that, if the conditions for life are right, it will happen, given that we cannot identify any unique conditions on Earth which gave rise to life and would not be found anywhere else? If we do not want to assume this, then we must postulate some set of unknown conditions which will prevent abiogenesis elsewhere. And that contravenes general scientific procedure, namely that you can only assume things for which you have evidence and which are necessary to explain the phenomena at hand.

THE OUTCOMES OF EMERGENCE

The elements which any exoplanets and exobeings will consist of will be the same as those on Earth. Not only that, but the properties of the various elements, such as metals, halogens, gases, determined by the numbers of electrons and the shells they form around atomic nuclei, will be the same. The combinations they can enter, the bonds they form and the reactions they participate in will be exactly as those determined by the chemistry we observe on Earth. This means that any differences between ourselves and exobeings will manifest themselves on higher levels of emergence, above that of atoms, molecules and their interactions. The question is then: 'To what extent will the common foundation of physics and chemistry, shared across the universe, determine the outcomes of complex cycles of emergence?'

27.4 Habitat Independence and Flexibility

Consider this question: Could exobeings live in darkness and be night hunters, availing of infrared vision and supersonic echo location like bats? In theory, they could. But that would restrict them to certain types of habitats and thus probably limit their potential to construct advanced technological artefacts. Could exobeings use infrasonic sound (below 20 hertz) like elephants? Again, in theory, they could. But they would need a very large body volume with an internal organ capable of producing such low-frequency sounds. Certainly, this would have the advantage that such sounds could be perceived over long distances. With whales in water, which is a better conductor than air, sounds can be picked up by different whales at distances of tens of kilometres. However, this would make exobeings dependent on a specific habitat. An important key to our success on Earth is the extent to which we can adapt to virtually any environment, hot or cold, high or low, wet or dry, an ability which we have to a greater extent than other animals, not least because we construct housing which shields us from adverse environmental conditions. Of course, over time, species can spread to new habitats and different types can be found. For instance, the snow leopard lives in the alpine regions of Central and South Asia and is related to other members of the *Panthera* genus, tigers, lions, jaguars and leopards, which generally live in warmer climates. Additionally, we can control and manipulate the food chain, by farming plants and keeping livestock, an option which animals do not have, to anything like our extent.

27.5 To Recap: The Likelihood of Life

Life is tenacious as it can occur in many types of environments. Complexity will arise where the conditions are right. But just what are these conditions? The absolute minimum for an exoplanet with exobeings capable in theory of communicating with us are listed below (for (2) to (5) to be realised, language would be necessary). Would exobeings need to be land-dwelling creatures to develop advanced technology? They could be amphibious, but entirely marine creatures would experience great difficulty in constructing any equipment necessary for interstellar communication. Indeed, is it not unlikely that marine creatures would be aware of the universe above the surface of their water world?

When starting out from Earth one can posit three major stages for life in the long path from the outset to ourselves. First of all, we have single-celled prokaryotic life forms. Then, with the rise of complex cells, eukaryotes, with a membraned nucleus housing the organisms's DNA along with other structures within the cell, life moves forward to larger organisms, roughly those visible to the human eye, such as insects, birds and fish. Somewhere in the great diversity of this life we arose, from the higher primates on a lineage which began some six to seven million years ago. The genus *Homo*, with its various species including *Homo sapiens*, came into existence.

Putting these three divisions of life in a sequence one can offer an assessment of their likely occurrence throughout our galaxy (Table 27.1). Note that the sequential development outlined here in no way suggests that humans are at the top of a tree of life in some evaluative sense, as used to be posited in the long superseded pre-scientific notion of a 'great chain of being.'

Minimal requirements for exobeings on an exoplanet

1. A rocky planet with liquid water orbiting a stable star in its habitable zone, in a calm and stable stellar environment for several billion years; the star should not emit large flares and the planet should have a magnetic field and suitable atmosphere to shield it from cosmic rays
2. Exobeings with the manual dexterity and cognitive powers to build and maintain sophisticated technical artefacts
3. The transmission of specialised knowledge across the generations
4. Large-scale societies with a division of labour across a range of activities
5. The attainment of digital technology capable of communication beyond the exoplanet

---

Table 27.1 Likelihood of life, in decreasing order

Category	Occurrence	Some preconditions
1. Microbial life	Common	Cell development, maintenance and reduplication
2. Animal life	Fairly common	Complex cell forms
3. Human-like exobeings	Extremely rare	A 'runaway' brain and manual dexterity

The first category may well occur already within our Solar System, as single-celled bacteria in the underwater worlds of Europa or Enceladus or possibly still in the surface fissures of Mars. Assuming sufficient time and suitable conditions, there is no reason why bacteria could not develop into fish and later animals, assuming, however, that they manage the vital step from prokaryotes to eukaryotes, a step which took nearly two billion years on Earth.

My personal view is that microbial life is widespread across our galaxy and, by extension, in the other galaxies of the universe. The conditions under which it arose on Earth are the same as in any other comparable solar systems with water-world planets or moons. Furthermore, I think that animal life would also be fairly common on planets which share environmental parameters with our Earth.[4]

In my opinion the last step, from animal to exobeing, which on Earth was the evolution from hominids to early *Homo* species to *Homo sapiens*, is the real crux in the development of life with intelligence similar to ours. What selective pressures or environmental advantages led to a great increase in brain size, far beyond the requirements of our environment, after we began our development as the *Homo* species? In the various chapters of the present book this question has been pursued and answers have been suggested.

4 The presence of molecular oxygen in the atmosphere may be one of these factors. However, not all forms of life require oxygen and not all photosynthesise. But all those on Earth are carbon-based and require water.

The step from animal to human-like life forms would have further interesting aspects, involving, above all, the relationship of exobeings to their own animal world. Given that predation, and the carnivorous diet that this entails, was a strong factor in the evolution of higher cognition on Earth, is it likely that, in their evolution, exobeings were, and could well still be, carnivores? If this were the case, it would imply many additional factors: Would they breed animals for slaughter and consumption as we humans do? Would they have transitioned to a vegetarian diet? What would the biosphere of an exoplanet, with its plant and animal life, look like and how would it be managed by exobeings? Would they use animals for work and keep some as pets? How much of the space on an exoplanet would still be allocated to the animals from which exobeings would have evolved?

27.6 The Role of Serendipity

Given that all speculation about life in the universe stems from observations on Earth, we simply do not know how much serendipity,[5] fortuitous advantageous events/developments, which occurred throughout our evolution, might apply to exobeings in their evolution on their exoplanet (see Section 7.5 above). Some examples from Earth are obvious, such as the disappearance of the non-avian dinosaurs after the K–Pg extinction event (the asteroid strike about 66 million years ago). But do we know whether the step from prokaryote to eukaryote was a case of serendipity or a common, quick development on most exoplanets, which happened to have taken an excruciatingly long time on Earth (well over a billion years)? And then there is the good luck of not having been struck by a huge asteroid or sterilised by the intense radiation of a nearby supernova or by a direct hit from a gamma-ray burst.

5 The term 'serendipity' was apparently first used in a letter by the English novelist and man of letters Horace Walpole (1717–1797), with reference to an originally Persian folk tale *The Three Princes of Seredip* (Serendip is an old word for Sri Lanka), an Italian version of which was published in Venice in 1557 by one Michele Tramezzino. The eponymous characters of the tale were always making happy discoveries by chance. With time, the term came to refer to any unplanned fortunate happening.

The greatest example of serendipity in our time is the fact that the 100-metre-wide meteor which caused an explosion in the Earth's lower atmosphere on 30 June 1908 did so in probably the most uninhabited part of our planet, central-eastern Siberia. Only a few people died in the event which devastated an area of over 2,000 square kilometres. Imagine the destruction and loss of life if that explosion had occurred over a large city or, for that matter, over the sea, triggering a massive tsunami dwarfing that which hit the east Indian Ocean in December 2004.

27.7 Being Out of Sync

Much of this book has been concerned with evolution and palaeoanthropology. Now consider that we might come across an exoplanet for which these stages of its existence lie in a vastly more distant past than for us, say by their being already one million years into a digital age.

And what would an exocivilisation look like which was one million years into its digital age, let alone 10 million or 100 million years, assuming that such a civilisation had managed to survive that long? Just think of the changes to our technological environment in all realms of society in the last 100 or just 50 years. Now such an exocivilisation would have developed much, much further than we have in the past century or so. Just what the world of such an exocivilisation would look like is no longer a consideration for informed, scientific speculation, the procedure used in this book, but would represent pure speculation. However, evidence of such a civilisation might be forthcoming in the form of interstellar robotic probes, which could move autonomously through space, just as the two Voyager missions, launched from Earth in 1977, are still doing, drifting slowly into the vastness of space beyond our Solar System.

And what about the opposite direction? Imagine there is an exoplanet with exobeings, but they are all 10,000 years behind us in their technological development. In front of them might be a great future in digital technology, but they would not know it and we would not either. Not only that, but all the present-day terrestrial efforts to find life elsewhere in our cosmic neighbourhood might not detect such an exoplanet and hence not know of its inhabitants.

27.8 Post-Human/Post-Biological?

The literature on futurism contains certain buzz words. 'Post-human' is one and 'post-biological' is another. As a linguist I am sceptical of the use of such terms as they imply that they refer to reality. Just what does 'post-human' mean? In a scientific context it can mean a situation in which the uncertainty and unreliability of human behaviour and the vulnerability and fragility of the human body have been replaced by an altogether more predictable and consistent machine-like existence within the framework of artificial intelligence. The picture often painted in popular representations is one where artificial intelligence takes over humans and we are then dominated by supercomputers. 'Post-biological' refers to an extension of this thinking where our biological evolution comes to an end and we transfer to an entirely electronic, digital existence. Some respected scientists consider this development a distinct possibility, at least for exobeings far ahead of us time-wise.

In the relevant discussions, in literature and the media, the scenarios are appropriately vague. Just how we would shift from a biological to a digital existence is never clearly stated, but the transition is nonetheless presented as a real likelihood, nay a certainty. Of course, if we continue to exist for, say, another 10 million years, we will have developed in ways we cannot conceive of today and technology will have come to infiltrate every realm of our life; indeed, it has done that already. But whether we will have discarded all traces of our biological existence, and whether that would have happened on an exoplanet with several million years as a head start, is uncertain. The brain is inextricably intertwined with the sentient body which contains it. And it is the interaction of the feeling brain with its biological envelope which triggers the unique sensation of human consciousness. No machine can ever vie with this delicate interplay of thoughts and feelings which constitutes our inner life and which receives expression in language. Enhancements to the human brain are definitely an option, to counteract visual or auditory impairments, for example, but replacing the brain and doing away with biology in some future 'post-biological' existence is a very different matter. If we continue along the route of ever advancing digital technology, it is likely the enhancement of our biology rather than its replacement which will be characteristic of forms of human life in the foreseeable future.

28

What Might Exolife Be Like?

What might exobeings really be like? To begin answering this question consider the deep history of our own evolution. Would the evolution of exobeings show the same key turning points we find in our own evolution? Would they develop complex multicellular life forms at an early stage and then move on to become vertebrates with a central nervous system controlled by a brain, allowing them to move around freely in their surroundings?[1] Indeed, to what extent would what appear to us as preconditions, vertebrae and a skeleton to stabilise an animal's body, be necessary on an exoplanet?

The degree of variability on an exoplanet could be similar to that found on Earth, with a certain range for cognitive ability (types of intelligence), personality, aspects of physical appearance such as size, eye and hair colour or shape of skeleton (observable in body build, hands, feet, arms, legs, etc.).

1 It would be interesting to consider how their mental lives might pan out. Would they have psychiatric diseases, like schizophrenia, and possibly further clinical conditions like depression or continua of deviations from their norm, such as autism spectrum disorders? And, of course, exobeings may have further conditions or diseases which we do not have due to the anatomy and physiology of their brains and bodies.

Would there be two sexes or would exobeings be hermaphrodite, each capable of both inseminating another exobeing and of conceiving, bearing and giving birth to a new exobaby? Assuming they reproduce sexually – and that is a very big assumption – they might well display sexual dimorphism, with females different from males, maybe somewhat smaller, with less muscle substance, more body fat, etc. But in fact females might be larger than males, of course, as is the case with many animals on Earth, apart from mammals and birds.

Consider the issue of appearance: if exobeings lived on a planet with a cool, faintly glowing red star, they would probably need a greater light intake than if their star was like our Sun. However, they might well avail of infrared light coming from their star. Exobeings might have large eyes or even more than two. Imagine an exobeing with four large eyes in the centre of a face, each moving independently, nostrils on a forehead and a mouth under a chin. This would still be structurally close to Earth mammals but we might find it aesthetically disturbing because it would be beyond the limits of variation which we are aware of from our own world.

Would they have a circadian rhythm, assuming that the exoplanet in question would spin on its axis like Earth and be characterised by periods of light and darkness? This question is relatively easy to answer. We know that many planets orbiting red dwarf stars are very close to their parent star and would be tidally locked to it so that there would be no night and day,[2] just a twilight zone at the transition between the star-facing side and the dark side of the planet.

Another issue might be the number of moons a planet has. There is great variation in this respect. We have just one, but Jupiter has over 80. Now imagine than an exoplanet had not one but five moons. Then the figure five would have a much greater cultural significance than it has on Earth. Assuming that an exoplanet is a rocky planet with liquid water (like Earth), would the moon(s) cause tides in the oceans? If so,

2 This simple fact would have considerable ramifications for the cultural life of exobeings. Just think of what literature on Earth would be like without all the references to night and day, to light and darkness.

this phenomenon would be of cultural significance for exobeings on that planet, just consider our metaphorical use of phrases referring to tides. like 'ebb and flow'.

28.1 Lifespan for Exobeings

We are born, live our lives and die. For exobeings this would be the same. Otherwise they could not have evolved on their planet. Just like us, they would need a certain lifespan during which to go from birth through childhood to adulthood (characterised on Earth by sexual maturity), then have offspring and pass away at some later stage. Given the operation of natural selection and random mutations, along with genetic drift and gene flow between populations, life forms on an exoplanet would evolve in the Darwinian sense just as on Earth.

But what would the typical lifespan of an exobeing be like? We have long childhoods (compared to other animals) and use these to acquire a variety of personal and social skills, not least of which is language. We then get to puberty, in the early years of our second decade of life. By our mid-teens we are ready (biologically) to have offspring. Assuming Darwinian evolution for life forms on an exoplanet, exobeings would have a period of maturation between birth and their having offspring, what we call childhood. If they have exolanguages comparable to the languages on Earth, it may be fair to assume that their childhoods would be long compared to other animals on their planet. This assumption is bolstered by observations on Earth: animals with high cognitive abilities have extended childhoods, say a number of years. Elephants, for instance, do not reach sexual maturity until at least 10 years (somewhat earlier for females than for males). This is strikingly different from other forms of life, which often mature between spring and autumn within a given year; some smaller forms of life, such as insects, mature much more quickly.

It is more difficult to estimate what the adult lifespan of exobeings would be like. This would depend on a number of factors, not least of which would be their rates of metabolism, and cultural issues such as advances in medicine, which have greatly extended our own chances of reaching a ripe old age. Much of this is due to the successful fight against infectious disease and the treatment of systemic illnesses like heart disease and cancer. Nowadays, in the early twenty-first century, most people in the Western world can expect to live well into their seventies (women somewhat longer than men), assuming nothing untoward befalls them. For the advancement of science and technology this has advantages, as trained individuals can contribute to their field of expertise for a longer period of time.

If exobeings live in societies with language and sophisticated technology, can one automatically assume that their lifespans would be much greater than those of other forms of life on their planet? Considering forms on Earth, one can see that lifespans vary greatly, and so the answer is 'No.' Some animals, especially those with slow metabolisms, live to great ages. A famous example is the Greenland shark, which has been estimated to have a lifespan of anything between about 300 and 500 years. One moral of this story is that longevity does not correlate with intelligence in the animal world. Octopuses, for all their neuronal flexibility and learning ability, live pitifully short lives, between six months and a few years, normally dying after mating (males) or their eggs hatching (females).

So just how long could the lifespan of exobeings be? This is an open question. We know the factors which contribute to aging among humans, such as telomere shortening on the chromosomes of our DNA, a general deterioration in cell quality, most noticeable in changes to our appearance as we get older, and possibly changes to epigenetic factors on our DNA. But it is known from the famous HeLa cell line[3] that it is possible to start a

3 The label HeLa is an acronym from Henrietta Lacks, an African-American woman who died of cervical cancer in 1951. When admitted to the Johns Hopkins Hospital in Baltimore, Maryland, some of the cancerous tissue was removed and maintained as a culture (without seeking the consent of the patient or her family). The cells of this tissue proved to be remarkably durable and a line was derived from the original tissue sample. These were then disseminated among medical researchers and today there are billions of HeLa cells used in laboratories, all deriving from the original cancerous tissue removed from the patient some 70 years ago. This has basically shown that an 'immortal' cell line could be produced outside the human body.

cell line outside the body, which can basically endure indefinitely. If science and technology on an exoplanet were more advanced than on Earth, it is in principle possible that exobeings could discover a means of halting the aging process. How likely this is we do not know and such speculation is fruitless. However, it is worth noting that if exobeings succeeded in stopping, or significantly decelerating, physical aging, this would have a detrimental effect on their biological evolution, which would depend, as on Earth, on a high turnover across the generations, providing a favourable scenario for Darwinian natural selection.

28.2 What Would Their Average Size Be?

The average size of humans is very roughly halfway between that of our Sun and the sub-microscopic world of atoms and molecules. We live on a scale defined by the size of our bodies, our limbs, the distances we can walk with our legs, the size of things we can grasp with our hands and what we can see with our eyes. When dealing with objects outside our normal scalar range, we use tools and machinery, whether to lift an object weighing tons, when loading containers, for instance, or when manipulating the tiny moving parts of a small gadget like a wrist watch.

What would the normal scalar range be for adult exobeings? This would be determined by a number of factors. Certainly, the supply of oxygen and nutrients would affect their average body size, as would gravity on an exoplanet. This would also determine things like the weight and resilience of a skeleton because gravity would not only affect the weight of an exobeing's limbs but things like the damage falling could have, or the relative weight of lifting other bodies, such as animal prey.[4]

The scale of living things on an exoplanet would define what elements of their planet they would intuitively recognise: to begin with they would

4 It is no coincidence that the largest animals on Earth are found in water, given the greater buoyancy it offers compared to air. Blue whales are the largest animals ever to have existed and can weigh well over 150 tons with a length of nearly 30 metres.

conceptualise objects on the level of their approximate body size. Then they would have to discover the world of the very small, that of chemistry and physics, by the slow advancement of science,[5] just as happened on Earth. And, if they became space travellers, they would have had to master the technology necessary to attain the escape velocity of their planet and reach space beyond it.

28.3 Alternative Ecologies and Behaviours

An exoplanet orbiting a red dwarf would probably be quite close to this star as the habitable zone (where water would exist on its surface in liquid form) would be in a tight orbit around the star. Leaving aside the dangers of a planet being struck by solar flares at that small distance from a star, there is the issue of tidal locking. A celestial body is tidally locked to another if it always shows the same face to the gravitationally attracting body, as is the case with our Moon vis à vis the Earth.

This question would be of great significance for exobeings. If tidally locked, they would have no diurnal rhythm determined by the relatively quick rotation of their home planet on its axis as we do on Earth. So how they would rest would probably be different from the manner in which we sleep during the night. Their bodies would most likely not have a hormone regime which would trigger tiredness when darkness falls and wakefulness when light returns after night.

However, sleep[6] is a behavioural feature which varies across a spectrum. Some animals, like marine mammals, can experience sleep for one hemisphere of their brain and then the other, allowing them to be alert for

5 Science on an exoplanet might have gone down several wrong paths, as on Earth, believing in things like the four temperaments (personality types), the belief in ether (a substance filling all space), in miasmas (poisonous air responsible for disease), in vitalism (a non-physical life force in all beings), etc.

6 Our sleeping patterns divide into five phases: 1, light sleep; 2, deeper sleep (about half of the time slept); 3 and 4, very deep sleep; and 5, REM sleep. If we are deprived of REM sleep, our ability to concentrate and learn diminishes rapidly.

predators. Some animals can sleep standing, like horses, which shift the legs they stand on; some birds can sleep during flight, resting one hemisphere of the brain, then the other.

No one can go without sleep for more than a day or two without being disturbed physically and psychologically. Sleep is essential for us to regenerate. Prolonged lack of sleep can damage our health considerably. If exobeings were to evolve in a roughly comparable manner to ourselves, would periodic rest also be necessary for them to maintain consistent levels of alertness, concentration and general health?

A close orbit to a parent star would mean that what exobeings would see as a year would be much shorter than ours (the time taken to complete one orbit around their star). To have seasons, their planet would need a tilt like our Earth (about 23.5 degrees), which results in the Sun being higher or lower in the sky at any given location and hence imparting more or less heat to the Earth depending on its position at a given time of year.

During winter many animals hibernate, reducing their metabolism and hence conserving energy. This might be an option practised by exobeings who could in fact go further and suspend their metabolism almost entirely if conditions on a planet got very inhospitable, for example if a planet had a markedly elliptical orbit leading to it leaving the habitable zone of its parent star for a certain period of its astronomical year.

Varying temperatures on an exoplanet would have several consequences for exobeings. They might have body fur to cope with the cold. Would they be endothermic (warm-blooded) or exothermic (cold-blooded)? Would they sweat and smell as we do? And, if the latter, could they communicate via pheromones, at least secondarily?

Alternative Biologies

One could maintain that exoplanets have some very different biology to ourselves. Yes, possibly. But what could that be? On Earth, all our biology is based on carbon (atomic number of 6), the universal molecule-making element. There have been suggestions that one might have life based on

silicon (atomic number 14), which can form large molecules in a manner not unlike carbon but which are less stable overall.

The conditions for life could be radically different. For instance, on Titan there are lakes of methane (CH_4) despite the extremely low temperatures which would not allow liquid water anywhere. But combinations of water and ammonia could exist and remain liquid at much lower temperatures in other worlds, so that is a possibility.

There are also elements which are relatively rare across the universe, but nonetheless important for life on Earth. For example, phosphorus is generated in supernovae and is unevenly distributed.[7]

The natural world on an exoplanet does not necessarily have to be green, the colour which is not absorbed by chlorophyll during terrestrial photosynthesis, hence the green colour of plants and foliage. Planets around red dwarfs, for instance, could tend towards red, blue or even black colours during photosynthesis. If an exoplanet orbited around a red dwarf, the amount of visible light reaching it would be considerably less than what we experience on Earth. This in turn would mean that life forms on such a planet would have to maximise the absorption of light and so might end up being black because a (perfect) black body reflects no energy away from itself as well as absorbing light in the infrared spectrum. So we might not be looking for a blue and green climate like Earth: the atmosphere could be reddish due to the red glare of the star and the plants and animals could be dark to absorb most energy.

Recent work to detect photosynthesis on an exoplanet has sought to examine the spectrum of reflected light from the surface of the planet. On Earth there is a noticeable increase in red and near-infrared emissions by photon-absorbing organisms, a signature of photosynthesis known as 'red-edge'. Such spectral prominences might hold for other wavelength regions (colours) for photosynthetic organisms on other planets, and hence provide evidence of the phenomenon for terrestrial scientists examining reflected surface light from an exoplanet.

7 Arsenic has been suggested as an alternative to phosphorus. Its chemical behaviour is similar, but this suggestion does not enjoy much support among astrobiologists.

Alternative Phenotypes

There is a natural tendency to imagine that exobeings are minimally different from us. For popular films it has been enough to portray them as having swollen, hairless heads, bulbous eyes, large ears with small mouths and pointed chins. Paint them green and you have off-the-shelf aliens. Now to us such constructed beings will still look fairly familiar because they exhibit the same type of bilateral symmetry we find in humans and other vertebrates.

Basically, animals are, in their bodies, either (i) radial, like jellyfish, or (ii) bilateral, with left-right symmetry of parts, like the other forms of life we know, including mammals and reptiles. We also know that radial symmetry in the natural world preceded bilaterality, which arose about 550 million years ago, just before the beginning of the Cambrian era.

With radially constructed animals, you can take a point at the centre and observe all parts of their bodies arranged symmetrically around this point. With bilaterally constructed animals, you can draw a line down the middle of the body and every part is arranged symetrically on either side of this imaginary line.[8] Each part of the body, which comes in twos, will be on the left and right of this line: two halves of your brain, eyes, ears, arms, hands, legs and feet. Internally, this also applies: two lungs, kidneys, etc.

Where there is only one part this is centrally aligned to the imaginary line down the middle: one mouth, tongue, throat, windpipe and food pipe; one brain stem and one spinal cord. The same holds for the male and female reproductive organs. The exceptions to this central alignment are (i) the heart, which lies slightly to the left in the human torso and (ii) the digestive system with the stomach, liver, intestines, which are folded in on themselves to maximise surface area but with a centrally aligned exit for waste.

8 We are unconsciously aware of this bilateral symmetry and find a lop-sided face, for instance, unaesthetic or even disturbing if it is very pronounced.

28.4 Feeling Like an Exobeing

Will exobeings show the range of different personalities, desires, actions, motivations which characterise ourselves? A common science-fiction representation of exobeings, at least in film, is to represent them as feelingless robotic beings, all doing the same thing, like trying to destroy humanity. But maybe they would be beset by similar levels of disagreement which characterise humans on Earth. That is probably the more likely scenario: exobeings can only arise, or have arisen, via the same type of complex inter- and intraspecies variety which we see on Earth. If they are cognitively complex, there will be many ways in which they could vary amongst themselves. Like us, they would probably have more differentiated personalities than the simpler forms of life from which they would have evolved, just as we do vis à vis many other animals.

On a more speculative level one can wonder whether exobeings would experience emotions as we humans do. Would they laugh and cry in the same way that we do (when experiencing happiness or sorrow)? Would they experience any or all of the six primary emotions: fear, anger, sadness, joy, surprise and disgust? Would they remember emotional events particularly well? Would they experience romantic or kinship love? Would they be curious? They probably would, given that curiosity was the driving force for scientific discoveries on Earth and would be necessary for analogous discoveries on an exoplanet. And some individuals are likely not just to be curious, but driven, especially those who reach high levels of achievement in their respective fields.

Would they enjoy pursuing a goal more than experiencing the goal once attained. 'Getting there is half the fun' as the phrase has it,[9] so in fact striving after something is frequently more enjoyable than attaining it. The more hype there is about a goal – achieving perfect romantic love – the greater the disappointment when we realise that it does not fundamentally change our lives and we return to our normal, middle-of-the-road

9 Or as Shakespeare wrote centuries ago 'All things that are, are with more spirit chased than enjoyed' (Gratiano in *Merchant of Venice*, Act II, Scene VI).

emotional state. Of greater relevance to our daily lives is the neuromodulator oxytocin, which is thought to further feelings of trust and love essential to human bonding. And certain neurotransmitters, such as dopamine, are assumed to generate positive emotions (addictive drugs like heroin and cocaine cause an oversupply of dopamine – the brain's reward substance – swamping the brain in the process). At the other end of the scale, we could ask if exobeings would suffer from anxiety, even existential angst? Would they have nightmares, or at least dream like us, assuming that they periodically entered a state of rest comparable to our sleep? There may be evolutionary grounds for assuming that exobeings would experience at least some emotions similar to those we humans experience. Disgust is a primitive emotion, already recognised as such by Darwin, stemming from the need for animals to recognise when food is safe to eat. Fear ensures that we do not enter into situations of obvious danger, for instance staying in the vicinity of powerful predators who could harm us. Positive emotions such as fondness and love further attachment and bonding between individuals and hence are beneficial to both individuals and groups, increasing their chances of survival.

28.5 What About Free Will and Morality?

A complex society can only arise and continue to develop if its members make decisions for different types of actions. This in turn depends on intelligent beings making decisions of their own accord, that is to say, on exercising free will. Whether this exists, independent of the physical substrate of our brains, is a much-discussed issue in both neuroscience and philosophy today (see Chapter 17). The question is whether we are genuinely free to do what we wish or whether our decisions for action are ultimately determined by configurations of our brains at certain moments. At present, there is absolutely no way of knowing the exact configuration of all our neurons and how they are firing at any one moment, so free will cannot, in practice, be predicted from the neuronal state of the brain. Whether it ever will be, remains an unanswered question.

Irrespective of how one sees the status of free will with us humans, we do make decisions concerning future actions. If we did not, there would be no room for individuals to engage in novel activities and hence trigger progress for their social group. A complex society consists of many different individuals with varying personalities. For an exosociety, and more so, for an exocivilisation to arise, this variation would also have to exist. Personalities of individuals would probably arise from a combination of genetic endowment and early development – a classic mix of nature and nurture – with memory and continuity of individual identity yielding a sense of self for exobeings.

Linked to free will is the question of moral behaviour, resting on inherent notions of good and evil. Whether exobeings would have a sense of morality is somewhat uncertain but the chances are good, seeing as how morality stems from extended social cooperation and altruism as a feature which, while costing an individual something at one point in time, can be helpful in the long run to both that person and the group they live in.

Both philosophers and developmental psychologists have discussed the issue of morality in detail. The Canadian–American philosopher of mind, Patricia Churchland, sees empathy among humans as a keystone in morality and as what lies behind such expressions as 'do unto others as you would have done unto yourself'. The American developmental psychologist Michael Tomasello regards the shared intentionality among humans as a chief source of a sense of morality. Already with young children we can observe cooperative behaviour and a sense of commitment to such collaboration. This is further evidenced in studies in which children appear to feel guilty when they make a one-sided decision to abandon some action or goal agreed upon with others. Morality is thus taken to be a quality which arose in social groups and performs the function of in-group regulation. It provides a brake on wrongful actions and promotes those who act in a manner favourable to the community.

What about empathy? Do animals not show it? This is difficult to assess. When wolves kill and devour the calf of a moose in front of her eyes, do they feel responsible or even sorry for her losing her young? When chimpanzees tear monkeys apart to eat them, do they regret having killed these creatures? Even more, when several chimpanzees gang up on a single one from another community and rip the animal to pieces, do they feel guilt at the unwarranted killing they just carried out?

Is empathy an epiphenomenon of our highly developed cognition and theory of mind? One might posit a theory of emotion where humans assume that other humans have similar feelings to ourselves and hence would similarly experience pain and sorrow. In this context, the complexity and cognitive ability of life forms matters. Female frogs do not have a personal relationship to the tadpoles they beget. Rather, they deposit their spawn in suitable water conditions and leave it there. Like many animals, they do not even see their offspring, let alone know them. Sea turtles deposit umpteen eggs in sand to hatch out on their own, with the young turtles making their way down to the sea.[10] Mammals, who give birth to live young, seem to have a closer relationship to them. Female chimpanzees, for instance, have been observed grieving the death of one of their young, carrying the dead body around for days afterwards. Elephants also mourn their dead, holding vigil on the corpse of a deceased member of the herd.

On an exoplanet, could one expect an equally distributed theory of mind/emotions, a sense of empathy and ultimately of morality as we observe here on Earth with humans? The nature of humans as highly social creatures, dependent on group cooperation at every turn, would suggest that exobeings, capable of similar technological achievements to ourselves, would be organised into complex societies based on collaboration and cooperation, with a sense of empathy and morality as guiding principles in the dense networks typical of advanced societies.

Finally, one should add that the existence of morality does not, of course, mean that all individuals would be led by it in their group behaviour. Quite the contrary, many individuals can and do choose to ignore morality for their own advantage: one can blunt one's sense of empathy by continuously flouting it in the striving for power and domination, as we can see by even a cursory look at our world and its history.

10 This is a dangerous undertaking during which many of them are lost on the way because predators are waiting for them on the path of the newly hatched turtles across the sand.

28.6 What Are Exobeings Likely to Share with Us?

The outset for any consideration of what exobeings could be like begins on the simplest level, that of molecular and cellular organisation. The following sections list some of the common features which exobeings might share with us humans, especially concerning their physical make-up and the mechanisms by which they would have evolved. The eight possible make-up features listed here could only possibly apply to exobeings still in a biological phase of development; but if they were beyond this, the features would still hold for their biological past.

Possible make-up features of exobeings

1. The use of DNA, or an equivalent mechanism, to pass on genetic information through the generations.
2. The development of a system akin to sexual differentiation, allowing recombination of DNA to produce offspring with different sets of genes from two parents.
3. The development of a central nervous system with a concentration at one end of the body, later yielding a brain in a head. This would go hand in hand with a system of food intake at one end and waste outlet at the other, with the brain at the other end of the body. This brain would likely be a predicting machine, trying to make sense of the constant input of unordered data from the sensory organs (eyes, ears, body's nervous system responsible for limb movement, touch, etc.).
4. The development of vertebrate bodies, which allow for good locomotion on land, ultimately yielding limbs similar in function to those we have.
5. The development of eyes from light-sensitive cell assemblies, which allow optical signals to be processed by a brain. If an exoplanet orbits a red dwarf star, the light sensitivity of its exobeings might be in the infrared range of the electromagnetic spectrum.

6. The development of hearing with ear-like organs to pick up acoustic signals in the central range of about 20– 15,000 hertz. Hearing in this range would help exobeings avoid natural dangers and predators in their surroundings.
7. The development of a tactile and possibly an olfactory sense.
8. The existence of proprioception (awareness of one's own body).

Assuming exobeings have brains in a bony case, along similar lines to humans, they would have to have sensory organs, which would feed external stimuli to their brains that process them. After all, that is the procedure for all animals on Earth. We cannot know any details of such sensory organs, for instance whether they would they have a projecting nose like us, but the principle would have to be the same as with us humans.

The really big question is whether DNA replication would evolve on an exoplanet. If this arose, or a similarly powerful mechanism for the transfer of genetic information, then one could posit a complete biosphere for a planet, from the most simple form of life, single-celled prokaryotes, given enough time.

Exobeings could theoretically be hermaphrodites, that is not be distinguished for sex and have a form of self-fertilisation. It is debatable whether that form of reproduction would hold for intelligent, highly developed life forms, if the situation on Earth is anything to go by. On our planet, only simple forms of life show hermaphroditism: slugs and snails, for instance, but hardly any vertebrates, except for a few fish. Plants, of course, are hermaphrodites (but with the give and take of pollen) and could hardly be anything else, as free mobility would be a prerequisite for sexual reproduction, which arose on Earth about 1.2 billion years ago.

If exobeings are similar to us in their cognitive and physical abilities, some further characteristics might be assumed with some certainty.

Further possible characteristics of exobeings, assuming the gradual evolution from single-celled life, probably over hundreds of millions of years

1. Development of hand-like limbs with high levels of manual dexterity (with opposing thumb-like structures and nails on their hands, rather than claws).[11]
2. A type of locomotion comparable to our bipedality.
3. Development of a large brain for higher cognitive functions, the 'runaway' brain, which would distinguish exobeings from other exofauna.
4. Relative independence of exobeings from their initial environment, with management and manipulation of their later environments.
5. The management of fire (only with an oxygen-rich atmosphere).
6. Development of cooked food, providing for high-energy intake in short periods and promoting social cooperation, assuming (5) applies.

28.7 How Smart Might They Be?

What might the cognitive abilities of exobeings be like? This is a difficult issue to approach, given that we do not know how intelligence might manifest itself with exobeings and if the yardsticks we apply on Earth would be appropriate in the context of exobeings. And there is already considerable variability on Earth, just compare vertebrate brains with those in cephalopods like octopuses. But if there were manifestly considerable differences,

11 *Notharctus* (an early primate living about 50-40 million years ago in Europe and North America, remains of which were first discovered in Wyoming Territory in 1870) had a divergent thumb, which could be brought into opposition with the other fingers; it also had nails rather than claws. Such hands allowed the animals to move securely along thin ends of branches in trees, where fruit and flowers were to be found, and so gave them an advantage in the procurement of food.

exobeings might not have the technology to communicate with us, though they could have advanced social and/or emotional intelligence. Just consider that a couple of million years ago, when we were much closer to the last common ancestor between us and chimpanzees, our cognition had not evolved so far, assuming an approximate (but not uncontested) correlation between brain volume (evident in the size of skulls) and cognitive power. We might just strike on a planet where exobeings were at a comparable stage of development.

Contained in the current considerations is an issue which is very relevant to the theme of this book. What is the maximum level of intelligence which could be evident in exobeings? What is the upper limit for cognition with exobeings? Is it similar to humans or much higher? Assuming that there are different groups of exobeings on different planets, what would the average range of cognitive ability look like? And would our conception of intelligence be the same as theirs? We do not know, is the straightforward answer.[12] Nonetheless, we can approach this question from an anatomical perspective. The ratio of our brains to body size is unique in the animal world. The only animals which have brains larger than us, for example, elephants and whales,[13] have massively bigger bodies. Furthermore, the size and structure of the cortex is key to the level of general intelligence an animal displays. For exobeings, the size of the brain would probably be limited by the size and form of their bodies (as it is for us humans). A brain twice the size of ours would require a very different body from that which the average human has. There would be additional difficulties, such as giving birth to beings with such large brains. But even 50 per cent more cells in the cortex at roughly the same body size could represent a great increase in cognitive power vis à vis us humans. It would be interesting to know how such exobeings would use this extra intelligence, hopefully for the benefit of science and the advancement of knowledge – maybe they know what dark energy and dark matter are and have solved the enigma of quantum entanglement – but there is no guarantee of that.

12 Some scholars, like the British philosopher Colin McGinn, have pointed out that exobeings could be significantly more advanced than us in terms of cognitive ability.

13 The sperm whale has the largest brain of any mammal, almost eight kilograms in weight.

For the current discussion, one should furthermore bear in mind that intelligence does not correlate directly with brain size. There is a connection, but a comparison of humans with Neanderthals shows that this is not always so straightforward. From measurements of fossilised skulls, we know that Neanderthals had a brain volume at least as large as ours, if not indeed somewhat greater. So were they as intelligent as we are? True, they were more advanced than previously assumed but they hardly developed their technical abilities during the entire period of their existence. They made spears with tips of stone, indeed they did, but they never thought of developing a bow and arrow. And only at the very end of their period did they engage tantalisingly in cave art. So no matter how much we may now look on them favourably there is no denying that they did not have the flexibility and creativeness which characterises *Homo sapiens*, to anything like the same degree. This conclusion means that there may well be exoplanets on which *Homo*-like species may have evolved but not have the adaptive and creative qualities which characterise *Homo sapiens*. Any exoplanet without such beings would be unable to establish any contact beyond their world and hence probably remain unknown to us.

Extending the Mind

To a certain extent we have already gone beyond our own minds by outsourcing many activities which we formerly performed ourselves. Just think of how we use calculators and spreadsheets to perform mathematical operations which our recent forebearers would have done manually. Many researchers see in this type of activity an extension of our human minds. It is true that we can use computers to perform calculations which we simply could not do ourselves for reasons of complexity and the sheer size of many tasks, but to claim that a smartphone is an extension of our minds does not make much sense; it performs tasks automatically which we would not care to do manually anymore, like keeping lists of telephone numbers. Such 'extensions of the human mind' never make autonomous decisions or show self-awareness and certainly do not exercise free will.

For an exocivilisation, which is at least thousands of years into its digital era, the situation might be different. Its beings might have devised ways of directly coupling their brain activity with digital devices, which could perform tasks and provide direct feedback into the brain, thus enhancing their cognitive powers. On Earth we can only dream of that today, but on an exoplanet such technology might have been realised already.

28.8 How Would They Count?

An exocivilisation would not exist without a counting system with which it could manipulate numbers. Indeed, we could probably assume that exobeings would be able to count and have neuronal circuits responsible for counting, if the evidence for humans is anything to go by.[14] On Earth, from the early history of humankind, we know that counting systems are as old as the first attempts to record language in some permanent form, in fact the two activities are clearly related.

Counting systems which are cumbersome would limit the technological range and scope of an exocivilisation. A good example for this is the Roman numeral system. It uses more than one symbol for higher numbers: V, VI, VII and VIII are 5, 6, 7, 8 in the Arabic numeral system (itself derived from an earlier Indian system) and was really only of any use in representing positive integers. But the real drawback of Roman numerals is that there is no zero symbol (its invention is traditionally attributed to the seventh-century Hindu scholar Brahmagupta). You might think, going on an everyday common-sense understanding, that zero is nothing and so you do not need it when counting, which is what a numeral is for in the first place. But far from it: without zero you cannot have fractions. You cannot represent half or a quarter in the Roman system whereas in the Hindu–Arabic system this is simply 0.5 and 0.25 respectively. You can also use the zero to indicate multiple of tens: 10, 30, 90, 400, 1,000. And, importantly, it is very difficult to do calculations without zero, for example $((7 \times 3) + 4) - ((2 \times 10) + 5)$, where the result is zero.

14 Stanislas Dehaene has demonstrated that the intraparietal sulcus is active during arithmetic operations; see Dehaene (2011) for a book-length treatment of this matter.

The base of a numeral system is a further consideration. The most common base across the languages of the world is 10, yielding a decimal system, though there are some noticeable exceptions where a language uses (or used in the past) a base 20, giving a ventigesimal one (Basque, French, Irish). Where does this come from? Quite simply: we have 10 fingers which we use for counting, so 10 is an obvious number. We also have 10 toes, so if we use them to count as well, we have a system with base 20. Because we are not as conscious of our toes as of our fingers (even in the days before we wore shoes) the decimal system is more common.

A base-10 counting system is intuitive to us humans given the number of fingers we have, but other systems are equally possible and might exist if exobeings had 8 fingers or 12. Furthermore, science may use bases which are not derived from our environment: in computer programming, base 16, the hexadecimal system, is common. The simplest system of all is the binary one where there are only two digits, 0 and 1, or some other manifestation, such as on or off (current flowing or not), below a certain threshold or above it. If an exocivilisation were capable of interplanetary communication, they would have to have computers, and it is likely that they would have a binary system for these devices. It would also be intriguing to know what form of mathematics exobeings might have, especially in light of the discussion here on Earth about whether mathematics is constructed or discovered, revealing realities about the universe which are not perceptible to us with our given senses.

28.9 Would They Have a Sense of Time?

In the Einsteinian universe, time is real,[15] it is a fourth dimension along with the three dimensions of space to form space–time, which can bend and curve as it does in the presence of large celestial bodies, this being the phenomenon we know as gravity.

15 There is much discussion of the reality or not of time in the scientific community. For instance, the British physicist, Julian Barbour, has claimed that it is an illusion and the American physicist, Lee Smolin, has adopted the opposite position, pleading for its real existence.

But does time in the sense of duration exist? Consider a clock: it moves its parts and hands and appears to be keeping track of time as it passes. But does it really? What a clock is based on is a conventionalised rate of movement for its parts which we then call seconds, minutes, hours, days, months and years. But does the clock represent time which exists independently, in some background grid in the universe, in which time is moving forward continuously? If it did, time would be fundamental, like electromagnetic energy. But our sense of time passing might rest on an extrapolation from our perception of change in the world around us. This would make time dependent on human observers and not an independent property of the universe.

Imagine you have a closed chamber containing helium and you manage to cool the gas to absolute zero (0 K or –273 °C) and maintain it indefinitely at that temperature (this is not actually possible, but just imagine it for this thought experiment). Motion on the atomic level has reached a minimum and more or less stopped (though not on the quantum level). Does it now make sense to talk of the state of the chamber an hour ago, yesterday, or to talk of its state this evening or next week? You may have an uncomfortable feeling that it does not, so no change means no passage of time. And we actually have an intuitive idea that time is change. Consider expressions in language such as 'like time had stood still,' 'like time was frozen,' 'as if time had passed it by,' etc., when confronted with some situation or location in which little or no perceptual change has taken place.

Conceptions of past and future are generated in our brains by memories of previous states (the past) and our conception of states which have not occurred yet (the future), along with hypothetical states which might arise (expressed in grammar via the conditional). But what is the ontological status of such notions as past and future – do they really exist separately from our cognition? Might it be more accurate to describe the past as the set of states the universe has been in and the future as the set which are possible for the universe and which may arise given that change is continually happening? Time would then denote the movement from one state to another, there would then be no time independent of the system within which it is supposed to manifest itself.

What About the Arrow of Time?

There is a further issue[16] in this context, that of the arrow of time: do states change in a particular order which gives us the impression that time is always moving forward? For instance, we all grow older, we do not grow younger. A fried egg cannot be reversed back to the unbroken egg in the shell, a cup of coffee with cream and sugar in it cannot revert to the state when neither substance had been added to it, etc. There is a large body of literature on this question, which involves the notion of entropy, the number of configurations in which a (closed) system can be arranged. These configurations tend to increase as other factors come to influence the system. In the example of the egg, applying force to the shell will break it and the shell is then in a more complex state than it was when it was whole. When heat is applied to the contents of the egg, the white loses its transparency as the proteins it consists of are broken up by the heat. So the overall state of the egg has become more complex; entropy has increased for the egg. In general, you cannot reduce the configurations for a system; you cannot undo a broken glass which has fallen on the floor and smashed into innumerable shards, though, importantly, there is nothing in the laws of physics which forbids this.

However, one should also add that the increase in entropy is not linked to the passage of time in any deterministic way. Consider a cup of coffee into which you put creamer: the moment of greatest complexity of the system is when the creamer is distributing itself throughout the coffee. Once you stir the coffee it becomes evenly distributed, with somewhat higher entropy and lower complexity, although this happens after you put the creamer in, later in time.

Our intuition for the fact that certain states, once reached, cannot be undone, gives us the feeling that time moves forwards and not backwards. This is confirmed for us in all areas of life; for instance, an adult cannot

16 The questions do not actually stop here. One which occupies particle physicists is whether time is granular, an assumption made in a theory called quantum loop gravity. Does it proceed in discrete steps of Planck time, 10^{-43} seconds (below which it makes no sense to talk of time), or is there an infinite number of time points in any given interval of time, say three seconds?

'ungrow' back to a small child and, perhaps most obviously, no one can be 'unborn'. So certain is this conviction for us that the forward-moving arrow of time is a central aspect of the human experience.[17]

Our brains evolved in an environment in which predictions based on experience were important for survival through the ever-present current moment of time. Our experiences form memories and, given that the experiences are no longer present for us, we see them as confined to what we call the past. The future consists of our predictions of what might yet happen, and which has not yet occurred. Of course, the conscious brain's perception of time is flexible and can be distorted by neuromodulators which heighten our awareness on certain occasions, for example of great danger, nervousness, special significance for an individual. Say, for instance, you get a cramp while swimming some distance off the shore and then feel, when trying to reach land again, that you will never get there. That has to do with your subjective experience of time[18] in that situation, not with an actual dilation of time.

Telling Time on an Exoplanet

The basic divisions of time made on Earth derive from the length of time the Earth requires to rotate once on its axis – a day – and the amount of time it needs to complete one orbit of the Sun – a year. The subdivision of month is derived roughly from the phases of the moon and yields 12 units per year (approximately, there are 13 lunations per year). Divisions of the year into seasons is largely a tradition of the Northern and Southern hemispheres (outside the tropics) and results from climatic variations during the year, largely due to the orbit of the Earth around the Sun and its axial tilt of about 23.5 degrees. Divisions of the day into 24 hours,

17 Nowhere is this more eloquently expressed than in the opening lines of Shakespeare's Sonnet No. 60: Like as the waves make towards the pebbled shore, So do our minutes hasten to their end; Each changing place with that which goes before, In sequent toil all forwards do contend.

18 Our first-person intuitions of time have been the subject of much philosophical work, notably in the early twentieth century when the French philosopher Henri Bergson (1859–1941) discussed his dissenting views with Einstein, see Canales (2015).

and not, say, 36 hours, are artificial, as are the further subdivisions into minutes and seconds.

So what might one find on an exoplanet? The first question would be: 'Is the planet tidally locked, as are many planets orbiting very close to their parent star?' If so, the same side faces the star continuously, so there is a permanent day side and night side. Next is the size of the planet's orbit around its sun. Close to a star, planets would need much less time, possibly just a few of our weeks. But if further out than our Earth is from our Sun, an exoplanet would need much longer, from a few to many years as we measure them (Neptune, the most distant solar planet, has an orbit of nearly 165 years). Seasonal divisions, as we know them, might only exist if there are regular patterns of climate variation which could serve as guidelines.

29

Looking for Signs of Life

29.1 Biosignatures and Technosignatures

In discussions about the search for life beyond Earth two basic possibilities have been proposed: either we will discover signs that biology exists on other planets through atmospheric analysis or we will detect an unambiguous radio (or maybe laser) signal from outer space which does not have a natural source and hence can be assumed to originate from an exoplanet. The former discovery would be via a biosignature, such as abundant oxygen in the atmosphere of an exoplanet. The reason why this is a good bio-indicator is that free oxygen (as a diatomic gas, O_2) is highly reactive and if it existed in the atmosphere of an exoplanet it would disappear quickly by forming molecules like carbon dioxide or water, or rust with iron, unless it were continuously replenished by some biological source, like trees and plants on Earth, which release oxygen into the atmosphere during photosynthesis. This is why the Earth's atmosphere continues to have about one-fifth oxygen: the tropical forests and green foliage in general keep replenishing the oxygen which is lost in the atmosphere through various molecule formation processes. However, this could also be produced by photolysis, a process in which ultraviolet radiation can break down two water molecules into four positively charged hydrogen atoms, four electrons and an O_2 molecule. This could occur on a planet surrounding an active red dwarf. If the planet did not have a strong magnetic field, it would be directly subject to ultraviolet radiation from its parent star, promoting photolysis.

The second type of discovery would be via a technosignature, such as a beam composed of electromagnetic waves, which might be directed towards our Solar System, if exobeings were deliberately trying to contact another planet, or it could be due to 'leakage', the collateral diffusion of radio waves into space due to radio and television broadcasting on that planet. Such leakage has been occurring from Earth for almost a century now (that is, for about 0.000002 per cent of the age of our planet).

The scientific assessment of biosignatures versus technosignatures plays out in the projects which are currently planned by the major players in the field of astronomical investigation, NASA along with the National Science Foundation and the Department of Energy, Office of Science. NASA has two possible items on its agenda, the Habex (Habitable Exoplanet Observatory) and the Luvoir (Large Ultraviolet Optical Infrared Surveyor) telescopes, which are both geared towards finding biosignatures. It would seem that the search for technosignatures is a gamble with a too uncertain outcome and has thus been left to private enterprises such as the Breakthrough Listen project. There is also the issue of persistence: biosignatures would be around much longer than technosignatures so the likelihood of finding them is, in principle, greater.

Biosignatures could tell us that there is at least oxygen-producing life on other planets, where photolysis can be excluded. Strong traces of oxygen, water and, possibly, methane in the spectroscopies of their atmospheres would be indications of this. It would mean that (i) the oxygen must be produced by some biological mechanism and (ii) the supply must be maintained constantly, otherwise the free oxygen would be bound by common processes of oxidisation and quickly disappear from the atmosphere.

But whether the long path from microbes to exobeings would have been traversed successfully is something which could only be answered, if at all, by continuously scanning nearby solar systems for planets which look as if they could harbour complex forms of life. Another possible indication of such life, apart from the electromagnetic signals just mentioned, could be atmospheric disequilibrium caused by pollutants, such as carbon compounds released from fuel burning or their use in industries of various kinds. That would presuppose that exosocieties would have passed through a stage in their development comparable to the Industrial Revolution on Earth.

Scanning the space around us for technosignatures was undertaken by Project Ozma in 1960, led by Frank Drake, and as of 1984 by the SETI institute (www.seti.org). In recent years, space scanning has been mostly transferred to the Breakthrough Listen project located at the University of California, Berkeley. While this undertaking has garnered much knowledge about how to search for exolife, of its very nature it is something of a gamble: either one or more digitally advanced exocivilisations[1] on exoplanets exist out there and can be detected or they can not. Anything other than that, such as a culturally advanced non-technological planet could not communicate with us, nor could we even accidentally pick up a signal from such a planet. When looking for signals from beyond our Solar System, SETI astronomers have in the past concentrated on a region known as the 'waterhole' – a term devised by Bernard M. Oliver (1916–1995) as a reference to where people would traditionally congregate and talk together. The frequency range of this band is from 1420 to 1662 megahertz and its lower and upper limits are defined by the spectral lines of the hydroxyl radical (OH) and hydrogen (H), respectively. Thus the waterhole is a region of the electromagnetic spectrum where naturally occurring signals are at a minimum, where noise from celestial objects would be slight. There are tens of signals found in the waterhole per day and they need to be examined carefully. Fortunately, more and more powerful computer algorithms are being devised which automate much of this sifting through mountains of data, and some of these tasks can and have been outsourced to citizen

1 It may be that astronomers on exoplanets might only have some very rudimentary ideas about the universe, based on simple observations, like registering nearby supernovae, as did Chinese astronomers in 1054 when they saw the supernova whose remnant is now termed the Crab Nebula, over 6,500 light years from Earth.

scientists around the world. The analysis shows that most of the narrow-band signals are from terrestrial radio waves, reflected back down to Earth from the upper atmosphere, or which stem from any of the hundreds of satellites orbiting our planet today. There have been proposals about how to get around this issue. One would be to put a telescope on the dark side of the Moon to exclude any electromagnetic 'noise' produced here on Earth.

The search for signals has since moved away from the waterhole band to cover a much wider range, and advances in artificial intelligence have allowed scientists to examine large quantities of data within reasonable lengths of time. There are various possibilities regarding signals from afar. A signal might be accidentally picked up, and not be directed at us. In such a case we would simply eavesdrop on an exoplanet's communication. Or it might be a deliberate attempt at communicating (i) with any planet which, for them, might be 'out there' or (ii) specifically with our Earth. The second scenario is less likely than the first. After all, just how would exobeings on a planet, say, 100 light years from us know that there is intelligent life here on Earth?

29.2 The Nature of a Signal

To decipher a signal astronomers would examine its wave form and look for patterns which emerge over stretches of the signal, for example possible encoding via wavelength and/or amplitude. The latter would only be relative, as signals from a distant star would be weak and many of the amplitude differences could be levelled out by travelling over great distances. So the quality of the signal would be all important. Another question is whether a signal is one of a sequence. Would the signals rain down on us thick and fast? If the messages sent from Earth are anything to go by, the answer is probably 'No', unless some exobeings were certain there was intelligent life on Earth and tried to contact us with repeated messages.

A further issue would be the supposed content of a signal. Assuming that scientists broadcast a signal from an exoplanet, they are likely to send a message with scientific content, perhaps with additional information about life on their planet (as humans did with the Voyager discs, see Section 9.4 above). But could we decode such messages? Would exobeings supply a cipher to help us understand their messages? What sort of concepts would be encoded in the signals they send out? Of course, we would not have the context of their language, so deciphering would be devoid of the connection with an environment which we have when examining our languages. Nonetheless, the kinds of assumptions about exoplanets and exobeings discussed in this book would provide us with an approximate framework within which to attempt deciphering an exolanguage.

We should also consider the question of the medium used to broadcast a message. The default assumption might be that the message is probably encoded in language, that is, information in the communication system of the exobeings. But the message might use a translation medium to counteract degradation of the signal over great distances. Such a medium might be like Morse code, which can be broadcast as sounds or beams of light. This code consists of sequences of 'dits' and 'dahs,' short and long pulses, patterns of which encode the letters of the alphabet. It is a type of binary code, with only two values, and so is inherently resistant to signal degradation. Again, we would need to recognise this and work out the cipher used in this binary encoding of an original signal from an exoplanet. Astronomers on Earth picking up a signal would have to be alert to the possibility that a sequence of just two types of pulse, long or short (wavelength), high or low (amplitude), was actually an encoded message, at one remove from the content of a message. After that, the astronomers would need to discover if the signal contained linguistic or other information, such as scientific data, sounds from nature or cultural data, like music, all of which would require very different kinds of interpretation.

If it could be determined that a received signal probably contained language, its meaning would have to be ascertained. Any language will likely have building blocks with individual meanings, which are linked up into larger units to express more complex meanings. Assuming that this also applied in principle to exolanguages, one could attempt to decipher the composite meaning of the message in a signal. What would exobeings wish to communicate to us? Are they asking us (or anybody 'out there') to return the message? Even if this could be determined, a signal could be from so far away that it would be impractical to engage in any type of dialogue, if, say, the turnaround time for a message went beyond 10 years. At the end of the day, the result might just be that we know there is another planet with intelligent, sentient beings in our galactic neighbourhood.

29.3 METI: Trying to Get in Touch

In July 2015, an organisation was founded in San Francisco by the American scientist Douglas Vakoch, called Messaging Extraterrestrial Intelligence International, or METI for short. The goal of this non-profit research organisation is to send what it considers appropriate messages, encoded as radio signals, into space in the hope that one or more of these might be picked up by exobeings, who would then know that we exist and maybe contact us in return. The timescales involved in this enterprise are considerable. For instance, one of the messages beamed into space was directed at Luyten's Star (in the constellation Canis Minor), a red dwarf about 12 light years away. There is a super-Earth (Gliese 273b) orbiting this star in its habitable zone and if this contains a digitally advanced civilisation, it should be able to pick up the METI message. If the planet did harbour such a civilisation and they replied immediately, this would mean that between sending a message and receiving a reply there would be a time lapse of over 24 years. For stars further away, the intervening period between message and potential reply would be much greater.

The scientific community is divided on the issue of actively trying to communicate with life on exoplanets. Some point out the potential dangers of drawing attention to ourselves while others regard the benefits of finding exobeings and engaging in exchange with them as far outweighing any possible hazards, especially given the distances involved, which would act as a natural barrier to any possible hostile actions.

29.4 Would They Want to Know Us?

If the first exobeings we find were on a planet around a star in a globular cluster with, say, 20,000 other stars in relatively close proximity, those technologically advanced beings might already know of several other planets like theirs in their immediate neighbourhood and not be greatly interested in getting to know us. This might be the case with a globular cluster like (Messier) M4 (some 7,200 light years from Earth). These stellar structures are very old (M4 is over 12 billion years old) and may well have already witnessed the rise (and demise) of intelligent life on the planets orbiting their stars.

Say an exocivilisation knew of 30 other exoplanets, would they really be that interested in a 31st? If a civilisation was, say, 10 million years into its digital age, it might have no interest in exobeings not yet in this age, or just starting it, like us. Memories of the pre-digital age on such a planet would be very distant. And it might just be the archaeologists and palaeoanthropologists on that planet who would lobby for attempting contact with us. But, for the following chapters, we can think about how we would establish initial contact with exobeings and consider a number of scenarios.

30

The Issue of First Contact

First contact has been a topic of discussion among astronomers and other scientists, such as social scientists and philosophers, as well as providing subject matter for science fiction, for some considerable time. In order to discuss it reasonably, the issue needs to be broken down into a set of possible scenarios, some of which are conceivable but well beyond our reach and are likely to remain so as far as we can see.

30.1 Some Scenarios

Scenario 1. Scientists detect a highly modulated signal which does not appear to have a natural source, like a quasar, but to contain information which could come from an exoplanet. Several questions arise at once. Was the signal directed at us deliberately, or at least broadcast into space by exobeings, looking for others on what would be exoplanets for them, outside their solar system? Or was the signal simply leakage from their planet, like all the radio and television signals which have dissipated away from Earth in the past hundred years? The latter is unlikely as such leakage would be far too weak to be picked up light years away. The signal would have to constitute a strong beam like the radio signal directed in 1974 towards the star cluster M13 and known as the Arecibo message (see below).

Scenario 2. We detect an artefact, constructed by the inhabitants of an exoplanet, when it enters our Solar System. Many scientists regard this as a distinct possibility and much was made of the extra-solar object 'Oumuamua, discovered in October 2017 (see Section 30.3 below). Nonetheless, this scenario remains worthy of consideration, although any such artefact would have spent hundreds of thousands, if not millions, of years travelling through interstellar space so that the exobeings which launched the object would, in all probability, have ceased to exist in the meantime. The value of such a discovery would be to show that we are/were not the only technologically advanced civilisation in the Milky Way and that would greatly relativise the way in which we think of ourselves and our planet.

Scenario 3. We are visited by exobeings. This is highly unlikely unless they have developed some method of overcoming the limits of space travel at subluminal speeds. If wormholes exist and represent a viable means of space travel or if Alcubierre drives (see Section 9.1 above) could be built and work, the distance limitations just might be overcome, but at present this is the realm of science fiction.

Scenario 4. We travel to an exoplanet and come into contact with exobeings on that planet. This is the least likely of all scenarios and will not be possible as long as rockets have to travel via conventional means, using engines powered by fuel found on Earth. If we develop techniques such as light sails, driven, if not by the Sun, then by powerful laser beams, radiation pressure on the sail could be used to accelerate the rocket up to a significant fraction of the speed of light. There are enormous technical problems associated with such mechanisms, some of which seem insurmountable at the moment. Just think of what would happen to a spacecraft leaving our Solar System at 1 per cent of the speed of light – 3,000 kilometres per second – when it hit a tiny piece of rock or ice in the asteroid belt, Kuiper Belt or further out in the Oort Cloud. The latter is largely spherical in shape so a spacecraft could not fly over the region, unlike the Kuiper Belt. Furthermore, the spacecraft would have to be able to reduce its speed when approaching a celestial object of interest and, in the near vacuum of space with no friction, energy would be required to slow down an object like a spacecraft. And also bear in mind, that at one per cent of the speed of light[1] it would take 450 years to reach the nearest star of the several hundred billion in our galaxy; that would just be the beginning of a journey to exoplanets. The upshot of these considerations is that, until we overcome conventional means of travel, if we ever do, there is no question of us just travelling out into space to take a look at some exoplanets.

Of the above four scenarios, the first (signal detection) is the most probable, so it is worth considering for a moment. An observatory on Earth detects a signal from space, which does not have a natural source; then a scenario arises for which some people are prepared. The American astronomer Seth Shostak has produced a document entitled 'Contact: What happens if a signal is found?'. The SETI Institute has 'Protocols for an ETI Signal Detection', placed on file with the International Academy of Astronautics. And, of course, various authors, such as Michael Michaud, have concerned themselves with the matter (see his 2006 book *Contact with Alien Civilizations: Our Hopes and Fears about Encountering Extraterrestrials*).

If there were an extraterrestrial signal detection, there would be a cascade of contacts. The Central Bureau for Astronomical Telegrams, now based at Harvard University, would presumably alert the scientific world. The United Nations (UN) secretary general would also be informed, apparently. And there is the UN Office for Outer Space Affairs (located in Vienna, see www.unoosa.org) along with a formal agreement among states, formerly called the 'Treaty on Principles Governing the Activities of States in the Exploration and Use of Outer Space' (1967), now known under the shorter label 'Outer Space Treaty', to which 110 countries worldwide are

1 To put this figure in perspective, consider that at one per cent of the speed of light you could get from London to New York, a distance of over 5,500 kilometres, in less than two seconds.

signatories. A host of government agencies, not least their intelligence services, would click into action.

So what would we then do? Who would have the right to lead the way in dealing with the situation? Would the old rivalries between the USA and Russia come to the fore? Would the Chinese or the European Union insist on being involved in any decisions concerning the contact? The face humans might present to the exobeings could be one of disarray and internal conflict. The best hope would be for international scientific collaboration to investigate the matter further.

There are scientists of the opinion, 'If the cosmic phone rings, do not answer'; the physicist Stephen Hawking was among them. They claim we do not know what we might be letting ourselves in for in contacting an exoplanet.[2] The geographer and historian Jared Diamond has pointed out that the history of contact with outside groups has always been to the detriment of the others. Depending on who is more advanced, not least in a military sense, the encounter could be disadvantageous to the less advanced group. But that would only hold if there was physical contact; in the near future we would probably only be dealing with contact via electromagnetic waves, if any. And the reality is that if we were to detect a signal it would be weak and from a considerable distance. So military considerations are probably not likely to be highest on the agenda. Furthermore, the assumption that exobeings would wish to annihilate or enslave us does not rest on empirical evidence. As the anthropologist Kathryn Denning has said, the manner in which we envisage contact with exobeings in fact says more about us than about a potential 'them'.[3] One can only hope that the first reaction of the largest nations on Earth will not be to put their military on red alert (as is always done in science-fiction films).

The question can be turned around. If we knew there was an exocivilisation on another planet in our cosmic neighbourhood could we afford *not*

2 In order to communicate the significance of exoplanet contact a scale – the Rio Scale – was developed in 2000, with a range of 1–10, which is supposed to assist non-scientists in judging the significance of the contact. Recently, proposals for an updated version of the scale, Rio Scale 2.0, have been put forward.

3 For a succinct and competent review of the 'transmission debate', see Denning (2010).

to contact it? The study of life would now have two instances to draw conclusions from: the set of one would become a set of two. The consequences for many areas of human activity on Earth would be considerable. Take the established religions, the Abrahamic religions, as a specific example. If exobeings had no notion of a single god, which has existed for all time and created the entire universe, if they had never heard of established religions and had no counterparts to them, what would that mean for the religions on Earth, which claim to represent universal truth for all time and in all places? This question can be posed even if exobeings did have their own, very different religions – either way there would be a problem. On a broader level, it would be fascinating to learn how intelligent life developed on an exoplanet, the stages of its evolution, the biosphere of such a planet, what type of flora and fauna it had and a myriad other interesting aspects of life beyond our Solar System.

30.2 How to Contact Them: Language-Independent Messages

Composing a message which can be understood without language is not an easy task, but attempts have been made in the past, notably with the Arecibo message, composed by two American astronomers, Frank Drake (1930–2022) and Carl Sagan (1934–1996), and broadcast into space on 16 November 1974, from the refurbished Arecibo radio telescope on the Caribbean island of Puerto Rico (which collapsed early on 1 December 2020 before plans to decommission it could be put into effect). The message was sent in a specific direction, towards the star cluster, labelled M13 in the famous catalogue of some 100 objects listed by Charles Messier (1730–1817) as nebulae or star clusters. The M13 cluster is more than 22,000 light years distant from our Solar System so it is a very moot point whether it will ever be picked up by exobeings. It was quite short, only 1,679 binary digits (arranged as a matrix of 73 rows by 23 columns) and took three minutes to send.

For our discussion the message is interesting in that it contains no language, just seven pieces of information.

Contents of the Arecibo message

1. The numbers 1 to 10
2. The atomic numbers of hydrogen, carbon, nitrogen, oxygen and phosphorous, the elements of which DNA strands are composed
3. The formulas for the sugars and bases in DNA nucleotides
4. The number of DNA nucleotides and a graphic representation of the DNA double helix
5. A figure, constructed with pixels, which was supposed to represent an average human being
6. A line of pixels showing the Sun and the planets, with Earth raised above the line to indicate where the message originated
7. A stylised image of the Arecibo radio telescope

Whether exobeings could decipher such a message is questionable but, assuming curiosity on their part similar to that typical of most humans, they could certainly try. Like the many instances in which linguists deciphered very early texts, often available only as poorly conserved inscriptions on stone, they could perhaps reconstruct the message successfully.

In August and September 1977, NASA launched two spacecraft, Voyager I and II, from Cape Canaveral, Florida, which were intended to fly towards the outer reaches of our Solar System and beyond. They both contained a Golden Record (designed by a committee chaired by Carl Sagan) along with a record player, containing greetings in 55 languages from planet Earth, as well as 116 encoded images and some other audio and video files. There was also a set of pictorial diagrams on a plate beside the disc intended to explain to exobeings how to play the double-sided record to find out what it contains.

Both the Voyager spacecraft are currently hurtling through space at about 60,000 kilometres per hour but have only recently left the Solar System (in 2012 for Voyager I and in 2018 for Voyager II), which means that it will take tens of thousands of years before they might reach the region of even the nearest stars where they just might possibly be detected. The same is true of the plaques contained on the Pioneer 10 and 11 probes, launched in 1972 and 1973, respectively. These plaques contain pictures of a man and a woman, our Solar System, the trajectory of the probes, the astronomical location of Earth and a depiction of two hydrogen atoms, showing the hyperfine transition of hydrogen's electron. These depictions could help exobeings work out the dimensions of the other items on the plaque (the same procedure was used on the Voyager Golden Records).

The Pioneer and Voyager plaques/discs are like messages in bottles cast into outer space in the hope that someone somewhere sometime might find them and figure out where and who they came from.

30.3 A Messenger from Beyond?

Ignoring the vastness of interstellar space and the pitiful slowness of spacecraft traversing it for a moment, imagine a spacecraft entered our Solar System and got from the Kuiper Belt to Mars in a few days and was heading towards Earth. Our radio telescopes should begin to detect it fairly soon and from that moment onwards all the telescopes in the world would be pointed towards the unannounced visitor. Such a spacecraft would most likely not contain exobeings but have been sent by them from their planet as an unmanned probe. Now imagine that we could miraculously capture the craft and bring it down onto Earth unscathed. By examining such a spacecraft we could tell a lot about the exobeings who built it. The spacecraft would be evidence that exobeings exist/existed, and are/were in a position to construct and launch an interstellar spacecraft. In a way, that is what was intended with the Pioneer and Voyager spacecraft: for exobeings they would be messengers from beyond.

'Oumuamua

There is a small number of scientists who think that the situation just described may have actually happened, at least something like it. 'Oumuamua is the Hawaiian word for 'scout' and is the name given by humans to the interstellar object detected within our Solar System in

October 2017, by the Canadian astronomer Robert Weyrk, working at the Haleakaiā Observatory in Hawaii. The object is red in colour and unusual in shape: oblong (rather like a cigar or maybe a stretched pancake) and tumbling through space on a trajectory which implies that it is not a Solar System object but is being drawn into an inner orbit by the gravity of the Sun. ʻOumuamua does not have the characteristic tail of a comet trailing behind it. So it may not consist of dust and various ices like other well-known comets (though there are different opinions on this) such as Halley's comet, which flies by Earth every 75 years. Nor is it a rogue comet (from outside our Solar System) like 2I/Borisov, discovered in 2019 by a Russian amateur astronomer after whom it is named.

The appearance of ʻOumuamua inside our Solar System caused some astronomers, like the Americans Andrew Fraknoi and Avi Loeb (from Harvard University) to consider that it might be an artificial probe with biological material on board (assumed to be in the form of long-living microbes), sent deliberately by exobeings from an exoplanet attempting to spread their forms of life to other star systems in an exercise labelled 'directed panspermia.' In fact scientists, such as Carl Sagan, speculated back in the 1960s that life on Earth may have arisen by such an extrasolar probe arriving here and spreading the life forms it contained on our planet. Going in the opposite direction could also be an option worth considering: some scientists believe it would be ethically acceptable for humans to send out probes containing microbes from Earth to ensure that the life forms which have developed here would have a chance of seeding themselves elsewhere, thus preserving life if it died out on our planet.

Returning to ʻOumuamua, there is no evidence for its coming from an exoplanet and the idea is definitely not supported by astronomers generally. But we have to be open-minded and pay attention to anomalies suggesting such astronomical objects might be artefacts. And there might just be self-replicating probes travelling through space. These are often called von Neumann probes after the Hungarian–American scientist John von Neumann (1903–1957), who suggested that these might exist. It is unclear how such probes would construct themselves and where the material would come from for the task. Such probes would not grow organically and hence would need to be constructed by machines in space – a major challenge, to say the least. At any rate, given the vastness of space, any such probe, if found, would be merely a trace of an exocivilisation from which it came and which might have long since ceased to exist.

30.4 Communicating without Meeting Them

At present, the Breakthrough Listen project is constantly monitoring incoming signals via radio telescopes and looking for an irregularly modulated signal, which could be a message from beyond our Solar System, sent from an exoplanet. The radio signals received from natural sources – stars, quasars, etc. – are regular and would not show the degree of modulation indicative of a message encoding information. But, if such a message were picked up, could we interpret it? This really depends on how the message was organised. If exobeings sent a message in their exolanguage with no indication of how it could be interpreted, we would be at a loss to understand it. However, if such exobeings possessed intelligence on a level comparable to our own, they would also consider sending a message which we on Earth – exobeings for them – could decipher, in principle.

What would we want to know from such communication? Just that life exists somewhere else? This would answer the question, 'Are we alone'? But we might wish to communicate with them, to find out more about them and so relativise many key aspects of our earthly life. If exobeings were like ourselves, as in having brains and limbs like us, a sex distinction, comparable childhoods and lifespans and, above all, languages similar in structure to those on Earth (though not in actual manifestation), we could conclude that the manner in which human life evolved here followed preferred paths of biological evolution and would be likely to have arisen elsewhere in the universe as well. Indeed, if we discovered one other exoplanet with exobeings similar to ourselves in our cosmic neighbourhood, there are likely to be millions of such planets with similar exocivilisations in the galaxies of our universe.

To conclude, we can consider all the implications we would be justified in making about life on an exoplanet if we were to pick up a radio signal from one of these.

Reverse implication scale for life on exoplanets

If there were intelligent beings on an exoplanet:

1. These beings must have evolved from simpler forms of life over hundreds of millions of years; the beginning of life there must have started with the simplest of cells.
2. They must have developed intelligence far greater than what is required to survive in their natural habitat (Wallace's puzzle); just how this would have happened is unclear to us.
3. These beings would likely have complex emotional lives with curiosity as a key feature driving their existence; otherwise, they would not progress very far.

If they could transmit radio waves into interstellar space, it is likely that:

1. They would have mastered the necessary digital technology to do this.
2. They would have harnessed electricity and have an understanding of the fundamental forces governing the universe; this would include knowledge of the atomic and subatomic domains or at least an understanding of the effects which derive from these.
3. They would have sciences corresponding to astronomy, physics, chemistry, biology, geology and geography.
4. They would have an efficient counting system and have developed an equivalent to our mathematics along with its various branches.
5. They would have systems, comparable to human language, allowing them to communicate with each other and importantly to transfer knowledge across generations; some system akin to schooling would have to exist for this to be realised, otherwise each generation would have to start more or less from scratch.
6. Their planet would contain complex societies with divisions of labour and possibly class differences, as seen traditionally on Earth.

30.5 And If We Find One, What Then?

If we humans discover exobeings on an exoplanet, human history will be divided into two blocks: the time before and the time after that event, and it is worth speculating on this issue for a moment. Discovering an exocivilisation would have the most far-reaching consequences for all domains of life on Earth. It would have consequences for all the natural sciences, especially for astronomy, geology, evolutionary biology and palaeoanthropology, and of course for the study of language.[4] Of course, any 'discovery' would have to be indisputably confirmed and that may well not be so easy to do. But if it were, that would be a major game changer. Even if most people just went on with their daily lives as before, we could not ignore the fact that we would then know that we are not alone.

But for a moment, consider what the situation would be like for the exobeings: they might not be that interested in getting in touch with us for a number of reasons. They might lack natural curiosity, though that is unlikely, given that their own civilisation would only have arisen if they were interested in constructing one. They might already know other exocivilisations on other planets and simply not have the resources or interest to investigate yet another one. They might regard us as an inferior form of life, not worthy of investigation. They might be afraid that we would attack them. Given the aggressive behaviour of humans in the past and present, and the kinds of leaders which many countries on Earth have, the last assumption would not be at all unreasonable.

A technologically advanced civilisation on an exoplanet would likely consist of several hundred million exobeings, at least. Anything smaller would probably not have the societal and technical infrastructure necessary to reach a high level of technological advancement – just think of the number of firms and companies behind the space industry on Earth.

4 But there would be consequences for other domains of life on Earth, such as the religions of our world, which might have to relativise many of their dogmas concerning humanity and spirituality. Whether exobeings would have religions comparable to those on Earth, and a notion of a god which created the universe, is a completely open question.

Assuming that an exoplanet would have a mixture of water and land, it would have a geography in principle similar to what we have on Earth.[5] The development of regions on such an exoplanet could vary, if the significance of the Mediterranean area with Greece and Italy for the development of Western civilisations or of India, China and Japan for Asian civilisations, for instance, is anything to go by. Some regions could be more developed (in a technological sense) than others and it is with the populations of such parts that contact would probably be made, as it would only be those areas where digitally advanced technology would be most prevalent, or indeed present at all, again if the situation on Earth is typical.

One consequence of these considerations is that exocivilisations might be in competition with each other on an exoplanet, similar to the way in which species compete in evolution. There could well be a similar kind of grouping of societies into nation states as we have on our Earth, with language playing a key role in the formation of identity and the bonding of individuals in such states, as we see on Earth. Furthermore, is it likely that these states would be not just in competition but locked in war with each other? If one considers the role of violence on Earth, both in the animal and human spheres, this could well be true of exoplanets as well. Indeed, a completely peaceful, placid and technologically advanced exoplanet is unlikely, as competition is a key aspect of human interaction and does in fact stimulate advances, including those in science. Equally, it is probably unlikely that an exoplanet would be inhabited by entirely egalitarian groups without any social hierarchies. All forms of life vary: animals and humans are no exception. So some will be stronger than others, more powerful due to privileged access to resources, favourable geopolitical circumstances, etc. (as was true of Europe in recent human history). This leads to hierarchical structures in any group and so would probably be true of an exoplanet as well. Historical developments on an exoplanet would likely lead to some groups/societies/states attaining more power

5 With ever more powerful telescopes it may one day be possible to directly image the surface of an exoplanet. We might then be able to recognise features comparable in size to the Richat Structure on Earth. This consists of a series of concentric rings in a depression about 40 kilometres in diameter, located in Mauritania and clearly visible from space.

than others and these will most probably be those which are at the forefront of technology, just as on Earth today. Any assumption that exosocieties would be egalitarian with peaceful governments would need to offer robust arguments for how such societies might arise.

We do not know whether the cognitive gap between exobeings and other forms of life on their planets will be as significant as between us and chimpanzees on Earth. On an exoplanet there might be a more continuous evolutionary line among animals with higher cognitive abilities, which in itself would be of interest to us here on Earth: comparing evolutionary developments would throw light on the development of life on Earth.

Space Exploration and Contact Again

On any exoplanet there could be different attitudes to exploration: some might want to find exoplanets with exobeings and communicate with them. Others might simply wish to continue their lives on their own planet undisturbed.

However, our assumption may be that exobeings would probably wish to communicate with us. Consider that, behind all our work on discovering exoplanets, is human curiosity. If we were not quintessentially curious beings, we would not have a technologically advanced civilisation, because all the steps along the way, in terms of fundamental research and engineering developments, would not have been undertaken and so the foundations of our modern societies would be missing.

30.6 Predicting Reactions

We cannot predict how exobeings would react to us if contact were established. Depending on the size of an exoplanet, it might be a Super-Earth, up to twice the size of our planet, and depending on the distribution of land masses and oceans on its surface there could well be great differences in the developmental stages of societies on such a planet. Again, human history on Earth provides many examples of this. Members of more technologically advanced societies might wish to extend their sphere of influence beyond the planet to surrounding space. There are already people in key positions of the governments of countries like the USA, Russia and China thinking along these lines. True, in 1967, the Outer Space Treaty specifically excluded the placing of weapons of mass destruction in space and, by implication, prohibited the militarisation of space. But are all superpowers today prepared to keep to this? And what would the situation on an exoplanet be like? We might inadvertently become embroiled in strife on an exoplanet, depending on who we first established contact with.

However, when one considers the enormous distances involved and the present absence of interstellar travel, any fears of being the object of aggression from an exoplanet are currently unfounded. Nonetheless, it is still a legitimate question to ask if exobeings might view us negatively. Opponents of this view would maintain that we are extrapolating from patterns of our behaviour to theirs. Others consider the risk involved in trying to establish contact with exobeings, who might turn on us, too great (see Section 30.1 above). After all, we have no idea whether such exobeings would have a sense of morality and be guided by principles of mutual cooperation.

31

Language Beyond Earth

> I think it is fruitful to consider language as an evolutionary adaptation, like the eye, its major parts designed to carry out important functions.
>
> Steven Pinker, *The Language Instinct* (1994)

Language is a unique property of humans. It is located in our brains and is intimately connected with our experience of consciousness. Our interaction with other humans via language is the main means by which we can be sure that others experience levels of consciousness like ourselves. However, many animals have communication systems which in principle are similar to language, that is, they are used to convey information between members of a species, though not always by means of sounds. For instance, bees use a special set of movements in which information about a source of nectar, its size and distance from the hive, is transmitted by movements in space by the bees, their 'dance'. Whales use noises sent out beneath the water to other whales. Other senses can and have been used for communication. For example, many insects exude pheromones, scents with a certain signal value for a member of a species, normally to attract females. These systems

– however fascinating they may seem to us because they are so different to ours – are all quite limited in the amount of information they can convey and are furthermore generally rigid. Except for the song of some imitative birds, like certain types of thrush, and a few other cases, animal communication systems are generally restricted and inflexible. The kind of information conveyed by animals has largely to do with food, danger of predation, sexual selection and the like. Humans on the other hand show the highest level of cognitive flexibility and can use language to talk about literally anything, whether true or false, existent or not. We also have the capacity for reversal learning, in general adapting quickly to contingent circumstances in our environment, indeed to changes to the environment itself.

Recall that early modern humans, the species *Homo sapiens*, started to evolve after at least 400,000 years ago, possibly as much as 800,000 years ago, in eastern and southern Africa. From the point of view of anatomy, these were modern humans: they had the long necks associated with speech production and a fully developed brain with a large prefrontal cortex. *Homo sapiens* started spreading from Africa to other parts of the world. By 60,000 years ago they appear to have reached Australia, having dispersed through Asia. In Europe, the first *Homo* species turned up shortly after their appearance in Africa, having moved through the Levant into the Eastern Mediterranean. These groups consisted of Neanderthals and *Homo sapiens*. Although the former were spread across all of Europe and into Siberia, the Neanderthals did not survive, the last having died out about 35–40,000 years ago. Among the *Homo sapiens* who overlapped with the Neanderthals before their demise were those called Cro-Magnon (35,000–45,000 years ago), from the region of the Dordogne in south-west France where important finds were made. They were anatomically modern humans, with high foreheads (indicating a large prefrontal cortex) and an appearance similar to humans today. The Cro-Magnon were spread across Europe and even in the north of Russia; they lived in caves, used complex tools and engaged in wall painting.

Does the number of people who existed during evolution play a role? Not really, there were times when the numbers dropped considerably. For instance, there was a population crash around 75,000 years ago when

Mount Toba on northern Sumatra, Indonesia exploded and caused a worldwide environmental catastrophe which led to widespread starvation due to the lack of plant growth after the kilotons of volcanic ash, spewed high into the Earth's atmosphere, blocked out sunlight for years and led to a cooling of the atmosphere, possibly for centuries, with only a slow recovery of the world's population afterwards.

31.1 Research on Evolution

At present, the early twenty-first century, there is a huge body of knowledge on the evolution of life forms – plants, animals and humans. Ever since Charles Darwin's epoch-making *On the Origin of Species* (1859) there has been intense scientific work on just how life on our planet evolved (the question of abiogenesis, the trigger for the start of life, is one which Darwin, however, did not consider). Evolutionary biology is now a central field in natural science, based on an enormous amount of comparative data which has led to established insights into the nature and progress of evolution on Earth.

When it comes to language the situation is unfortunately not so clear-cut. There are two reasons for this. First, language leaves no trace in the fossil record, which means that linguists must look to ancillary sciences, like palaeoanthropology, for evidence of language evolution.[1] Second, language is now only found in *Homo sapiens*, ourselves, the only remaining species of the genus *Homo*, so that we cannot do comparative work across species, as we could in the development of such physical features as legs, feet, wings, fins, beaks, teeth, eyes, etc.

However, a few qualifications are required here. Palaeoanthropology can help with the skeletal make-up of early hominins. This shows an

1 It should be mentioned that there are computational models to determine how languages could evolve given some initial input data. Caution is required when generalising from such models; nonetheless, insights can be gained about the possible trajectories which human languages might take and hence assist researchers considering how exolanguages could arise.

ever-increasing size of the skull (more precisely of the brain case), which is an indication of increasing intelligence, assuming that the size of the brain, especially of the prefrontal cortex, correlates (approximately) with intelligence and that brain size matches body size favourably. Other anatomical features, which can be linked to language, are the shapes of the oral and respiratory tracts, various bones around these areas, especially the hyoid bone, and the probable existence of specialised nerves which control the speech apparatus, above all the hypoglossal nerve (from Greek *hypo* 'below' and *glossa* 'tongue'), a major cranial nerve (the twelfth and last of these) which passes through a canal from the base of the brain towards the front. A certain amount of comparative work, see Fitch (2010: section 3.9), between humans and other animals, involving abilities such as vocal control among birds, has shown that our own ability to produce the sounds of language had precursors or related functions in the animal kingdom (see Chapter 26 for more discussion).

The work just alluded to is important in forming a mosaic around the rise of language in humans, and in recent years this mosaic has slowly but surely been increasing in both size and internal differentiation. This is a fortunate development, given that research into the evolution of language was for a long time frowned upon, certainly since the famous ban by the Linguistic Society of Paris in 1866 on all discussions and debates on the topic,[2] something which stifled relevant research for a century or so.

31.2 When Did It All Start?

By the time *Homo sapiens* appears in the fossil record their brains were in the range of 1,300 cubic centimetres or more, so comparable to, or in some cases even somewhat greater than what we have today. *Homo*

2 Arguments from design, such as those put forward by Christian apologists like William Paley (1743–1805), are not, and have not been, used in linguistics. Paley gave the example of someone who found a watch on a path and who would have naturally assumed that it had a 'creator.' His error lies in confusing human artefacts (his watch) with evolutionary phenomena. Of course, at his time, nothing was known about how evolution works.

sapiens show a typical frontal cranial bulge so we can postulate that this housed the prefrontal cortex, which distinguishes our brains from those of other animals closely related to us. Now, assuming that the present human language faculty developed not earlier than 200,000 years ago (about 8,000 generations ago), this leaves about 200,000 years when we had our present-day brain size but without language in our modern sense. This period would correspond to that assumed by the proponents of the discontinuity hypothesis in which we had large language-ready brains, seemingly waiting for the 'spark' to ignite our language faculty with complex grammar. That spark would be the sudden and unmotivated mutation which allowed humans to produce non-linear, hierarchically ordered, potentially recursive syntactic structures.[3] However, those linguists who favour the continuity hypothesis assume that language gradually developed over a long period before it reached the structural and functional complexity which we observe today. This issue will be explained and discussed in more detail in Section 32.7 below.

Other scholars have assumed that the earliest forms of language reach back much further in time. Bickerton (2009) maintained that proto-language was the type of communication system used by *Homo erectus* about one and half million years ago. Other scholars are sceptical about pushing back the origin of language so far (Fitch 2010: 405). Bickerton's view would mean that for the development of modern language from proto-language we would have well over a million years: from 1.5 million to about 200 thousand years ago.

It is also uncertain just how language evolved: as a slow–quick–slow development (technically called an S-curve); by leaps and bounds; or as longer stable periods with sudden fits of evolution – known in biology as 'punctuated equilibrium', a term stemming from the American evolutionary biologist Stephen Jay Gould (1941–2002).

3 This preference of humans for tree structures, not just in language, but also in music, has been labelled *dendrophilia* (lit. 'love of trees' from Greek) by W. Tecumseh Fitch.

31.3 Where and Why Did It Start?

The last exodus of *Homo sapiens* out of (eastern) Africa took place about 80,000 to 60,000 years ago, across to the Arabian peninsula, and then in two movements, one up to the Levant and on towards Europe and another over to South Asia, and then towards East and South-East Asia. Because all the design features of language (see Section 20.4 above) are found in all languages of the world it must be assumed that these applied to the forms of language used by those late *Homo sapiens* who left Africa.

A traditional view of the rise of *Homo sapiens* is that the Great Rift Valley, which runs inland along the coast from the Red Sea down to the Strait of Madagascar in eastern Africa, separated early hominins into two groups, one which continued to live in the rainforests of central and western Africa, and another group which was left in the semi-arid savannah regions to the east of the continent. The inhospitable conditions led here to the necessity to develop intelligence. These conditions did not just consist of an arid climate but probably involved fluctuations in climate, which required continual adaptation by early hominins. Part of this adaptation to the eastern environment would have been the development of a system of communication which ultimately evolved into human language. The American linguist Derek Bickerton (1926–2018) regarded the selective pressure to communicate in situations where food is out of the visual range of others as providing an impetus for the development of signs with iconic value to refer to things not present, thus forming an essential characteristic of human language.

There is evidence for the east of Africa location for early humans and hence for language: chimpanzees are found only to the west of the Great Rift Valley. However, there is probably more to the story than this. Early fossil remains of hominins have been found extensively in southern Africa and across the north of Africa. It is unclear whether this is due to migration from eastern Africa or to the rise of different *Homo* subspecies in diverse areas of Africa.

The Out of Africa Hypothesis for Language

The basic principles of human language today must have been established before the spread of *Homo sapiens* from Africa as all extant languages show the same principles. Duality of patterning, referential displacement, the arbitrariness of the sound-concept relationship, the productive, open-ended nature of syntax, are features found in all the world's languages and so must have been present at the time of the last dispersal of *Homo sapiens* out of Africa (see above). Various reasons have been put forward for why *Homo sapiens* began to disperse out of Africa. Long periods of drought in Africa could have driven humans to coastal regions from where they crossed to adjacent land masses, for example across the Red Sea to the Arabian Peninsula. The dispersal was a slow process, perhaps a few kilometres per hundred years (see Section 13.4 above). It certainly did not have the character of a campaign of conquest comparable to what we see later, as with the spread of the Mongols throughout Asia, the Middle East and eastern Europe in the early- to mid-thirteenth century. The slow dispersal led to the gradual geographical separation of hominin groups, often with contact between them in later movements. These are factors which linguists now know lead to language diversification.

An additional linguistic argument has been put forward in support of the Out of Africa hypothesis for human language by the New Zealand linguist, Quentin Atkinson. He has suggested that there is a correlation between the original region for early language and the amount of distinctive units in the sound system of languages (technically called phonemes). The languages of Africa, as the oldest languages, have a high number of phonemes whereas the languages of Oceania, the region of the world last populated by humans, are known for having very few phonemes (Rotokas, spoken on the island of Bougainville, east of Papua New Guinea, holds the record with only six consonants and five vowels in the central dialect of this language). The click sounds in the native San languages of Namibia, from where they spread to Bantu languages (the Nguni subgroup) in South Africa, such as Xhosa and Zulu, may also be a conserved feature of very early languages.

Language Evolution Just Once?

Did language evolve multiple times, like eyes? This is not a trivial question because such developments can often show striking structural similarities. They are technically known as analogous evolutionary traits – similar features, which arise separately from each other, as opposed to homologous ones, which are connected in their evolutionary past. If that were also true of language, this could possibly mean that the key structural features of human language, such as non-linear order in syntax and duality of structure, were independently developed features.

In the context of the present book, any analogous developments have strong diagnostic value because they increase the likelihood that they could also arise under similar sets of circumstances on an exoplanet. Features which only arose once, even if they are widespread on Earth, are just a statistical set of one, and hence conclusions about the probability of such features developing on another planet cannot be drawn.

31.4 Primary and Secondary Functions in Biology

In the course of evolution, many primary bodily functions acquired a secondary function which was superimposed on the first. An example is urination, the process of ridding the body of liquid waste. Because of the individual smell of urine it came to be used by many animals for delimiting their territory vis à vis others of their species, a case of what is often called exaptation, the acquisition of a function for which something did not originally evolve.

Many features of human anatomy point to an adaptation for language, that is, to a secondary function. For instance, the larynx of humans is particularly low in the neck. This has a physiological disadvantage in that we could theoretically choke when we eat, as we must close off the top of the trachea each time we swallow. However, the advantage for speech is that we can produce a large number of sounds with a considerable volume. The sounds of human languages – the great range of vowels and consonants – is only possible given the shape of our mouths, the flexibility of our tongues and the additional use of the nasal cavity and the switching on or off of vocal fold vibrations, an essential feature for the consonants of all languages.

Human tongues are highly flexible bundles of muscle, capable of making tens of movements per second. To fulfil the primary function of moving food around the mouth and swallowing it, our tongues would not have to be so dexterous – just think of the tongues of dogs or cats which are quite limp but nonetheless sufficient for this primary purpose. So it would appear that the agile tongue muscle of humans developed because of the secondary function of speech production (Table 31.1).

A further example of secondary functionalisation connected to language concerns the vocal folds, which have the ability in humans to close and suddenly release, yielding the sound known technically as a glottal stop, a common sound in many languages of the world. It is found, for instance, in the pronunciation of *t* in London English in a word like *water*, often described as a 'catch in the throat' and indicated by an apostrophe in writing: *wa'er*.

Table 31.1 Primary and secondary functions with respect to speech

Body part	Primary function	Secondary function
Larynx	Pipe leading to lungs	Resonance chamber for sound
Vocal cords	Closing mechanism for lungs	Production of sound through vibration
Teeth	Chewing	Dental sounds
Tongue	Moving food around mouth	Production of a wide variety of sounds
Velum	Valve for closing off nose	Production of velar sounds
Nasal cavity	Breathing through nose	Production of nasal sounds

PRIMARY AND SECONDARY FUNCTIONS WITH EXOBEINGS

In animal evolution on Earth there are repeated examples of parts of the body acquiring a secondary function in addition to an earlier primary one. There is no reason not to assume that in the evolution of exobeings this would have taken place too. Secondary functionalisation may well also alter the phenotype of the part of the body affected, as with the human tongue.

Secondary functionalisation extends beyond the physical sphere. In (externalised) language, one can recognise not only a primary function in communication but also a clear secondary function in the expression of social identity and perspective. This is the main reason why dialects still exist in countries with standard forms of language. Although dialects tend to be viewed negatively and as not acceptable in official public usage they are maintained by their speakers as an expression of their social and linguistic identity, strengthening internal cohesion within local groups.

32

How Human Language Arose

> I foresee that the field will converge on a biologically driven decomposition of the human capacity to acquire language into a set of well-defined mechanisms, and the interfaces between them.
>
> W. Tecumseh Fitch, *The Evolution of Language* (2010)

32.1 Looking for a Beginning

There is probably no one reason for the rise of language but rather an ensemble of factors which all played a role in the gradual increase in sophistication from simple noises to the flexible system of communication we know today. Among linguists there is much discussion not only of the triggers for language but also of the steps involved and the manner in which *Homo sapiens* moved from one stage in language development to the next. As has been stated at several points in this book, one has to account not only for the structures of the attested languages across the world but also for the rise of the language faculty, internal to our brains, which makes languages possible in the first place.

Niche Construction

The idea of niche is a given in biology: organisms and animals evolve to occupy all the niches which an environment offers, given enough time. Take a rainforest as an example. Every niche is occupied: the forest floor is filled with bacteria, insects and animals, which move around the forest. The tree canopy is occupied by different animals, as are the levels below this. There are behavioural niches: hunters prey on animals and eat their flesh, scavengers, from vultures to hyenas, feed on what is left over. When flora die naturally, the remains are dealt with in a similar way, fungi and bacteria decompose vegetation and the cycle of nutrients continues. All of this happens because, for every niche in nature, there is some life form doing specialised work there. In a nutshell, if there is a niche anywhere in an environment, it will be filled by life, sooner rather than later.

Niches can be actively created as well. In the view of Derek Bickerton, niche construction results from the interaction of genetic evolution and cultural behaviour. Early hominins would have created a niche for themselves in the animal world after the divergence from the common ancestor with chimpanzees, and would then have begun to compete with other animals and assert themselves, first through confrontational scavenging and later directly through hunting in small groups. Bickerton sees the first stirrings of language in this early behaviour of the *Homo* species.

When Did It Begin?

Just when exactly language began depends on how one defines it and how one evaluates the tentative evidence in the fossil records of early *Homo* species. In the sense of communicating with acoustic signals, language was likely always present with the various species of *Homo*; early chimpanzees probably also used sounds for communication, panting and hooting as do their conspecifics today. In the sense of a structured symbolic acoustic system using words strung together into grammatical units, human language is much younger, with 200,000 years ago a generous figure for language comparable to present-day language in complexity and flexibility.

By 50,000 years ago, language probably had finally reached a stage where it had the range of structure and function which we know from languages today. This means that by the time recorded history starts - not more than 5,500 years ago - human language had long since developed all the structural characteristics and functions of languages today. The societies of 5,500 years ago were certainly less complex than present-day ones but the languages used were not less sophisticated - they were fully suited to the needs of the societies which used them.

32.2 Some Early Triggers

In any science, the scholars who are perceived most clearly by others are those who adopt a single stance on a topic. For instance, it is much easier to remember a scholar who represents the gestural theory of language evolution, another who holds to the song origin or a further one who believes gossip was central rather than someone who supports all three possibilities to varying degrees.

But when dealing with a matter as complex and multi-facetted as the evolution of language the truth is probably not to be found in one standpoint and one alone, but rather in a combination. So, in adopting this position here, it is not from an unwillingness to nail my colours to the mast but from a recognition that the truth is not black or white but somewhere in between.

Below, a number of factors in the rise of language are introduced. They are associated with certain authors who have highlighted a particular factor or group. The following factors are listed separately but they are interrelated and probably operated together.

Grooming

Higher primates, such as chimpanzees and gorillas, are known for grooming within their groups. Group size with animals is normally fairly limited, a group being a collection of animals which interact with each other and maintain internal relationships. Thus, a group is different from a flock of birds, for instance, the members of which simply congregate together and may have very limited interaction, like flying in formation. Grooming has the primary function of removing parasites from the skin/hair/fur but the secondary function of strengthening bonds between members of the group. The British anthropologist Robin Dunbar, in his 1996 book *Grooming, Gossip, and the Evolution of Language*, also proposed that human language may have its origin in a form of verbal grooming carried out in social groups. As these got larger it became impracticable to engage in physical grooming and so members of the group resorted to grooming via words instead of touch, as it were. This would be cheap, in the sense that an animal would not have to invest precious resources, such as time and attention, in its realisation, and nonetheless could reach the goal of strengthening social bonds, but only if language was accepted as a reliable type of group-internal commitment, an issue which has been the topic of research and critical assessment of the grooming hypothesis.[1]

The clear advantage of verbal, as opposed to physical grooming, is that it can encompass a much larger group of people. Verbal grooming is less intense but has a greater range. For supporters of the social brain hypothesis, this interaction within a larger group would have provided an additional stimulus for an increase in cognitive powers. Certainly, it would have provided an incentive to develop, or expand an existing 'theory of mind' by which humans could work out what others are thinking and thus interact more successfully with them.

Gossip

Gossip can be seen as a subtype of verbal grooming in which the amount of factual information is minimal and the expression of opinions about others comes to the fore. Robin Dunbar more specifically suggested that gossip was also significant in the early rise of language increasing bonding through this type of social interaction.

1 The topic of social communication, bonding and language has been expanded in recent literature, see Dunbar (2012), Knight and Power (2012).

Helping in the Group

The Swedish physicist and linguist, Sverker Johansson, sees in help and assistance key features in the origin of language. In his book *The Dawn of Language* (2021), he writes: 'Human language presupposed helpfulness. Language cannot have evolved, therefore, before we became helpful, which means the evolution of helpfulness is one of the keys to the origin of language.' Certainly, cooperation is essential for any group to survive and prosper, and organising this cooperation involves continual communication between the group's members. In this sense such help is yet another type of social interaction which would favour the continual development of language.

Contact Calls

Most animals have a means of contacting their young via acoustic signals should they become separated, a common occurrence in the wild. The pressure to have contact calls between mothers and offspring was obviously operating in evolutionary history and so these may well have contributed to language, especially with novel modulated calls developing over time. Predator calls are used to warn close relatives and/or group members of danger. An increasing complexity with such calls could have contributed overall to linguistic complexity.

Gestures and Pointing

Humans to this day engage in gesturing while speaking. In some cultures, such as in Italy, this is particularly marked. The link between gestures and sounds – the two are often coordinated in the animal world as well – has led linguists to see in gesturing a likely source of, or at least an area of common ground with early language. In its turn gesturing was preceded by posturing, as with a combat posture, for instance, used to intimidate other animals of the same or another species.

Just how advanced gestural proto-language was – hands first, then speech – is hard to say, though it is unlikely that internal language already existed with this system of gestures. Language and manual gestures (Corballis 2002, 2017; Goldin-Meadow 2012) could have further developed into a system of pointing (technically called deixis), where early *Homo sapiens* would have used the fingers of their hands to draw the attention of others to something in their environment.

Initially, gestures could have represented meaning (semantics) and then have developed grammar (syntax), which ended up by transferring to the modality of speech.[2] If this is the case, the transition issue is about how early hominins moved from a gestural system to one of vocal sounds with attendant meanings. One solution is to postulate that any gestural system would have been accompanied by sounds from the very beginning and that the sounds came to dominate in the system, with the gestures assuming an increasingly secondary role. Any hypothesis without sound from the start needs to provide a convincing account of how a communication system would adopt sound as an additional modality alongside an already existing gestural one. Sounds along with gestures provided a flexible two-tiered interconnected system: sounds could be left out on occasions, silent signalling can be very useful when hunting, for instance, and individuals could refrain from gestures, using sound alone on other occasions, such as in close proximity, in poor lighting, when using both hands for some other purpose, etc.

Supporters of gesturing in both the phylogeny and ontogeny of language, such as the American cognitive scientist Elizabeth Bates, have highlighted the fact that infants use pointing in their pre-linguistic phase of life. This would seem to support the view that gesturing is a remnant of pre-linguistic communication among humans.

However, if the internal language faculty we have is independent of sensory expression, and it can manifest itself in sounds or in gestures, this independence cannot have been there from the beginning unless one assumes that human language was originally purely gestural and the auditory expression arose as a later, independent development.

2 Note that gestures and sign languages are very different things. Gestures have a scaffolding effect during language acquisition and later in language use (Fitch 2010: 437) but sign language is an entirely different modality of language externalisation.

Expanding the System of Pointing

In human language, an important component involves pointing backwards or forwards form the point of time in speech. For this we use tenses and words referring to time, such as *Yesterday I was in Cork, Tomorrow I am going to Galway*. Such usages are ultimately abstractions from physical locations: *The river behind us; The mountain in front of us*. It is thus justified to see the incorporation of gestures into speech as a process in the genesis of human language. Indeed, the system was considerably expanded, with whole series of 'pointing elements' arising in languages, such as demonstrative pronouns – *this man, that girl, those people* – or personal pronouns used to link up sentences in a stretch of speech, for example, *My Spanish teacher is from Chile. She came to Ireland a few years ago. Computers are useful, everyone has one [of them] nowadays.*

Were Mirror Neurons Involved?

Neurophysiologists at the University of Parma, Italy, under the leadership of Giacomo Rizzolatti, conducted experiments on macaque monkeys in the 1990s, during which they discovered that certain neurons in the monkeys' brains fired not just when they were performing an action but also when they were observing this action being performed by others. The neurons involved were labelled 'mirror neurons' and equivalents were later shown to exist in humans. While not accepted by all scholars, this research sparked wide interest in different fields of science, notably in psychology and linguistics. For instance, new views of empathy arose, grounded on the triggering of a feeling in someone merely by observing it in another person, like sadness and unhappiness or joy and elation.

The view that language evolution could be causally connected with the mirror neuron system has been put forward by Michael Arbib, and his reasoning is as follows: mirror neurons help us in imitating the actions of others and internalising the steps involved. Hence, language production could have evolved out of this system, given that the area where mirror neurons are located in macaque monkeys is homologous with a region of the left temporal frontal cortex, close to Broca's area, largely responsible for language production. While this view is worthy of consideration it leaves many questions unanswered, for instance how it would connect with the origin of syntax.

Language and Song

Doubtless there are parallels between language and song. Singing involves modulation of the voice, in pitch, amplitude and duration, and these elements are present even with songs which do not have words to them.[3] Furthermore, in most languages today, pitch, amplitude and duration play a central role.[4] This role might be part of the language system in the strictest sense, that is, variations along these three parameters can result in changes in meaning as in tone languages such as Vietnamese or the various Chinese languages (technically called Sinitic languages). Even where pitch is not so important, as in English, amplitude and duration are significant, for instance with stressed versus non-stressed syllables, yielding contrasts like *record, convert, remake* (noun, stress on first syllable) and *record, convert, remake* (verb, stress on second syllable). Duration is important with vowels in English: the contrast between long and short vowels is the basis for word distinctions like *bit* (short vowel) versus *beat* (long vowel). In some languages, like Swedish and Italian, there are also length distinctions for consonants, for example Swedish *vit* 'white' (adjective, short t) versus *vitt* 'white' (noun, long t) or Italian *tono* 'tone' (short n) versus *tonno* 'tuna' (long n).

The use of pitch, amplitude and duration, known technically as prosody, for meaning distinctions in the languages of the world strongly suggest that these elements are deeply conserved from the earliest days

3 There are examples from jazz, known as scat singing, and in classical music the Bachiana Braziliera No. 5 by Hector Villa-Lobos contains a well-known example of singing without words. In fact there is a minor musical genre, known as a vocalise, of sounds without words, a well-known example being that by Rakhmaninov.

4 Humans have direct cortical motor connections with the medulla (at the top of the spinal cord) through which the muscles in the larynx are innervated, allowing for sensitive vocal control. We are the only primates which have this direct connection.

of pre-language. They are essential elements of music found also in the vocalisations of singing birds. The comparison with such birds has led several linguists concerned with the origin of language, notably the American W. Tecumseh Fitch, to look to other species for parallels to our vocalisations in the period of pre-language. Fitch points out that birdsong and human speech share a common ability: prosodic vocal communication. This could have led to complex, learned vocalisations, comparable to birdsong, which would have provided the basis for the sounds of later human language and would have preceded both semantics (meaning) and syntax (grammar). Fitch also sees the early development of children as recapitulating this evolutionary development: children use nonsense rhymes in their play, which rely solely on prosody, for example the counting rhyme 'eeny, meeny, miney, moe, catch a tiger by the toe.' Furthermore, the chant used in ritualistic language depends heavily on prosody (see below).

Another important aspect of animal vocalisations is that many species go through a process of learning these. Such vocal learners require direct connections from high in their brains to the muscles controlling vocalisations. This is true for singing birds which control their syrinx in this way. The syrinx is an inverted Y branch at the bottom of the bird's trachea (windpipe), which has muscle membranes around it allowing the bird to contract these at will, producing modulated sound; birds do not have the vocal folds of mammals (necessary for human voice generation). Some birds also engage in the imitation of sounds from their environment, which can include human speech in some cases, as with parrots or mynah birds. This is pure sound imitation and is not semantic: the sounds have no meanings for the birds.

Language and Music

Singing is the production of prolonged sound which varies in tone over a much higher pitch range than spoken human language. The sounds produced over this range are known as musical notes, which can also vary in duration and rhythm. This kind of sound production can also be realised by musical instruments, which can vary greatly in size, shape, method of generating sound and thus in the overtones they typically produce.

Darwin, in *The Descent of Man* (1871), said 'Primeval man, or rather some early progenitor of man, probably first used his voice in producing true musical cadences, that is in singing.' He saw musicality as a highly conserved feature in humans and noted that all cultures have musical traditions of one sort or another.[5] However, singing is not now a primary means of human communication and generally is used for largely artistic purposes.

How far does music, and singing in particular, go back in time? We do not know about singing as there is no means of determining via the fossil record whether humans sang in their early days. But what about musical instruments? Here the evidence is tantalising. For instance, there are early flutes, made from bone, by drilling holes at intervals. This is the case with the Geissenklösterle flutes, *c.* 37,000 years old (or possibly older) and made by Cro-Magnon *Homo sapiens*. And there is the bone flute, made from a cave bear femur, found in the Divje Baba Cave (north-west Slovenia) from about 43,000 years ago. It is regarded as having been created by Neanderthals in the area (though some scholars think it was made by Cro-Magnons). The 'flute' consists of a stretch of bone, some 13.3 centimetres long, with two holes and possibly two more, broken off at both ends. There is no agreement in the scholarly community about whether it was used as a musical instrument; if it was, it would be the oldest known example.

The Role of Sexual Selection

It is clear from many observations in the animal world that the males of several species go to considerable lengths to win a female. Some birds, such as the various birds of paradise found in Papua New Guinea and Eastern Australia, are known to perform complex dances when trying to convince a female to mate with them. Displays of colours are also typical, for instance in the plumage of birds which are sexually dimorphic, like

5 On the relationship of music and language, see further Mithen (2012), Fitch (2006). For other scientists, such as Steven Pinker, music is a spandrel, that is, an evolutionary byproduct.

pheasants or mandarin ducks, where the male has much more colourful feathers than the female. The antlers of male deer provide another example, the Irish elk with heavy (40-kilogram) antlers holding the record – carrying this weight permanently on one's head just to impress females shows a firm dedication to successful mating. Making an impression on a potential mate can involve other means. Again, there are quite fascinating examples in nature. One of the most intriguing must be that of the white-spotted puffer fish (*Torquigener albomaculosus*), found in the waters around the southern islands of Japan and only discovered in 1995. This little creature (only about 10 centimetres in size) goes to inordinate lengths to create a pattern of rings in the sand on the sea floor, about two metres in diameter, in an attempt to lure his loved one into the centre for mating. But before that might take place, she inspects the entire structure of rings and only if it meets with her approval will she reach a decision in favour of the hopeful puffer fish male.

Among the many strategies to charm the opposite sex was also song, and Darwin believed that it had its origin in sexual selection strategies among many animals and early humans. He thought that music arose out of song similar to that which birds and some other animals have, and therefore, very early on in evolution, animals learned to control pitch and amplitude of sounds. Rhythm was also important and often matched muscular movements made at the same time, just as it does for many performers of music today. This would have formed the basis for language according to the *musical proto-language hypothesis.*

This view is also held by the American evolutionary psychologist Geoffrey Miller, who suggested that female-partner achievement was a force for the development of song and, later, language, a view partially shared by others, see Fitch (2006), for example.

Ritualistic Language

A central feature of much ritualistic language is chant, a drawn-out delivery of language, somewhere between speech and song. It is unlikely that language arose solely from primitive chant, but the latter is an aspect of the rituals of many cultures to this day, and it would have supported the continuing development of language. Rituals are types of communal practices and would presuppose the existence of social groups which increased their internal bonding via such practices. In pre-scientific cultures, where so little was understood of natural phenomena such as lightning, flooding or volcanic eruptions, rituals arose with chanted formulaic language intended to placate the agents assumed to be behind so many dangerous natural phenomena. Thus, this view sees language in the wider context of human symbolic behaviour, which is a part of every culture, though widely varying in form and substance.

Practising Deception and Resolving Tension

The issue here has to do with a use to which human language was put and one which would have furthered its continued development. An essential feature of language is that words are cheap, they do not require an investment of effort or physical commitment on the part of the speaker. Hence trust is necessary between speakers, otherwise it can be abused. Linguistic signals are intrinsically unreliable as they can very easily not refer to existing situations, but rather deceive others into thinking that the meanings of statements are true.

If, however, as W. Tecumseh Fitch has suggested, languages developed in families to begin with, the high degree of inherent trust in these social units might have led to an acceptance of the unreliability of linguistic signals in favour of the other advantages which they have. But it is precisely this potential use for deception which some scholars, such as Chris Knight in the 1970s, proposed as a key feature in the development of language as it provided its users with an advantage in a group.

Nonetheless, language also provided advantages for social communities. In increasingly sophisticated social groupings, language presented a means of resolving issues, or at least of venting anger without recourse to physical violence among individuals. This was inherently advantageous as it reduced injuries within the group and helped to maintain levels of physical readiness vis à vis outsiders.

32.3 From Proto-Language to Language

All human languages show features which were obviously not there from the start. Hence there are issues surrounding the transition from stages without these key features to ones where they were present. I have chosen to highlight three of these features as without them there would be no language in our present-day sense.

From Direct Reference to Symbolic Reference

At the very beginning of language evolution, sounds would have referred directly to things. If an individual used a set of sounds and pointed to a cave, these sounds referred directly to this object. However, modern languages work indirectly: our words refer to concepts of things, rather than directly to the things themselves, see Section 20.5 above. So the question is how human language got from situation (A) to situation (B), shown in Table 32.1 (see also section Section 20.5).

For situation (B), humans would have required concepts, like the concept of river or forest while thinking about it. There is evidence that this was already present, at least embryonically, in the animal world. The British evolutionary linguist James Hurford gives the example of a hunter following its prey. This will be out of sight sometimes, for instance when it dashes

Table 32.1 Transition from direct to symbolic reference

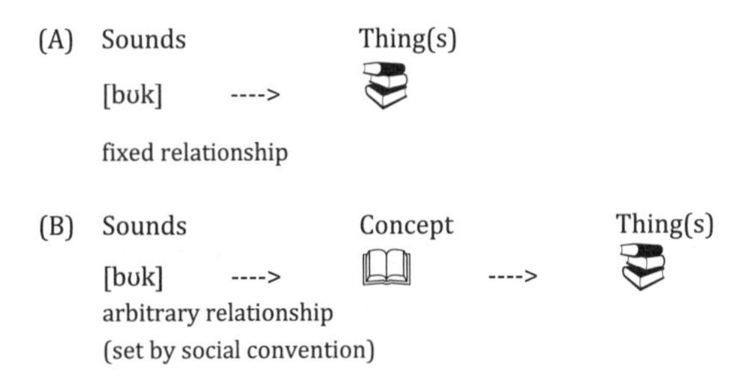

behind a bush or down a burrow, but the hunter still has an idea of the prey, so it must have a concept in its mind of the animal it is pursuing, called 'object permanence' in psychology (Hurford 2014: 64). For animal cognition, one needs to posit concrete concepts, such as things in the natural world. The fox chasing a rabbit realises that this is an animal, an animate object, self-aware, moving and delimited from its environment. If the rabbit makes it to the burrow and disappears underground, the fox realises this, often pushing its snout into the hole and trying to get the rabbit out; so the fox has a concept of the rabbit and recognises that it still exists, even if it cannot see it. Such cognitive concepts would have continued in early humans, indeed would have been greatly expanded with the increase in cognitive powers. Thus, concepts would have bootstrapped language into situation (B) in Table 32.1. There are also abstract concepts which early humans would have had, such as strength, weakness, tiredness, alertness. Once the relationship of sounds and concepts became arbitrary, as in situation (B), the way was open for the use of language for abstract thought.

Among the many features of human language, compared to animal communication systems, is that of trust. What does one mean here by 'trust'? It refers to the extent to which communication signals between members of a species are genuine and reliable. Consider bee communication for a moment: those bees returning from a reconnaissance flight will convey to other bees in the hive where a source of nectar and/or pollen is via their special 'waggle dance'. When doing this, bees never 'lie', they always convey correct information so other bees can 'trust' them. Now consider human language. Do you trust everything you hear? Obviously not. Human language allows us to tell tales, to lie and indeed to create great literature which stems from the imagination of writers. This poses a serious question for the origin of language: animal communication systems are as a rule genuine and not mendacious,[6] for instance primate calls which have intrinsic meaning are reliable within a species because they are difficult to fake and hence deceive others. So how did the shift to a system which was potentially false and unreliable take place and what kind of advantage

6 There are a few notable exceptions where an animal pretends to be another for its own advantage, for example some cuckoos which cheat other birds into feeding their young in the latter's nest.

could there have been, in the beginning of language evolution, for such a system to be favoured by the early *Homo* species? One reason would certainly be that because we do not have sabre teeth and sharp claws we must work more cooperatively together, especially previously in environments filled with dangerous predators. This fact alone would have stimulated social bonding among humans to a greater extent and furthered the use of language to reach this goal.

Another reason would be that, in a community with high levels of public trust, a system which has great communicative power, but is potentially unreliable, would nonetheless have been accepted and would have continued to evolve. The lack of a necessary connection between language and the reality it can describe did in itself add to its inherent power, as it allowed for planning into the future and to the unleashing of creative potential in individuals. The net effect of high levels of accountability in communities, probably starting from family units, was that their members accepted the cheap signals of language in its early phases. This then represented an evolutionarily stable strategy, which persisted.

Complete unreliability was later avoided by humans making a default assumption that statements in language are true, and that speakers have good intentions, unless there is evidence to the contrary. In the mid-twentieth century these aspects of language use were formulated as a set of rules for successful communication by the British philosopher, H. P. (Herbert Paul) Grice (1913–1988), who devised a series of conversational maxims, including the cooperative principle, which he regarded as central to human communication.

The 'Mother Tongues' Hypothesis

Within a community, some individuals could have been especially prone to using language, despite its potential unreliability. A view put forward by W. Tecumseh Fitch is that language use began between mothers and their children – the so-called 'mother tongues' hypothesis. Here the inherent reliability of mother-to-child communication would have outweighed the potentially unreliable nature of language and hence the latter feature would not have acted as a brake on the spread of language to all members of a community.

Qualitative Leaps in the Development of Language

In the development of human language there must have been a number of leaps in evolution which led to language attaining new qualities. Some of these have been mentioned above, see Section 31.2. Here the concern is with the rise of abstract reference and the attendant complexification of language, aspects which distinguish it in principle from all systems of communication in the animal world.

Independence of the Here and Now

The issue of potential unreliability just outlined is connected with another development, indeed one which in a way resulted from it. This is independence of environment by which language came to be dissociated from the present physical environment of its users. This feature is known as 'displacement' (one of Hockett's features, see Section 20.4 above). By this is meant referring to objects in the world when they are not being perceived by the senses in the 'here and now'.[7]

A later development, made possible by the independence of thought and communication from the here and now, is the modular system of language in which each subsystem achieved quasi-autonomy but with well-defined interfaces. This represents a kind of optimisation, seen in other bodily functions, such as the immune system, achieved in the course of the slow evolution of language.

From the Concrete to the Abstract

The release from the here and now led to a movement towards the more abstract and symbolic. This shift has been described as a move from 'cued' to 'detached' brain activity (Hurford 2014: 103) where the former is directly connected to the current situation and the latter is not 'directly stimulated by the immediate situation'. Detached meanings would have led to cases

7 The American neuroanthropologist Terrence Deacon has stressed that the symbolic nature of our communication frees us from the here and now.

like non-present reference, as in *Me fish later,* which would have resulted ultimately in a system of tenses referring to the past (what has happened) and the future (what is planned/projected).

Such developments are most likely connected to the structure of human cognition and may have been dependent on advances in this area leading to the ability of language to show locational and temporal displacement, referring to places and times other than the here and now. The ability to abstract away from the present is associated on a linguistic level with irrealis constructions, which we realise in English with *if*-clauses, as in *If find wood, me make fire,* or verbs like *suppose, imagine,* etc. A further extension of such hypothetical structures would be provision for conditions which might arise, as for negatives, for example *If not snow next day me fish.*

The meanings of individual words also became less concrete, allowing for the rise of abstract nouns which testify to a common path in language development. Just consider the abstract use of parts of the body, such as *head* for leader of a group, *foot* for the bottom of a structure, *back* for the notion of carrying, *arm* for the notion of force. Some of these abstractions are culturally determined; for instance, the use of *right* (as a term in law) and *sinister* (Latin for *left*) stem from the use of the right hand in legal actions and the idea that things from the left were somehow dark portents of misfortune.

The move from the concrete to the abstract is a typical path of change for many systems. Another confirmation is given by the development of pictographic writing systems, where the symbols (pictographs) were originally little drawings of objects or beings and became increasingly abstract so that they are no longer immediately recognisable as direct representations. This happened, for instance, in the development of the Chinese writing system: consider the word for 'person, man,' 人; 'timber, tree,' 木; or 'mountain,' 山; these symbols can still be recognised as (very) abstract representations of their meanings.[8]

8 In the earliest form of Chinese writing from the end of the second millennium BCE (the Oracle Bone Script), the symbols, from which the modern ones are ultimately derived, were much closer visual representations of the objects they stood for.

The Rise of Complex Structures

Biological systems do not remain static over time but tend to develop greater complexity and thus become more efficient. In the evolution of language this can be seen where, in its initial stages, sounds would have come to have had meanings attached to them. There must have been a developmental trajectory like the following, going from the smallest units to the largest ones:

(1) sounds → words → sentences

But it is unclear whether words were assembled from single units of sound, like building blocks, or whether there were first chunks of language which were later broken down into individual units. And, on the next highest level, the question is whether simple sentences occurred as blocks or whether they were constructed by joining words together. In technical terms, the issue is whether proto-language was holistic or atomistic (Hurford 2014: 104); the following shows examples of each:

(2)

a.	holistic [woman gives child food]	b.	atomistic [woman] [gives] [child] [food]

Whatever happened first, whole chunks or building blocks, the result was the same: groups of sounds came to exist in specific combinations which formed words with meaning.[9] Then these combinations were strung together to yield basic sentences. If chunks were there first, one could imagine one like [I'm going fishing now], realised as say *kusant*, then later with some additional meaning, such as [I'm going fishing now

9 It is uncertain whether the first words were intended to somehow represent the meanings directly via sound. Instances where this is apparently the case, such as the word *woof* for the barking of a dog, or *meeow* for the sound produced by cats, are regarded as examples of onomatopoeia, a Greek word for 'sound suggesting meaning.' The examples are usually taken from animal sounds and vary across languages. One or two may refer to objects, like *tick-tock* for the sound of a mechanical clock. In modern languages onomatopoeia is not of importance.

to the river], say *kusant wep,* as opposed to [I'm going fishing now to the lake], *kusant hod* (of course, niceties like tense or prepositions would not have existed to begin with). Either way there is no syntax to begin with. There would have been what linguists call 'lexical proto-language' – single words/chunks used in succession, which represent units of form-meaning mappings. The pronunciation of these words would become linked to the meaning they carried (technically called semantic mapping). So at this stage there would have been sounds combined to form words with associated meanings; sentence grammar (syntax) would come later.

Once single words were available, they could be combined into simple strings. Consider a very simple language with the words *uk* 'me', *nesk* 'fish', *dunu* 'apple', *kalu* 'water', *pente* 'drink' and *atoh* 'eat'. These words could be combined to produce different basic sentences like the following:

(3)	a.	*uk nesk atoh*	'me fish eat'
	b.	*uk dunu atoh*	'me apple eat'
	c.	*uk kalu pente*	'me water drink'

These are simple phrases where elements can be exchanged, as with the nouns in the first two examples above. So far there are only nouns and verbs along with some pronouns. But consider adding the words *aku* 'head', *dop* 'hand', *tek* 'foot' for parts of the body and a word like *fint* for 'sore'. Now you have an adjective and can say something like *uk tek fint* 'me foot sore'. In the course of many centuries, single words combining multiple meanings can arise so that something like *uktekfint*, or simplified to *utefint* (both *k*'s deleted), might become the way to say 'foot is hurting'. Many languages, like native languages of North America and some in eastern Siberia, are known for compressing multiple meanings into single 'words' (such languages are called 'polysynthetic' in linguistics). But they did not start out like this.

Additional features can develop naturally in languages, which would add to their expressive power. Say there is a word *polod* roughly meaning 'before' or 'yesterday' in the hypothetical language. Then one could form a phrase like *uk nesk atoh polod* 'me fish eat before'. With time, the word *polod* could become attached to any verb and to mark it as past tense, so *uk pentepolod* would mean 'I drank' (from 'I drink before'). This kind of development is very common across the world's languages and is technically known as 'grammaticalisation,' meaning that a full word becomes an ending on another word (and is usually reduced in form so something like *pentelod*, from *pentepolod*, would mean 'drank'). An example where this has happened in English would be *kingdom*, with the ending *-dom* coming from an earlier independent word meaning 'position, condition.' When *dom* became an ending, it was then used with many other stems yielding words such as *wisdom*, *freedom*.[10]

Conventionalisation: Agreement in Communities

For the early development of language an essential question is how agreement arises in a community about what strings have what meanings. Bear in mind that this is a gradual process. It would have taken some time for words to achieve a final form. Take the invented examples presented in the previous section. For example, the word for 'me' might have consisted of a variety of forms, like *uk* ~ *ok* ~ *ek* and settled down in time to *uk*. The word for 'apple' could have been *donu* ~ *duno* ~ *dano* before it became reduced to just *dunu*. So, at the very beginning, different individuals in a community would have used somewhat different pronunciations. But in the important process of conventionalisation, speakers came to agree on just one form for each word. Such agreement would hold for one speech community. Neighbouring communities might come to different agreements and a contrast like *uk* ~ *ok* ~ *ek* might come to distinguish varieties of an embryonic language, which later would become dialects. However, keeping to one form in one speech community would serve the double function of strengthening linguistic identity within the community on the one hand and delimiting it from further groups on the other.

10 Languages always show evidence of earlier formation patterns, because remnants of older structures and features co-exist with newer ones at any one time; for example, plurals from an earlier stage of English like *foot – feet*; *man – men*; *mouse – mice*; *child – children* (actually a double plural), alongside the myriad regular formations in final *-s*, as in *cars*, *bikes*, *planes*, *birds*, *cats*, *dogs*, etc.

Conventionalisation of linguistic forms is most favoured in small, closely knit communities where it is easiest to reach tacit, unconscious agreement on what elements comprise the language spoken. Where speech communities are in contact with each other, new hybrid forms could arise and have done so in the histories of so many languages (Hickey 2020).

32.4 The Evolving Levels of Language

The scientific analysis of human speech in the past two centuries or so has led to divisions being made which apply to all the world's languages. These divisions are called 'levels of language' (for details, see Section 22.2 above).

Language consists of various levels involved basically with (i) sounds, which combine to form the acoustic shape of words, (ii) the specific ordering of these words and (iii) the attachment of meaning to them. Students of linguistics will know these levels as phonetics/phonology, morphology/syntax and semantics. The levels of language are normally presented in works on language in the order: sounds, words/grammar and meaning.

In evolutionary terms, sounds must have come first and formed simple words with basic, literal meanings. Combining words into larger units, our sentences, would have come later. Or initially blocks of sound were used, which later were subdivided into units that could be recombined in different ways to yield sentences (see holistic versus atomistic distinction in Section 32.2).

Human language covers two broad domains, the cognitive domain, where thoughts are encoded as sequences of words, and the physical domain, concerned with the production of sound. The latter is the modality through which human language is expressed (externalised) in the vast majority of cases, signing (see Section 23.2 above) being another common modality.

How Does a System Expand?

Once a principle is established, like the arbitrary relationship between words and their meanings, the system can expand rapidly across a very few generations. Children at this stage would have grasped a principle in the language they were acquiring: the lack of a necessary connection between the sounds of words and their meanings; that is, the sound shape of a word does not determine its meaning. This fact then became an essential characteristic of human language. And at some stage it became part of our genetic code, passed on to every new generation as part of their biological endowment. Just how this principle, and many similar ones in language, became genetically encoded is not known. But it did happen because there is no language on Earth which does not have this principle (arbitrary relationship between words and their meanings) and children do not have to learn that there is no connection between words and their meanings. They start on this assumption and keep to it all their lives.

32.5 Sounds and Sound Systems

All the world's languages exploit a subset of the possible sounds which humans can produce, somewhere between the vocal folds in the throat and the lips of the mouth. In language evolution, those sounds which humans in a community uttered came to have particular significance for distinguishing words from each other. Consider the following 16 English words; they are all different, but they all share the final element /-e:l/ (technically called a syllable rhyme consisting of a nucleus and a coda.). Not all possible combinations need be attested, for example there is no word *zail* in English. And the same pronunciation can have two meanings, for instance *pale* can be 'stake' or 'lacking in colour'. There may also be a different spelling with the same pronunciation, such as *pail* 'bucket'. Furthermore, the absence of a sound can also be used to distinguish words, consider the first word, *ale*, below and all the others:

(4)

ale	*bail*	*dale*	*fail*
gale	*hail*	*jail*	*kale*
mail	*nail*	*pale*	*rail*
sail	*tail*	*veil*	*wail*

The spelling can vary, for example *(g)ale, (f)ail, (v)eil,* but this is an orthographical not a linguistic issue, in that it involves spelling not pronunciation. Only the beginning of the words (their syllable onsets) differ. What we can see here is the *phonemic principle* in action: the sound contrasts distinguish the meanings of the various words. The above words consist of individual sounds, each occupying one of three slots:

(5)

slots	sounds	spelling
1:	/f, b, m, g, v … /	*f, b, m, g, v …*
2:	/e:/	*ai, ale, eil*
3:	/l/	*l*

Only one slot needs to vary to get a new word. In the examples above it is the first slot, but it could be slot 2, as in *fail # feel # fool # foal # fall,* or slot 3 as in *fail # feign # fair # fate # fade # face # faze # fake # fame*. A slot may be empty and hence contrast with a filled one, as in *feel # fee* (no *l* in second word), *fail # ale* (no *f* in second word). Slots can also be filled by more than one element. The onset of a syllable has a number of possibilities, for example *fail* can contrast with *snail, quail, flail* in slot 1 (the onset). Now note the following words:

(6)

tap	*sap*	*rap*
trap	*strap*	*crap*

There are a number of combinations which we do not find: *stap* does not exist, nor does *srap*. But whereas it is just an accident of English phonetics that there is no *stap* – there is *step* and *stop*, for instance – the sequence *srap* is not possible in principle: only the sibilant *sh* /ʃ/ can occur before *r* in English, as in *shrimp, shrift, shrew, shrink*. So there are restrictions in any language with regard to what combinations of sounds are permissible and the set of restrictions are not always the same; for instance, /sr-/ occurs in Irish, as in *sraith* 'series,' *sreang* 'wire.'

There are other ways in which a language system can utilise sound differences to distinguish words. Very common among the world's languages are distinctions in tone, found especially often in languages of East and South-East Asia. In such languages, vowels are produced with typical tones – falling, rising, high, low, falling-rising, rising-falling, etc. – called contour tones as there is a pitch change during their articulation (some languages use register tones with fixed pitch levels). The words in which tones vary have different meanings; for example in Mandarin Chinese there are four tones (and a neutral tone), which occur as follows in the commonly cited sample word *ma*: *mā* (high) 'mother, mamma' (first tone), *má* (rising) 'hemp' (second tone), *mǎ* (falling-rising) 'horse' (third tone), *mà* (falling) 'chide, scold' (fourth tone). There can also be a word with neutral (non-specific) tone, which is weakly accented, as in the interrogative particle *ma*, added at the end of questions.

Some languages have many more tones, such as Cantonese, spoken largely in the southern province of Guandong in China. It has six tones (a mixture of register and contour tones), sometimes as many as nine are posited, though the additional three can be predicted by position in a word. This last point is important: a language can have more sounds than the number which are used to distinguish meaning, as these additional sounds are usually predictable by the position they occur in. For instance, most varieties of English have a 'clear' /l/ (somewhat light-sounding) at the beginning of a word, as in *leap, let, laugh,* but a 'dark' /l/ (somewhat hollow-sounding) at the end of a word, as in *field, sale, cool.* These two pronunciations of /l/ do not distinguish meaning but simply result from their relative position in a word (most accurately in a syllable).

Yet another distinction is stress on different syllables of a word, as this can be used to distinguish meanings. Certain sets of nouns and verbs in English exhibit this phenomenon, where the nouns are stressed on the first syllable but the verb is on the second syllable, as in *a 'rethink, to re'think; a 'record to re'cord; 'perfect* (adj.), *to per'fect.* These examples share the same basic meaning, but differences in stress may also be seen across words which lie further apart in meaning, as in '*differ* and *de'fer*.

Sounds, when produced on land, have a further possible source: whistling. In whistling languages, sound is produced by bringing the lips together to form a small hole and then blowing air out (sometimes inhaling for short periods, using a so-called ingressive airstream). Alternatively, two fingers (from one or both hands) can be held just inside the lips and air exhaled. In both cases, a single sound (the whistling sound) with distinctive harmonics is produced in the mouth, which acts as a resonance chamber. This can be varied in tone, duration and amplitude. Similar sounds can be produced by cupping the hands and forming a small slit between the

SOUNDS IN EXOLANGUAGES

One might imagine that exolanguages could use very different mechanisms for producing sounds and hence for distinguishing meaning. How could we argue for or against such an assumption? One way would be to reflect on how sounds are produced by various animals on Earth and extrapolate from there. Consider that with all primates vocalisation involves air escaping from a facial opening, the mouth. However, animals in the sea, such as toothed whales (odontocetes like sperm whales and dolphins), can produce vocalisations, which sound like clicks or in some cases continuous tones, which vary in pitch and are produced in the head and emitted through the bulge found above their mouths. This is clearly recognisable in dolphins which have phonic lips, roughly the equivalent of our vocal folds leading to one or two blowholes on the top of their heads. Baleen whales produce sounds in their larynx maybe using a U-fold feature parallel to airflow for this; they do not exhale while making sounds. Many whales and dolphins use their sounds not only to echolocate potential prey but also for communicating across long distances, as visibility in water is much more limited than on land, as is smell (though some fish, like sharks, have a good sense of smell). The assumption for this book is that exobeings would not live in a marine environment so the vocalisation patterns of cetaceans are not considered a possibility for languages spoken by technologically advanced exocivilisations.

thumbs, which are held together. In this case the resonance chamber for the sound is the interior of the cupped hands. Whistling languages cannot reach the level of differentiation of normal language and so are used for limited purposes, as between hunters, herders or farmers when communicating information relevant to their typical activities. A well-known example is the *Silbo Gomero* 'whistle of Gomera', traditionally used on the Canary island of La Gomera for communication across the deep valleys there. The variations in whistling attempt to represent (in crude form) the sounds of Spanish, essentially the vowels.

Concepts, Words and Meanings

In the beginning were concepts. These probably arose by recognising objects, delimiting things in the environment and observing similarities in shape and/or behaviour, which would have led to stored representations of them in the mind. This would furthermore have led to the realisation that there are classes of things. Take a simple example like bears. People would have realised that there are big bears, small bears, peaceful bears, angry bears. From this comes the realisation that there is a class of such animals and that, when one is confronted with one in the forest, this is an exemplar of the class of bears. Now combine the notion of class with displacement – the act of talking about something which is not present in one's immediate surroundings, either with past or future time reference. We now have a concept detached from an external stimulus: People would have warned others of bears,[11] first when one or more was seen while hunting or foraging in the forest. But later warnings could have been preemptive, as for children: warnings were issued before these predators came into sight, a class of animals not present right now, but perhaps in the immediate future.

11 The word for *bear* is typical of Germanic languages (English, German, Dutch, Swedish, etc.) and is most likely derived from the word meaning 'brown'. In Latin the word for bear is *ursus* and in Greek *arktos* while in Russian the word for this animal is *medved*, roughly 'the honey eater' (also found in other Slavic languages). The Greek word is assumed to be closest to the Indo-European original, whose stem would have contained the consonants *rkt* bolstered by a vowel or two; it is related to the English geographical term *arctic* from the Great Bear constellation in the northern sky.

Scholars acknowledge that the idea of 'concept' is contentious (Hurford 2014: 61). To begin with, do only humans have concepts? In the abstract sense, like the concept of 'beauty', 'laziness', 'stubbornness', 'diligence', this may well be true as we have no unambiguous evidence pointing to the higher primates having such cognitive distinctions and classifications, though there may be indirect behavioural evidence pointing in this direction. But at least the concept of a concrete object or being is probably present in most animals. A hawk probably has a concept of small rodents, such as mice, or other birds, like ducks, which it can hunt as a source of food. Nearly all animals would have a concept of size, often within their own species: this is essential when considering whether to challenge another to a fight or attempt to take food away from it.

Bickerton (2009) and Fitch (2010: 400–403) assume that animals have concepts on a basic level but they do not have a representational form of language for these concepts. Their calls serve the purpose of direct communication, as with the various calls of vervet monkeys, rather than being an expression of abstract concepts. Furthermore, such calls cannot be segmented into elements and then recombined to generate different new calls.

It is hard to say whether animals have a concept of 'class of objects' and, if so, how nuanced this concept is for them. Certainly, animals recognise others of their species and in general only mate with each other or species very closely related, yielding, for instance, a mule, which is a cross between a male donkey and a female horse, or a liger, which is a cross between a lion and tiger.[12] But beyond species recognition animals would seem to recognise others as just not the same as themselves.

Grammar: How Did Compositionality Arise?

Once words had developed, either from a holistic or atomistic precursor stage, combinations arose which had specific meanings depending not just on the meanings of individual words but on their order in a series.

12 Cross-species breeding can be successful but the outcome is usually infertile.

With that, the sentence was born, in essence the expression of thought as a group of words in a definite order with a beginning and an end. Combining words in a specific order, which has a meaning as a whole, is labelled 'compositionality' and is the central feature of syntax in all human languages.

The meaning of a sentence is derived (i) from the meanings of the individual words and (ii) the rules which determine how these words can combine in a given language, such combinations of words in sentences with specific aggregate meanings being instances of compositionality. For instance, the meaning of the sentence *Fiona spoke to Fergal* is determined by the meanings of the individual words and the manner in which they are combined to yield a well-formed sentence of English: noun + verb + preposition + noun. Here, the noun before the verb is the subject, the initiator of the verb's action; then we have the verb; and after that the object, the person who experiences the verbs's action, here the person being spoken to. In addition, other word orders can be used within a language for different functions: for questions in German or Italian, for example, the verb is put before the subject, *Kommst Du morgen?* (come-you tomorrow?), *Vieni domani?* (come-you tomorrow?).[13] With time, basic sentences would have expanded by allowing for further elements which added aspects of meaning, as in *Fiona spoke to Fergal on the bus this morning* or, by means of insertions, *Fiona, our new colleague from Cork, spoke to Fergal on the bus this morning*. All of this would have happened with linguistic forms very different from those of modern English, of course.

Structure and Iterated Learning

The question posed in the heading of the previous section, of how the members of early embryonic speech communities arrived at linguistic systems which showed regularity in their syntactic structures, still requires an answer. To come closer to providing one, scholars – above all the British cognitive scientist Simon Kirby with colleagues in Edinburgh, such as Kenny Smith – have considered a key aspect of how individuals

13 English is a bit more complicated in this respect as it needs an additional verb together with the main verb: *Are you coming tomorrow?*

acquire knowledge in social contexts: through iterative learning.[14] What this means is that information is passed on from individual to individual and, in this process, each new transfer leads to additional regularity being added to the system of form-meaning mappings. Individuals can guess the underlying meanings of utterances when presented with enough instances where they are also aware of the communicative intentions of others. By this means an irregular, stochastic (random) input can, in a relatively short series of transfers, result in an output which is regular and internally structured. This is achieved by the individuals in each generation (transfer event) adding regularity to a system they inherit from others, thus providing an intergenerational positive feedback loop.

Simon Kirby maintains that structure and iterated learning are related. The key insight is that linguistic structure emerges due to transmission over generations, something which is also true of new sign languages (see Section 23.3 above) and of pidgins and creoles (see Section 25.9 above). Linguistic structure is the solution that cultural evolution finds to the problem of how to make random input learnable.[15]

And it is not just structure which arises but conventions as well. Speakers who inherit increasingly structured input become aware of the conventions and norms governing the structure in this input, that is, they begin to pay attention to grammatical rules. And sometimes this is conscious (Kirby has documented examples of his test persons 'correcting' themselves in iterated learning experiments).

Viewed from the vantage point of language evolution we can see that iterated learning across generations in a community would lead, within

14 This model has been presented as an alternative to explaining how children acquire language as native speakers when exposed to so little input from their surroundings. The iterated learning model has been explicitly contrasted with Chomsky's innateness model, see Smith et al. (2003) as well as Kirby et al. (2014).

15 Emergent structure can be observed in other human activities, noticeably in the genesis of games. These are good examples of relatively rapid cultural evolution; for example, tennis, football, cricket, chess are all instances of games which arose from simple beginnings through a multi-step process of codification to reach their modern forms. In many cases the codification is on-going, just think of the changing rules of association football (English soccer).

a relatively short period of time, to regularity among the permitted combinations of words, and it would lead to agreement on the syntax of the evolving language. This is possible, as Kirby and his associates have shown in both computational modelling and laboratory experiments, because learners (unconsciously) impose order on the data they are exposed to without distorting the input too much. This kind of cultural evolution is much quicker than the slow and almost imperceptible biological evolution by random mutation and natural selection.

32.6 What Is Morphology?

If you have had even the slightest brush with linguistics you will have heard the term 'morphology'. Derived from Greek roots, the word means the study of forms. It is found in many sciences, such as geology and biology, and in linguistics it has a specific meaning: the study of word forms. Morphology in linguistics can be divided into two broad sections: that which is concerned with the formation of words, called lexical morphology, and that which is concerned with endings on words (possibly involving additions to the beginning or middle of words), called inflectional morphology.

Word formation is fairly straightforward (in English at least) and studies how new words are composed of existing words or parts of words, as in *flowerbed* from *flower* + *bed; waterfall* from *water* + *fall; unsettling* from *un* + *settle* + *-ing*, etc. For the purpose of this book, word formation is not that relevant and so no details of this area of morphology will be pursued.[16]

Inflectional morphology is concerned with endings added to word stems to indicate such categories as tense, number and person (with verbs) – *she sees, they were, we hiked, I ran*, etc. – or case and number (with nouns) – *Fiona's coat, new phones, one meal*, etc. It also covers the set of words which cannot stand on their own but derive their meaning from co-occurring with other words. Instances of these would be prepositions, as in *over the city, to the station, under the table*, etc. or articles like *a, the*

16 There are many introductions to word formation and the area is dealt with in textbooks on linguistics, see Harley (2006) or Plag (2018), for example.

or demonstrative pronouns like *this, that, those*. A cover term for all these word types is 'function word' or 'grammatical word.'[17]

When considering the origin of language one should bear in mind that neither function words nor grammatical endings existed at the very beginning. They need time to appear in a language: function words arise from full words, so-called lexical words, by being continuously used in a certain context. The words then lose their full, lexical meaning and become indicators of some grammatical category. Take the small particle *a* in English. Historically, this is derived from the cardinal number 'one': 'a person' originally meant 'one person.'[18]

An important source of endings in a language is the process called grammaticalisation (see the section 'The Rise of Complex Structures' above). The idea here is fairly simple: one word is increasingly interpreted by speakers as representing grammatical rather than full, lexical meaning. A good example is the future tense with *go* in English, as in *I'm going to Belfast next week*. Originally such sentences meant just that, someone is going somewhere. Later they came to mean that something will happen in the future, such as *I'm going to read that book*, which does not involve movement of any kind.

How Languages Differ in Morphology

Even a cursory look at the major European languages shows that these vary greatly in the number of endings, found mainly on nouns and verbs, in

17 These are the most common words in any language, with content words following at some distance. The distribution of word frequency appears to follow what is called Zipf's law (after the American, George Zipf 1902–1950) whereby the cline from the most common to the least common item follows a power law: the second most common word is half less common as the first, the third most common is a third less common than the second, the fourth a quarter less common than the third, etc. Because of this, the 100 most common words in English account for about half of the words in the language as used by its speakers.

18 This is still visible in German where the form *ein* can mean 'a' or 'one' as in *ein Haus* 'a/one house,' *ein Auto* 'a/one car,' etc. The interpretation of *ein* depends on the context in which it occurs, as in *Dort war ein Auto geparkt* 'A car was parked there'; *Sie haben nur ein Auto* 'They have only one car.'

their grammars. English and Swedish, for instance, have very few whereas Russian, Finnish, Hungarian and German have a lot. When learning one of the latter languages you might wonder what the purpose of so many grammatical endings is. Well, they allow speakers to keep track of what elements in a sentence belong together. Consider the following sentences in Italian:

Le ragazze tedesche sono andate in Italia.
'The (fem) German (fem) girls (fem) have gone (fem) to Italy.'

Lo zio della mia madre è morto.
'The (masc) uncle (masc) of-the (fem) my (fem) mother (fem) is dead (masc).'

Agreement in gender between nouns, adjectives and certain verb forms adds cohesion to a sentence and helps in real-time processing of sentences as they are uttered (technically called 'reference tracking'), often in situations which are less than optimal, for instance in a noisy environment. Grammatical gender is a system of morphological correlations used for reference tracking in an inflectional language such as Italian, German or Russian with the traditional labels 'masculine', 'feminine' (Italian, Spanish) and 'neuter' (German, Russian) being used for both persons and objects.[19]

Languages differ greatly in the manner in which their grammars are organised and over time this organisation can change. Linguists see in long-term language change cycles in which grammars can lose or gain endings and move from one type to another in the course of many centuries; the study of such change and the classification of languages by their grammatical organisation is called 'linguistic typology' (see Section 22.3 above).

19 So to say the Germans think of the Sun as female (because it is 'feminine', *die Sonne*) and the moon as male (because it is 'masculine', *der Mond*) is nonsense. Equally, juxtapositions of the Germanic and the Romance character, because gender assignment is the reverse in the latter languages (cf. Italian *il sole* 'the Sun' and *la luna* 'the Moon'), are also nonsensical.

VARIATION AND CHANGE AMONG EXOLANGUAGES

There are two basic types of linguistic change, internal and external. The former results largely from children creating regularity with sets of forms during first language acquisition, such as plurals of nouns or tenses of verbs, and/or by constructing an internal grammar for their native language which is slightly different from that of their parents. External change is triggered by socially motivated variation in a community, for instance the use of 'fashionable' variants by young adults, which leads to shifts in a language. Similar variation with change might well occur in an exolanguage, given an extended childhood for acquiring a native language and complex societies with groups jostling for key positions and others striving to imitate those in such positions.

32.7 Syntax: The Grammar of Sentences

For many scholars, syntax is the holy grail of linguistics. It can be approximately equated with grammar in the triad of divisions for language: pronunciation, grammar and vocabulary. Syntax provides the ordered structures into which words are inserted to render what we say meaningful. In that sense, syntax is at the inner core of language. By 'inner' is meant here the deep-seated, unconscious part of language, which we use continuously to construct sentences but of which we are not normally aware.

We acquire the syntax of our native language during our early childhood in a process which is unconscious and instinctive. No child refuses to internalise the syntax of their native language just as no child refuses to walk or use their eyes for binocular vision.

There is no real equivalent to syntax in the communication systems of animals. Syntax is unique to humans and would be to exobeings on their planet assuming that they would be the only beings with language. Syntax is not ordered in a linear fashion: the sequence of elements in a sentence in any language does not have to correspond to the order in which things happen in the outside world. Consider the following sentence, in which the bracketed elements in bold with the subscript *a* belong together semantically but are separated by two clauses:

(7) **[Are the cookies]**$_a$ [which Mammy baked yesterday]$_b$ [before Daddy came home]$_c$ **[in the tin]**$_a$?

Present-day languages allow for such word orders. However, in the view of many linguists, such ordering was not permitted in the early evolution of syntax and only emerged gradually. According to some other linguists, this type of ordering arose suddenly with the appearance of hierarchical, non-linear syntax in *Homo sapiens*, not more than 100,000 years ago. This single fact has engendered an intense debate among linguists in recent decades about how we humans came to have a system of syntax which is free from the contingencies of linear ordering in the time-dominated, sequential world we live in.[20]

How Did We Come to Have Syntax?

In the scholarship on the origin of language, there are basically two camps: one which believes that syntax arose slowly, in tandem with pronunciation and vocabulary. This view sees syntax as proceeding from a basic initial form of linear language, in which the sequence of words corresponded one-to-one to the sequence of events being referred to. The second view sees syntax, in our modern sense, as arising suddenly due to a chance mutation in a single individual. The two views are traded under the labels (i) the *continuity hypothesis* and (ii) the *discontinuity hypothesis*, respectively. The latter is a decidedly minority view but because it has been proposed by the founder of modern linguistics, Noam Chomsky, it is always considered and quoted.

20 There are present-day languages which appear to have only a linear order of the type seen in such slogans as *Don't drink; don't drive; stay safe.* The Indonesian language Riau, investigated by the linguist David Gil, and Piraña, investigated by the anthropological linguist Dan Everett, are putative examples of such languages (the syntactic nature of Piraña – see Everett (2008) – is contested and lacks independent corroboration by other scholars).

The case for the continuity hypothesis hardly needs to be stated in detail as it assumes language evolved relatively slowly and along the same lines as other human faculties would have developed (more on this below). The discontinuity hypothesis is an entirely different story. To begin the discussion consider the following quotation from a study by Chomsky:

> Within some small group from which we are descended, a rewiring of the brain took place in some individual, call him Prometheus, yielding the operation of unbounded Merge, applying to concepts with intricate (and little understood) properties ... Prometheus's language provided him with an infinite array of structured expressions. (Chomsky 2010: 59)

The key element in this passage is the notion of Merge (spelt in the literature with a capital 'M'). According to Chomsky this is an operation in the syntactic module of our language faculty (in our brains) and allows us to build sentences which have an internal hierarchical order and which have no pre-set limits on their size as the Merge operation can apply recursively (again and again, with the outcome of itself as the input to a new cycle of Merge). The abstract form of Merge would be two objects A and B merging to a set {A, B}. This operation can be applied repeatedly so that {A, B} + {C, D} → {{A, B}, {C, D}}. An example would be the following: *Fergal* + *Fiona* → {*Fergal, Fiona*}; *like* + *coffee* → {*like, coffee*}; {*Fergal, Fiona*} + {*like, coffee*} → {{*Fergal, Fiona*}, {*like, coffee*}}, which would be externalised as a sentence via speech or signing (different modalities). This simple example leaves out a lot of detail, but the principle is clear: sentences are built by repeatedly combining elements[21] by the same operation of merging two parts on the left to a single one on the right.

When two objects are merged, one of them dominates and becomes the head in the resultant phrase; for example, *Fiona sings* is about an action – the verb *sings* is the head and the noun *Fiona* is the modifier. In the case of *Italian song* the noun *song* is the head and the adjective *Italian* is the modifier. The operation by which one of two merged elements becomes

21 Chomsky talks of concepts merging because he assumes the Merge operation arose before actual words existed in language.

the head of the resulting structure is called 'labelling'. The two structures just mentioned can now be merged to yield *Fiona sings an Italian song*. The generation of such structures can yield an unbounded array of hierarchically organised structures. The system thus contains units in structures (the sets resulting from Merge), a feature of the syntax of all human languages.

Songbirds can string chunks of vocalisation together but neither they nor any non-human primate can produce hierarchical units (with heads and modifiers) and then merge these into complex syntactic structures; their systems are linear without an internal 'vertical' structure.

To recap: Chomsky believes that syntax is based on a single potentially recursive operation Merge and that the ability to perform this operation arose with a single individual, thus claiming that this feature of all the languages in the world resulted from the rewiring of the brain of one person. You might ask, why should Chomsky assume such a scenario for which there is no evidence? In addition, he offers no explanation of how and when this rewiring took place. Was it a gene mutation and how was this expressed? Did the possession of Merge bestow a survival advantage on Prometheus during his lifetime (Chomsky assumes this individual was a male)? How did his siblings and friends react to this feature he possessed? With bewilderment or with deep appreciation of the linguistic options which he had and they did not? If no one else in his environment possessed Merge, this syntactic ability of Prometheus would have been of no value to him as no one would have understood the hierarchical structures he generated with it. And was Merge passed on to following generations by Prometheus' children (who he must have had), who all inherited this favourable mutation? This problematic issue does not arise with the continuity view: the transition from linear grammar to hierarchical grammar was gradual and the difference between any two generations would not have been so drastic as to impair communication or cause confusion. This transition could be compared to language change, which never proceeds so rapidly as to be the cause of misunderstandings between generations.

It is true that syntax is not temporally ordered and has recursion[22] as a key characteristic, but why should Chomsky claim that this arose through the appearance of an operation Merge in a single individual '[w]ithin some small group from which we are descended'? To attempt an answer here one must understand that Chomsky had consistently developed a theory of syntax since the mid-1950s, known as generative syntax, and that in the four decades up to the 1990s he increasingly insisted that syntax must be based on very simple operations, indeed ideally on a single one.[23] This led him to postulate Merge, put forward most clearly in his 1995 book *The Minimalist Program*, in which he described this line of linguistic inquiry.

To sum up, scholars who disagree with Chomsky's discontinuity hypothesis are critical of three main elements of his view on the origin of syntax.

Chief elements of the discontinuity hypothesis

1. It posits an unspecified rewiring of the brain (just what was this?)
2. It assumes that this occurred with a single individual (Chomsky's Prometheus)[24] from whom it apparently spread to everyone else in subsequent generations.
3. It assumes that complex language suddenly arose internally, independent of any external use of language.

In terms of evolutionary biology, point (1) can only be a chance genetic mutation in DNA, which led to a change in the structure of the brain. True,

22 Some scholars, like Michael Corballis (see Corballis 2011), think that the incorporation of recursion into other cognitive domains antedates its incorporation into language.

23 As clearly stated (Berwick and Chomsky 2016: 7), Chomsky was concerned with reducing syntactic operations to the simplest possible form, which would represent 'a narrower language phenotype' that would apply to all humans.

24 Chomsky does not say why he chose this name, nor why he regards the originator of complex syntax as a man and not a woman, but given the fact that Prometheus in Greek mythology was a Titan god who stole fire from Zeus and created humanity from clay, it might seem a fair name to use. Prometheus also came to symbolise humanity's striving to achieve scientific knowledge, so he could, at a stretch, be linked to linguistics.

if this led to more flexible syntax, it would have a fitness advantage, but this would assume that it was present in many individuals in a population and that these individuals with their mutation came to dominate over very many generations. But Chomsky states that Merge appeared with one individual, so it was a single chance mutation with far-reaching consequences for humanity. Are there instances in genetics of similar mutations which have such wide-ranging effect? Maybe, consider the single point mutation on gene *FGF5* in certain dogs which results in longer hair with dogs of different breeds, like pomeranians and collies. Could the Merge operation be traced to a single point mutation like that for the coats of dogs? This is difficult to ascertain because the Merge operation does not have a physical expression like length of hair, and so we may be comparing apples and oranges. There is perhaps one further source of Chomsky's Merge operation: an epigenetic change in Prometheus. Such a change would have resulted from methylation, where methyl groups[25] latch onto a part of the DNA, repressing gene transcription at that point. Another means for epigenetic change is histone modification whereby an epigenetic factor latches onto the tail of a histone (a disc-shaped protein around which DNA is wound), preventing DNA around a group of histones from being transcribed by blocking the unwinding of the DNA sequence, which is necessary for transcription into RNA and later translation into a functionally specific protein. These changes can happen during the lifetime of an individual and can, according to some biologists, be transferred to the next generation.

In later publications, Chomsky continued to remain non-committal on the cause of Merge, cf. '[p]erhaps it was an automatic consequence of absolute brain size ... or perhaps some minor chance mutation.' (Berwick and Chomsky 2016: 65). Incidentally, if it was an *automatic consequence of absolute brain size* (emphasis, RH) then Neanderthals, who had brains at least as large as *Homo sapiens*, should also have experienced the Merge mutation.

25 A methyl group is a derivative of methane by the loss of one hydrogen atom resulting in CH_3.

The second element of the discontinuity hypothesis maintains that the sudden rewiring only occurred with a single individual, which is consistent with the view that Merge arose internally without being previously determined by other developments. There is no further argument given for this development so scholars had to decide there and then whether to accept this assumption or not.

A Question of Direction: Externalisation or Internalisation?

Chomsky's hypothesis for the origin of language sees it as having arisen as a means of organising thought, which was then externalised as speech (see point (3) above). There is a clear direction implied here: language developed from the inside out. The majority of scholars working on the origin of language adhere to the continuity hypothesis: they see the development of language from simple concrete forms to more abstract complex ones, all the while being realised as speech sounds, perhaps accompanied by gestures. The abstract use of language is closely associated with the process of internalisation. By internalised language is meant its use in the organisation of thought and its independence of any external use as speech or signs. Once this independence was achieved the structures of language no longer had to correspond to the temporal sequence in which the external events they referred to occurred.

Chomsky and his followers see language's sequential temporal ordering as imposed by the demands of externalisation (Berwick and Chomsky 2016: 12) whereas those who support the opposing view see language as developing from temporally bound sequences of words, which gradually became increasingly more abstractly organised and internalised with this process, leading to the non-temporal syntax of human language. Internalisation led to a decoupling from the environment, opening up new structural options for the human language faculty. So the disagreement is not about whether we have a non-linearly organised language faculty but how we got to this point in our evolution.

Language may have been internalised at an early stage and then this released it from strictly chronologically arranged syntax. Later, this internalised language had an influence on the syntax of externalised language.

Cognitively motivated factors such as topicalisation strategies, (used to add focus to elements in a sentence) led to a discrepancy between chronology and syntax in externalised language: for example, *Fergal bought a new car after he got a better job* (he got the job first and then bought the car, but the focus is on the latter and hence it appears first in the sentence).

But there is a major problem here: if externalisation (the rise of speech to express thoughts) appeared after Merge arose through a single mutation, the secondary specialisation of the organs of speech must have arisen afterwards. There is no way by which the anatomical development of the tongue with its control through the hypoglossal nerve, the hyoid bone and the round shape of our oral cavity, allowing intricate vocalic distinctions, could have been suddenly achieved *after* the appearance of Merge. Hundreds of generations would be required to produce the linguistically agile tongue characteristic of modern humans. This issue is not discussed by Chomsky and his collaborators because, although concerned with the evolution of language, they were not concerned with how phonetics developed (let alone how it came to be attached to mental representations of a sound system, and then linked to phonology). The following quotation illustrates the assumption of sudden appearance which underlies the discontinuity hypothesis.

> The language faculty is an extremely recent acquisition in our lineage, and it was acquired not in the context of slow, gradual modification of preexisting systems under natural selection but in a single, rapid, emergent event that built upon those prior systems but was not predicted by them. (Bolhuis et al. 2014)

The above could only apply to syntax, if at all. The specialisation of the tongue for the production of sounds (the means of externalising *i*-language in Chomsky's terms) would need a considerable amount of time to go from the form it had with the earliest *Homo* species to that of modern humans.

Merge Outside Language

When considering central features of language and their probable origin it is beneficial to ask whether these features can also be recognised in other realms of human cognition and behaviour, in part or wholly. Recall that children perceive their parents as a set of two beings linked together in their function as parents. This could be formulated as a Merge operation: {man} + {woman} → {husband, wife}, later {mother, father}. Children also perceive of their grandparents as two sets of the same objects: {maternal grandmother, maternal grandfather} and {paternal grandmother, paternal grandfather}. Admittedly, the relationship across the generations is not one of Merge, in Chomsky's sense, because the two sets of grandparents come first and then generate the parents by each providing just one member of the latter set. But nonetheless children perceive that there are two levels in their wider family, one with a single set and one with two sets of the same type, so that the next step in recognising that – in principle – two sets of beings could generate one combined set of beings is cognitively a small one.

> In the literature on recent generative syntax much emphasis is put on digital infinity, which is taken as characteristic of human language:[26] 'Language is, at its core, a system that is both digital and infinite. To my knowledge, there is no other biological system with these properties .... ' (Chomsky 1991: 50).

This notion of digital infinity can be deconstructed. Let me take 'infinity' to begin with. Chomsky is using infinity in the sense of 'open-ended', 'with no predefined limits.' In this sense, virtually any human activity, not just language, is infinite. If I sit in an armchair and start to tap my hand on the armrest, there is no predefined limit on the number of times I can do this. The number of times would be constrained by external factors: my hand might get sore after a while, I might have other things to do or I just get bored and stop. Equally, there is no pre-set limit to the number of words I can put in a sentence (through concatenation) or to the number of sentences I can say. The language faculty does not limit the number or size of sentences in advance. Chomsky rightly points to the open-ended nature of human language, which he, however, regards as arising suddenly, it did not proceed in increments. We cannot postulate that the development of Chomskyan digital infinity for language arose through stages such as (i) 3-word sentences, (ii) 10-word sentences, (iii) 20-word sentences, (iv) open-ended sentences.

26 As Chomsky noted, Galileo in his *Dialogo* (1632) remarked on the ability of human language to produce a potentially infinite array of constructions (sentences) from a finite set of elements (phonemes [sounds], Galileo refers to letters).

However, it should be stressed that language was *always* open-ended in the length and number of sentences possible. If it were not, the burden of proof would lie with the person maintaining that there were pre-set limits.

Now what about the 'digital' in 'digital infinity'? What is meant here is that each element of language is discrete, meaning it is clearly separated from others without transitions or grey areas in between. This notion is like binary numbers in computer technology: a value is zero or one but not something in between.[27] This notion of discreteness is something which we have for many areas in life, for instance, we are discrete as beings, clearly separated from all others. We recognise forms of life as discrete entities in their environments. There are also multitude examples on the socio-cultural level: for example, you are either married or not, there is no legal state of being 'half-married', or 'one-quarter divorced'. And many systems which we create are 'digitally infinite', a mundane example being model construction sets with a finite number of pieces: strips, plates, blocks, interlocking parts from which one can construct any number of objects. There is no pre-set limit on the number of units you can build (assuming you re-use the elements at will, just as we use words again and again as we require them). Granted, human language is vastly more complex than a model construction set, but both share the principle of discrete elements out of which one can construct, in theory, an unlimited number of units, so both are potentially infinite in the sense of 'open-ended'.

THE STRUCTURE OF AN EXOLANGUAGE

To enable all the intricate interactions of a complex civilisation with advanced science and technology, a powerful language, characterised by open-endedness of its structures, would be a precondition. Whether it would show recursiveness like human language is an open question. But exobeings, having evolved from earlier forms of life, would have an innate understanding of hierarchies, much as humans do, and this fact suggests that exolanguages would also have hierarchical structures.

27 This system can be realised in practice via threshold values; for example, anything above five volts is treated as '1' and anything lower than five volts is '0'.

32.8 A Possible Parallel: The Immune System

Chomsky regards the sudden popping into existence of modern syntactic ability with a single individual as something unique in biological evolution. In the form in which he expresses it – a 'sudden rewiring' of the brain – it is impossible to prove. However, for argument's sake, one can consider another biological system which definitely took the route of ever greater abstraction and developed an increasingly modular organisation in the process: this is the immune system.

There is no evidence for the evolution of the immune system available from palaeoanthropology as no traces of this system are found in the fossil record. Nonetheless, by comparing immune systems in different forms of life today, an approximate history can be reconstructed.

The first point to note is that all forms of animal life have means of distinguishing self components from non-self components and they will reject the latter by phagocytosis (ingesting and destroying foreign material in the body). All vertebrates have cells in their blood streams which constantly monitor for invasive foreign bodies. They also have immunoglobulins (antibodies) which, in aggregate, form a record (immunological memory) of previous infections to which the organism is then, as a rule, immune. True, the particular lymph nodes present and the immunoglobulin types available vary, and they are increasingly differentiated in more complex life forms.

The Innate and Adaptive Immune System

There two major parts to the immune system, the innate immune system and the adaptive or acquired immune system.[28] The first part is non-specific and includes the skin and slightly acidic mucous membranes at openings in the body, in the nose and mouth, for example. These ward off all kinds of germs and foreign particles, like dust or dirt, which might

28 See Flajnik and Kasahara (2010) for detailed information on the origin and evolution of the adaptive immune system.

enter the body. Acid in the stomach will also kill many pathogens in food while bacteria in the gut (intestinal flora) will render it difficult for germs to gain a foothold. The white blood cells (leukocytes), which are constantly circulating in the body and react non-specifically to intruders, are part of the innate immune system.

The innate immune system exists to provide early defence against pathogen attack, and to alert the adaptive immune system to the fact that pathogen invasion has occurred. This dual function appears to operate through a very ancient signalling pathway, the Toll pathway, responsible for regulating antimicrobial peptides (known as 'AMPs'), that long predates the adaptive immune system, and is present in the fruit fly (*Drosophila melanogaster*), vertebrates and, most probably, in plants as well. Another component of innate immunity, phagocytic cells (see above) such as macrophages that scavenge incoming pathogens, could have their origins in unicellular amoeba-like eukaryotes.

The adaptive immune system consists of many components, above all the T cells (so-called because they mature in the thymus gland), which are present in lymph nodes (hence their alternative label lymphocyte). T cells are on constant alert to attack by any intruders in the body. They are made aware of intrusion by dendritic cells, which are in contact with the outside world, usually through the lungs and nose, and responsible for recognising a pathogen (cell-mediated immunity). The dendritic cells provide cues called antigens, molecular structures on their surface. The immune system is thus activated and responds in a manner appropriate to the foreign bodies by marshalling different types of response, specifically by passing the antigen information on to the T cells. The dendritic cells move from an infection site to the nearest lymph node where T cells are located, ready for deployment in defence, and so act as a bridge between the innate and adaptive immune system. Those T cells which come into contact with dendritic cells in the lymph nodes are a subpopulation called armed effector T cells and most move to the infection site to determine what needs to be undertaken to combat the infection.

The T cells attack infected body cells to destroy them and thus prevent pathogens from co-opting the body's cells into propagating them. B cells, so-called because they are produced and mature in the bone marrow, use the information conveyed to them about the antigens on the pathogen in question to produce antibodies (antibody-mediated immunity), special kinds of proteins which can latch onto the pathogens on a lock-and-key principle and hence neutralise the pathogens, which are then ingested by macrophage cells, the latter recognising pathogens with antibodies on their surface.

A further subpopulation of T cells, called memory cytotoxic T lymphocytes, remain in the body and maintain information about specific antigens, so that if the latter were to enter at some later date they would be recognised and neutralised quickly (long-term immunity).

The innate immune system also includes natural killer cells (known as 'NKCs'), which can detect foreign bodies or even cancerous cells and neutralise them. They can detect if a cell in the body has developed an abnormality by registering changes in proteins coded for by the major histocompatability complex. They then bind to the cell and release certain chemicals which induce apoptosis, automatic cell death. These cells are similar to the cytotoxic T cells, which will kill any body cell that is too heavily infected to recover. Other cells of the immune system die after ingesting pathogens, such as neutrophils, the majority type of white cells, contained in the pus which develops at a wound.

The responses of the immune system include a rise in body temperature, which renders multiplication of the pathogens more difficult. Localised infections, such as an insect bite or infected wound, will also show swelling due to fluid leakage by infected cells. The swelling attracts phagocytes, which ingest both the pathogens and damaged cells of the body.

Language Faculty and Immune System: Shared Principles of Organisation?

What has the immune system got to do with language? In functional terms, nothing. However, the manner in which it has arisen and in which it is organised shares *principles of organisation* with the human language faculty. While it is uncertain just how the immune system of primitive life forms began, it is assumed that there was an external trigger (pathogen intrusion, namely infection) which kick-started the process. That began a process of *internalisation*, whereby the body built up a complex and efficient system, consisting of various purpose-specific modules, for dealing with intrusion by pathogens.

Consider also that an organism could store pathogenic information in its DNA from all the pathogens it has encountered.[29] This would be very cumbersome and all that would be needed is for a new pathogen to come along

---

Origin and organisation of the human immune system

External trigger:

1. Organism attempts resistance to pathogens. One method is by storing information about a certain harmful virus/bacterium to avoid being infected by a more serious variant of that pathogen, technically called superinfection exclusion.

Internalisation, expansion and genetic encoding:

2. A system develops in the body for dealing with pathogens *in principle*. Two subsystems arise: innate and adaptive immunity, with further subcomponents, for instance, neutrophils, B and T cells, which are further differentiated.
3. The adaptive immune system contains memory T cells and B cells, which allow the individual organism to maintain a record of the pathogens which the body has been exposed to. This permits the individual to react flexibly to whatever pathogens they might be subsequently confronted with.
4. The innate immune system is genetically encoded, and hence available to all following generations, just as the language faculty is.

---

29 Our DNA contains lots of bits and pieces of viruses we have come in contact with during our evolution, as much as 8 per cent of our genome. These 'pieces' are generally insertions of code from RNA viruses which occurred during cell infection at some stage. However, the totality of this genetic code is not organised as a systematic and coordinated means of combating pathogens.

and the organism would have no defence against it.[30] For the immune system, it is far more efficient to have a means by which it will recognise and remember those pathogens the organism has been exposed to during its lifetime. With language, we have the same principle: individuals inherit a language faculty, rather than information about specific languages. This faculty permits the individual to acquire whatever language(s) they are exposed to in their early childhood.

32.9 Language and Thought

Thoughts are the cognitive chunks into which we organise our mental lives. Language uses sentences, which in their simplest form can be said to express single thoughts. Such thoughts can express a positive or negative proposition, such as *Fiona is a clever girl, Fiona is not a nervous person.* And from a logical point of view sentences can have a truth value, for example, *Hydrogen is the simplest element* (true), *The world is flat* (false). Another feature of sentences in human languages is equivalence as can be seen in the following: *Fergal washed the car, The car was washed by Fergal* (active ~ passive); *Fiona is a good runner, Fiona can run well* (attribute expressed by a noun or a verb). This shows the flexibility of language in expressing thoughts.

There is a view held by some linguists and a very few philosophers, like the American Jerry Fodor (1935–2017), that language and thought are inextricably linked and that in evolutionary terms language evolved to formulate thoughts. Such a view sees thoughts as consisting of concepts, which are ordered in a specific manner and causally connected to words in the sentences of language. This would mean that we think by using the language of thought. Linguists holding this view would then postulate that this language of thought was later externalised as speech, or a combination of speech and gestures, which would be secondary and derivative.

30 In language, the equivalent would be storing all the sentences you have ever heard. That strategy would break down the moment you had to process new, not previously heard sentences.

How could one show that the language of thought is somehow primary and speech secondary? Is there a set of independent criteria which could be used here? Consider the linguistically accepted view that spoken language is primary and written language secondary. There are three good arguments for this.

Arguments for primacy of spoken language

1. From phylogeny: written language arose much later than spoken language and is derived from it.
2. From ontogeny: when children acquire language they speak initially; if they do not consciously learn how to write, they never will.
3. From poverty of representation: writing does not capture all aspects of language, for instance, prosody (pitch, amplitude, intonation, rhythm) is not expressed in written language.[31]

We can apply this set of criteria in a consideration of whether the language of thought is primary and speech secondary. (1) Chomsky has stated that the argument from phylogeny holds: the language of thought existed before speech. But that is an assumption for which there is no empirical evidence. The two could have developed in tandem, for instance. (2) Whether an argument from ontogeny would apply is not known as we cannot assess whether children have a language of thought before they begin to speak. (3) This argument could be used to claim that the language of thought is more nuanced that externalised speech, but how much any greater internal differentiation is due to cognition rather than language would be well-nigh impossible to establish, especially as we have no definite way of ascertaining to what extent our thoughts go beyond what language can express.

31 There is also an argument from biological endowment: you need an instrument to write, such as a pen, pencil, painting brush or whatever. However, this argument would not apply to the current discussion.

There is an apparent contradiction in the 'language for thought' assumption: if there is a language of thought, and language is necessary for thought, then one could conclude that language came first and then complex thought. Or at least that in the course of our cognitive development, language overtook thought and came to dominate it. These alternative possibilities could be put in the following form:

A. Did advanced cognition make language possible?
B. Was language responsible for advanced cognition?

Linguists disagree amongst themselves about options A and B. For example, among those scholars who have studied the origins of language, Derek Bickerton was of the opinion that language enabled complex thought (B). Michael Tomasello, on the other hand, believes that there was a change in the mode of thought, which distinguished us from the great apes and that this new mode enabled language (A).

Of course, it could have been a mixture of both (A) and (B). Charles Darwin[32] argued that, in evolution, language and thought were inextricably bound to each other. That is doubtless still the case with our species, though language is not a necessary precondition for thought as we can think without language; for instance, when planning a journey, you conceive of the route from the outset to the goal by imagining yourself travelling the way without the need to 'recount' the journey to yourself with language.

Tomasello thinks that a 'theory of mind' – joint cooperative perspectives and meta-awareness of social behaviour – arose maybe up to 400,000 years ago. This could have led to the development of a proto-language which was primarily gestural. The next important step was the development of sanctioned group norms (Tennie et al. 2009). The linguistic expression of these sanctioned norms was regulated, and increasingly complex language appeared. In a stage previous to this, both Bickerton and Tomasello see cooperative foraging as a strong selective pressure in favour of higher

32 Cases of birds imitating human speech, as with parrots, do not constitute animal language. Nonetheless, there have been impressive instances like the African grey parrot, Alex, which could use about 50 words and combine these sensibly.

cognitive abilities. Both think apes have 'displacement' (a central feature of human language in the Hockettian list, see Section 20.4 above) and that planning for the future is a key aspect of their behaviour. They show a degree of meta-awareness but these aspects of their cognition are competitive and not cooperative as with humans.

Finally, it might be useful to consider whether thought with modern humans is possible without language. One approach is to see whether humans can engage in mathematics without language. Dehaene (2011: 247) has shown that, when processing numbers, the intraparietal sulcus (slightly to the side on the top of the head, far away from the language areas of the brain) is active, implying that any mathematical operations we might perform are independent of language. The opposite view would see the origins of mathematical capacity in an absraction from linguistic structures, something which has not been demonstrated by fMRI investigations. Perhaps a quotation from Einstein (from a letter to the French mathematician Jacques Hadamard) might be insightful in the present context.

> The words of language, as they are written or spoken, do not seem to play any role in my mechanism of thought. The psychical entities which seem to serve as elements in thought are certain signs and more or less clear images which can be "voluntarily" reproduced and combined .... From a psychological viewpoint this combinatory play seems to be the essential feature in productive thought .... Conventional words or other signs have to be sought for laboriously only in a secondary stage ...

Introspection and Language

Introspection, the examination of one's own thoughts and feelings, would appear to be a uniquely human characteristic. In all probability animals cannot introspect to anything like the level possible with humans. After all, if they could, they would behave differently, and begin to do things other than eating, mating and sleeping: they would be communicating the results of this introspection to others. For this, language, or some nuanced communication system, would be necessary. It is scarcely conceivable that animals could have rich and complex mental experiences and not be able to communicate these to others. Imagine the frustration of being a dog and not being able to convey the thoughts of one's rich inner life to other dogs, let alone to humans. The ability to introspect implies a strong notion of self. However, animals appear not to have this as they do not possess the ability to recognise themselves, for example in a mirror. Only chimpanzees and some other higher primates pass what is called the 'positive mirror test', which is often taken as a marker of self-awareness.

What Is Inner Speech?

The relationship of language to our inner mental lives has preoccupied not just linguists but psychologists as well. William James (1842–1910), the founding father of American psychology, saw thought as involving the 'I', the aspect of the self which is manifested in the continuous personal identity which we feel, given our continuous consciousness. This he viewed as different from the 'me', which is a reference to one's concept of oneself, often realised via language as in 'I thought it was me who came up with the idea'.

The early-twentieth-century Russian psychologist, Lev Vygotsky (1896–1934), carried this thinking further, positing that we continually engage in inner speech.[33] In a collection of essays, *Thinking and Speech,* published in the year of his untimely death, he outlined his belief that young children go through a series of phases, starting with social speech (with others around them) to private speech (talking to themselves) to inner speech (without any articulation, no phonetic activity), calling this process 'internalisation'.

Neuroscience has established that the same brain areas are activated in inner speech as in externalised speech. Contentwise, inner speech can be about self-regulation, planning, problem solving, etc. This inner speech can often consist of dialogues with imaginary individuals. For instance, we can rehearse

33 This should not be confused with subvocalisation, which is the use of one's organs of speech without producing sounds, typical of some individuals, for instance when reading a text to themselves.

the arguments for and against a certain decision or course of action and frequently we motivate ourselves by going through the positive reasons for something in inner speech. We can also repeat things in inner speech to bolster our short-term memory, like repeating a telephone number to yourself.

32.10 The Evolution of the Language Faculty

The language faculty is part of our biological endowment, in this case, a module of the human brain which allows us to acquire and use language. It is, to use a somewhat simplistic metaphor, the hardware which allows the software of language to run. However, the metaphor is not complete because the language faculty also generates language, in other words the hardware makes the software.

The framework within which human languages did and can develop is determined by the language faculty by setting the limits on structural variation, which means it imposes constraints on possible languages. On a physical level it permits the range of sounds which humans can make with their organs of speech (vocal folds, throat, oral cavity, tongue, lips, etc.). On a cognitive level it permits collections of sounds to form as words with meaning, and then collections of words as sentences, again with meaning. But importantly, the language faculty licenses certain structural features of grammar: it allows the recursive embedding of elements to create hierarchical, non-linear structures, which humans can produce and decode in real time when listening to others. The structural options of sentences are highly constrained by three main factors.

How Complex Can a Language Be?

The languages in nearly all of Europe, parts of the Middle East as well as most of South Asia, the so-called Indo-European languages, belong to a large family which probably arose about 4000 BCE in the area of Ukraine/southern Russia/northern Caucasus and dispersed in all directions from there.

Sources of structure in human language

1. Functional elements: subject, verb, object (or objects). These arise from our environment in which we (subjects) act (verbs) and affect things or beings (objects). Certain word classes develop from the manner in which we perceive our environment and actions: *The big tree, the deep river* (adjectives: *big, deep*). *The woman ran quickly to her children. The hunters advanced slowly on the prey* (adverbs: *quickly, slowly*).
2. Information organisation: simple declarative sentences reflect direct happenings in our environment: *The man killed the boar*. More complex options allow the presentation of information in different ways: *The man dragged the boar, which he had killed, back to the cave.* The linear order of sentence elements does not need to correspond to the temporal order: *The man dragged the boar back to the cave after he had killed it*. (Temporal sequence: he killed the boar then dragged it.)
3. Discourse organisation: at an advanced stage, human language developed means for providing cohesion among sets of sentences or parts of a sentence: *The man tried to cross the river but he slipped on a wet stone and Ø fell into the water*. (The word *he* refers back to the man and Ø indicates that the pronoun is assumed and not specified.) *The woman cut the meat, pounded it with a stone and placed it over the fire.* (Here, *it* is a pronoun referring back to the meat.)

An apparent paradox can be observed with these languages. The earliest written forms in this family were highly complex, with very intricate grammatical patterns, these growing simpler as the centuries passed. English is a good example of this: at the earliest period of its attestation, before 1000 CE, it had many endings on nouns and verbs and grammatical gender similar to present-day German or Russian. Equally, Classical Latin was more complex grammatically than the modern Romance languages such as French or Italian. True, complexity can be a cyclic matter, with elements of grammar being lost and new ones arising, and at different rates in different places, but the overall trend for all the Indo-European languages is the same: complexity has been diminishing over time.[34] This fact also holds for older versus modern forms of languages from other families: Classical Arabic vis à vis modern varieties of Arabic, Classical Hebrew vis à vis Modern Hebrew.

This situation has led many linguists to consider whether we can define the structural envelope of complexity for human languages. It has turned out to be notoriously difficult to do this. Languages do not require complex grammars, many get along quite well without this, not just English (largely), but languages as different as Chinese, Vietnamese and Afrikaans. The insight from these considerations is that native speakers can manage language systems with great formal complexity, such as Finnish or Hungarian, because they acquire them with ease in their early childhood. For instance, you do not need a threeway distinction in gender for all nouns of a language, as in German or Russian, but if it exists children will master it effortlessly if exposed to the language in the first years of life. For us humans it would seem that we can manage much more complexity, if this is demanded of us when acquiring a specific language. But if it is not required, we do not acquire it.

COMPLEXITY IN EXOLANGUAGES

The current issue is of relevance when considering possible exolanguages. Some of these may be more complex than others and complexity on certain levels may vary across languages, as it does on Earth, with some languages, like Georgian or Irish, having complex sound systems, and others, like Hawaiian or Malay, having much simpler ones. Indeed, it has been suggested that languages far from East Africa, where the original languages of the late 'Out of Africa' hominin dispersal were spoken, like those in the Pacific area, are simpler, at least in their sound systems, as a result of their relative distance to the original source area. With exolanguages, we might not notice the variation in complexity because the very parameters across which this varies would be unknown to us, so that even a relatively simple exolanguage might appear especially complex to humans from Earth, given the nature and scope of our cognition.

34 For reasons of space is not possible to discuss all the factors involved in this development, but contact between disparate groups of speakers (Hickey 2020) certainly plays a role with small, closely knit, low-contact communities having languages of high formal complexity.

32.11 Language, Evolution and Innateness

Language is not a human artefact; it is not subject to aesthetic considerations, which are often superimposed by humans on the objects they create. Instead, it must fulfil its ultimate function of communication, which may be done even if a residue from earlier stages is still present or where there are fuzzy edges to categories.

Like other evolutionary phenomena, language became genetically encoded over a very long period of time. In this respect it can be compared to the way in which most animals have experience from previous generations of their species encoded in their genes, allowing this experience to be transmitted to following generations. This is why animals know how to behave - for instance how to fly, what plants to eat, how to build a nest or spin a web - without being instructed by their parents. For humans, language experience would seem to have led to the genetic encoding of universal features which are common to all languages but specific to none. The details of single languages are what are acquired by children in the first five or six years of their lives. This view of (i) an inherited core and (ii) an additional amount gained from one's immediate surrounding in childhood is the essence of recent linguistic theory which is concerned with determining what elements of language structure are universal - common to all languages - and hence most likely to be innate.[35]

How Did It Become an Instinct?

If the language faculty became part of our biological endowment, can it be regarded as an instinct? The answer would seem to be 'Yes,' if it fulfils certain conditions.

35 The American scientist Terrence Deacon sees it as a conundrum that language is passed on socially, and that without social interaction there is no language acquisition. However, because language arose in social groups there was no selection pressure for language acquisition to work without social interaction.

Particular (externalised) languages are acquired (learned unconsciously) during early childhood by virtue of the human language faculty. It is this which enables children to construct an internal grammar of the language they hear around them with great speed and accuracy, without any instruction, despite the fragmentary and unorganised nature of the data children are confronted with. They dissect the sound stream they hear around them, correctly working out the structures behind sounds and sentences and internalising these structures for later use when speaking this language. The linguistic terms used in this connection are 'performance', the actual use of a language, and 'competence', the ability of an individual to speak and accurately judge this language. Children construct the competence of their native language from the performance of those around them (usually their parents). Needless to say, this will not be completely identical in all details: the slight variations in competence between generations represents a key source of language change.

Criteria for the language faculty as an instinct

1. It is encoded in our DNA, hence present at birth.
2. It is fixed as a framework for language externalization.
3. It is common to all humans, unfolding within a given time period.
4. Its expression in language acquisition is triggered by the growth of individuals.
5. It determinines the level of acquisition attained by an individual (full native speaker status reached before puberty).

The view of the language faculty as a genetically encoded instinct is not uncontested in linguistics. Some linguists refute the notion and would appeal more to something like the Baldwin[36] effect in language acquisition: good learners of language had a survival advantage over others and this created a

36 Called after the American scientist James Mark Baldwin (1861–1934), who first proposed the idea which is now commonly accepted as a valid pathway in evolution.

positive feedback loop, with these good learners turning into super-learners in later generations, who would seem to have an ability like an instinct. Note that the Baldwin effect is not Lamarckian – this is not the incorporation of features acquired in an organism's lifetime into its DNA. Opponents of this view would argue that the Baldwin effect only applies to external behaviour, such as birds learning to open refuse bags to get at their edible contents. The Baldwin effect cannot explain the internalisation of behavioural patterns, linguistic or otherwise. In addition, evidence for genetic encoding of the language faculty comes from the pathological condition known as specific language impairment (see Section 24.4 above) where the genetic mutation of the *FOXP2* gene inhibits normal and complete acquisition of language. Furthermore, if (first) languages were learned, rather than acquired, there would be individuals who would be not so successful at this, that is, there would be variation in the quality of native language acquisition across a community. But this is not the case, as shown by the acquisition of grammatical gender. All Russians, Germans, French, Italians or Spanish know the genders of all words in their respective languages and virtually never make mistakes in this area. If this was learned linguistic behaviour, some individuals would be better at it than others. This is precisely the situation one has with second language speakers, who consciously learn such languages after puberty, by which time the instinctual acquisition of language no longer operates to anything like the extent it does during childhood.

THE EXOLANGUAGE FACULTY

Given time, evolving biological systems tend to release themselves from the concrete environment which triggered them and go through a process of internalisation. This leads to abstract structures, which can be applied to multiple new situations, and renders the system more efficient. An instance of this is the immune system discussed in Section 32.8. If language evolved slowly for exobeings, we could expect that they would also come to have a language faculty, an internalised, abstract ability to acquire, process and use languages within limits determined by their version of universal grammar. If this were not the case, young exobeings would have to learn the entire language of the previous generation from scratch. This would greatly restrict the flexibility and range of exolanguages and would probably hinder them in developing digital technology, with which they could communicate with beings on other planets.

32.12 Language and the Physical Brain

In his 2013 book *Language, Cognition and Human Nature*, Steven Pinker stated that any theory linking language directly to physical processes in the brain 'faces a formidable set of criteria: it must satisfy the constraints of neurophysiology and neuroanatomy, yet supply the right kind of computational power to serve as the basis for cognition' (Pinker 2013: 85). Despite the seemingly daunting nature of this task several scholars in recent years have grappled with the issues surrounding it. Apart from articles, there have been two monographs dealing with language and the physical brain: Murphy (2020) and Friederici (2017). The goal of this work is not just to recognise approximate regions which 'light up' during language use, but to identify in as much detail as possible the neuronal mechanisms by which language is realised.

Much of this recent research has centred on neural oscillations, often colloquially termed 'brain waves.' Neural oscillations emerge from the activity of neurons (on the smallest scale) and of neuronal ensembles (functional clusters on a larger scale consisting of millions of neurons). They vary along the following dimensions: (i) frequency (number of waves per unit of time), (ii) amplitude (height of a wave from crest to trough divided by two) and (iii) phase (the location of the wave on the horizontal axis). These oscillations occur due to the fact that when information is transferred across the synapse between two neurons an extracellular current is generated at the dendrites of the receiving neuron.

The assumption of scholars working in this field, such as the British scientist Elliot Murphy, is that the 'the functions of the brain are manifested via oscillatory activity and [that] information transmission across the brain can be measured as a function of phase coherence', or: 'the computational operations of language can be explained through the oscillatory activity of the brain and how our species-specific oscillatory profile evolved' (Murphy 2020: 12).

Clusters of neurons can fire together, leading to a peak in the electrical wave, which spreads outwards and can be detected on the scalp over the braincase of an individual via electroencephalography (EEG) recordings. The synchronous and rhythmic firing of neuronal clusters produces a regular wave form, which has both frequency and amplitude, the former being determined by the rhythm of the current causing the waveform. The neural oscillation can be picked up by another neuronal cluster (in a different part of the brain) with which the first is physically linked, and the firing of both these clusters can be recognised in the observable waveform. The postulation made by neuroscientists is that the oscillations represent the physical basis for the communication between regions of the brain. Because oscillations, with amplitude increase or decrease, can occur together, with the execution of some cognitive process, they are said to index the latter, that is, they represent the physical signal that this process is taking place.

Neural oscillations reflect synchronised fluctuations in neuronal excitability and are grouped by frequency as shown in Table 32.2.

Table 32.2 Types of neural oscillations

Name	Frequency (hertz)	Association
Delta	0.5–4	Deep sleep, declarative memory
Theta	4–8	Spatial memory
Alpha	8–12	Attention modulation[37]
Beta	12–30	Forms of prediction
Gamma	30–150	More local forms of processing

37 Lower frequencies, as in the alpha range, are known to synchronise distant cortical regions, perhaps suggesting that these connections may represent the neural substrates for connections between different modules in language.

Oscillations can also fluctuate in amplitude and do so in a gradual (phasic) or rapid (tonic) manner, with this behaviour potentially reflecting coordinated computations. Importantly, they can combine with different phases (starting points on the horizontal axis) and amplitudes, leading to a complex wave form consisting of different component waves.[38] Wilson and Bower (1992) already postulated that the phase and frequency of cortical oscillations may reflect the coordination of general computational processes within and between cortical areas.

According to Elliot Murphy, the 'language network' in the brain extends to substantial parts of the superior and middle temporal cortex, inferior parietal cortex, and also subcortical areas such as the basal ganglia, the hippocampus and the thalamus.[39] Oscillations can thus travel across the brain, forming interfaces between different regions. The model he proposes assumes that the data structures involved in linguistic computation (syntactic and semantic features, such as gender, animacy, person, case) are indexed by discrete gamma cycles, which are 'embedded' within the slower theta and delta cycles. Oscillations are used to cluster these features together, as when the operation Merge (see Section 32.7) is assumed to cluster certain features together within a local set during a syntactic derivation. Aspects of the delta phase and its coupling with cross-cortical gamma oscillations, together with parahippocampal and temporal delta oscillations, are proposed to be responsible for indexing syntactic categorical information.

32.13 Language and Memory

Memory is a multi-staged process which consists of at least three parts: (i) the acquisition of information, triggered by an experience (internal or external); (ii) the encoding of this information in a set of neurons in

38 Phase-amplitude coupling between theta and gamma neural oscillations have been causally implicated in memory retrieval and maintenance.

39 Compare the following claim: 'a previously unknown principle in neuroscience [is] thalamic control of functional cortical connectivity', indicating that the role of the thalamus in cognition is much wider than has been typically assumed, see Schmitt et al. (2017).

different parts of the brain, something which may be a preferential activity during sleep; and (iii) the later retrieval of this information by the firing of these neuronal sets together, as a neuronal ensemble. This is known technically as an engram, a term which was introduced by the German evolutionary biologist Richard Semon (1859–1918).

Which cells are used to store memories is determined by the hippocampus, which sorts the information input to the brain, or produced by the brain in the case of an idea, which forms a memory. The hippocampus assigns cells in different regions of the brain to encode the memory, this then accounting for why memories can encompass an auditory, visual, tactile, olfactory or a general emotional aspect.

The persistence and vividness of a memory depends on the strength of the synaptic connections between these cells. If they are strong and constantly used – scientists speak of long-term potentiation – then the memory is also strong. Memories also require consolidation, which appears to involve a process called hippocampal replay; this may in fact take place unconsciously and re-enact the encoding process in compressed form (this has been shown to happen in rodents).

The connections for information pertaining to one's native language(s) show a strength second to none. It is impossible to forget one's native language, though it may become inactive, or dormant, if you do not use it, maybe after emigrating to a new country and switching to a new language. One's native language can always be reactivated.

During the age-appropriate decline in cognitive abilities, language also suffers but to varying extents depending on linguistic level. The ability to form sentences according to the norms of one's native language remains virtually unimpaired to the end of one's life, no matter what age one reaches. The retrieval of words to form sentences shows considerable impairment as age advances, already setting in during one's 60s, though this varies from person to person. The pronunciation of one's native language can also suffer as muscle coordination and control decline in advanced age.

Types of Memory

Memories can be classified into different types. The first division is that between declarative (explicit) and procedural (implicit) memory, declarative memory is further subdivided into semantic and episodic memory. Semantic memory encompasses factual information, such as 'how fast is the speed of light' and, of course, the meanings of words, such as 'what does "sycophant" mean?' Episodic memory, as the adjective implies, refers to events, your eighteenth birthday, the day you graduated from college, got your first job, etc. Procedural memory covers a number of subtypes, of which an important one is how to perform a certain activity, like buttoning a shirt, driving a car, etc. Such memory consists of the information for a coordinated number of steps or actions. Knowledge of the syntax of your native language is encoded as procedural memory: it consists of those steps necessary to construct well-formed sentences in your native language. Furthermore, it is unconscious knowledge; you do not usually think about the abstract structure of a sentence when expressing a thought, though you will, most probably, consider what words to use.

MEMORY WITH EXOBEINGS

Memory is a key part of our existence: the store of memories we build up during our lives provides us with a sense of self and of unique personal identity. But memory refers to much more, particularly to abilities we have, how to ride a bicycle (unconscious – muscle coordination and balance adjustment) or how to play chess (conscious – the permissible moves of pieces and their effects). Without memory, many of our faculties, such as language – a combination of largely unconscious with a degree of conscious knowledge – would simply not function. This would apply equally to exobeings: advanced intelligence would be based on the ability to store information and retrieve it later, at whatever level of consciousness. This ability would be an essential precondition for any communication system, that is, for an exolanguage.

33

The Language of Exobeings

Would exobeings have language? The answer to this question without a doubt is: 'Yes.' They would have language in the sense of a powerful and flexible means of communicating thoughts and ideas between individuals. Why? Complex societies arise through continual differentiated interaction among their members. While many non-human animals do live in communities whose members engage in considerable interaction this does not reach anything like the level characteristic of humans. Furthermore, no beings can acquire all the knowledge of a complex society from scratch on their own. Each generation of a society builds on existing knowledge, which is transferred from generation to generation by being documented using language. To build a technologically advanced society, language would need to be documented in some fixed form, which on Earth means using one of the many writing systems, captured physically, usually on paper, or digitally as bits and bytes in computer storage. A functionally similar system would be required by exobeings and their societies.

Furthermore, any language used by exobeings will have evolved, like the beings themselves, over a long period of time, during which the language will have developed key structural features, optimising its internal use in exobeings' brains and its external application as a means of communication.

33.1 What Might Their Language Be Like?

Recall that human speech is produced by air being exhaled slowly from our lungs, past our vocal folds, which vibrate to produce sound, then through the mouth (and possibly nose as well), with the tongue adopting different configurations (shapes) and touching parts of the mouth to articulate various sounds. These are produced by the exhaled air vibrating in a frequency range, generously given as between about 80 hertz and 8,000 hertz (this varies for children, male and female adults, and there is individual variation as well; sibilants, S-sounds, show the highest frequency).

For exobeings to produce sounds in a like manner their anatomy would have to be similar to ours, in principle. They would have to have a respiratory system and an oral tract with tongues comparable to ours. In addition, they would need the neural control mechanisms to manipulate the various parts of their anatomy necessary for speech. Whether this would be the case is unknown, but consider the following.

An important part of the present-day analysis of potential Earth-like exoplanets consists of examining the spectral lines of their atmospheres for signs of oxygen (see the discussion of biosignatures in Section 29.1 above). This means that if we find an exoplanet with a lot of oxygen in its atmosphere, there must be some biological source continuously producing it, which also means that other organisms on such a planet could avail of this molecular oxygen as an energy source, just as aerobic forms of life on Earth do. So there is a distinct likelihood that beings on such a planet would have developed a means of intaking oxygen and, given that their biologies would be most likely carbon-based, they could give off carbon dioxide as a byproduct of utilising the oxygen intake. Now this intake could be realised in different ways. The exobeings might take in the oxygen through some type of diffusion on their surface (for instance, through skin cells as frogs sometimes do) or they might have gill-like organs to capture the oxygen passing through the gills.

Or they may simply have lungs. If the latter were the case, they would have a respiratory tract comparable to ours and one which could exhale air and which potentially could be used to produce sounds for language.

What Would Exobeings Hear?

The above remarks refer to the production of language. Exobeings would probably also need to hear language to be able to process what others produce. There are good reasons for assuming that exobeings would be able to pick up sounds in the range of our languages. The reason is that common sounds in nature – wind, the rustle of leaves, the sounds of feet and hooves, water splashing or a rock falling – would all be in the frequency range of language. Perceiving these sounds would be vital to exobeings, just as this is for us on Earth where we can hear sounds and noise between at maximum 20 hertz and 15,000 hertz (the high end is attenuated with increasing age). Hence it is a reasonable working hypothesis to assume that exobeings would have evolved language in which sound is produced in a range similar to that which holds for humans.

Relative Loudness of Speech

We have a range from whispering (speech without phonation, i.e. without vibration of the vocal folds) to shouting, but the normal mid-range of speech is around 60 decibels, which can be heard over a range of a few metres in the absence of too much ambient noise. This has advantages given our social organisation. Sixty decibels allows us to communicate easily with one or a few people and to address a larger group by slightly raising one's voice, if necessary. Exobeings with a complex social structure would also communicate with a similar number of others, from a single individual to small groups to larger ones, on occasions. Ambient noise might be an issue for them depending on their societies and how much machinery they might have in their immediate surroundings, as on a loud factory floor, on a busy street or in comparable situations on their planet. There might also be frequent loud noises from nature, like strong winds, which could affect the noise level used for their language.

33.2 Could We Understand Them?

A staple fare of science-fiction films is the landing of an alien spacecraft on Earth. The beings inside such an aircraft are often just really somewhat displaced versions of humans:[1] they may have extra-large ears, bulging eyes, long thin limbs, reptile-like skin or whatever. Equally, the language they speak may just be standard American English with a few twists: sometimes they speak with a clipped, staccato voice, sometimes with shifted syntax, for instance by placing verbs at the beginning of all sentences.

However, for argument's sake, let's assume that a spacecraft from an exoplanet landed on Earth and that the exobeings[2] in the craft were benevolent and interested in having an exchange with us, how would this proceed?

Imagine the following scenario. A group of humans marvel at the wonderful spaceship which has just landed near them. They walk up to the strange craft, a door opens and a ramp is let down. A few exobeings step out. They pause for a moment. Each side considers the other with some curiosity and bemusement. We approach them and they utter sounds apparently coming from mouth-like structures in the front of their heads. We, the humans, point to the spacecraft and the exobeings utter something that sounds like 'apsidolok.' This, the humans think, must be the word for 'spaceship.'

Whether a situation like that will ever happen is very unlikely; it belongs more to the realm of fantasy than any kind of reality that we will experience. But if we did have some kind of contact with exobeings and wished to communicate sensibly with them, the question of medium would be central. Let us consider four basic questions to begin with.

1 Needless to say, we could certainly not interbreed with exobeings, no matter how similar they might appear to us.

2 What representatives of their world would the exobeings on the spacecraft be? Would they speak for their entire planet? Consider the competition between the major powers of our Earth (USA, Russia, China) and now transpose our situation to an exoplanet. Those who might contact us would probably be a selection from the major political units (states), or just one, as the case may be, on that planet.

Contact and communication with exobeings

1. **Could we understand their language on first contact?**
   No.
2. **Could we work out their language, given some time to do so?**
   Probably, on the condition that it was internally structured and of a form which we could grasp. We would have to recognise the modality of its expression – sounds or signing. Then we could perhaps figure out how it was organised and how it functioned.
3. **Could we learn their language and communicate successfully with them?**
   Possibly, if their language was similarly structured to ours, with equivalents to nouns and verbs, for instance, and organised into sentences with meanings similar to those found in human languages, which we could grasp.
4. **What general cognitive features would be reflected in the structure of an exolanguage?**
   (i) The recognition of objects and beings, expressed in grammatical categories similar to nouns in terrestrial languages. Distinctions between one and more than one (singular versus plural), between definite and indefinite (this noun, a noun). The ability to refer to classes of objects (stars, planets, spaceprobes).
   (ii) An extension of this into an abstract domain with concepts, feelings, attitudes, etc. In general, the means to introspect and consider the nature of their life and language. A notion of self, of an interlocutor in discourse and of others about whom one could talk (first, second and third person in human languages).
   (iii) The reflection of spatial distinctions in diverse word classes such as prepositions or deictic particles (positional elements like *on, in, over, under, behind* and pointing elements like *this, that* and *those*).
   (iv) The reflection of temporal divisions as past, present or future with a class of words similar to verbs which express states or actions. Such words would also be used to express hypothetical situations (irrealis constructions), vital to planning in an exosociety (*we could try and contact them, decipher their language*, etc.).

(v) A class of evaluative terms similar to our adjectives and adverbs, with scalar distinctions, such as the range from good to bad, big to small, fat to thin, wet to dry, dark to light, etc.

Depending on the manner in which exosocieties and their languages would have developed (with internal variation and differences), it would be likely that some forms of language would have more status than others. Indeed, there could well be preferred, codified forms of language, equivalent to our standard languages, and speakers might well be prescriptive in their attitudes to others who speak differently to themselves. Such factors would be powerful driving forces for language change in exosocieties.

33.3 Could They Understand Us?

Turning the tables around for a moment, one can ask: could they understand us? Most definitely not. There are over 6,000 languages on Earth, all of which evolved over thousands of years, with pronunciation, grammar and vocabulary of their own. Just consider the English words in Table 33.1.

Table 33.1 Some common words of English and their (standard) pronunciations in phonetic transcription

Written form	Pronunciation	Written form	Pronunciation
man	[mæn]	*take*	[teɪk]
woman	[wʊmən]	*bring*	[brɪŋ]
child	[ʧaɪld]	*up*	[ʌp]
see	[si:]	*down*	[daʊn]
hear	[hɪə(r)][3]	*front*	[frʌnt]

3 The *r* at the end of this word is not pronounced by all speakers. Generally, in British English and varieties of English in the Southern Hemisphere, it is not. In forms of English in North America and in other varieties, such as Scottish and Irish English, the *r* is always pronounced after a vowel. This matter is actually quite complicated but for simplicity's sake the statement here will do. For more information, see Hickey (2023).

Table 33.1 (cont.)

Written form	Pronunciation	Written form	Pronunciation
touch	[tʌʧ]	*back*	[bæk]
come	[kʌm]	*big*	[bɪg]
go	[goʊ]	*small*	[smɒ:l]

Even assuming that exobeings had languages with phonetics like ours, none of the above words would be the same in an exolanguage, except by the unlikely event of coincidental development of their vocabulary.[4] This means that if we were confronted with exobeings, we could not understand anything in their language, nor could they understand ours, to begin with. And it is not just a question of vocabulary: we organise the words of our languages into sentences with a specific syntax (a meaningful arrangement of words in a string which we then speak or process when listening to someone else). It is extremely unlikely that their syntax would be sufficiently similar to ours for us to understand it or for them to understand ours, without any training.

In a nutshell, it is out of the question that an exobeing would be able to immediately decipher any stretch of human language. However, with time and professional help from linguists, we humans could come to understand an exolanguage, such as a message which we might receive, and we might just be able to assist exobeings in understanding our language and establishing equivalents between any given human language and the particular exolanguage we are confronted with (assuming there was time and opportunity to do this). Linguists would begin by trying to establish how the language is encoded (with a non-spoken message), searching for patterns and variations, which might offer clues to the contents. Stretches of speech, if they were ever presented to us, would be that much easier because they would (probably) not be encrypted. Of course, we would not have the exact context in which a language had arisen and came to be spoken. But assuming that any exolanguage would reflect the conditions and

4 An interesting question for historical linguistics would be whether exobeings would know what their Proto-World language, the first on their planet, was or was like.

societies on an exoplanet and the nature of the beings using it, it would in theory be possible to decipher their language and establish correspondences between a human language and an exolanguage. There are examples of similar undertakings here on Earth which have been successful, as shown briefly in the following section.

How to Decipher a Written Code

In 196 BCE, a decree was issued in Memphis, Egypt on behalf of the then king, Ptolemy V. In 1799, a French officer in the Napoleonic army in Egypt, one Pierre-François Bouchard (1771–1822), came across a large granite-like stone in the Egyptian port town of Rosetta (Arabic: Rashid), with one flat side on which the decree was to be found in three different languages. The first two were forms of Ancient Egypt, a high, courtly form and a low, vernacular form, so to speak. The third language was Ancient Greek. After the defeat of the French by the English, the stone, which came to be known as the Rosetta Stone, was taken to London and has been on continuous display there since 1802. Because the text on the stone was also in Greek, which is well known, it has been possible to decipher the Ancient Egyptian (whose later continuation was Coptic, not Arabic).

Other spectacular decipherments have been made, one of which also involved Greek. During the Mycenean period (that of the oldest forms of Greek) remnants of a language, labelled Linear B and found mostly at Knossos on Crete, were deciphered as an early form of Greek by the English architect and self-taught linguist Michael Ventris (1922–1956). Other famous examples in the history of linguistics are, for instance, the recognition of the Indo-European origins of Sanskrit, Greek, Latin, Gothic, Celtic, etc. by the English polymath and judge in India, Sir William Jones (1746–1794), and the decipherment of Hittite (an Indo-European language spoken in central/eastern Turkey) by the Czech linguist Bedřich Hrozný (1879–1952).

But attempts at decipherment are not all success stories. The Voynich Manuscript is a book compiled in Italy in the early decades of the fifteenth century (going on the carbon dating of the vellum on which it was written).

It is named after Wilfrid Voynich (1865–1930), a Polish antiquarian, who bought it in Italy in 1912. At present it is housed in the Rare Manuscripts section of Yale University Library, having changed hands a few times. The book consists of about 240 pages of text and illustrations, which seem to suggest that it is a botanical treatise or a description of paradise, given the many dancing human figures it contains. The text, written without a single mistake, is in an unknown language using an unknown script, and all attempts at determining the language have remained fruitless. It may be in a cipher, that is, a deliberately displaced representation of a known language, such as Latin, but attempts to decipher it on this basis have not met with any success either.[5] The Voynich Manuscript should be a warning to us that deciphering a message from outer space might not be as simple a task as we might imagine, though it would certainly be worth the effort.

Deciphering Messages from Beyond

The successful decipherment of ancient languages on Earth has been achieved by the techniques of comparative linguistics: the scholars take unknown elements in the language being deciphered and try to link these to known elements in other attested languages. If we were to successfully decipher a message from beyond Earth, our tried-and-tested comparative techniques would only be of limited value. We would have to try working out the meanings of the message and then try to work out the system of representation used by exobeings. This is just what any exocivilisation will have to do with the Voyager space probes' Golden Records, should anyone beyond Earth ever find this and try to decipher the contents.

5 One reviewer actually suggested that the manuscript is not written in a language or a cipher of a language at all and hence is undecipherable. What the purpose of the manuscript would have been in that case is entirely unclear.

34

Looking Forward: The Basic Questions Again

Normally the final section of a book on language matters would simply be labelled 'Conclusion,' where the author reviews the main arguments in the book, draws the threads together and presents the results and insights in summary form. However, with this book there can be no definitive conclusion as the subject matter is speculative. But what one can do is summarise the possibilities of exolanguage as a series of questions with tentative answers. Admittedly, the following may be regarded by some readers as unduly anthropocentric, too heavily reliant on what we humans are like. However, in keeping with the principle applied throughout this book, the speculative sections begin with what we know from our existence on our Earth and then move in careful steps to consider what might be the case for exobeings on an exoplanet.

Question:

Would exobeings have language in principle similar to human language?

Answer:

Yes, if:

(i) They have speech organs similar to ours, specifically, a breathing apparatus with vocal folds which vibrate, producing sound which can be modulated in various ways, and a resonance chamber like our larynx and mouth.[1] This would make it possible to articulate vowels. Consonants would be trickier as the parts of the body which we use for them, the lips, the teeth, the alveolar ridge (behind the teeth), the palate, the velum and above all a flexible and agile tongue might not be present in their physical make-up or, if analogous structures existed, it is unsure whether they would have been specialised for the secondary function of language production, as has happened with humans. To be an efficient and flexible speech mechanism their apparatus would have to be independent of their general body movement and their breathing would have to show fine control with slow exhalation.

Exobeings would also require some kind of auditory organ, along the lines of our ears, to pick up the sound waves emanating from their speech apparatus.

There could be variation in the modality of language externalisation: exobeings might use other means than sound, such as signs or just maybe touch, to communicate with each other. However, there are strong arguments in favour of sound as the preferred modality. The sounds of human languages are in the same frequency range as common sounds in nature, as discussed above. Furthermore, you do not have to face an individual or touch that person for sound comprehension – this would be advantageous when working in poor light or hunting near others while not in their line of vision.

(ii) They have brains similar to ours, i.e. a controlling organ for a nervous system attached to muscles used to articulate sounds. Furthermore, if the exobeings had the ability to produce sounds, it stands to reason that their evolution would also have provided them with a means

1 Birds use a structure called a syrinx, shaped like an inverted 'Y' and located just above their lungs. They produce vocalisations by contracting muscles in the walls of the syrinx. This structure developed separately from the vocal cords which we, and other non-human primates, show. See remarks in Section 32.2.5.

of interpreting sequences of these sounds as sense units, that is, words strung together to form sentences which express thoughts arising in their brains. This is what linguists would call the semantic interpretation of the phonetic stream: the assignment of meaning to the sequences of sounds produced by speakers.

(iii) Their brains allow for unconscious knowledge which would encode both (a) general principles of language organisation (innate) and (b) structural knowledge of a specific language (acquired early on in life). It might well be, given a long period of evolution, that (a) became part of their genetic endowment by being encoded in the information passed from one generation to the next as part of their biological make-up and that (b) was acquired during a key phase like early childhood. Assuming that exobeings arise through some system of birth by parents, they would have childhoods in which they grow and mature toward adulthood during which they were capable of reproduction. In the early part of their lives the language or languages of their surroundings would be internalised by them, thus providing the basis for the effortless use of language in later years, and resulting in the unconscious acquisition of a native language (or languages).

Question:

Would exobeings have several languages on their planet?

Answer:

If the assumption that exobeings would use language for secondary functions like social or group identification is correct, one can assume that any populations geographically and/or demographically spread across a planet would show differences in speech, first resulting in different dialects of a language. Given enough time and geographical dispersal, these differences could lead to a lack of mutual comprehensibility, resulting in separate languages. If the exoplanet in question was organised geographically into a system of individual units, similar to states on our Earth, there could well be a process of standardisation by which each state developed a single form of language, or just a few, which would serve for communication in official, public contexts.

Question:

Would exobeings have a notion of 'native speaker' of their own language?

Answer:

Assuming that there are a number of languages on an exoplanet and that not all exobeings speak all languages, there would be individuals who acquired one language rather than another during their childhoods. Furthermore, assuming that exobeings could not speak several languages with equal degrees of competence, there would be different 'native speakers' of different languages. An unanswered question would be whether the ability to acquire a language with the competence of children would decline with exobeings when they reach pre-adulthood, after puberty, as is the case with humans on Earth.

Question:

Could exobeings speak several languages?

Answer:

There is no a priori reason to suppose that the cognitive instruments for reaching a goal – here, the language faculty for acquiring a first language – could only be used in one instance. For example, we use our legs to walk, but we can also skip, run and dance. We can use our feet to shift, turn and kick things, to swim in water, etc. Exobeings could be similarly flexible in their cognition as we are.

Given that exobeings and their exolanguages would have evolved, it might well be that their innate language faculty would have developed to allow for the acquisition of more than one language in childhood.

Furthermore, it might well be that the innate language faculty of exobeings would allow for the acquisition of more than one language in childhood and/or the learning of a second or third language in adulthood. If their language faculty were independent of a specific language, there should be no reason why it could not be applied to more than one language by one and the same exobeing, given the right environmental circumstances. Indeed, if they are cognitively more advanced than we are, maybe they could master several languages easily.

Question:

Would the subjective experience of language be the same for exobeings as for us?

Answer:

This question depends on whether their language(s) would have an emotional component such as we experience when reading poetry or, with some people, when they read religious texts, often accompanied by chant. We do not know how likely it is that exolanguages would be entirely cerebral with no feelings attached to them. Indeed, we do not know to what extent exobeings would experience feelings and how such feelings might affect or perhaps determine the way they behave, including contacting beings on other planets to theirs.

Question:

What elements of their environment would be reflected in exobeings' language?

Answer:

Exobeings would have to negotiate the geometry of three-dimensional space in manners similar in principle to those employed by humans (walking, running, jumping, moving in various ways). Furthermore, one can assume that exolanguages would express the processes involved in this negotiation in a manner similar to the means employed in human languages. They could also be expected to experience a categorical distinction between objects and actions, and this would be encoded in their language as nouns and verbs (see relevant sections in Part V above).

Question:

Would their language(s) be subject to change?

Answer:

Yes. Just as exobeings would have evolved, so too would their language(s) have changed and developed over long periods of time. Furthermore, their language(s) would be subject – in principle – to the same type of variation and change which one finds with languages on Earth: both internal variation, connected with the transmission of language across the generations, and external variation, resulting from preferences for pronunciations and structures in language(s) determined by interactions among speakers in their societies.

Question:

Could exobeings learn our language?

Answer:

In principle, yes, assuming they could process the sounds we make, and that they could interpret sets of these sounds (words) as having specific meanings and arranged in specific orders (sentences). If their language was founded on the use of sentences consisting of words, they could, in theory, decipher ours. The organisation of their grammar into units like our sentences would presuppose that the cognition of exobeings would consist of a continuous string of thoughts as with human consciousness. Whether that would be the case is less certain. It would definitely be an efficient way to organise one's internal cognitive world, but that does not mean that they would do this in the same way as we do.

Question:

Could we learn their language?

Answer:

Again, in principle, yes, assuming that we had some way of deciphering it, this allowing us in turn to reconstruct it for ourselves. Assuming they use sound for language, there is no reason why they should not utilise the possibilities of tone for meaning distinctions as do about half the world's languages.[2] How good we might be able to get at using their language is another question. It would depend on the precise manner in which their language exploited the options provided by their cognition and anatomy. If this happened in a fashion in principle similar to how it did on Earth, we should be able to learn their language. And it might well be that some of our children could acquire their language with a greater level of competence. But that would be a question for later on, not at the time of first encounter, and would depend on factors like the duration and quality of contact with exobeings. But most likely we could communicate with them and, again assuming that exobeings were interested, we could reach a level of sensible exchange with them on which we could impart knowledge of our world to them and in return gain information about the world they came from.

Question:

Could we use some other means of communication, such as pictures or signs we might draw?

Answer:

In theory yes, but this could be more cumbersome and the exobeings might not understand what information we are trying to convey to them, or gain from them. If their language faculty could select the modality for communication, say select either sounds or signs, as the human language faculty can do, we might be able to use alternatives to spoken language with them.

2 If an exolanguage used tone in its sound system, this would put Chinese scientists at an advantage in deciphering it.

35

Some Final Thoughts

The considerations in this book have been largely about the nature of human life and language with a view to assessing (i) what exolife forms and exolanguages might be like and (ii) the chances of our communicating successfully with such exobeings, should we ever come into contact with them. Let us remind ourselves of four general preconditions for beings on an exoplanet in order for them to be capable of engaging in such communication.

General preconditions for exobeings

1. Darwinian evolution on an exoplanet would be necessary to give rise to beings with cognitive powers far beyond what they would require to survive in their environment – like humans, they would possess 'runaway' brains. Evolution on Earth did not generally favour the development of life forms like us humans, otherwise there would be more groups of animals with similar levels of intelligence to us. Assuming that the situation on an exoplanet would be any different would require incontrovertible evidence that this is the case.

2. Hand in hand with increased intelligence would go the rise of higher consciousness. As on Earth, such consciousness could only be generated by a physical substrate. For us, it is the continuous interaction of tens of billions of neurons in our brains, which engenders consciousness. Given that exobeings would have a physical existence and possess brain analogues (either central or distributed), their consciousness would result from the activity of these brains.
3. Beings would need to develop hand-like limbs with a precision grip enabling them to construct things. Otherwise they would be like dolphins: intelligent creatures without the ability to make anything. They would also need to be able to move freely in their environment on leg-like limbs, which would imply a skeleton-like structure with muscles and other types of bodily tissue.
4. Constructing machinery and technical objects would demand a level of interpersonal cooperation which could only be reached if the beings in question had a language-like communication system. This would also be necessary to store information already gained in a society, which would allow each generation to build on the knowledge attained by its predecessors.

The likelihood of such beings arising is probably small, very small. But this has happened: we humans testify to this. The universe 'allows' intelligent, sentient beings to arise, who can then reflect on the nature of the universe and largely comprehend it. Although the likelihood of such beings arising is slight there are so many Earth-like planets, probably in the region of 50 billion in our galaxy alone, going on conservative estimates, that this has most likely occurred more than once, maybe many times across the galaxy. However, the chance of our coming face to face with such beings is extremely slim in the foreseeable future, given the enormous distances even within the small corner of the galaxy we inhabit. And any beings matching the preconditions (1) to (4) above would also need to have mastered the technology necessary to broadcast their existence to others within the Milky Way.

35.1 Deciphering a Signal

Languages are constructed from building blocks with meaning, which are put in specific orders when we speak. This is the manner in which information is transferred between individuals via speech. It is likely that similar information transfer would take place between exobeings, otherwise they could not have evolved successfully, forming communities and, later, societies. Information encoding in any signal we might receive could only be done in blocks, otherwise there would be no way to decipher it. This is the principle used in the famous Arecibo message broadcast nearly 50 years ago and is that still employed by the practitioners of METI (Messaging Extraterrestrial Intelligence). So the task for linguists, working with astrophysicists, would be to recognise the chunks of information in any signal received. Of course, there are various issues involved here. How was a signal encoded? Has the signal become distorted due to interference from interstellar radiation over the long journey from its source to Earth? What range of the electromagnetic spectrum was used to broadcast a signal? How was the signal modulated, via amplitude or frequency? The latter is really the only sensible method, as any amplitude modulation would degrade during a signal's long journey through space.

We might receive not only a composite signal consisting of language but also other information, for instance, about the broadcasting planet, its inhabitants and their scientific knowledge. How would we know where the language starts and ends in such a signal? And, of course, what arrives in Earth might well be only part of a signal, degraded to such an extent that we could recognise it as artificial but not be able to decipher it in any meaningful way.

35.2 A Last Comment

The above are all practical considerations which will play a role as we move forward in the search for exoplanets with advanced forms of life capable of communication with us. Throughout this book I have, at intervals, discussed the likelihood of our finding life beyond our Earth. In this final section, I would like to put the probability of discovering signs of advanced life in something like an order of descending likelihood.

Events implying the existence of exobeings

Event 1	**An artificial signal is received from beyond our Solar System**
Caveat	The signal might have some alternative explanation not hitherto considered, even if a terrestrial source can be excluded categorically.
Significance	Assuming the caveat can be dismissed, we could conclude that the signal was sent by beings on an exoplanet with digital technology. If successfully deciphered, or partially deciphered, information could be acquired concerning the senders of the signal.
Comment	This would be an instantaneous confirmation of intelligent life elsewhere. But there will doubtless be difficulties in locating the source of a signal. Pinning it down to a single planet around a distant star would be a challenge but, with a persistent signal, interference patterns gained from different telescopes, could aid in locating the source, and parallax measurements could determine the distance. Variations due to Doppler effects could determine the orbital and rotational properties of the planet with the senders of a signal.
Event 2	**With our ever more powerful telescopes we discover an exoplanet on which we can indisputably posit the existence of life**
Caveat	The signals would have to allow of no alternative explanations and several independent signals – both biosignatures and technosignatures, for example – would be necessary to eliminate any possibility of a false positive.
Significance	Assuming all caveats can be rejected, then we could conclude that a planet probably contains life. Just what form this would

	take and whether it was technologically advanced would depend on how we could interpret the various signals. Strong technosignatures would imply that the planet hosted life with appropriate technology. Communicating with such life forms would be another matter.
Comment	Confirming that intelligent life exists on an exoplanet would be slow and painstaking. Many likely candidates would fall by the wayside during the ascertainment process. But, in time and with increased precision of instruments, more accurate data will be gathered, forming a mosaic of information about exoplanets comparable to our Earth.
Event 3	**An extrasolar artificial object is detected within our Solar System**
Caveat	It would have to be established beyond all doubt that the object did not have a natural origin.
Significance	Assuming the caveat can be dismissed, we could conclude that there is/was a civilisation which is/was capable of constructing such an object and sending it into interstellar space. If we could capture such an object and if it contained information (like the Golden Records on the Voyager spacecraft) then the gain in information would be considerable.
Comment	This is a waiting game. We have only been in a position to closely monitor movements in the space around us for some decades so we have no idea whether objects from beyond our Solar System appear commonly in our cosmic backyard or not.
Event 4	**A spacecraft with exobeings enters our Solar System and approaches or indeed lands on Earth**[1]
Caveat	It would have to be ensured that any such report is not a hoax. Incontrovertible objective evidence for such an event would be required. Vague uncorroborated reports about any such landing can be dismissed out of hand.

1 As the English physicist Brian Cox has noted, the most noticeable thing for exobeings approaching our planet would be all the city lights spread across the continents in key concentrations, which can be seen during nighttime on Earth.

Significance	Assuming that the event was genuine, it would change the history of the world in an instant.
Comment	Unless exobeings (if they exist) had some way of travelling safely close to the speed of light, they will never appear within our Solar System. Period.
Event 5	**Human beings travel to an exoplanet and contact exobeings there**
Caveat	This option is nowhere near the reach of current or foreseeable technology.
Significance	Such an event would be epoch-making for us, but it can be regarded as clearly unfeasible.
Comment	See previous comment.

In my opinion only Event 1, possibly Event 2, is at all likely to happen in the foreseeable future. For Event 1, the great advances made in our knowledge of how human languages are structured should put us in a position to make reliable statements about the language(s) of exobeings, were a detected signal to contain what we could recognise as language. Whether we could initiate a phase of exchange with an exocivilisation would depend crucially on how far the source of the signal was from Earth. Any distance greater than, say, 20 light years would make exchange an intergenerational undertaking and very slow to reach significant levels. Whether such an exchange would be desirable for both sides and whether it would lead to some kind of cooperation between our worlds remains to be seen. And, most likely, any initial euphoria would probably yield quickly to pragmatic realism, given the nature of interstellar distances. Nonetheless, the most important consequence of such contact would be the certainty that we are not now alone in our corner of the Milky Way. It would be a beginning, confirming the existence of advanced life in our cosmic neighbourhood. In fact, we could advance our search for exoplanets to other possible locations and, in time, increase our knowledge of various forms of extrasolar life and discover what spectrum of variation it covers. In such a situation we could at the very least approach the language question in a realistic and goal-oriented manner, which could help us considerably in making decisions concerning our involvement with forms of intelligent life beyond Earth, should they ever be discovered.

Appendix A

A Possible Roadmap to Exobeings

The following is a brief summary of the probable preconditions and developments which would characterise the pathways to exobeings on exoplanets.

Facts which hold for all exoplanets:

Laws and constants of physics and the fundamental forces of the universe will apply as on Earth.

The range and properties of elements will largely be the same as on Earth.

Atoms and molecules will show the same tendency to aggregate as they do on Earth.

The phenomenon of emergence, both weak and strong, will be found on exoplanets depending on different levels and scales.

Electricity can, in principle, be generated as on Earth, converting mechanical into electrical energy (by moving a coil in a magnet field). Assuming an oxygen-rich atmosphere, energy sources, which rely on combustion and produce heat, will be also available on an exoplanet.

Planet configuration, necessary or at least desirable:

A circular orbit around a stable star.

A magnetic field to deflect harmful radiation from the parent star.

Moon(s) providing orbital stability for a planet.

No tidal locking, to ensure axial rotation independent of the parent star.

An average temperature allowing for liquid water.

A mixture of land and water on the planet's surface.

The presence of molecular oxygen in the atmosphere (requires a mechanism for continually releasing oxygen, e.g. photosynthesis).

The presence of greenhouse gases like carbon dioxide and/or methane in the atmosphere to ensure a roughly equal distribution of temperature across the planet.

Energy gradients on land and in sea to provide dynamic conditions which would stimulate the evolution of life forms.

Abiogenesis:

Life must have been triggered through a coincidence of right conditions.

Cells must be self-maintaining and engage in continual replication.

Eukaryotic life forms (cells with differentiated internal structures) must exist to allow biological complexity to evolve.

Movement from sea to land (marine to terrestrial life, assuming abiogenesis first occurs in a planet's oceans).

Predation may have arisen, leading to an increase in intelligence, as on Earth.

Evolutionary biology:

A framework for life forms to evolve increasing complexity by equivalent processes to natural selection, gene flow and genetic drift. Exobeings will have developed along this trajectory, no matter how advanced they are in a digital age.

No life forms, which would prevent the evolution of intelligent beings, can dominate on an exoplanet, as dinosaurs did on Earth before the K–Pg extinction event.

Survival through extinction events:

The active geology on a planet's surface must not be so great as to constitute extinction events from which life could not recover.

Freedom from asteroid strikes, nearby supernovae, gamma-ray bursts and other catastrophic disturbances from beyond the planet.

Anatomical and physiological developments:

Development of hand-like limbs with a power and precision grip, necessary for the construction of artefacts.

Great increase in size of their brain-compatible structure, far beyond what is required to survive in their environment.

Evolution of some mechanism for producing modulated sound, functionally comparable to our vocal folds, which would permit the externalisation of language.

Cognitive development:

Rise of the ability to engage in abstract thought, to plan for the future, to conceive of artefacts which might be constructed, etc.

Environmental and cultural developments:

Management of fire; perhaps with the advent of cooking (not essential, assumes exobeings are heterotrophs).

Perhaps housing, to allow the maintenance of suitable surroundings independent of environment.

Perhaps medicine, to maintain health and prolong life expectancy.

Expansion of science, advanced technology:

Rise of societies with divisions of labour and large-scale cooperation between groups, to realise technological achievements based on scientific insights.

Appendix B

A Possible Roadmap to Exolanguage

The manner in which exolanguage might arise, and what basic characteristics it might have, is outlined below. These postulations derive from what we know about the evolution, development and diversification of language on Earth. The reference is to exolanguage in general; individual exolanguages would likely appear with increasing separation of social groups through geographical dispersion and/or internal divisions.

Evolution of language:

The first precondition would be a physical substrate - brain or comparable structure generating advanced cognition - with dedicated region(s) for the production and reception of language.

A slow development in which sounds are initially combined to form words and possibly used with gestures for communication in small social groups. Word combinations then arise which are the precursors of sentences in modern language and word classes, above all equivalents to nouns and verbs, would begin to appear. These developments are antecedent, or at best parallel to rises in cognitive ability.

A sophisticated communication system would also be integral to abstract thinking and would allow large-scale and detailed cooperation on a social level, ensuring the transmission of accumulated knowledge across generations.

General nature of language:

Language faculty:

Genetically encoded; provides maximum flexibility; allows acquisition of specific languages.

Instinctual: unfolds in early childhood, simply by time proceeding and a being growing.

Modality:

Sound in the 20–15 kilohertz range (corresponding to sounds in nature, evolutionary preference for this); production of sound: vocal folds (precondition: pulmonic airstream or equivalent).

Signing: would presuppose the presence of hands, fingers, mouths for facial expression.

Structure of language:

Production: cerebral mechanism for constructing linguistic units, roughly comparable with our sentences.

Reception: cerebral mechanism for deconstructing these units in real time via Bayesian inference based on knowledge of language structure and the world speakers live in.

Rise of phonemic principle – words consisting of sounds, which in themselves do not possess, but distinguish meaning.

First words: possible on the basis of sound blocks (very early stage).

Later rise of morphology (grammatical endings on words); grammaticalisation as a force in language change – perhaps, depends on language type.

Basic word classes:

1. Nouns – for objects in three-dimensional space.
2. Verbs – for actions on a time axis or states at one time.
3. Adjectives/adverbs – qualifiers for (1) and (2).
4. Prepositions/cases – relations in space–time.

Distinction between classes and exemplars (*trees : this tree*)/definite and indefinite classifications (*the tree : a tree*).

Developments in meaning:

Distinctions in number – *one*, [*two, a few*], *many* (singular [dual, paucal], plural); quantifiers – positive: *some, all*, negative: *none, nothing.*

Deictic terms ('pointers') – various pronouns (personal, demonstrative, etc.), space and time reference (*here, there*; *yesterday, tomorrow*).

Concrete–abstract distinction.

Use of metaphorical language, application of the concrete to the abstract – would depend on cognition of exobeings.

Development of means to refer to the past and the future and degrees of distance from the present (time of discourse).

Negation and counterfactuality – say what is not the case or what could be the case; interrogation – request information from hearer.

Rise of syntax:

Initially linear, bound to temporal sequences, later independent of external time frame; development of hierarchical structures which can recur within sentences, which are, in principle, open-ended in their number.

Pragmatics:

Language in use, developing out of conventions of interaction; negotiation of meaning between speakers, fuzziness of language as flexibility, indirectness; norms of politeness (perhaps, depending on nature of exosocieties).

Diversification of language:

This would be the outcome of language change which in turn would result from two basic sources:

1. internal – due to shifts when children construct the competence of their native language from the performance of those around them;
2. external – determined by social preferences in the use of variants in language (mainly pronunciation, but also grammar and vocabulary).

Geographical spread over an exoplanet, as well as contact between various speakers, would lead to language splitting into different varieties, which in turn would become separate languages when mutual comprehensibility no longer holds.

Glossary

General

The following is a list of common terms used in the body of the book. In order to avoid repetition, some terms are not included here as they are defined and discussed in the main text. If you cannot find a definition here then please consult the index, which contains the number(s) of the page(s) where the terms can be found in the book.

abiogenesis The crucial process by which life – animate forms – arose from inanimate matter triggering the evolution of biology on Earth.

adenosine triphosphate (ATP) An organic compound which provides the energy needed for many processes in the body such as muscular contraction or nerve conduction.

allele A variant form of a gene. Different alleles can result in different characteristics in the phenotype, the organism as it manifests itself on the highest scale.

amino acid An organic compound which consist of an amine and a carboxylic acid group. Some 20 of them are relevant to the making of proteins in the human body, proteinogenic amino acids.

amygdala A part of the limbic system of the brain, located at the base, deep in the medial temporal lobe, and shaped like a pair of almonds. It has been implicated in both the formation of memory and the emotions of fear, anxiety and aggression.

anaerobic A reference to life forms which do not require oxygen or cannot tolerate it, such as some types of bacteria, or to conditions without oxygen. Initially life forms on Earth were anaerobic.

analogous In evolutionary biology, the term used to refer to convergent traits which are similar but do not share a common source in the past. See *homologous*.

anthropic principle A belief, which in its strong form, states that the universe shows fine-tuning *in order that* humans can exist. The weak form just states that the universe shows the properties necessary for life, a simple truism. There are many, widely discussed fomulations of this principle.

Anthropocene A proposed epoch, the present one, which is centrally characterised by humans and our effect on Earth's biosphere. A variety of starting points have been suggested, ranging from the beginning of farming to the mid-twentieth century when environmental degradation and species elimination became acutely obvious.

archaea A domain of life consisting of microscopic single-celled, prokaryotic organisms, similar to but distinct from bacteria.

axon A thread-like extension, technically called a cytoplasmic protrusion, from the body of a neuron along which electric signals are transmitted to other neurons and which is insulated by a myelin sheath. When a signal reaches a synaptic terminal it is passed to the next neuron by neurotransmitters. See *dendrite*.

bacteria Microorganisms consisting of only one cell and often without a nucleus in a membrane (prokaryotes). Bacteria are the most common form of life and are found in all environments, including within our bodies where they play an important role in many biological processes, such as digestion.

basal ganglia A collective term for a number of structures at the base of the human brain. In evolutionary terms these are primitive parts present in most vertebrates and all primates.

biosignature The tell-tale composition of a planet's atmosphere, visible in the spectral lines it produces. These betray whether it contains free oxygen, which implies that this is replenished regularly from some biological source, such as plants giving off oxygen during photosynthesis.

biosphere The totality of all ecosystems on Earth.

bipedalism Moving forward on two feet. Humans are eminently bipedal, using legs in which the upper and lower parts are fully distended, forming a line from the feet to the hips. This contrasts with those animals which move forward on all fours, with the ends of the front limbs formed like hands with bent fingers. Higher primates, like chimpanzees and gorillas are such 'knuckle-walkers.' This kind of movement cannot be maintained for very long because the knees of the hind legs are bent, making effective movement difficult.

bottleneck, evolutionary A metaphoric term for a period in evolution when the variety of life in general, or at least the size of particular populations, was severely curtailed, usually through natural catastrophes, such as a major volcanic eruption or a large asteroid impact.

Cambrian explosion A period of sudden expansion of life forms, as evidenced in the fossil record, with a strong rise in predation. It began around 540 million years ago.

carbon dating A technique for dating organisms and objects containing organic matter by measuring the ratio of the carbon-14 isotope to carbon-12, a stable form, in a given volume of organic material. The method yields reliable results as carbon-14 has a half life of 5,730 years and this can be measured accurately up to about 10 cycles. More recent dating methods can achieve greater time depth.

cell division A fundamental process in all biology, which is the basis for the proliferation of life forms. It is a complex process consisting of several stages from DNA replication to cytokinesis, the actual splitting of one cell into two. Most cells undergo mitosis (cell reduplication) while sex cells also undergo meiosis (fourfold division with *gametes*).

Cenozoic era Lit. 'new life', this is the current geological era, extending from 66 million years ago (after the K–Pg extinction event) to the present day. It is also known as the Age of Mammals, given the rise of mammals, which greatly diversified later, leading to primates and ultimately to species of *Homo*.

chromosomes Packets of DNA strands wound together tightly and present in all cells. Humans have 46 chromosomes; two particular chromosomes are essential in distinguishing sex: females have two X chromosomes while males have one X and a much smaller Y chromosome.

clade A grouping of organisms or animals which all have a common ancestor and are thus connected by descent. A clade can be a species or an entire family. Also known as a monophyletic group.

consciousness A reference to the sensation of awareness which all humans have during their waking life. It is intimately connected to the sensation of self and the feeling of continuity of being throughout one's life. The experience of consciousness is interwoven with the language of thought, in which we construct a continuous mental narrative for ourselves.

contingency In sciences dealing with the past, contingency refers to unpredictable, chance events/developments as opposed to those which were predetermined and thus bound to happen. Assessing the role of contingency is a central concern in many sciences, for instance in evolutionary biology.

cortex The outer layers of neurons on the cerebellum (upper part of the brain). It is arranged as a series of ridges (pl. *gyri*, sg. *gyrus*) and grooves (pl. *sulci*, sg. *sulcus*), which greatly increase its surface area. The cortex contains over 16 billion neurons, which are responsible for all higher-order cognitive abilities, such as language, thought and most likely consciousness in general.

dendrite An extension from the body of a neuron, which is used to receive signals from other neurons. Each neuron can have several hundred, if not a few thousand, such connections with other neurons. See *axon*.

determinism A stance in philosophy which maintains that all events are fully determined by previous states of systems and forces operating within them. With reference to human decisions, determinism negates the existence of free will.

DNA An abbreviation for deoxyribonucleic acid. It is a very large molecule held in the cell nucleus of all eukaryotic life forms and has encoded in it the information to produce the entire organism which carries it. DNA is arranged as a spiralling double helix, a kind of twisted ladder with strands on either side, forming a sugar phosphate backbone, and rungs connecting across. The latter consists of certain combinations of four bases – cytosine (C), guanine (G), adenine (A), thymine (T) – as A and T or C and G pairs – with sequences of these pairs encoding information. In humans, long strands of DNA are wrapped tightly to form 46 chromosomes in the nucleus. To reproduce itself, the DNA molecule separates the strands of the original double helix, copies it and closes off the open sides left open after separation. By that means, two identical strands are gained from one source strand. See *gene, genome; transcription and translation.*

Earth-like A general term used to refer to exoplanets which would share a large number of key features with our Earth and thus be potential sites for exolife. See Section 8.8 for details.

echo location An ultrasonic system used by tooth whales and dolphins to locate objects/beings in their surroundings. They produce clicks in their nasal passages, which are then focused forward through a bundle of tissue called the melon. Fatty tissue in the lower jaw receives the return signal bounced off objects/beings and transmitted back to their ear.

electroencephalography A non-invasive procedure for measuring electrical currents in the brain. It involves placing electrodes on the scalp, which pick up the flow of electricity between neurons, and the wave-like pattern is displayed on a computer screen and/or printer.

empirical A term which refers to evidence reached by objective collection and analysis of data. To be empirical is a basic demand of all scientific theories, in as much as this is possible.

encephalisation quotient The relation of brain size to body size. In the animal world, bottlenose dolphins have the highest value, locating them close to humans.

endothermic A scientific term for the generation of heat by an animal (known colloquially as 'warm-blooded'). See *exothermic*.

ethology The study of animal behaviour, especially in a comparative sense, examining under which circumstances different types of behaviour evolved.

exaptation A development in which a feature is co-opted for a function which it was not originally intended for. Feathers, originally for keeping an animal warm by trapping air, but later used for flight, are a good example. Another example would be webbing between the toes of some animals, such as Newfoundland dogs, enabling them to swim efficiently.

exobeings A term to refer to intelligent beings on an Earth-like planet beyond our Solar System, who would have a communication system comparable to human language and who could, in principle, engage in interstellar messaging.

exothermic A label for those animals which do not generate the heat they require internally but absorb it from their environment (colloquially known as 'cold-blooded'), for example lizards, which need the sun's heat to maintain a high body temperature. See *endothermic*.

extremophile A class of organisms which can survive in conditions not previously thought possible, such as at temperatures over 100 °C (boiling point of water) or less than 0 °C (freezing point of water) or in environments which are highly saline.

fMRI An abbreviation for functional Magnetic Resonance Imaging, a technology developed to image blood flow in the brain because blood is released to active neurons at an increased rate, which can be detected magnetically. Given this correlate of neuronal activity, areas of intense blood flow are also those of increased neuron firing and can be associated with specific functions which individuals carry out, such as visual or auditory tasks. Recent investigations have engaged in precision fMRI of individuals to determine the neural substrates of cognitive control.

FOXP2 A gene responsible for vocal control and the acquisition of vocal sequences, which in its human form constitutes an integral part of our genetic ability/predisposition for language. Neanderthals had a similar, though not identical, FOXP2 allele to ours, so that they most likely had a comparable degree of vocal control which would have been a precondition for language. FOXP2 is an abbreviation of 'Forkhead Box Protein P2.'

gamete The reproductive cells of an organism, termed eggs with a female and sperm with a male. These combine during fertilisation, resulting in a new combination of chromosomes (50 per cent from each parent) and leading to pregnancy for the female.

gene A section of DNA associated with a particular function in the construction of the organism containing it. Genes are expressed, by which is meant that the information is read off and used in making one or more proteins, which then trigger further functions in the organism. There are several steps in gene expression – transcription, RNA splicing, translation and later modification. Genes can be suppressed, 'turned off', by the processes of methylation during which methyl groups are attached to DNA or by the proteins around which they are wrapped – histones – being modified such that they stop the gene from being transcribed. The human genome consists of about 20–25 thousand genes, each of which can be anywhere from 300 to 1,000,000 nucleotide pairs long.

genome The complete DNA in the nucleus of a cell (occasionally in the cytoplasm surrounding the nucleus) required for the construction of the organism containing it.

genotype The totality of what an organism inherits from its progenitor, its heredity. See *phenotype*.

genus A term in the classification of life forms, which is above the level of species. Humans belong to the genus *Homo* and the species *sapiens*. The system of labelling life forms via genus + species was devised by the Swedish botanist Carl Linnaeus (1707–1778).

glial cells Support cells for the neurons in the brain, supplying them with oxygen and nutrients, and removing waste. There are about as many glial cells as there are neurons in a human brain, between about 85 and 90 billion.

haplogroup A group of gene variants (alleles) inherited from a single parent showing high levels of conservation, for instance in the Y chromosome, found in males, and in the mitochondrial DNA, passed on unchanged through the female line. Haplogroups are thus useful in tracing human lineage.

heterotroph Organisms which have an external intake of food, such as terrestrial animals which feed on plants or other animals.

Holocene The geological epoch in which we are currently located. It began about 11,700 years ago, after the end of the last glacial period (ice age) when farming first developed. It was preceded by the Pleistocene and together with this it forms the Quaternary period. See *Anthropocene*.

homologous This refers to features and/or structures found among species having common ancestry but with different realisations. The forelimbs of vertebrates (the part from the elbow to the fingers in humans) vary greatly from the arms of bats to the front flippers of whales to the forearms of humans. See *analogous*.

ion An atom or a molecule which has lost or gained one or more electrons. This changes the charge of the atom, it becomes positive on the loss of electron(s) or negative when an electron is gained. See *redox reactions*.

Lamarckism A view in pre-Darwinian biology, propounded by the French scientist Jean-Baptiste Lamarck (1744–1829), which assumed that characteristics acquired during the lifetime of an organism could be passed on to a following generation. While this is certainly not true in any simple form, there is evidence that epigenetic modifications can be transferred to offspring, thus complementing Darwin's notion of evolution by natural selection.

mRNA An abbreviation for 'messenger ribonucleic acid', a single-stranded molecule containing the code for a single gene, read off the DNA in a cell nucleus and transported out of the nucleus into the cytoplasm (fluid contained in the cell) to be translated by a *ribosome* into the protein which the original gene codes for (simplified representation of the process).

natural selection A process, first formulated by Charles Darwin, by which organisms, which have traits beneficial to them in their environment, have a greater chance of survival and so contribute to the overall evolution of their species.

neuroplasticity A characteristic of human brains which allows them to form new connections and hence encode new information relating to knowledge and behaviour. Although this ability declines after puberty it is nonetheless taken to persist throughout adult life.

neurotransmitters Chemicals which are activated when an action potential (an electrical charge) arrives at the synaptic gap between two neurons. These jump the gap of the cleft and lead to the electric charge being transferred to the opposite side and hence to the next neuron. Dopamine and serotonin are neurotransmitters which affect mood and desire.

nucleotide A structural unit of DNA consisting of a nitrogenous base (the 'rung' of the double helix) as well as a sugar molecule and a phosphate group (part of the 'backbone' of the double helix). The base is one of the following: adenine, cytosine, guanine or thymine. Bases are also present in RNA but uracil takes the place of thymine.

ontogeny The development of the individual as opposed to that of the group or species. See *phylogeny*.

palaeoanthropology The study of the early stages of the genus *Homo* (and its predecessor *Australopithecus*). Due to new fossil discoveries and advances in genome sequencing techniques the field is highly dynamic, and standard wisdoms are being constantly revised.

phenotype The totality of an organism's traits observable on the highest scale. The phenotype of a human is the individual as that person is observed by others in shape, features and behaviour. See *genotype*.

phylogeny The development of an entire group or species as opposed to the development of an individual. See *ontogeny*.

plate tectonics The movement of the continents on the molten rock just below the Earth's crust, discovered by Alfred Wegener (1880–1930) but not widely accepted until the mid-twentieth century. This movement is responsible for volcanic activity, sea-floor spreading and earthquakes on the surface of the Earth.

prebiotic An adjective referring to organic material which existed before life forms arose. The related term 'probiotic' refers to organic compounds which provide a direct input to life forms, for example the amino acids used in the cells of human bodies.

proteins Large biomolecules assembled from amino acids into specific three-dimensional structures. Proteins fulfil a huge number of functions in all organisms, from metabolism and growth to the immune system.

punctuated equilibrium A notion put forward by the American evolutionary biologist Stephen Jay Gould (1941–2002) that evolution can be relatively slow and stable over long periods and suddenly go through a rapid phase of transition.

redox reactions A commonly occurring class of chemical reactions in which electrons are gained by an element/molecule, called reduction, or lost, called oxidation.

ribosome A structure in cells which reads off the instructions in *mRNA* and constructs a specific protein. Its output is the protein coded for by a particular gene on the DNA. However, because the ribosome is outside the cell nucleus, the gene on the DNA must be first *transcribed* into mRNA, moved out of the cell and then *translated* by the ribosome into the target protein.

RNA An abbreviation for 'ribonucleic acid', a type of single-stranded molecule that consists of a long chain of nucleotide units, each of which contains a nitrogenous base, a ribose sugar and a phosphate group. RNA transmits the genetic information from DNA in the nucleus of cells to the protein-encoding devices (*ribosomes*) in the cytoplasm (liquid of a cell containing the nucleus and various organelles). Both DNA and RNA are considered essential building blocks of life.

RNA World The view that RNA molecules were the earliest which could self-replicate (preceding DNA). Later, DNA developed, consisting of two interwoven strands with bases attached to the backbone as opposed to the single strand of RNA.

senescence Accumulated errors in reproduction, accumulated degeneration. With senescence, chromosomes fail to replicate accurately, or change their structure somewhat or hang around that bit too long. Phenotypically, it is registered as 'getting old'.

species A classificational division in the animal world which is below that of genus. Thus, humans are of the genus *Homo* and the species *sapiens*.

synaptic gap A gap at the end of an *axon* leading to a neuron. The signal passing towards the neuron jumps this gap via neurotransmitters which are released for this purpose (and then withdrawn during reuptake).

taxonomy A system of classification used to render divisions and structures in a system visible, for example animal forms, languages, etc.

tetravalent An atom with four electrons in its outer orbit, which can be shared with other atoms in chemical (covalent) bonding. Carbon is tetravalent, which accounts for its great flexibility in forming those molecules which make up all organic matter.

thalamus A structure located on top of the brain stem under the cerebrum. It plays a pivotal role in relaying sensorimotor signals to and from the body.

transcription and translation Two related processes by which proteins are formed in cells on the basis of information in DNA. First, a protein (transcription factor) binds to a locus in the DNA, controlling the transcription of a section of the DNA into mRNA, which then exits the nucleus and is translated by ribosomes in the cytoplasm within the cell into the required protein.

vertebrate A reference to all animals which show a spinal cord along their backs with a nerve centre at one end (the brain). Vertebrates have bones which form a skeleton offering support to the soft parts of the body. Birds, reptiles and mammals are examples of vertebrates.

zygote The cell which results from a female egg being fertilised by a male sperm. The zygote goes on to form an embryo, which in time leads to a fetus, then to the entire animal/human at birth.

Linguistics

This section of the glossary is intended as an aid to those readers who are not au fait with linguistics. The definitions given relate to the topics discussed in the body of the book. Due to the orientation of this book certain areas of linguistics are not centrally represented, such as sociolinguistics or pragmatics.

acoustic phonetics One of the three divisions of phonetics, which is concerned with the physical properties of sound and its transmission through air (default). See also *articulatory* and *auditory phonetics*.

acquisition The process whereby children internalise linguistic information unconsciously, using it later when they wish to speak the language in question, their native language. Acquisition is largely unguided and shows a high degree of completeness compared to second-language learning.

adjective A word class which generally qualifies a noun, as before the noun: *the dry snow* or after a form of the verb 'to be': *The snow is very dry*. Adjectives can themselves be qualified by adverbs (as with *very* in the example just given).

adverb A word class comprising those elements which qualify verbs/verb phrases (*she smiled slyly*) or nouns/noun phrases (*a remarkably good linguist*). Some adverbs can qualify a clause or an entire sentence as in *Surprisingly, Fiona left for home*.

alphabet A system of letters intended to represent the sounds of a language in writing. For all west European languages the Latin alphabet, with some modifications, for example with additional signs over or under letters, has been the outset for their writing systems. There are, however, very different alphabets for other languages, such as Russian, Georgian, Armenian, Hebrew, Arabic, etc.

alveolar A classification of sounds which are formed at the alveolar ridge (the bony plate behind the upper teeth). Alveolar sounds are formed with the tip or the blade of the tongue. Examples are /t, d, s, z, l, n/ in English, among the most common sounds in all languages.

analytic A term used for a language which tends to use free morphemes (roughly equivalent to separate words) to indicate grammatical categories. Examples are Modern English and French, to a certain extent. Other languages, such as Chinese or Vietnamese, are very clearly analytic. See *synthetic*.

apex The tip of the tongue; adjective: *apical* or *apico-*.

aphasia A general term for language malfunctions which result from brain damage, through accident or disease. The malfunctions can affect production or understanding and may involve grammatical or semantic levels of language or both. Occasionally, aphasia can be inherited due to a genetic defect as with specific language impairment.

arbitrariness The lack of any necessary relationship between linguistic signs and their referents, that is, objects in the outside world. For instance, the relationship between the sounds of the word *sky* and the concept of sky is arbitrary but fixed by social convention. The equivalent convention in French specifies *ciel* as the appropriate word, in German *Himmel*, in Russian *nebo*, etc.

arcuate fasciculus A bundle of nerve fibres connecting the two main areas of the brain which are crucial to language, Broca's area (in the inferior frontal gyrus) and Wernicke's area (in the superior temporal/inferior parietal lobe).

articulation The set of muscular movements necessary to produce a specific sound. There is a distinction between manner and place of articulation.

articulatory phonetics One of three divisions of phonetics, which concerns itself with the production of sounds (compare *acoustic* and *auditory phonetics*).

auditory phonetics One of the three divisions of phonetics, which is concerned with the perception of sounds. See also *articulatory* and *acoustic phonetics*.

babbling phase A phase during infancy, from about the middle to the end of the first year of life. During this phase children often produce sounds which are not necessarily part of the sound system of the language they are acquiring.

Bayesian inference A procedure used by humans when listening to language: the hearer takes in the first word(s) and constructs possible continuations a sentence might have, going on their knowledge of the world, the topic being discussed and their acquaintance with the structure of the language being heard. In real time, hearers continually revise their hypotheses about the sentence, as new words are heard which predict a continuation of sentence structure and meaning in a certain direction.

Broca, Paul (1824–1880) A French doctor who investigated patients who had experienced a physical trauma to the left frontal lobe of the cortex and who had difficulties in producing speech as a consequence of this.

Broca's aphasia A form of aphasia in which the production of speech is impaired. This may or may not involve difficulties in understanding language. See *Wernicke's aphasia*.

Broca's area A part of the brain – approximately above the left temple – called after its discoverer the French doctor Paul Broca and which is responsible for speech production. See *Wernicke's area*.

Chomsky, Noam (1928–) The main figure in modern linguistics, Chomsky is responsible for the great increase in theoretical linguistics which set in during the cognitive revolution of the 1960s, which he was instrumental in initiating. Among his many ground-breaking tenets is the assumption that children are born with a predisposition to acquire language and have unconscious structural knowledge of language in general, so-called 'universal grammar' which enables them to acquire their native language (or languages with bilinguals) quickly, very well and without any formal instruction.

closed class Any linguistic set whose elements form a relatively small and fixed number. For instance, phonemes (distinctive sounds), morphemes (grammatical endings) and syntactic structures (sentence types) are all closed classes but the lexicon (vocabulary) is an open class as it is continuously expanding. See *open class*.

competence According to Chomsky, this is the abstract ability of an individual to speak the language which they have acquired as native language in their childhood. The competence of a speaker is unaffected by such factors as nervousness, temporary loss of memory, speech errors, etc. Competence also refers to the ability to judge if a sentence is grammatically well formed; it is an unconscious ability.

computer language A formal language which is used to create computer programs. Such languages are highly restricted and rigid, with no exceptions to the rules of their syntax and semantics. They are thus not really comparable to human language. There is a general division into *procedural* languages (such as Python or C and its derivatives) and *declarative* languages (such as Lisp or Prolog). The former are used to define the steps to be taken to solve a problem or reach a predefined goal. The latter are used when the user states the goal which should be arrived at and leaves it to the program to determine the quickest and most economical 'path' to reach this.

conduction aphasia A type of aphasia where patients fail to connect what they hear to what they say due to damage to the arcuate fasciculus.

consonant One of the two main classes of sound. Consonants are formed by a constriction in the supra-glottal tract (or occasionally at the vocal folds as with the glottal stop, shown as [ʔ] in phonetics). Consonants contrast with vowels in their relatively low sonority (fullness of sound) and are hence found typically in the margins of syllables, that is, in onsets and codas as in *stopped* /stɒpt/.

continuity vs. discontinuity hypothesis A set of contrasting views on how language evolved with humans. The continuity hypothesis assumes that language evolved slowly through the later species of the genus *Homo* whereas the discontinuity hypothesis, put forward by Noam Chomsky, posits a sudden point at which modern language, with hierarchical syntax, arose due to a 'rewiring of the brain'.

convention An agreement, reached unconsciously by speakers in a speech community, that relationships are held to apply between linguistic items (words in the language) and concepts relating to the outside world.

creativity A feature of human language manifest in speakers' ability to produce and to understand an open-ended number of sentences.

critical period A period in early childhood in which language acquisition is most effective (roughly the first six years). If exposure to a language begins later, acquisition does not usually result in native-like competence. The watershed for successful acquisition is puberty after which it is nearly always incomplete. The periods of greatest sensitivity to language can vary according to level: it would appear to be earliest for phonology (the sound system).

deixis In the broadest sense, the phenomenon of 'pointing' in human language, either spatially or temporally or across sentences in discourse. Deixis is realised by many means: pronouns for nouns, tenses with verbs, adverbs of place in locational deixis, such as *here, there, beyond,* etc.

demonstrative pronoun A pronoun whose function is to indicate position in discourse, in relation to the speaker (*this, that*). Demonstrative pronouns frequently express definiteness, as in *I am not interested in this book.*

diachronic A view of language over time, in terms of historical development. See *synchronic.*

digital infinity A characteristic of human language, given this label by Noam Chomsky. It specifies that language consists of discrete, separate units (words, clauses, sentences) and that these can combine in potentially infinite ways, meaning that language, i.e. sentence generation, is open-ended.

discontinuous Any linguistic unit which has been split by the insertion of another, as with the main clause in *Her husband, whose name I always forget, rang this morning.*

displacement One of the key characteristics of human language which enables it to refer to situations which are not in the here and now, such as *Fiona studied in Belfast when she was young.*

duality of patterning A structural principle of human language whereby larger units consist of smaller building blocks, the number of such blocks being limited but the combinations being open-ended, that is, potentially infinite. For instance, all words consist of combinations of a limited number of sounds, say about 44 in English (depends on variety). Equally, all sentences consist of structures from a small set, with different words occupying different points in the structures, which can be arranged to produce any number of novel combinations.

epiglottis A part of the human anatomy which closes off the larynx when one swallows, thus preventing food or drink from entering the lungs and causing one to choke.

etymology An area within historical linguistics which is concerned with the origin and development of the form and meaning of words, and the relationship of both these aspects to each other.

exolanguage A term used in this book for a language spoken by exobeings.

generative linguistics The main school of linguistics since the middle of the twentieth century, which assumes that speakers' knowledge of language is largely unconscious and essentially rule-governed. The models used are intended to generate – properly describe – how underlying syntactic structures are mapped onto actual sentences.

grammar A level of language concerned with how words combine together to form sentences. It also refers to speakers' knowledge of how to produce well-formed sentences, in which case it is an ability, part of speakers' competence, in their native language.

grammaticalisation An historical process consisting of a change in status from lexical to grammatical for certain elements, frequently due to loss of lexical meaning (semantic bleaching). For instance, the (archaic) adverb *whilom* 'formerly, erstwhile' derives from a dative plural of the Old English word *hwilom* 'at times,' which was with time not felt to be an inflected noun but a different word class, an adverb. It later moved on the pathway to Modern English *while*, both in the sense of 'during' and 'although.'

haptic communication Communication via touch; this is not an established modality for human languages, though touch can be used in unusual circumstances, such as when communicating with someone in locked-in state.

hierarchy In syntax, hierarchy refers to a vertical, non-temporal order of elements rather than a purely linear one, the latter reflecting the temporal order of events referred to in a sentence.

historical linguistics The study of how languages develop over time as opposed to viewing them at a single point in time. This was the major direction in linguistics up until the advent of structuralism initiated by Ferdinand de Saussure (1857–1916) at the beginning of the twentieth century. See *synchronic* and *diachronic*.

holophrase A sentence which consists of just one word. This is a common feature of early language acquisition when children utter a noun or verb and intend an entire sentence. The next stages are two and then multi-word sentences, which gradually lead to adult-like speech production.

iconic A term for the correspondence between linguistic form and meaning where the former is suggestive of the latter, as in *teeny* for very small, *thud* for a blunt fall. It is debatable whether iconicity played a role in the evolution of language.

inflectional language Any language which relies heavily on the use of inflections (grammatical endings) to indicate grammatical relationships, such as Latin or German.

innate A term meaning roughly 'given by birth,' referring to endowments which are part of our genetic make-up. The human language faculty is assumed to be innate, enabling children to acquire the language(s) in their surroundings in a very short time (the first few years of life) without instruction.

innateness hypothesis The view that children are born with a predisposition to acquire language. It contrasts with the view that knowledge of language is gained solely by experience (assumed by behaviourism, up until the late 1950s).

instinct A reference to features of behaviour not acquired but genetically encoded. Flight with birds or weaving a cobweb for spiders are examples, as these activities are not learned by observing others. The extent to which language is an instinct is debated among linguists.

internalisation (i) A process in which abstract knowledge of the language a child is acquiring, above all its syntax, is stored unconsciously in long-term memory. (ii) In evolutionary terms, a gradual process whereby externally determined language became increasingly abstract and encoded in the brain as internal knowledge, decoupled from its external source.

International Phonetic Alphabet A means of transcribing sounds of language, devised in the nineteenth century by French and English phoneticians. The principle is simple: each symbol stands for just one sound and so can never be ambiguous or misinterpreted; for example, /x/ stands for the differently written sound in German *Bach* <ch> 'stream,' Spanish *trabajo* <j> 'work,' Russian *dukh* <kh> (in Latin transliteration) 'spirit.' Abbreviated to IPA.

interrogative A sentence type which represents a request for information on the part of the speaker. The syntax of such sentences is usually different from declarative ones, for instance, in Italian by inverting subject and object, *Conosci Amelia?* 'Do you know Amelia?' and in English by the use of *do* as an auxiliary verb, *Do you like linguistics?*

intonation That part of the sound system of a language which involves the use of pitch to convey information, for example to denote interrogative structures.

intuition The feeling native speakers have about what structures are well-formed and acceptable in their language and what are not. The ability to intuit one's native language derives from unconscious language acquisition in early childhood together with innate universal grammar, which sets the envelope of possible variation in human languages.

iterated learning A type of learning, investigated empirically by the British cognitive scientist Simon Kirby, in which learners increasingly introduce structure and rules into data they are repeatedly exposed to. This is done by segmenting the data into recognisable parts and sections and allowing these to be recombined into new structures. Kirby and his associates maintain that this process ostensibly makes such data more learnable and that it holds for early language acquisition.

language acquisition The process by which children acquire knowledge about their native language in the first years of life in a rapid and unconscious manner. Acquisition is distinguished from learning, which refers to gaining explicit knowledge of a second language in later life (after puberty).

language change A process by which developments in a language are introduced and established. Language change is continual in every language and is largely regular. However, the rate of language change varies considerably across languages. It depends on a number of factors, not least on the amount of contact with other communities on the one hand (this tends to further change) and the degree of standardisation and universal education in a speech community on the other (this tends to inhibit change). Language change can also take place during language acquisition when children construct their internal grammars from external linguistic input.

language faculty A predisposition for acquiring language which all humans are born with and which enables children to acquire any language in a remarkably short period of time, despite fragmentary input from their surroundings. See *universal grammar*.

language of thought hypothesis The view, propounded by the American philosopher and cognitive scientist, Jerry Fodor (1935–2017), that thought has properties similar to language, in that the former combines concepts into hierarchical structures similar to those found in the latter. This view is close to that of Noam Chomsky, who maintains that the language of thought is primary and spoken (or signed) language is a secondary externalisation of the former.

language pathology A branch of linguistics which studies disorders of language: (i) with a view to gaining insights into the nature of human language in general; and (ii) with the goal of improving the rehabilitation chances for patients, as with persons who have suffered a stroke. See *aphasia*.

larynx A part of the human anatomy located in the neck and colloquially called the 'voice box' as it houses the vocal folds, where voice is generated by their vibrating in air escaping from the lungs.

larynx, descent of the A stage in the evolution of humans during which the larynx, with the vocal folds, was lowered in the throat, thus providing more resonance in the production of voice (this lowering takes place after birth with infants as part of their growth). A lowered larynx has been shown for many other animals and can be reached dynamically by certain animals who wish to exaggerate their prowess in order to impress or ward off others of their species.

lexicon The vocabulary of a language. It can refer to the book form of a dictionary (usually with an alphabetic listing of words) or the assumed lexicon which speakers possess mentally. The nature and organisation of this mental lexicon is radically different from a conventional dictionary.

linguistic determinism The view, propounded by Benjamin Lee Whorf (1897–1941), that language determines the way in which people think. Also termed the *linguistic relativity hypothesis.*

linguistics The study of language. As a scientific discipline built on objective principles, linguistics did not develop until the beginning of the nineteenth century, during which time the approach was largely historical. In the early twentieth century, linguistics came to focus on language structure at one point in time (structuralism). The mid-twentieth century saw the rise of a different approach – generative linguistics – which stressed unconscious knowledge and underlying grammatical structures to be found in all languages.

manner of articulation A reference to how a sound is produced. There are three basic types: (i) stops (or plosives), which have a complete blockage of the pulmonic airstream, e.g. /p, t, k/ as in *pan, tan* and *can* respectively; (ii) fricatives, in which there is a constriction producing turbulence in the air stream, as in /s, z, ʃ, ʒ/ in *sink, zinc, assure, azure,* respectively; and (iii) affricates, which consist of a stop and a fricative, e.g. /ʧ, ʤ/ *church* and *judge* respectively. In all these cases the active articulator is the tongue. See *place of articulation, voice.*

Merge The central operation in syntax in the model put forward by Noam Chomsky in the *The Minimalist Program* (1995). This operation takes two elements {a} and {b} and forms a set consisting of both {a, b}. This output can then serve as the input for another cycle of Merge in syntactic operations which are recursive in nature. Chomsky has maintained that all operations in syntax can be explained using this simple process and that it arose suddenly with a 'rewiring of the brain' with a single individual not more than 100,000 years ago. See *continuity vs. discontinuity hypothesis.*

metaphor The application of a word to another with which it is figuratively but not literally associated, such as *food for thought.*

modality A means by which language is externalised, that is, manifest outside the brain. Speech is by far the most common modality, but signing is a viable alternative, particularly for deaf individuals, and has its own grammar comparable to that of spoken language.

noun One of the major parts of speech, which refers to objects and also to concepts or ideas that are regarded as forming entities parallel to real-world objects, for instance by showing the property of countability, as with *feelings, ideas*.

onomatopoeia The use of human sounds to putatively imitate what they represent, such as animal sounds, as in *meeow* for a cat, *woof* for a dog. Onomatopoeia has not played any significant role in the development of vocabulary.

open class A term denoting a class which does not have a predetermined number of members, such as vocabulary. See *closed class*.

organs of speech Parts of the human anatomy around the mouth and throat which are used in speech production, such as the glottis (gap between the vocal folds), uvula, velum, palate, alveolar ridge, lips and the tongue. From an evolutionary viewpoint, organs of speech represent secondary adaptations and specialisations of organs which have some other primary function.

performance The actual production of language as opposed to structural knowledge of one's native language. See *competence*.

pharynx That section of the throat which lies immediately above the larynx.

phonaesthesia A phonetic phenomenon in which there is an apparent connection between sound sequences, such as /fl-/ in English, and specifiable semantic contents, such as 'fast flow of liquids; quick movement' (*flow, flush, flux; fling, flip*). Also termed sound symbolism.

phonation The vibration of the vocal folds during the production of a sound. By placing your thumb and index finger against your throat you can feel this vibration of the vocal folds.

phoneme In traditional phonology, the smallest unit in language which distinguishes meaning, such as /k/ and /g/, as in *coat* and *goat*. Each phoneme has one or more realisations, called allophones.

phonology The study of the sound system of one or more languages. Phonology involves the classification of sounds and a description of the interrelationship of these elements on a systemic level.

pitch A reference to the relative frequency used to produce vowels. A rise in pitch is a common means to realise prominence in a syllable, for instance with the first syllable of a word in Finnish.

place of articulation The point in the vocal tract at which a sound is produced. This can be anywhere from the lips at the front to the glottis (the gap between the vocal folds) at the back of the vocal tract. The most common place of articulation is the alveolar ridge just behind the upper teeth, using the tip of the tongue as with /s, z, t, d, n, l/. See *manner of articulation, voice.*

plural A category in all languages which refers to more than one object. Languages have a particular means for expressing this category, frequently by using a characteristic inflection (especially in synthetic languages).

poverty of linguistic input An assumption that children in their early years do not receive enough stimulus for them to acquire their native language fully without assuming a genetic head start in this process due to their inherited language faculty and universal grammar.

preposition A grammatical word which occurs in conjunction with a noun or phrase and which expresses the relation it has to other elements in a sentence, as in *on the roof, under the table.* In an analytic language like English, prepositions play a central role in the grammar, as in *Fiona explained the matter to Fergal.*

principles and parameters A model of generative linguistics, proposed by Noam Chomsky in the early 1980s, which assumes that everyone is born with an unconscious knowledge of what constitutes a basic language, in other words what essential *principles* it embodies. The term *parameters* alludes to those sections of language structure which receive special values (within a given range) from the particular language acquired by speakers in early childhood.

pronoun A grammatical element which refers to a noun previously mentioned; thus, it has a deictic function pointing backwards in discourse, as in *The lecturer was here and he spoke to us on a special topic.*

proposition A statement which can be assessed as being true or false, such as *The sun is shining,* which contains a proposition and in any given situation its truth value can be assessed.

prosody A term which refers to such properties of language as pitch, stress, loudness, tempo and rhythm.

Proto-World A postulated language assumed to be the first humans spoke and from which all others are derived. In general, historical linguists do not think that it is possible to reconstruct such a proto-language, given the great time depth involved and all the changes which later languages have undergone.

pulmonic A reference to the lungs as in 'pulmonic airstream'.

qualifier Any element which specifies some other element by rendering its meaning more specific, as in *a doubtful claim.*

quantifier Any term which serves to indicate an amount, such as *all, some, a few*, or the set of numerals in a language.

recursive The repeated application of a rule or process whereby the output of one cycle provides the input to the next.

S-curve A visual representation of change, which is characteristic of languages (and other systems). This kind of change begins slowly, has a quick middle section and a slow final phase.

semantics The study of meaning in language. This is an independent level and has several subtypes, such as word, grammatical, sentence and utterance meaning.

sense relations The semantic relationships between words as opposed to those which hold between words and concepts, which is termed *denotation*.

sentence The basic unit of syntax. A structural unit, which contains at least a subject and a verb, possibly with other complements, and which may occur with various clauses or which may be concatenated with other sentences. Sentences are used for making statements and expressing thoughts and would appear to be universal, i.e. common to all languages.

sign language A communication system in which people use their hands (and sometimes mouthing as well) to convey signals. Sign languages have come to be regarded as fully fledged systems, comparable to natural languages, mostly used by those individuals who are congenitally deaf and who learn sign language in childhood. There are many forms of sign language, some of which have arisen spontaneously in deaf communities.

singular A grammatical category which indicates a single occurrence of something. This is taken as the unmarked or default instance in language. The plural, and even more so the dual [two] or paucal [a few], are marked forms, usually with special inflections characterising them.

speech community Any identifiable and delimitable group of speakers who use a more or less unified type of language.

specific language impairment (SLI) A general term for language development disorders, which involve a spectrum of conditions, only some of which may be due to defective genetic input. The disorders generally involve production difficulties, late speech onset, difficulties in forming sentences or lexical issues, such as problems storing and retrieving words.

stochastic A reference to any process or change which is random and not the predetermined outcome of some initial, previous conditions. The appearance of the language faculty in humans is regarded by Noam Chomsky as stochastic.

structure A network of connections between elements of a system, for instance sentence structure is the set of relations which exist between parts of a sentence.

syllable An important structural unit in phonology. A syllable consists of a series of sounds which are grouped around a nucleus, usually a vowel. A closed syllable is one which has a consonant or consonants after the nucleus, an open syllable ends in a vowel, consider *got* /gɒt/ and *go* /gəʊ/ respectively.

synchronic A view of language at one particular point in time, usually the present.

syntax The possible combinations of words in a language to form sentences. A sentence consists minimally of a verb and a subject and maximally of a string of clauses, usually in a specific relationship to each other. Syntax is governed by rules of well-formedness, which specify which combinations are acceptable to native speakers and which are not. It is the task of a syntactic theory to determine these rules and relations.

synthetic A language which is characterised by an extensive inflectional morphology, such as Latin or Russian. Centuries of change, during which words fused together to give compound forms, can be seen as the trigger for this type arising. For this reason, young languages, like creoles (see Section 25.9), are not synthetic in type. See *analytic*.

token A reference to non-unique words in a text or stretch of speech, that is, to all words. For instance, in the sentence *The young girl spoke to the older girl*, there are eight tokens but only six types because the words *the* and *girl* occur twice.

tone language Any language in which variations in pitch have semantic significance, and are used to distinguish meanings. This principle is found in many languages of Africa and East and South-East Asia. In such languages, words may be segmentally the same but distinguished by tone only.

tongue The most frequently used active articulator in all languages. The tongue can be divided into these areas: the tip (Latin *apex*), blade (Latin *lamina*), back (Latin *dorsum*). The tongue may also show a groove, for instance with palato-alveolar fricatives such as /ʃ, ʒ/, as in *fission* and *fusion*, respectively. The tip can be made to roll in the escaping airstream, as is the case with the apical rolled /r/ of many Romance languages.

type (i) A term used in classifying languages according to their basic grammatical structure, see *typology*. (ii) A reference to a unique word in a text, see *token*.

typology The description of the grammatical structure of languages independently of genetic relationships. Languages which occupy a geographically delimited area, for instance the Balkans, may come to share structural properties, irrespective of historical background or genetic affiliation.

universal Any feature or property which holds for all languages. The number of these is limited though near-universals – those which hold true for the majority of languages – are more common, such as a distinction between masculine and feminine in the third person (not true of Finnish *hän* 'he/she', however).

universal grammar A term referring to the assumed common core of all languages, knowledge of which is genetically encoded and innate for every individual. Such knowledge provides constraints on what internal grammars children can construct and can account for the speed and proficiency of child language acquisition.

verb One of the two major lexical categories – the other is that of nouns – which is used to express a state or an action. The set of inflectional forms of a verb is termed a *conjugation* (parallel to *declension* with nouns). Verbs are usually distinguished for person and number along with tense and mood and frequently for aspect (the manner in which an action takes place) as well.

vocabulary The set of words in a language. These are usually grouped into word fields, like furniture, clothing, food, etc., so that the vocabulary can be said to show an internal structure. The term *lexicon* is also found and has two meanings (the words of a language and the mental storehouse of the words one knows).

vocal folds The two folds which are at the bottom of the throat and which open and close at great speed, producing the effect of 'voice', which is characteristic of nearly all vowels and many consonants. The vocal folds can come together in a single movement producing a glottal stop [ʔ], as for the <t> in a colloquial London pronunciation of *water*; they can be drawn close together – without closure – to articulate a glottal fricative [h], as in *hat*. The term *glottis* refers to the gap between the vocal folds.

voice The vibration of the vocal folds in the production of certain sounds: for example, English *bold* consists of four sounds all of which are voiced.

vowel A sound which is formed with voice (nearly always) and without any obstruction of the airstream in the oral cavity. Because such sounds have a high degree of sonority (sound volume) they tend to form the nucleus of syllables.

Wernicke, Carl (1848–1905) A German scientist who, in the second half of the nineteenth century (1874), discovered the area behind the left ear where language is processed as it is heard.

Wernicke's aphasia A type of aphasia in which understanding language is impaired because the region in question, Wernicke's area, has been damaged. See *Broca's aphasia, conduction aphasia*.

Wernicke's area A part of the brain which is taken to be responsible for the comprehension of language. It is located just above the left ear (see *Broca's area*).

whisper Speaking without the vibration of the vocal folds. In whispered speech all sounds – consonants and vowels – are voiceless.

word A morphological form which is internally stable, can stand on its own and which can in principle be moved to a new position in a sentence. Words can be internally complex, as with compounds consisting of a base and one or more endings or prefixes, such as *un-like-ly, un-think-able* or of two or more bases: *book-case, dance-hall, food-counter*.

word class A group of words which are similar in their grammatical characteristics: the kinds of inflections they take, their distribution in sentences and the relations they enter with other words. Typical word classes are nouns, verbs, adjectives, adverbs and prepositions.

word order The arrangement of words in a linear sequence in a sentence. There is normally an unmarked, or 'canonical', word order in a language – such as subject–verb–object (S–V–O) in English, V–S–O in Irish, Welsh, Tagalog, S–O–V in Turkish, Basque, Japanese – but usually alternative word orders exist, particularly to allow for emphasis in a sentence such as placing elements at the front to highlight them.

writing A means of representing language in a relatively permanent, non-spoken, non-signed form, by means of an alphabet or syllabary. Because of the importance of writing in many societies, it enjoys high status, which leads to non-linguists appealing to written language as the locus of 'correct' usage.

Timelines

The following timelines attempt to show key events leading to modern humans from the very beginning. The emphasis is on developments which were beneficial to our appearance and survival on Earth and might be shown one day to have parallels to those on exoplanets with exobeings. N.B.: The dates given are approximate.

History of the Universe and Our Solar System

Years in past	Events
13.8 bya	Sudden, enormous expansion, known colloquially as the Big Bang, from an infinitely dense and hot origin. Time and space comes into existence. 380,000 years later the universe has cooled enough for atoms to form, mostly hydrogen with some helium.
13.6–13.5 bya	Vast clouds of gas and dust show slightly irregular distributions due to quantum fluctuations at the very beginning, which are exaggerated during a period of exponential inflation (during a tiny fraction of first second after the 'bang'). The collapse of these clouds leads to the first galaxies, including the Milky Way, which probably later absorbed other galaxies with which it collided. It is one of about 100,000 galaxies in the Laniakea Supercluster (this also subsumes the Virgo Supercluster, see Figure 4.2).
5 bya	Our Solar System – the Sun, planets and asteroids – evolves from a huge disk of dust and gas (nebula), which had developed clumps and increased due to gravitational attraction (similar nebulae can still be observed within our galaxy). Nuclear fusion begins in the centre of the Sun, leading to it emitting light and heat. Our Sun is a third-generation star, which evolved from the gas and debris of previous stars that already contained heavy elements.

	Our Solar System is located in the Orion–Cygnus arm far away from the centre of our galaxy (over 26,000 light years). The Earth develops a magnetic field due to its outer core of molten iron continually moving and thus produces a dynamo effect. The magnetic field shields the Earth from harmful radiation from the Sun (solar wind) and from space (cosmic rays).
4.5 bya	The Moon forms due to the collision of the Earth with a planetary body (called Theia) about the size of Mars; this hurls huge quantities of material into space, which ends up orbiting Earth and clumping to form the Moon. Because of its size, about a quarter of the diameter of the Earth, and its position (about 384,000 kilometres from the Earth), the Moon has a stabilising effect on the rotation of the Earth.

History of Our Earth

Years in past	**Events**
4.55 bya	The Earth forms from rocks orbiting fairly near the Sun (like Mercury, Venus and Mars), which stuck together, exerting a gravitational pull that led to them accruing more material. The gas and ice giants (Jupiter, Saturn; Uranus and Neptune) develop and maybe migrate to their present positions. Planets also have moons which arise from local concentrations of material. In some cases, large amounts of water (already existent in the universe) are added to planets and moons and captured in comets, which orbit the Sun along long irregular pathways.
4 bya	After the period of Late Heavy Bombardment, the Earth is free of serious impacts from space. Prokaryotes (simple cell microorganisms) arise. The Earth is anaerobic (lacking oxygen) to begin with but this changes when vast quantities of cyanobacteria release substantial amounts of oxygen into the atmosphere.

2.5–2 bya	The first eukaryotes arise through the incorporation of one type of prokaryote into another. The resulting complex cells still have mitochondria, stemming from the incorporated prokaryotes with their own, now inactive DNA providing evidence of this process. Cells develop membraned nuclei and various organelles, different functional parts within a cell. Complex life forms begin to arise. Early forms might have developed near hydrothermal vents (such as black smokers) on the ocean floor.
1.2–1 bya	Sexual reproduction appears. Instead of simple cell divisions with identical results (mitosis), a system of meiosis develops. Here cells resulting from division (one to two to four) only contain half the chromosomes of the first cell. These cells are called gametes, two of which fuse together during fertilisation. As gametes come from different sources, one from each parent, a genetic mixture occurs which is typical of all sexually reproducing eukaryotes and ultimately of humans, where the large gamete is the egg of the mother and the smaller gamete is the sperm of the father.
1 bya	Forms of life evolve with a bone structure running through the centre of the body (vertebrates). These have a brain-like structure at one end where the mouth (for food intake) and major sensory organs are located. At the other end is an anus permitting waste to pass from the body. The brain controls the nervous system, which extends down the back (spinal cord) and branches out into the various limbs.
540 mya	Cambrian 'explosion': a sudden increase in life forms, with skeletons and hard bodies (shells), is apparent in the fossil record. Several unique life forms date to this period, above all those involving predation.
370 mya	Life forms move from the sea to the land. The source of these are lobe-finned fish, which had leg-like, jointed fins and primitive lungs. The latter probably developed to cope in aquatic situations with low oxygen, such as in shallow swamps. These fish are the ancestors of all tetrapods – four-limbed creatures. (Sea mammals resulted from a re-migration from land to the sea about 50 mya.)

252 mya	Permian–Triassic extinction event, known as the Great Dying, leads to the extinction of over 80 per cent of all marine life and about 70 per cent of all terrestrial life. This is the most severe of all extinction events; its cause is uncertain, but asteroid impact, supervolcanoes, sudden depletion of oxygen in the air and sea have all been put forward as possible reasons.
240 mya	The age of the dinosaurs begins. They come to dominate animal life across Earth.
66 mya	A major asteroid impact occurs near the Yucatán peninsula (present-day Mexico). The resulting ecological catastrophe (fires, tsunamis, blocking of sunlight for decades) eradicated the dinosaurs and nearly all other forms of life. Some animals survive, like reptiles and birds. The demise of the dinosaurs opens the way for other vertebrates, above all mammals, to evolve unhindered.

History of *Homo* Lineages

Years in past	**Events**
10 mya	Earliest apes evolve. Somewhat later gorillas and chimpanzees split.
6–7 mya	The tribe hominini split with the last common ancestor between chimpanzees and the genus *Australopithecus* 'southern ape' (precursors of *Homo* species).
4–1.5 mya	Habitual bipedality develops early on. Primitive tool making appears with *Homo habilis* (*c* 2.4–1.6 mya). *Homo erectus* (*c.* 2–0.12 mya) is the next major species which spreads throughout Africa and Eurasia even reaching east Asia as the remains of *Homo floresiensis* in Indonesia probably show. By about 0.5 mya *Homo heidelbergensis* had reached Europe and *Homo naledi* is documented for South Africa by about 0.3 mya. By about 60 kya Oceania and Australasia are settled by anatomically modern humans.

1.5–1 mya	Management of fire achieved, affording protection from predators and furthering social bonding. Cooking of food follows shortly afterwards, greatly improving energy intake in a short period of time. This is taken as furthering brain volume with an attendant reduction in gut size. The reduction in the size of the teeth (especially the molars) and the jaw muscles may be related to the increasing size of the brain case.
600–300 kya	First modern human beings (*Homo sapiens*) evolve in eastern and/or southern Africa. An exact dating of their appearance is not possible.
100–70 kya	Human languages among *Homo sapiens* have gained modern structure. This must have happened before the last dispersal, as the internal organisation of all human languages is the same today and so must derive from a common source (parallel independent development extremely unlikely).
70 kya	Last wave of dispersal of *Homo sapiens* out of Africa via the southern Arabian peninsula up into the Near East and on to Europe also across to South, East and South-East Asia. There was a limited amount of interbreeding with the following groups.
35–40 kya	Neanderthals (*Homo neanderthalensis*), who had emerged in Europe and Western Asia about the same time as *Homo sapiens*, become extinct as do other poorly documented species such as the Denisovans (central Asia). This leaves *Homo sapiens* as the only extant human species.

Human History

Years in past	**Events**
17 kya	Wall paintings were made in a cave complex at Lascaux (Dordogne, France) which are early examples of *Homo sapiens* art.

15 kya	*Homo sapiens* groups cross from Siberia through Beringia not later than 15 kya (probably earlier) and quickly spread down into North and South America.
12–11 kya	Göbekli Tepe, an archaeological site in Anatolia, Turkey, is early evidence of permanent settlement and buildings. Flourishing of Fertile Crescent – modern Iraq, southern Turkey and the Levant (present-day Syria, Lebanon, Israel).
11.7 kya	End of the last glacial period of the Quaternary period. Ice sheet recedes exposing the European land mass. Rise of agriculture with the domestication of plants and animals and gradual transition from hunter-gatherer groups to sedentary farming communities.
11–7 kya	Beginning of Neolithic period characterised by stone artefacts and implements found in archaeological sites.
10 kya (i.e. 8000 BCE)	Beginning of Yellow River (Huanghe) civilisation in China.
5000–4000 BCE	Indo-European groups, living in the steppe region north of the Caucasus, begin to disperse. The dialects of these groups gradually become branches of the later Indo-European family, which still later diversify into the individual languages of groups such as Indo-Aryan, Hittite, Tocharian, Hellenic (Greek), Italic (later Romance), Celtic, Germanic, Slavic, Baltic, most of which are attested to this day (but not Hittite or Tocharian).
3500 BCE	Cuneiform script was developed in the Near East and used first for Sumerian, spoken in Mesopotamia (present-day Iraq), and for the unrelated languages Akkadian and Hittite (*c* 3.7 kya).
3300 BCE	Beginning of Indus Valley civilisation.
3000–1000 BCE	Bronze Age, a designation for a period during which an alloy of copper and tin was widely used and traded throughout Europe and the Near East.
1200–500 BCE	Iron Age, initiated by the so-called Late Bronze Age collapse in the Eastern Mediterranean and the Near East (including the Nile valley), during which widespread incursions and invasions disrupted communities in this large area.

1000 BCE	The Phoenicians employed an alphabetic script (from an earlier form called Proto-Sinaitic); this was later adopted by the Greeks, which provided the impetus for later modern alphabets in Europe. In alphabets, letters stand for sounds, not for objects.
800 BCE	Classical antiquity begins with the rise of Greek city states.
500 BCE	Roman Republic is founded, continued as the Roman Empire from 27 BCE onwards.
400 BCE	Detailed records of astronomy observations begin to appear in China.
200 BCE	Paper invented in China.
100 BCE	Antikythera mechanism constructed in Greece.
O CE	Beginning of the Common Era (references from here on are to centuries).
3–9 CE	Classical Period of Mayan civilisation.
4 CE	The Roman Empire disintegrates into an Eastern and a Western half (395 CE). The Western Roman Empire continues to decline.
5–6 CE	Flourishing of Indian astronomy, which later influenced Chinese and Islamic astronomy.
5–15 CE	The Middle Ages in Europe follow, a period in which theology dominates most spheres of scientific investigation.
late 7 CE	Printing is invented in China.
8–13 CE	Islamic Golden Age, during which science flourishes.
14–15 CE	Aztec civilisation flourishes in areas of present-day Mexico.
14–16 CE	Inca civilisation flourishes in areas of present-day Peru.
14–17 CE	The Renaissance, a period of renewed interest in classical antiquity and of scientific activity, begins in Italy and spreads to all of Western Europe.
1610	Galileo builds the first telescope and discovers the larger moons of Jupiter.

1687	Isaac Newton publishes his *Philosophiæ Naturalis Principia Mathematica*, laying down the foundations of classical (deterministic) physics.
1799	Alessandro Volta develops the first battery (voltaic pile), thus harnessing electricity.
1831	Michael Faraday discovers electromagnetic induction and how to use it to generate electricity.
1865	James Clerk-Maxwell succeeds in unifying electricity and magnetism in a theory of electromagnetism.
1900	Max Planck introduces the notion of quantum, the smallest packet of energy, which is later empirically attested and forms the basis for quantum mechanics.
1905/1916	Albert Einstein publishes his theory of special and general relativity, respectively, showing that the speed of light was an absolute constant, that matter and energy were equivalent and that gravity was manifested as the curvature of space–time, an assumption empirically demonstrated some years later.

Abbreviations: bya = billion years ago; mya = million years ago; kya = thousand years ago; BCE before the common era; CE common era (references here are in centuries).

Figure credits

Figure 1.1 – drawn by the author.

Figure 4.1 – Source: NASA.

Figure 4.2 – Source: NASA.

Figure 8.1 – Source: NASA.

Figure 8.2 – Source: NASA.

Figure 13.1 – drawn by the author.

Figure 13.2 – Source: S. López, L. van Dorp and G. Hellenthal 2015. Human Dispersal Out of Africa: A Lasting Debate in *Evolutionary Bioinformatics Online*. Libertas Academica Ltd., pp. 57–68.

Figure 13.3 – Source: https://ungo.com.tr/2021/06/hayvanlar-neden-insana-benzer-bir-dile-sahip-degil/.

Figure 13.4 – Source: image from Christopher Walsh in J. Bradbury 2005. *Molecular Insights into Human Brain Evolution*. PLoS Biol 3, p. 50.

Figure 13.5 – Source: with permission of Kevin D. Hunt.

Figure 13.6 – Source: Vecteezy.com.

Figure 13.7 – Source: Zina Deretsky, National Science Foundation.

Figure 15.1 – Source: Vecteezy.com.

Figure 15.2 – Source: Bruce Blaus.

Figure 16.1 – Source: with permission of Bryan Kolb. Bryan Kolb and Bryan D. Fanitie 2008. Development of the Child's Brain and Behavior in Cecil R. Reynolds and Elaine Fletcher-Janzen (eds) *Handbook of Clinical Child Neuropsychology* Third edition. Heidelberg: Springer.

Figure 20.1 – drawn by the author.

Figure 20.2 – drawn by the author.

Figure 20.3 – drawn by the author.

Figure 20.4 – drawn by the author.

Figure 22.1 – drawn by the author.

Figure 22.2 – Source: Theresa Knott.

Figure 22.3 – Source: Shila Alcala.

Figure 22.4 – drawn by the author.

Figure 22.5 – drawn by the author.

Figure 24.1 – Source: https://web.archive.org/web/20070713113018/http://www.nia.nih.gov/Alzheimers/Publications/UnravelingTheMystery/Part1/NeuronsAndTheirJobs.htm.

Figure 24.2 – Source: https://open.oregonstate.education/aandp/chapter/14-3-the-brain-and-spinal-cord.

References

Barnes, Luke A. and Geraint F. Lewis 2020. *The Cosmic Revolutionary's Handbook (or: How to Beat the Big Bang)*. Cambridge: Cambridge University Press.

Barrat, James 2013. *Our Final Invention. Artificial Intelligence and the End of the Human Era*. New York: St Martin's Press.

Berlin, Brent and Paul Kay 1969. *Basic Color Terms: Their Universality and Evolution*. Berkeley: University of California Press.

Berwick, Robert C. and Noam Chomsky 2016. *Why Only Us? Language and Evolution*. Cambridge, MA: MIT Press.

Bickerton, Derek 2009. *Adam's Tongue. How Humans Made Language, How Language Made Humans*. New York: Hill and Wang.

Bloomfield, Leonard 1933. *Language*. New York: Henry Holt.

Bolhuis, Johan J., Ian Tattersall, Noam Chomsky and Robert C. Berwick 2014. 'How could language have evolved?' *PLOS Biology* 12:8: e1001934: https://doi.org/10.1371/journal.pbio.1001934

Bostrom, Nick 2014. *Superintelligence: Paths, Dangers, Strategies*. Oxford: Oxford University Press.

Bowling, Daniel L., Jacob C. Dunn, Jeroen B. Smaers, Maxime Garcia, Asha Sato, Georg Hantke, Stephan Handschuh, Sabine Dengg, Max Kerney, Andrew C. Kitchener, Michaela Gumpenbersger, W. Tecumseh Fitch 2020. 'Rapid evolution of the primate larynx?', *PLOS Biology*, August 11, 2020, https://doi.org/10.1371/journal.pbio.3000764.

Canales, Jimena 2015. *The Physicist and the Philosopher: Einstein, Bergson and the Debate That Changed Our Understanding of Time*. Princeton: Princeton University Press.

Chomsky, Noam 1991. Linguistics and cognitive science: Problems and mysteries. In Asa Kasher (ed.), *The Chomskyan Turn*. Oxford: Blackwell, pp. 26–53.

Chomsky, Noam 1980. *Rules and Representations*. New York: Columbia University Press.

Chomsky, Noam 1995. *The Minimalist Program*. Cambridge, MA: MIT Press.

Chomsky, Noam 2010. Some simple evo devo theses: how true might they be for language? In Richard K. Larson, *Viviane Déprez and Hiroko Yamakido The Evolution of Human Language: Biolinguistic Perspectives*. Cambridge: Cambridge University Press, pp. 45–62.

Corballis, Michael C. 2002. Did language evolve from manual gestures? In Alison Wray (ed.) *The Transition to Language*. Oxford: Oxford University Press, pp. 161–179.

Corballis, Michael C. 2011. *The Recursive Mind: The Origins of Human Language, Thought, and Civilization*. Princeton, NJ: Princeton University Press.

Darwin, Charles 1869 [1859]. *On the Origin of Species by Means of Natural Selection; Or, The Preservation of Favoured Races in the Struggle for Life*. Fifth Edition, with Additions and Corrections. London: John Murray.

Darwin, Charles 1871. *The Descent of Man, and Selection in Relation to Sex*. London: John Murray.

Dehaene, Stanislas 2011. *The Number Sense: How the Mind Creates Mathematics*. Second edition. Oxford: Oxford University Press.

Denning, Kathryn 2010. Unpacking the great transmission debate, *Acta Astronautica* 67: 1399–1405.

Deudney, Daniel 2020. *Dark Skies: Space Expansionism, Planetary Geopolitics, and the Ends of Humanity*. Oxford: Oxford: Oxford University Press.

Dunbar, Robin 2012. Gossip and the social origins of language. In Maggie Tallerman and Kathleen R. Gibson (eds) *The Oxford Handbook of Language Evolution*. Oxford: Oxford University Press, pp. 343–345.

Emery Nathan J. and Nicola S. Clayton 2001. Effects of experience and social context on prospective caching strategies by scrub jays, *Nature* 22.414: 443–446.

Fedorenko, Evelina and Rosemary Varley 2016. Language and thought are not the same thing: Evidence from neuroimaging and neurological patients, *Annals of the New York Academy of Sciences* 1369 1. 132–153.

Fellows Yates, James A., Irina M. Velsko et al. 2021. The evolution and changing ecology of the African hominid oral microbiome, *Proceedings of the National Academy of the Science of the USA*, May 18, 2021 118 (20) e2021655118 https://doi.org/10.1073/pnas.2021655118

Fisher, Ronald A. 1930. *Genetical Theory of Natural Selection*. Oxford: Clarendon Press.

Fitch, W. Tecumseh 2006. The biology and evolution of music: A comparative perspective, *Cognition* 100: 173–215.

Fitch, W. Tecumseh and David Reby 2001. The descended larynx is not uniquely human, *Proceedings of the Royal Society B* 268: 1669–1675.

Fitch, W. Tecumseh 2010. *The Evolution of Language*. Cambridge: Cambridge University Press.

Flajnik, Martin F. and Masanori Kasahara 2010. Origin and evolution of the adaptive immune system: genetic events and selective pressures, *Nature Reviews Genetics* 11(1): 47–59.

Frankish, Keith (ed.) 2017. *Illusionism as a Theory of Consciousness*. Special issue of *Journal of Consciousness Studies*. Exter, UK: Imprint Academic.

Friederici, Angela D. 2017 *Language in Our Brain. The Origins of a Uniquely Human Capacity*. Cambridge, MA: MIT Press.

Ginsburg, Simona and Eva Jablonka 2019. *The Evolution of the Sensitive Soul: Learning and the Origins of Consciousness*. Cambridge, MA: The MIT Press.

Ginsburg, Simona and Eva Jablonka 2022. *Picturing the Mind: Consciousness through the Lens of Evolution*. Cambridge, MA: The MIT Press.

Goldin-Meadow, Susan 2012. What modern-day gesture can tell us about language evolution. In Maggie Tallerman and Kathleen R. Gibson (eds) *The Oxford Handbook of Language Evolution*. Oxford: Oxford University Press, pp. 545–557.

Goodall, Jane 1990. *Through a Window: 30 years observing the Gombe chimpanzees*. London: Weidenfeld and Nicolson.

Gould Stephen Jay 1990. *Wonderful Life. The Burgess Shale and the Nature of History*. New York: W. W. Norton & Company.

Harari, Yuval Noah 2018. *21 Lessons for the 21st Century*. London: Jonathan Cape.

Harley, Heidi 2006. *English Words: A Linguistic Introduction*. Oxford: Blackwell.

Heller, René and John Armstrong 2014. Superhabitable worlds. *Astrobiology* 14.1. doi: 10.1089/ast.2013.1088.

Hickey, Raymond (ed.) 2020. *The Handbook of Language Contact*. Second edition. Malden, MA: Wiley-Blackwell.

Hickey, Raymond 2023. *Sounds of English Worldwide*. Malden, MA: Wiley-Blackwell.

Hockett, Charles F. 1960. The origin of speech, *Scientific American* 203(3): 88–97.

Hooke, Thomas 1665. *Micrographia or, Some physiological descriptions of minute bodies made by magnifying glasses*. London: J. Martyn and J. Allestry.

Hurford, James R. 2014. *The Origins of Language. A Slim Guide*. Oxford: Oxford University Press.

Innis Dagg, Anne 2009. *The Social Behavior of Older Animals*. Baltimore, MD: Johns Hopkins University Press.

Johansson, Sverker 2021. *The Dawn of Language. How we Came to Talk*. Translated from Swedish. London: MacLehose Press.

Kandel, Eric R. 2018. *The Disordered Mind. What Unusual Brains Tell Us About Ourselves*. New York: Farrar, Strauss and Giroux.

Kelly, David 2011. *Yuck! The Nature and Moral Significance of Disgust*. Cambridge, MA: The MIT Press.

Kirby, Simon, Thomas L. Griffiths and Kenny Smith 2014. Iterated learning and the evolution of language, *Current Opinion in Neurobiology* 28: 108–114.

Knight, Chris and Camilla Power 2012. Social conditions for the evolutionary emergence of language. In Maggie Tallerman and Kathleen R. Gibson (eds) *The Oxford Handbook of Language Evolution*. Oxford: Oxford University Press, pp. 346–349.

Kurzweil, Ray 2005. *The Singularity is Near*. New York: Penguin Group.

Lakoff, George and Mark Johnson 2003 [1980]. *Metaphors We Live By*. Chicago: University of Chicago Press.

Lenneberg, Eric H. 1967. *Biological Foundations of Language*. New York: Wiley.

Lewis, Geraint F. and Luke A. Barnes 2016. *A Fortunate Universe. Life in a Fine-Tuned Cosmos*. Cambridge: Cambridge University Press.

Marino Lori, Daniel W. McShea and Mark D. Uhen 2004. Origin and evolution of large brains in toothed whales, *The Anatomical Record. Part A, Discoveries in Molecular, Cellular, and Evolutionary Biology*, 281(2):1247–1255.

Matsuzawa, Tetsuro 2010. Cognitive development in chimpanzees: A trade-off between memory and abstraction? In Denis Mareschal, Paul C. Quinn and Stephen E. G. Lea (eds) *The Making of Human Concepts*. Oxford: Oxford University Press, pp. 227–244.

McGinn, Colin 2015. *Prehension: The Hand and the Emergence of Humanity*. Cambridge, MA: MIT Press.

McMahon, April and Robert McMahon 2012. *Evolutionary Linguistics*. Cambridge: Cambridge University Press.

Meyer, Matthias, Juan-Luis Arsuaga, Cesare de Filippo … and Svante Pääbo 2016. Nuclear DNA sequences from the Middle Pleistocene Sima de los Huesos hominins, *Nature* 531: 504–507.

Michaud, Michael 2007. *Contact with Alien Civilizations. Our Hopes and Fears about Encountering Extraterrestrials*. New York: Springer.

Mithen, Steven 2012. Musicality and language. In Maggie Tallerman and Kathleen R. Gibson (eds) *The Oxford Handbook of Language Evolution*. Oxford: Oxford University Press, pp. 296–298.

Montgomery, Stephen H., Jonathan H. Geisler, Michael R. McGowen, Charlotte Fox, Lori Marino and John Gatesy 2013. The evolutionary history of cetacean brain and body size, *Evolution*. 67(11): 3339–3353.

Mu, Liancai, and Ira Sanders 2010. Human tongue neuroanatomy: Nerve supply and motor endplates, *Clinical Anatomy* 23(7): 771–791.

Murphy, Elliot 2020. *The Oscillatory Nature of Language*. Cambridge: Cambridge University Press.

Ni, Xijun et al. 2021. Massive cranium from Harbin in northeastern China establishes a new Middle Pleistecene human lineage. *The Innovation* 2. https://doi.org/10.1016/j.xinn.2021.100130.

Pinker, Steven 1994. *The Language Instinct. The New Science of Language and Mind*. London: Allen Lane.

Pinker, Steven 2013. *Language, Cognition and Human Nature. Selected Articles*. Oxford: Oxford University Press.

Pinto, Yair, David A. Neville, Marte Otten et al. 2017. 'Split brain: divided perception but undivided consciousness,' *Brain* 1;140(5):1231–1237.

Plag, Ingo 2018. *Word-Formation in English*. Second edition. Cambridge: Cambridge University Press.

Powell, Lauren E. Robert A. Barton and Sally E. Street 2019. Maternal investment, life histories and the evolution of brain structure in primates, *Proceedings of the Royal Society B. Biological Sciences* B.28620191608.

Raup, David M. and J. John Jr. Sepkoski 1982. Mass extinctions in the marine fossil record, *Science* 215.4539: 1501–1503

Raup, David M. and J. John Jr. Sepkoski 1984. Periodicity of extinctions in the geologic past, *Proceedings of the National Academy of Sciences of the United States of America* 81.3: 801–805.

Sapir, Edward 2004 [1921]. *Language. An Introduction to the Study of Speech*. New York: Dover Publications.

Saussure, Ferdinand de 1959 [1916]. *A Course in General Linguistics*. Ed. Charles Bally and Albert Schechaye. Trans. by Wade Baskin. London: Peter Owen.

Schmitt, Ian, Ralf D. Wimmer, Miho Nakajima et al. 2017. Thalamic amplification of cortical connectivity sustains attentional control, *Nature*. May 11; 545(7653): 219–223. doi: 10.1038/nature22073.

Schulze-Makuch, Dirk and William Bains 2017. *The Cosmic Zoo. Complex Life on Many Worlds*. Berlin-Heidelberg: Springer.

Schulze-Makuch, Dirk, Abel Méndez, Alberto G. Fairén, Philip von Paris, Carol Turse, Grayson Boyer, Alfonso F. Davila, Marina Resendes de Sousa António, David Catling and Louis N. Irwin 2011. 'A two-tiered approach to assess the habitability of exoplanets,' *Astrobiology* 11.10: 1041–1052.

Shelley, Mary 2018 [1818]. *Frankenstein*. The 1818 Text. London: Penguin.

Smith, Kenny, Simon Kirby and Henry Brighton 2003. Iterated learning: a framework for the emergence of language, *Artificial Life* 9(4): 371–386.

Tennie, Claudio, Josep Call and Michael Tomasello 2009. Ratcheting up the ratchet: On the evolution of cumulative culture, *Philosophical Transactions of The Royal Society B Biological Sciences* 364(1528): 2405–2415.

Tremblay, Marc and Hélène Vézina 2000. New estimation of intergenerational time intervals for the calculation of age and origin of mutations, *American Journal of Human Genetics* 66: 651–658.

Trombetti, Alfredo 1905. *L'unità d'origine del linguaggio*. [The unified origin of language] Bologna: Luigi Beltrami.

Wilson, Matthew A. and James M. Bower 1992. Cortical oscillations and temporal interactions in a computer simulation of piriform cortex, *Journal of Neurophysiology*. doi.org/10.1152/jn.1992.67.4.981.

Wragg Sykes, Rebecca 2020. *Kindred: Neanderthal Life, Love, Death and Art*. London: Bloomsbury.

Wrangham, Richard W. 2009. *Catching Fire: How Cooking Made Us Human*. New York: Basic Books.

Bibliography

General

General Science Books

Armstrong, J. Scott and Kesten C. Green 2022. *The Scientific Method. A Guide to Finding Useful Knowledge*. Cambridge: Cambridge University Press.

Atkins, Peter 2003. *Galileo's Finger. The Ten Great Ideas of Science*. Oxford: Oxford University Press.

Canales, Jimena 2015. *The Physicist and the Philosopher: Einstein, Bergson and the Debate That Changed Our Understanding of Time*. Princeton, NJ: Princeton University Press.

Clegg, Brian 2013. *Dice World. Science and Life in a Random Universe*. London: Icon Books.

Davies, Paul 2006. *The Goldilocks Enigma. Why is the Universe Just Right for Life?* London: Penguin.

Gribbin, John 2002. *Science. A History 1543–2001*. London: Penguin.

Gribbin, John 2004. *Deep Simplicity. Chaos, Complexity and the Emergence of Life*. London: Penguin.

Gribbin, John 2011. *Alone in the Universe: Why Our Planet Is Unique*. Malden, MA: Wiley.

Gribbin, John and Mary Gribbin 2020. *On the Origin of Evolution. Tracing 'Darwin's Dangerous Idea' from Aristotle to DNA*. London: William Collins.

Marletto, Chiara 2021. *The Science of Can and Can't. A Physicist's Journey Through the Land of Counterfactuals*. London: Penguin.

McAllister, James W. 1996. *Beauty and Revolution in Science*. Ithaca, NA: Cornell University Press.

Schrödinger, Erwin 2012. *What is Life? With Mind and Matter and Autobiographical Sketches*. Cambridge: Cambridge University Press.

Wilczek, Frank 2021. *Fundamentals. Ten Keys to Reality*. New York: Penguin.

Evolutionary Biology and Genetics

Boyd, Robert 2003. *Evolution. The History of an Idea*. Berkeley, CA: University of California Press.

Brusatte, Steve 2018. *The Rise and Fall of the Dinosaurs. The Untold Story of a Lost World*. London: Macmillan.

Carroll, Sean B. 2020. *A Series of Fortunate Events: Chance and the Making of the Planet, Life, and You*. Princeton, NJ: Princeton University Press.

Coen, Enrico 2012. *Cells to Civilizations: The Principles of Change That Shape Life*. Princeton, NJ: Princeton University Press.

Darwin, Charles 1869. *On the Origin of Species by Means of Natural Selection; Or, The Preservation of Favoured Races in the Struggle for Life*. Fifth Edition, with Additions and Corrections. London: John Murray.

Dawkins, Richard 1986. *The Blind Watchmaker*. London: Longman.

Dawkins, Richard 2016. *The Selfish Gene*. 40th anniversary edition. Oxford: Oxford University Press.

Fisher, Ronald A. 1930. *Genetical Theory of Natural Selection*. Oxford: Clarendon Press.

Fitch, W. Tecumseh and David Reby 2001. The descended larynx is not uniquely human. *Proceedings of the Royal Society* B 268: 1669–1675.

Gould, Stephen Jay 1990. *Wonderful Life. The Burgess Shale and the Nature of History*. New York: W. W. Norton & Company.

Gould, Stephen Jay 2002. *The Structure of Evolutionary Theory*. Cambridge, MA: Harvard University Press.

Harrison, D. F. N. 1995. *The Anatomy and Physiology of the Mammalian Larynx*. Cambridge: Cambridge University Press.

Hazen, Robert M. 2019. *Symphony in C. Carbon and the Evolution of (Almost) Everything*. New York: W. W. Norton & Company.

Hopcroft, Rosemary 2018. *Oxford Handbook of Evolution, Biology, and Society*. Oxford: Oxford University Press.

Huxley, Julian 1942. *Evolution. The Modern Synthesis*. London: Allen & Unwin.

Kemp, T. S. 2005. *The Origin and Evolution of Mammals*. Oxford: Oxford University Press.

Jones, Stephen, Robert D. Martin, David R. Pilbeam, Sarah Bunney and Richard Dawkins (eds) 1994. *The Cambridge Encyclopedia of Human Evolution*. Cambridge Cambridge University Press.

Lane, Nick 2010. *Life Ascending. The Ten Great Inventions of Evolution*. London: Profile Books.

Lane, Nick 2015. *The Vital Question. Why Is life the Way It Is?* London: Profile Books.

Raup, David M. and J. John Sepkoski Jr 1982. Mass extinctions in thc marine fossil record. *Science* 215(4539): 1501–1503.

Raup, David M. and J. John Sepkoski Jr 1984. Periodicity of extinctions in the geologic past. *Proceedings of the National Academy of Sciences of the United States of America* 81(3): 801–805.

Ridley, Matt 2003. *Nature versus Nurture. Genes, Experience and What Makes Us Human*. London: Fourth Estate.

Roth, Gerhard 2013. *The Long Evolution of Brains and Minds*. New York: Springer.

Shermer, Michael 2012. *The Believing Brain: From Spiritual Faiths to Political Convictions. How We Construct Beliefs and Reinforce Them as Truths*. London: Robinson.

Shubin, Neil 2013. *The Universe Within: Discovering the Common History of Rocks, Planets, and People*. New York: Pantheon Books.

Tattersall, Ian 1999. *Becoming Human. Evolution and Human Uniqueness*. New York: Harcourt-Brace.

Tattersall, Ian 2012. *Masters of the Planet: The Search for Our Human Origins*. New York: St Martin's Press.

Wallace, Alfred Russel 1904. *Man's Place in the Universe. A Study of the Results of Scientific Research in Relation to the Unity or Plurality of Worlds*. London: Chapman and Hall.

Wallace, Alfred Russel 1905. *Darwinism: An Exposition of the Theory of Natural Selection with Some of Its Applications*. New York: Macmillan.

Zalasiewicz, Jan and Mark Williams 2012. *The Goldilocks Planet. The 4 Billion Year Story of Earth's Climate*. Oxford: Oxford University Press.

Zalasiewicz, Jan, Colin N. Waters, Mark Williams and Colin Summerhayes (eds) 2019. *The Anthropocene as a Geological Time Unit: A Guide to the Scientific Evidence and Current Debate*. Cambridge: Cambridge University Press.

Neuroscience

Bolhuis, Johan J. (ed.) 2000. *Brain, Perception, Memory. Advances in Cognitive Neuroscience*. Oxford: Oxford University Press.

Brown, Colin M. and Peter Hagoort (eds) 1999. *The Neurocognition of Language*. Oxford: Oxford University Press.

Cobb, Matthew 2020. *The Idea of the Brain. A History*. London: Profile Books.

Deacon, Terrence W. 2011. *Incomplete Nature: How Mind Emerged from Matter*. New York: W. W. Norton & Company.

Eagleman, David 2015. *The Brain. The Story of You*. Edinburgh: Canongate.

Eagleman, David 2020. *Livewired: The Inside Story of the Ever-Changing Brain*. New York: Pantheon Books.

Greenfield, Susan 2000. *The Private Life of the Brain*. London: Penguin.

Greenfield, Susan 2016. *A Day in the Life of the Brain: The Neuroscience of Consciousness from Dawn till Dusk*. London: Allen Lane.

Kandel, Eric R. 2018. *The Disordered Mind. What Unusual Brains Tell Us About Ourselves*. New York: Farrar, Strauss and Giroux.

Op de Beeck, Hans and Chie Nakatani 2019. *Introduction to Human Neuroimaging*. Cambridge: Cambridge University Press.

Striedter, Georg F. and R. Glenn Northcutt 2019. *Brains Through Time: A Natural History of Vertebrates*. Oxford: Oxford University Press.

Tononi, Giulio 2012. *Phi: A Voyage from the Brain to the Soul*. New York: Pantheon Books.

Human Cognition and Consciousness

Baars, Bernard 1997. *In the Theater of Consciousness: The Workspace of the Mind*. New York: Oxford University Press.

Baars, Bernard and Nicole M. Gage 2010. *Cognition, Brain, and Consciousness. Introduction to Cognitive Neuroscience*. Second edition. San Diego/London: Academic Press.

Baars, Bernard and Nicole M. Gage 2012. *Fundamentals of Cognitive Neuroscience: A Beginner's Guide*. San Diego/London: Academic Press.

Baggini, Julian 2012. *The Ego Trick*. London: Granta Books.

Barbey, Aron K., Sherif Karama and Richard J. Haier (eds) 2021. *The Cambridge Handbook of Intelligence and Cognitive Neuroscience*. Cambridge: Cambridge University Press.

Barrett, Louise 2015. *Beyond the Brain. How Body and Environment Shape Animal and Human Minds*. Princeton, NJ: Princeton University Press.

Birch, Jonathan, Simona Ginsburg and Eva Jablonka 2020. Unlimited associative learning and the origins of consciousness: a primer and some predictions. *Biology and Philosophy* 35: 56. doi.org/10.1007/s10539-020-09772-0.

Blackmore, Susan J. 2018. *Consciousness: A Very Short Introduction*. Second edition. Oxford: Oxford University Press.

Boden, Margaret 2008. *Mind as Machine. A History of Cognitive Science*. Oxford: Oxford University Press.

Chalmers, David 2010. *The Character of Consciousness*. Oxford: Oxford University Press.

Damasio, Antonio 2012. *Self Comes to Mind: Constructing the Conscious Brain*. New York: Vintage Books.

Damasio, Antonio 2018. *The Strange Order of Things. Life, Feelings and the Making of Cultures*. New York: Pantheon Books.

Dehaene, Stanislas 2014. *Consciousness and the Brain. Deciphering How the Brain Codes Our Thoughts*. New York: Viking Penguin.

Dehaene, Stanislas 2014. *Reading in the Brain: The Science and Evolution of a Human Invention*. New York: Viking Penguin.

Dehaene, Stanislas 2020. *How We Learn. The New Science of Education and the Brain*. London: Penguin Books.

Dennett, Daniel C. 2005. *Sweet Dreams: Philosophical Obstacles to a Science of Consciousness*. Cambridge, MA: MIT Press.

Dennett, Daniel C. 2017. *From Bacteria to Bach and Back: The Evolution of Minds*. London: Penguin.

Dowling, John E. 2018. *Understanding the Brain. From Cells to Behavior to Cognition*. New York: W. W. Norton & Company.

Edelman, Gerald M. 2004. *Wider Than the Sky: The Phenomenal Gift of Consciousness*. New Haven, CT: Yale University Press.

Feinberg, Todd E. and Jon M. Mallatt 2016. *The Ancient Origins of Consciousness: How the Brain Created Experience*. Cambridge, MA: MIT Press.

Feldman Barrett, Lisa 2017. *How Emotions are Made. The Secret Life of the Brain*. New York: Houghton Mifflin.

Frankish, Keith (ed.) 2017. *Illusionism as a Theory of Consciousness*. Special issue of *Journal of Consciousness Studies*. Exeter: Imprint Academic.

Frankish, Keith and William Ramsey (eds) 2012. *The Cambridge Handbook of Cognitive Science*. Cambridge: Cambridge University Press.

Fuentes, Augustín 2017. *The Creative Spark: How Imagination Made Humans Exceptional*. New York: Dutton.

Gazzaniga, Michael S. 2018. *The Consciousness Instinct: Unraveling the Mystery of How the Brain Makes the Mind*. New York: Farrar, Straus and Giroux.

Geary, David C. 2005. *The Origin of Mind. Evolution of Brain, Cognition, and General Intelligence*. Washington, DC: American Psychological Association.

Gennaro, Rocco J. (ed.) 2018. *The Routledge Handbook of Consciousness*. London: Routledge.

Ginsburg, Simona and Eva Jablonka 2019. *The Evolution of the Sensitive Soul: Learning and the Origins of Consciousness*. Cambridge, MA: MIT Press.

Ginsburg, Simona and Eva Jablonka 2022. *Picturing the Mind: Consciousness through the Lens of Evolution*. Cambridge, MA: MIT Press.

Gregory, Richard L. 2004. *The Oxford Companion to the Mind*. Second edition. Oxford: Oxford University Press.

Hameroff, Stuart and Steven Penrose 1996. Orchestrated reduction of quantum coherence in brain microtubules: a model for consciousness?, in Stuart Hameroff, Alfred W. Kaszniak and A. C. Scott (eds) *Toward a Science of Consciousness: The First Tucson Discussions and Debates*. Cambridge, MA: MIT Press, pp. 507–540.

Humphreys, Nicholas 2006. *Seeing Red: A Study in Consciousness*. Cambridge, MA: Harvard University Press.

Koch, Christof 2019. *The Feeling of Life Itself. Why Consciousness Is Widespread but Can't Be Computed*. Cambridge, MA: MIT Press.

Kriegel, Uriah (ed.) 2020. *The Oxford Handbook of the Philosophy of Consciousness*. Oxford: Oxford University Press.

LeDoux, Joseph 1996. *The Emotional Brain. The Mysterious Underpinnings of Emotional Life*. New York: Simon and Schuster.

Marcus, Gary 2003. *The Birth of the Mind. How a Tiny Number of Genes Creates the Complexities of Human Thought*. New York: Basic Books.

McGinn, Colin 1991. *The Problem of Consciousness*. Oxford: Blackwell.

McGinn, Colin 2015. *Inborn Knowledge. The Mystery Within*. Cambridge, MA: Harvard University Press.

Mercier, Hugo and Dan Sperber 2017. *The Enigma of Reason. A New Theory of Human Understanding*. London: Penguin.

Mithen, Steven 1996. *The Prehistory of the Mind. A Search for the Origins of Art, Religion and Science*. London: Thames and Hudson.

Miyake, Akira and Priti Shah (eds) 1999. *Models of Working Memory: Mechanisms of Active Maintenance and Executive Control*. Cambridge: Cambridge University Press.

Northoff, Georg 2014. *Unlocking the Brain*. Oxford: Oxford University Press.

Pereira, Alfredo, Jr and Dietrich Lehmann (eds) 2015. *The Unity of Mind, Brain and World Current Perspectives on a Science of Consciousness*. Cambridge: Cambridge University Press.

Redish, A. David 2013. *The Mind within the Brain*. Oxford: Oxford University Press.

Searle, John R. 1997. *The Mystery of Consciousness*. New York: New York Review of Books.

Tomasello, Michael 2014. *A Natural History of Human Thinking*. Cambridge, MA: Harvard University Press.

Tomasello, Michael 2019. *Becoming Human: A Theory of Ontogeny*. Cambridge, MA: Harvard University Press.

Zelazo, Philip David, Morris Moscovitch and Evan Thompson (eds) 2007. *The Cambridge Handbook of Consciousness*. Cambridge: Cambridge University Press.

Human Anatomy and Sexuality

Darwin, Charles 1871. *The Descent of Man, and Selection in Relation to Sex*. London: John Murray.

Darwin, Charles 1872. *The Expression of the Emotions in Man and Animals*. London: John Murray.

Gazzaniga, Michael S. 1992. *Nature's Mind. The Biological Roots of Thinking, Emotions, Sexuality, Language and Intelligence*. New York: Basic Books.

Lieberman, Daniel E. 2011. *The Evolution of the Human Head*. Cambridge, MA: Harvard University Press.

Lieberman, Daniel E. 2013. *The Story of the Human Body*. London: Penguin Books.

Miller, Geoffrey 2001. *The Mating Mind: How Sexual Choice Shaped the Evolution of Human Nature*. London: Anchor Books.

Symons, Donald 1979. *The Evolution of Human Sexuality*. Oxford: Oxford University Press.

Wind, Jan 1970. *On the Phylogeny and Ontogeny of the Human Larynx*. Groningen: Wolters-Noordhoff Publishing.

Evolution of Human Behaviour and Culture

Barrat, James 2013. *Our Final Invention. Artificial Intelligence and the End of the Human Era*. New York: St Martin's Press.

Bostrom, Nick and Milan M. Ćirković (eds) 2011. *Global Catastrophic Risks*. Oxford: Oxford University Press.

Christian, Brian 2011. *The Most Human Human: What Artificial Intelligence Teaches Us About Being Alive*. New York: Doubleday.

Christian, Brian and Tom Griffiths 2016. *Algorithms to Live By: The Computer Science of Human Decisions*. New York: Henry Holt and Co.

Dartnell, Lewis 2019. *Origins: How the Earth Made Us*. New York: Penguin Random House.

Diamond, Jared 1992. *The Third Chimpanzee. The Evolution and Future of the Human Animal*. New York: HarperCollins.

Dunbar, Robin, Chris Knight and Camilla Power (eds) 1999. *The Evolution of Culture*. Edinburgh: Edinburgh University Press.

Fuentes, Augustín 2008. *Evolution of Human Behavior*. Oxford: Oxford University Press.

Harari, Yuval Noah 2018. *21 Lessons for the 21st Century*. London: Jonathan Cape.

Hebb, Donald 1949. *The Organization of Behavior*. New York: John Wiley.

Jablonka, Eva and Marion J. Lamb 2014. *Evolution in Four Dimensions. Genetic, Epigenetic, Behavioral, and Symbolic Variation in the History of Life*. Revised edition. Cambridge, MA: MIT Press.

Laland, Kevin 2013. *Social Learning: An Introduction to Mechanisms, Methods, and Models*. Princeton, NJ: Princeton University Press.

Laland, Kevin 2017. *Darwin's Unfinished Symphony: How Culture Made the Human Mind*. Princeton, NJ: Princeton University Press.

Levinson, Stephen C. and Pierre Jaisson 2006. *Evolution and Culture*. Cambridge, MA: MIT Press.

Levinson, Stephen C. and N. J. Enfield (eds) 2006. *Roots of Human Sociality. Culture, Cognition and Interaction*. London: Bloomsbury.

McNeill, David 2015. *Why We Gesture. The Surprising Role of Hand Movements in Communication*. Cambridge: Cambridge University Press.

Ord, Toby 2020. *The Precipice. Existential Risk and the Future of Humanity*. London: Bloomsbury Publishing.

Rees, Martin 2003. *Our Final Century. Will Civilisation Survive the Twenty-First Century?* London: Arrow Books.

Rees, Martin 2018. *On the Future. Prospects for Humanity*. Princeton, NJ: Princeton University Press.

Székely, Tamás, Allen J. Moore and Jan Komdeur (eds) 2010. *Social Behaviour. Genes, Ecology and Evolution*. Cambridge: Cambridge University Press.

Taylor, Shelley E. 2003. *The Tending Instinct. Women, Men, and the Biology of Relationships*. New York Henry Holt & Co.

Tennie, Claudio, Josep Call and Michael Tomasello 2009. Ratcheting up the ratchet: On the evolution of cumulative culture, *Philosophical Transactions of The Royal Society B Biological Sciences* 364(1528): 2405–2415.

Workman, Lance, Will Reader and Jerome H. Barkow 2020. *The Cambridge Handbook of Evolutionary Perspectives on Human Behavior*. Cambridge: Cambridge University Press.

Wrangham, Richard W. 2009. *Catching Fire: How Cooking Made Us Human*. New York: Basic Books.

Wrangham, Richard W. 2019. *The Goodness Paradox. How Evolution Made Us More and Less Violent*. London: Profile Books.

Palaeoanthropology

Aiello, Leslie C. and Christopher Dean 1990. *An Introduction to Human Evolutionary Anatomy*. London: Academic Press.

Berger, Lee and John Hawks 2017. *Almost Human: The Astonishing Tale of Homo Naledi and the Discovery That Changed Our Human Story*. Washington: National Geographic.

Cameron, David W. 2004. *Hominid Adaptations and Extinctions*. Sydney: University of New South Wales Press.

Compton, John S. 2016. *Human Origins. How Diet, Climate and Landscape Shaped Us*. Cape Town: Earthspun Books.

De Waal, Frans 2005. *Our Inner Ape*. New York: Riverhead Books.

Finlayson, Clive 2019. *The Smart Neanderthal. Bird Catching, Cave Art, and the Cognitive Revolution*. Oxford: Oxford University Press.

Henke, Winfried and Ian Tattersall (eds) 2007. *Handbook of Paleoanthropology*. Springer: Berlin/Heidelberg.

Reich, David 2018. *Who We Are and How We Got Here. Ancient DNA and the New Science of the Human Past*. Oxford: Oxford University Press.

Stringer, Chris and Robin McKie 1997. *African Exodus. The Origins of Modern Humanity*. New York: Henry Holt.

Stringer, Chris 2011. *The Origin of Our Species*. London: Allen Lane.

Neanderthals and Denisovans

Pääbo, Svante 2014. *Neanderthal Man: In Search of Lost Genomes*. New York: Basic Books.

Wragg Sykes, Rebecca, 2020. *Kindred: Neanderthal Life, Love, Death and Art*. London: Bloomsbury.

Astronomy, Astrophysics and Astrobiology

Ball, Philip 2001. *Stories of the Invisible. A Guided Tour of Molecules*. Oxford: Oxford University Press.

Barbour, Julian 1999. *The End of Time. The Next Revolution in Our Understanding of the Universe*. Oxford: Oxford University Press.

Barnes, Luke A. and Geraint F. Lewis 2020. *The Cosmic Revolutionary's Handbook (or: How to Beat the Big Bang)*. Cambridge: Cambridge University Press.

Bertone, Gianfranco 2010. *Particle Dark Matter: Observations, Models and Searches*. Cambridge: Cambridge University Press.

Billings, Lee 2013. *Five Billion Years of Solitude. The Search for Life Among the Stars*. New York: Penguin.

Carroll, Sean M. 2016. *The Big Picture. On the Origins of Life, Meaning and the Universe Itself*. New York: Dutton.

Cockell, Charles S. 2015. *Astrobiology: Understanding Life in the Universe*. Malden, MA: Wiley-Blackwell.

Conrad, Pamela Gales 2016. *Planetary Habitability*. Cambridge: Cambridge University Press.

Conway Morris, Simon 2004. *Life's Solution: Inevitable Humans in a Lonely Universe*. Cambridge: Cambridge University Press.

Deutsch, David 1998. *The Fabric of Reality: The Science of Parallel Universes and Its Implications*. London: Penguin.

Gargaud, Muriel (ed.) 2011. *Encyclopedia of Astrobiology*. Heidelberg/New York: Springer.

Greene, Brian 2004. *The Fabric of the Cosmos. Space, Time and the Texture of Reality*. London: Penguin.

Greene, Brian 2012. *The Hidden Reality: Parallel Universes and the Deep Laws of the Cosmos*. London: Penguin.

Hawking, Stephen 1988. *A Brief History of Time. From the Big Bang to Black Holes*. London: Bantham.

Hawking, Stephen and Leonard Mlodinow 2010. *The Grand Design. New Answers to the Ultimate Questions of Life*. London: Bantham Books.

Kolb, Vera M. 2019. *Handbook of Astrobiology*. London: Routledge.

Krauss, Lawrence M. 2001. *Atom. A Single Oxygen Atom's Odyssey from the Big Bang to Life on Earth ... and Beyond*. Boston: Little, Brown and Company.

Krauss, Lawrence M. 2012. *A Universe from Nothing: Why There is Something Rather Than Nothing*. New York: Simon and Schuster.

Lewis, Geraint F. and Luke A. Barnes 2016. *A Fortunate Universe. Life in a Fine-Tuned Cosmos*. Cambridge: Cambridge University Press.

Liddle, Andrew 2015. *An Introduction to Modern Cosmology*. Third edition. Malden, MA: Wiley-Blackwell.

Penrose, Roger 2011. *Cycles of Time: An Extraordinary New View of the Universe*. London: Vintage Books.

Scharf, Caleb A. 2014. *The Copernicus Complex: Our Cosmic Significance in a Universe of Planets and Probabilities*. New York: Farrar, Straus and Giroux.

Scharf, Caleb A. 2021. *The Ascent of Information: Books, Bits, Genes, Machines, and Life's Unending Algorithm*. New York: Penguin Random House.

Tegmark, Max 2015. *Our Mathematical Universe: My Quest for the Ultimate Nature of Reality*. New York: Vintage Books.

Life Beyond Earth

Al-Khalili, Jim (ed.) 2017. *Aliens: The World's Leading Scientists on the Search for Extraterrestrial Life*. New York: Picador.

Barnes, Rory (ed.) 2010. *Formation and Evolution of Exoplanets*. Malden, MA: Wiley.

Carroll, Michael 2017. *Earths of Distant Suns. How We Find Them, Communicate with Them, and Maybe Even Travel There*. Heidelberg/New York: Springer.

Ćirković, Milan M. 2018. *The Great Silence: Science and Philosophy of Fermi's Paradox*. Oxford: Oxford University Press.

Cooper, Keith 2020. *The Contact Paradox. Challenging our Assumptions in the Search for Extraterrestrial Intelligence*. London: Bloomsbury.

Coustenis, Athena and Thérèse Encrenaz 2013. *Life beyond Earth: the Search for Habitable Worlds in the Universe*. Cambridge: Cambridge University Press.

Dartnell, Lewis 2007. *Life in the Universe: A Beginner's Guide*. London: OneWorld Publications.

Deeg, Hans J. and Juan Antonio Belmonte (eds) 2018. *Handbook of Exoplanets*. New York: Springer International.

Denning, Kathryn 2010. Unpacking the great transmission debate, *Acta Astronautica* 67: 1399–1405.

Dick, Steven J. (ed.) 2015. *The Impact of Discovering Life beyond Earth*. Cambridge: Cambridge University Press.

Feinberg, Gerald and Robert Shapiro 1977. *Life beyond Earth: The Intelligent Earthling's Guide to Life in the Universe*. New York: William Morrow & Co.

Goldsmith, Donald 2018. *Exoplanets. Hidden Worlds and the Quest for Extraterrestrial Life*. Cambridge, MA: Harvard University Press.

Heller. René and John Armstrong 2014. Superhabitable worlds. *Astrobiology*. 14(1): 50–66.

Henin, Bernard 2018. *Exploring the Ocean Worlds of Our Solar System*. New York: Springer.

Jayawardhana, Ray 2013. *Strange New Worlds: The Search for Alien Planets and Life beyond Our Solar System*. Princeton, NJ: Princeton University Press.

Kaufman, Marc 2011. *First Contact: Scientific Breakthroughs in the Hunt for Life Beyond Earth*. New York: Simon and Schuster.

Kichin, Chris 2012. *Exoplanets: Finding, Exploring, and Understanding Alien Worlds*. New York: Springer.

Kirschner, Marc W. and John C. Gerhart 2005. *The Plausibility of Life. Resolving Darwin's Dilemma*. London: Yale University Press.

Loeb, Avi 2021. *Extraterrestrial: The First Sign of Intelligent Life Beyond Earth*. New York: Houghton Mifflin Harcourt.

Mason, John W. (ed.) 2008. *Exoplanets: Detection, Formation, Properties, Habitability*. New York: Springer.

Meyers, Walter Earl 1980. *Aliens and Linguists: Language Study and Science Fiction*. Athens, GA: University of Georgia Press.

Michaud, Michael 2007. *Contact with Alien Civilizations: Our Hopes and Fears about Encountering Extraterrestrials*. New York: Springer.

Miller, Ben 2016. *The Aliens are Coming! The Extraordinary Science Behind Our Search for Life in the Universe*. New York: The Experiment.

Regis, Edward, Jr (ed.) 1985. *Extraterrestrials. Science and Alien Intelligence*. Cambridge: Cambridge University Press.

Schulze-Makuch, Dirk and William Bains 2017. *The Cosmic Zoo. Complex Life on Many Worlds*. Berlin/Heidelberg: Springer.

Schulze-Makuch, Dirk and Louis N. Irwin 2018. *Life in the Universe: Expectations and Constraints*. Third edition. Berlin/Heidelberg: Springer.

Schulze-Makuch, Dirk, Abel Méndez, Alberto G. Fairén et al. 2011. A two-tiered approach to assess the habitability of exoplanets, *Astrobiology* 11(10): 1041–1052.

Seager, Sarah 2011. *Exoplanets*. Tucson, AZ: University of Arizona Press.

Shostak, Seth and Alex Barnett 2003. *Cosmic Company: The Search for Life in the Universe*. Cambridge: Cambridge University Press.

Tasker, Elizabeth 2017. *The Planet Factory. Exoplanets and the Search for a Second Earth*. London: Bloomsbury.

Toomey, David 2014. *Weird Life. The Search for Life that is Very, Very Different from Our Own*. New York: W. W. Norton & Company.

Ward, Peter and Donald Brownlee 2000. *Rare Earth: Why Complex Life is Uncommon in the Universe*. New York: Springer.

Webb, Stephen 2002. *If the Universe is Teeming with Aliens. Where is Everybody? Fifty Solutions to the Fermi Paradox and the Problem of Extraterrestrial Life.* New York: Springer.

History of the Topic

Crowe, Michael 2003. *The Extraterrestrial Life Debate 1750–1900: The Idea of a Plurality of Worlds from Kant to Lowell.* Mineola, NY: Dover Publications.

Dick, Stephen J. 1982. *Plurality of Worlds: The Origins of the Extraterrestrial Life Debate from Democritus to Kant.* Cambridge: Cambridge University Press.

Dick, Stephen J. 2001. *Life on Other Worlds: The 20th Century Extraterrestrial Life Debate.* Cambridge: Cambridge University Press.

Dick, Stephen J. 2018. *Astrobiology, Discovery, and Societal Impact: Controversy and Consensus.* Cambridge: Cambridge University Press.

Shklovskii, L. S. and Carl Sagan 1966. *Intelligent Life in the Universe.* New York: Dell Publishing.

Language Beyond Earth

Freudenthal, Hans 1960. *Lincos: Design of a Language for Cosmic Intercourse.* Amsterdam: North-Holland.

McConnell, Brian S. 2001. *Beyond Contact: A Guide to SETI and Communicating with Alien Civilizations.* Cambridge, MA: O'Reilly.

Oberhaus, Daniel 2019. *Extraterrestrial Languages.* Cambridge, MA: Harvard University Press.

Ollongren, Alexander 2013. *Astrolinguistics. Design of a Linguistic System for Interstellar Communication Based on Logic.* New York: Springer.

Vakoch, Douglas A. (ed.) 2011. *Communication with Extraterrestrial Intelligence.* New York: State University of New York Press.

Vakoch, Douglas A. (ed.) 2014. *Archaeology, Anthropology and Interstellar Communication.* Washington, DC: NASA.

Space Exploration

Dawson, Linda 2020. *The Politics and Perils of Space Exploration: Who Will Compete, Who Will Dominate?* New York: Springer.

Deudney, Daniel 2020. *Dark Skies: Space Expansionism, Planetary Geopolitics, and the Ends of Humanity.* Oxford: Oxford University Press.

Launius, Roger D. 2018. *The Smithsonian History of Space Exploration: From the Ancient World to the Extraterrestrial Future.* Washington, DC: Smithsonian Books.

Logsdon, John (ed.) 2018. *The Penguin Book of Outer Space Exploration: NASA and the Incredible Story of Human Spaceflight Audiobook*. London: Penguin.

Schwartz, James S. J. 2020. *The Value of Science in Space Exploration*. Oxford: Oxford University Press.

Vakoch, Douglas A. (ed.) 2011. *Psychology of Space Exploration: Contemporary Research in Historical Perspective*. Washington, DC: NASA History Series.

Ethology

Attenborough, David 2022 [1990]. *The Trials of Life. A Natural History of Animal Behaviour*. Second edition. London: William Collins.

Avital, Eytan and Eva Jablonka 2000. *Animal Traditions: Behavioural Inheritance in Evolution*. Cambridge: Cambridge University Press.

Axelrod, Robert 1997. *The Complexity of Cooperation. Agent-Based Models of Competition and Collaboration*. Princeton, NJ: Princeton University Press.

Bekoff, Marc, Colin Allen and Gordon M. Burghardt (eds) 2002. *The Cognitive Animal. Empirical and Theoretical Perspectives on Animal Cognition*. Cambridge, MA: MIT Press.

Carroll, Sean B. 2016. *The Serengeti Rules. The Quest to Discover How Life Works and Why It Matters*. Princeton, NJ: Princeton University Press.

Cheney, Dorothy L. and Robert Seyfarth 2007. *Baboon Metaphysics. The Evolution of a Social Mind*. Chicago: Chicago University Press.

Goodall, Jane 1990. *Through a Window: 30 Years Observing the Gombe Chimpanzees*. London: Weidenfeld and Nicolson.

Kelly, David 2011. *Yuck! The Nature and Moral Significance of Disgust*. Cambridge, MA: MIT Press.

Reznikova, Zhanna 2007. *Animal Intelligence. From Individual to Social Cognition*. Cambridge: Cambridge University Press.

Animal Vocalisation and Communication

Bolhuis, Johan J. and Martin Everaert (eds) 2013. *Birdsong, Speech and Language, Exploring the Evolution of Mind and Brain*. Cambridge, MA: MIT Press.

Bradbury, Jack W. and Sandra L. Vehrencamp 1998. *Principles of Animal Communication*. Sunderland, MA: Sinauer.

Call, Josep and Michael Tomasello 2007. *The Gestural Communication of Apes and Monkeys*. Hillsdale, NJ: Lawrence Erlbaum.

Cheney, Dorothy L. and Robert M. Seyfarth 1990. *How Monkeys See the World. Inside the Mind of Another Species*. Chicago, IL: University of Chicago Press.

Connor, Richard C. and Dawn M. Peterson 1994. *The Lives of Whales and Dolphins*. New York: Henry Holt.

Griffin, Donald R. 2001. *Animal Minds: Beyond Cognition to Consciousness*. Chicago, IL: Chicago University Press.

Innis Dagg, Anne 2009. *The Social Behavior of Older Animals*. Baltimore, MD: Johns Hopkins University Press.

Kershenbaum; Arik 2020. *The Zoologist's Guide to the Galaxy. What Animals on Earth Reveal about Aliens*. London: Penguin.

Lonsdorf, Elizabeth, Stephen R. Ross and Tetsuro Matsuzawa (eds) 2010. *The Mind of the Chimpanzee*. Chicago: University of Chicago Press.

Maynard Smith, John and David Harper 2003. *Animal Signals*. Oxford: Oxford University Press.

Rogers, Lesley J. and Gisela T. Kaplan (eds) 2004. *Comparative Vertebrate Cognition. Are Primates Superior to Non-Primates?* New York: Kluwer Academic.

Rumbaugh, Duane M. and David Washburn 2003. *Intelligence of Apes and Other Rational Beings: Current Perspectives in Psychology*. New Haven, CT: Yale University Press.

Searcy, William A. and Stephen Nowicki 2005. *The Evolution of Animal Communication. Reliability and Deception in Signalling Systems*. Princeton, NJ: Princeton University Press.

Wynne, Clive, D. L. 2004. *Do Animals Think?* Princeton, NJ: Princeton University Press.

Artificial Intelligence

Armstrong, Stuart 2014. *Smarter than Us. The Rise of Machine Intelligence*. Berkeley, CA: Machine Intelligence Research Institute.

Barrat, James 2013. *Our Final Invention. Artificial Intelligence and the End of the Human Era*. New York: St Martin's Press.

Bostrom, Nick 2014. *Superintelligence: Paths, Dangers, Strategies*. Oxford: Oxford University Press.

Christian, Brian 2020. *The Alignment Problem: Machine Learning and Human Values*. New York: W. W. Norton & Company.

Kurzweil, Ray 2005. *The Singularity is Near*. New York: Penguin Group.

Kurzweil, Ray 2012. *How to Create a Mind: The Secret of Human Thought Revealed*. New York: Viking Books.

Russell, Stuart 2019. *Human Compatible: Artificial Intelligence and the Problem of Control*. New York: Viking Press.

Savulescu, Julian and Nick Bostrom (eds) 2018. *Human Enhancement*. Oxford: Oxford University Press.

Tegmark, Max 2017. *Life 3.0: Being Human in the Age of Artificial Intelligence*. New York: Random House Publishing.

Zarkadakis, George 2015. *In Our Own Image. Will Artificial Intelligence Save or Destroy Us?* London: Rider.

Language and Linguistics

Bloomfield, Leonard 1933. *Language*. New York: Henry Holt.

Hockett, Charles F. 1960. The origin of speech. *Scientific American* 203(3): 88–97.

Sapir, Edward 2004 [1921]. *Language. An Introduction to the Study of Speech*. New York: Dover Publications.

Saussure, Ferdinand de 1959 [1916]. *A Course in General Linguistics*. Charles Bally and Albert Schechaye (eds). Translated by Wade Baskin. London: Peter Owen.

Language and Mind

Aitchison, Jean 2007. *The Articulate Mammal: An Introduction to Psycholinguistics*. Fifth edition. London: Taylor and Francis.

Albert, Martin L. and Loraine K. Obler 1978. *The Bilingual Brain*. New York: Academic Press.

Andrews, Edna 2019. *Neuroscience and Multilingualism*. Cambridge: Cambridge University Press.

Bowerman, Melissa 2001. *Language Acquisition and Conceptual Development*. Cambridge: Cambridge University Press.

Berlin, Brent and Paul Kay 1969. *Basic Color Terms: Their Universality and Evolution*. Berkeley, CA: University of California Press.

Chafe, Wallace 2020. *Thought-based Linguistics. How Languages Turn Thoughts into Sounds*. Cambridge: Cambridge University Press.

Chomsky, Noam 1968. *Language and Mind*. New York: Harcourt Brace Jovanovich.

Chomsky, Noam 1980. *Rules and Representations*. New York: Columbia University Press.

Chomsky, Noam 1991. Linguistics and cognitive science: problems and mysteries, in Asa Kasher (ed.) *The Chomskyan Turn*. Oxford: Blackwell, pp. 26–53.

Chomsky, Noam and Andrea Moro 2022. *The Secrets of Words*. Cambridge, MA: MIT Press.

Cummings, Louise 2020. *Language in Dementia*. Cambridge: Cambridge University Press.

Curtiss, Susan 1977. *Genie: A Psycholinguistic Study of a Modern-Day 'Wild Child.' Perspectives in Neurolinguistics and Psycholinguistics*. Boston, MA: Academic Press.

Dehaene, Stanislas 2011. *The Number Sense: How the Mind Creates Mathematics*. Second edition. Oxford: Oxford University Press.

Friederici, Angela D. 2017 *Language in Our Brain. The Origins of a Uniquely Human Capacity*. Cambridge, MA: MIT Press.

Hauser, Marc D., Noam Chomsky and W. Tecumseh Fitch. 2002. The faculty of language: what is it, who has it, and how did it evolve? *Science* 298: 1569–1579.

Larson, Richard K., Viviane Déprez and Hiroko Yamakido 2010. *The Evolution of Human Language: Biolinguistic Perspectives*. Cambridge: Cambridge University Press.

Levinson, Stephen C. and David P. Wilkins 2006. *Grammars of Space. Explorations in Cognitive Diversity*. Cambridge: Cambridge University Press.

Lucy, John A. 1992. *Language Diversity and Thought. A Reformulation of the Linguistic Relativity Hypothesis*. Cambridge: Cambridge University Press.

Murphy, Elliot 2020. *The Oscillatory Nature of Language*. Cambridge: Cambridge University Press.

Obler, Loraine K. and Kris Gjerlow 1999. *Language and the Brain*. Cambridge: Cambridge University Press.

Pavlenko, Aneta 2014. *The Bilingual Mind and What it Tells Us About Language and Thought*. Cambridge: Cambridge University Press.

Rizzolatti, Giacomo and Corrado Sinigaglia 2008. *Mirrors in the Brain: How Our Minds Share Actions and Emotions*. Oxford: Oxford University Press.

Sharwood Smith, Mike 2017. *Introducing Language and Cognition. A Map of the Mind*. Cambridge: Cambridge University Press.

Steinberg, Danny and Natalia Sciarini 2006. *An Introduction to Psycholinguistics*. London: Pearson Longman.

Warren, Paul 2012. *Introducing Psycholinguistics*. Cambridge: Cambridge University Press.

Evolution of Language

Aitchison, Jean 1996. *The Seeds of Speech. Language Origin and Evolution*. Cambridge: University Press.

Allott, Robin 1989. *The Motor Theory of Language Origin*. Market Harborough: The Book Guild.

Anderson, Stephen R. and David W. Lightfoot 2002. *The Language Organ. Linguistics as Cognitive Physiology*. Cambridge: Cambridge University Press.

Andresen, Julie Tetel 2013. *Linguistics and Evolution. A Developmental Approach*. Cambridge: Cambridge University Press.

Arbib, Michael A. 2012. *How the Brain Got Language. The Mirror System Hypothesis*. Oxford: Oxford University Press.

Armstrong, David F., William C. Stokoe and Sherman E. Wilcox 1995. *Gesture and the Nature of Language*. Cambridge: Cambridge University Press.

Barnard, Alan 2016. *Language in Prehistory*. Cambridge: Cambridge University Press.

Berwick, Robert C. and Noam Chomsky 2016. *Why Only Us? Language and Evolution*. Cambridge, MA: MIT Press.

Bickerton, Derek 1981. *Roots of Language*. Ann Arbor, MI: Karoma Press.

Bickerton, Derek 1995. *Language and Human Behaviour*. Seattle, WA: University of Washington Press.

Bickerton, Derek 2009. *Adam's Tongue. How Humans Made Language, How Language Made Humans*. New York: Hill and Wang.

Bickerton, Derek 2014. *More than Nature Needs. Language, Mind, and Evolution*. Cambridge, MA: Harvard University Press.

Boeckx, Cedric and Kleanthes Grohmann (eds) 2013. *The Cambridge Handbook of Biolinguistics*. Cambridge: Cambridge University Press.

Botha, Rudolf 2020. *Neanderthal Language. Demystifying the Linguistic Power of Our Extinct Cousins*. Cambridge: Cambridge University Press.

Botha, Rudolf and Chris Knight (eds) 2009. *The Prehistory of Language*. Oxford: Oxford University Press.

Botha, Rudolf and Chris Knight (eds) 2009. *The Cradle of Language*. Oxford: Oxford University Press.

Bouchard, Denis 2013. *The Nature and Origin of Language*. Oxford: Oxford University Press.

Burling, Robbins 2007. *The Talking Ape. How Language Evolved*. Oxford: Oxford University Press.

Calvin, William H. and Derek Bickerton 2000. *Lingua Ex Machina: Reconciling Darwin with the Human Brain*. Cambridge, MA: MIT Press.

Carstairs-McCarthy Andrew 1999. *The Origins of Complex Language. An Inquiry into the Evolutionary Beginnings of Sentences, Syllables and Truth*. Oxford: Oxford University Press.

Cavalli-Sforza, Luigi L. 1997. Genes, peoples, and languages. *Proceedings of the National Academy of Sciences* 94: 7719–7724.

Chomsky, Noam 1987. *Language and Problems of Knowledge*. Cambridge, MA: MITPress.

Chomsky, Noam 1995. *The Minimalist Program*. Cambridge, MA: MIT Press.

Chomsky, Noam 2000. *The Architecture of Language*. Oxford: Oxford University Press.

Chomsky, Noam 2007. Biolinguistic explorations: design, development, evolution. *International Journal of Philosophical Studies* 15: 1–21.

Christiansen, Morten H. and Nick Chater 2016. *Creating Language. Integrating Evolution, Acquisition, and Processing*. Cambridge, MA: MIT Press.

Christiansen, Morten H. and Simon Kirby (eds) 2003. *Language Evolution*. Oxford: Oxford University Press.

Corballis, Michael C. 2002. Did language evolve from manual gestures?, in Alison Wray (ed.) *The Transition to Language*. Oxford: Oxford University Press, pp. 161–179.

Corballis, Michael C. 2011. *The Recursive Mind: The Origins of Human Language, Thought, and Civilization*. Princeton, NJ: Princeton University Press.

Corballis, Michael C. 2017. *The Truth about Language. What It Is and Where It Came From*. Chicago: University of Chicago Press.

Corballis, Michael and Stephen E. G. Lea 2000. *The Descent of Mind. Psychological Perspectives on Hominid Evolution*. Oxford: Oxford University Press.

de Boer, Bart 2001. *The Origins of Vowel Systems*. Oxford: Oxford University Press.

Deacon, Terrence W. 1997. *The Symbolic Species. The Co-Evolution of Language and the Human Brain*. London: Penguin.

Dediu, Dan 2015. *An Introduction to Genetics for Language Scientists*. Cambridge: Cambridge University Press.

Dediu, Dan and Stephen C. Levinson 2013. On the antiquity of language: the reinterpretation of Neanderthal linguistic capacities and its consequences. *Frontiers in Psychology* 4. https://doi.org/10.3389/fpsyg.2013.00397.

Dessalles, Jean-Louis 2007 [2000]. *Why We Talk. The Evolutionary Origins of Language*. Oxford: Oxford University Press.

Deutscher, Guy 2006. *The Unfolding of Language*. London: Penguin.

di Sciullo, Anna Maria (ed.) 2017. *Biolinguistics*. London: Routledge.

di Sciullo, Anna Maria and Cedric Boeckx 2011. *The Biolinguistic Enterprise. New Perspectives on the Evolution and Nature of the Human Language Faculty*. Oxford: Oxford University Press.

Dor, Daniel, Chris Knight and Jerome Lewis (eds) 2014. *The Social Origins of Language*. Oxford: Oxford University Press.

Dunbar, Robin 1996. *Grooming, Gossip and the Evolution of Language*. Cambridge, MA: Harvard University Press.

Dunbar, Robin 2012. Gossip and the social origins of language, in Maggie Tallerman and Kathleen R. Gibson (eds) *The Oxford Handbook of Language Evolution*. Oxford: Oxford University Press, pp. 343–345.

Emmorey, Karen 2002. *Language, Cognition, and the Brain. Insights from Sign Language Research*. Mahwah, NJ: Lawrence Erlbaum Associates.

Everaert, Martin B. H., Marinus A. C. Huybregts, Robert C. Berwick et al. 2017. What is language and how could it have evolved? *Trends in Cognitive Sciences* 21(8): 1–21.

Falk, Dean 2009. *Finding Our Tongues: Mothers, Infants and the Origin of Language*. New York: Basic Books.

Fitch, W. Tecumseh 2010. *The Evolution of Language*. Cambridge: Cambridge University Press.

Goldin-Meadow, Susan 2003. *The Resilience of Language: What Gesture Creation in Deaf Children Can Tell Us About How All Children Learn Language*. New York: Taylor and Francis.

Goldin-Meadow, Susan 2012. What modern-day gesture can tell us about language evolution, in Maggie Tallerman and Kathleen R. Gibson (eds) *The Oxford Handbook of Language Evolution*. Oxford: Oxford University Press, pp. 545–557.

Haiman, John 2020. *Ideophones and the Evolution of Language*. Cambridge: Cambridge University Press.

Håkansson, Gisela and Jennie Westander 2013. *Communication in Humans and Other Animals*. Amsterdam: John Benjamins.

Hauser, Marc D. and W. Tecumseh Fitch 2003. What are the uniquely human components of the language faculty?, in Morton H. Christiansen and Simon Kirby (eds) *Language Evolution*. Oxford: Oxford University Press, pp. 158–181.

Heine, Bernd and Tania Kuteva 2007. *The Genesis of Grammar: A Reconstruction*. Oxford: Oxford University Press.

Herder, Johann Gottfried 1996 [1772]. *Über den Ursprung der Sprache* [Essay on the origin of language]. Translated by John H. Moran. Stuttgart: Verlag Freies Geistesleben.

Hurford, James R. 2011. *The Origins of Grammar: Language in the Light of Evolution*. Oxford: Oxford University Press.

Hurford, James R. 2014. *The Origins of Language. A Slim Guide*. Oxford: Oxford University Press.

Jackendoff, Ray 2002. *Foundations of Language: Brain, Meaning, Grammar, Evolution*. Oxford: Oxford University Press.

Jackendoff, Ray 2012. *A User's Guide to Thought and Meaning*. Oxford: Oxford University Press.

Jenkins, Lyle 2000. *Biolinguistics. Exploring the Biology of Language*. Cambridge: Cambridge University Press.

Jespersen, Otto 1922. *Language. Its Nature, Development and Origin*. London: Allen & Unwin.

Johanson, Donald and Blake Edgar 1996. *From Lucy to Language*. New York: Simon and Schuster.

Johansson, Sverker 2005. *Origins of Language: Constraints on Hypotheses*. Amsterdam: John Benjamins.

Johansson, Sverker 2021. *The Dawn of Language. How we Came to Talk*. Translated from Swedish. London: MacLehose Press.

Kenneally, Christine 2007. *The First Word: The Search for the Origins of Language*. London: Penguin.

Kirby, Simon 1999. *Function, Selection and Innateness. The Emergence of Language Universals*. Oxford: Oxford University Press.

Kirby, Simon 2012. Language is an adaptive system: the role of cultural evolution in the origins of structure, in Maggie Tallerman and Kathleen R. Gibson (eds) *The Oxford Handbook of Language Evolution*. Oxford: Oxford University Press, pp. 589–604.

Kirby, Simon, Thomas L. Griffiths and Kenny Smith 2014. Iterated learning and the evolution of language, *Current Opinion in Neurobiology* 28: 108–114.

Knight, Chris and Camilla Power 2012. Social conditions for the evolutionary emergence of language, in Maggie Tallerman and Kathleen R. Gibson (eds) *The Oxford Handbook of Language Evolution*. Oxford: Oxford University Press, pp. 346–349.

Knight, Chris, Michael Studdert Kennedy and James R. Hurford (eds) 2000. *The Evolutionary Emergence of Language. Social Function and the Origins of Linguistic Form*. Cambridge: Cambridge University Press.

Lefebvre, Claire, Bernard Comrie and Henri Cohen (eds) 2013. *New Perspectives on the Origins of Language*. Amsterdam: John Benjamins.

Lenneberg, Eric H. 1967. *Biological Foundations of Language*. New York: Wiley.

Lieberman, Philip 2000. *Human Language and Our Reptilian Brain. The Subcortical Bases of Speech, Syntax and Thought*. Cambridge, MA: Harvard University Press.

Lieberman, Philip 2006. *Towards an Evolutionary Biology of Language*. Cambridge, MA: Harvard University Press.

Loritz, Donald 1999. *How the Brain Evolved Language*. Oxford: Oxford University Press.

MacNeilage, Peter F. 2008. *The Origin of Speech*. Oxford: Oxford University Press.

MacWhinney, Brian and William O'Grady (eds) 2014. *The Handbook of Language Emergence*. Malden, MA: Wiley-Blackwell.

McMahon, April and Robert McMahon 2012. *Evolutionary Linguistics*. Cambridge: Cambridge University Press.

McNeill, David 2012. *How Language Began. Gesture and Speech in Human Evolution*. Cambridge: Cambridge University Press.

McWhorter, John 2003. *The Power of Babel: A Natural History of Language*. New York: Harper Perennial.

Messing, Lynn S. and Ruth Campbell (eds) 1999. *Gesture, Speech and Sign*. Oxford: Oxford University Press.

Moro, Andrea 2016. *Impossible Languages*. Cambridge, MA: MIT Press.

Mufwene, Salikoko S., Christophe Coupé and François Pellegrino (eds) 2019. *Complexity in Language. Developmental and Evolutionary Perspectives*. Cambridge: Cambridge University Press.

Niyogi, Partha 2006. *The Computational Nature of Language Learning and Evolution*. Cambridge, MA: MIT Press.

Noiré, Ludwig 1917. *The Origin and Philosophy of Language*. Chicago and London: Open Court Publishing.

Piattelli-Palmarini, Massimo and Robert C. Berwick (eds) 2013. *Rich Languages from Poor Inputs*. Oxford: Oxford University Press.

Pinker, Steven 1994. *The Language Instinct. The New Science of Language and Mind*. London: Allen Lane.

Pinker, Steven 1999. *How the Mind Works*. New York: W. W. Norton & Company.

Pinker, Steven 2000. *Words and Rules: The Ingredients of Language*. London: Weidenfeld and Nicolson.

Pinker, Steven 2007. *The Stuff of Thought: Language as a Window into Human Nature*. London: Allen Lane.

Pinker, Steven 2013. *Language, Cognition and Human Nature: Selected Articles.* Oxford: Oxford University Press.

Pinker, Steven and Paul Bloom 1990. Natural language and natural selection. *Behavioral and Brain Sciences* 13(4): 707–727 (with open peer commentary, pp. 727–784).

Pinker, Steven and Ray Jackendoff 2005. The faculty of language: what's special about it? *Cognition* 95: 201–236.

Ritchie, L. David 2022. *Feeling, Thinking, and Talking. How the Embodied Brain Shapes Everyday Communication.* Cambridge: Cambridge University Press.

Rousseau, Jean Jacques 1781. Essai sur l'origine des langues [essay on the origin of languages], in *Collection Complète des Oeuvres* [complete collected works] 1780–1789, volume 8. Geneva.

Scott-Phillips, Thom 2014. *Speaking Our Minds. Why Human Communication is Different, and how Language Evolved to Make It Special.* Basingstoke: Palgrave Macmillan.

Smith, Kenny, Simon Kirby and Henry Brighton 2003. Iterated learning: a framework for the emergence of language. *Artificial Life* 9(4): 371–386.

Smits, Rik 2016. *Dawn. The Origins of Language and the Human Mind.* Translated from Dutch. London: Routledge.

Stamenov, Maxim L. and Vittorio Gallese (eds) 2002. *Mirror Neurons and the Evolution of Brain and Language.* Amsterdam: John Benjamins.

Stokoe, William C. 2001. *Language in Hand: Why Sign Came Before Speech.* Washington DC: Gallaudet University Press.

Stroik, Thomas S. and Michael T. Putnam 2013. *The Structural Design of Language.* Cambridge: Cambridge University Press.

Studdert-Kennedy, Michael and Herbert Terrace 2017. In the beginning: a review of Robert C. Berwick and Noam Chomsky's *Why Only Us. Journal of Language Evolution* 2(2): 114–125.

Tallerman, Maggie (ed.) 2005. *Language Origins. Perspectives on Evolution.* Oxford: Oxford University Press.

Tallerman, Maggie and Kathleen R. Gibson (eds) 2011. *The Oxford Handbook of Language Evolution.* Oxford: Oxford University Press.

Tattersall, Ian 2012. *Masters of the Planet: The Search for Our Human Origins.* London: Macmillan.

Tomasello, Michael 1999. *The Cultural Origins of Human Cognition.* Cambridge, MA: Harvard University Press.

Tomasello, Michael 2008. *Origins of Human Communication*. Cambridge, MA: MIT press.

Tomasello, Michael 2014. *A Natural History of Human Thinking*. Cambridge, MA: Harvard University Press.

Trabant, Jürgen and Sean Ward (eds) 2001. *New Essays on the Origin of Language*. Berlin: Mouton de Gruyter.

Trombetti, Alfredo 1905. *L'unità d'origine del linguaggio*. [The unified origin of language] Bologna: Luigi Beltrami.

Wray, Alison (ed.) 2002. *The Transition to Language*. Oxford: Oxford University Press.

Żywiczyński, Przemysław, Nathalie Gontier and Slawomir Wacewicz (eds) 2017. *Language Evolution: Focus on Mechanisms*. Special issue of *Language Sciences* 63: 1–130.

Żywiczyński, Przemysław, and Slawomir Wacewicz (eds) 2017. *The Evolution of Language. Towards Gestural Hypotheses*. Frankfurt: Peter Lang.

Language Acquisition

Bavin, Edith L. 2009. *The Cambridge Handbook of Child Language*. Cambridge: Cambridge University Press.

Bloom, Paul 2000. *How Children Learn the Meanings of Words*. Cambridge, MA: MIT Press.

Clark, Eve 2016. *First Language Acquisition*. Third edition. Cambridge: Cambridge University Press.

Curtiss, Susan 2014. The case of Chelsea: The effect of late age at exposure to language on language performance and evidence for the modularity of language and mind, in Carson T. Schütze and Linnaea Stockall (eds) *Connectedness. Papers by and for Sarah Van Wagenen*. Los Angeles: University of Los Angeles Working Papers in Linguistics 18: 115–146.

Curtiss, Susan, Victoria Fromkin, Stephen Krashen, David Rigler and Marilyn Rigler 1974. The linguistic development of Genie, *Language* 50(3): 528–554.

Fletcher, Paul and Brian MacWhinney (eds) 1994. *The Handbook of Child Language*. Oxford: Blackwell.

Lust, Barbara C. 2006. *Child Language. Acquisition and Growth*. Cambridge: University Press.

Lust, Barbara C. and Claire Foley (eds) 2003. *First Language Acquisition. The essential readings*. Oxford: Blackwell.

McCune, Lorraine 2008. *How Children Learn to Learn Language*. Oxford: Oxford University Press.

Meisel, Jürgen M. 2011. *First and Second Language Acquisition. Parallels and Differences*. Cambridge: Cambridge University Press.

O'Grady, William 2005. *How Children Learn Language*. Cambridge: Cambridge University Press.

Ritchie, William C. and Tej K. Bhatia (eds) 1999. *Handbook of Child Language Acquisition*. Oxford: Blackwell.

Rowland, Caroline 2014. *Understanding Child Language Acquistion*. London: Routledge.

Saville-Troike, Muriel 2013. *Introducing Second Language Acquisition*. Second edition. Cambridge: Cambridge University Press.

Saxton, Michael 2010. *Child Language. Acquisition and Development*. Los Angeles: SAGE Publications.

Tomasello, Michael 2003. *Constructing a Language: A Usage-Based Theory of Language Acquisition*. Cambridge, MA: Harvard University Press.

Sign Languages

Baker, Anne, Beppie van den Bognaerde, Roland Pfau and Trude Schermer (eds) 2016. *The Linguistics of Sign Languages. An Introduction*. Amsterdam: John Benjamins.

Brentari, Diane 2010. *Sign Languages*. Cambridge: Cambridge University Press.

Davis, Jeffrey E. 2010. *Hand Talk: Sign Language among American Indian Nations*. Cambridge: Cambridge University Press.

Le Guen, Olivier, Josefina Safar and Marie Coppola (eds) 2021. *Emerging Sign Languages of the Americas*. Berlin: de Gruyter Mouton.

Lucas, Ceil 2001. *The Sociolinguistics of Sign Languages*. Cambridge: Cambridge University Press.

Lucas, Ceil, Robert Bayley and Clayton Valli 2001. *Sociolinguistic Variation in American Sign Language*. Washington, DC: Gallaudet University Press.

Mallery, Garrick 2011 [1881]. *Sign Language Among North American Indians*. New York: Dover Publications.

Sandler, Wendy and Diane C. Lillo-Martin 2006. *Sign Language and Linguistic Universals*. Cambridge: Cambridge University Press.

Zeshan, Ulrike and Connie de Vos (eds) 2012. *Sign Languages in Village Communities. Anthropological and Linguistic Insights*. Berlin: de Gruyter Mouton.

Human and Animal Communication

Anderson, Stephen R. 2004. *Doctor Dolittle's Delusion: Animals and the Uniqueness of Human Language*. New Haven, CT: Yale University Press.

Savage-Rumbaugh, E. Sue 1986. *Ape Language. From Conditioned Response to Symbol*. New York: Columbia University Press.

Savage-Rumbaugh, E. Sue, Stuart G. Shanker and Talbot J. Taylor 1998. *Apes, Language, and the Human Mind*. Oxford: Oxford University Press.

Language and Music

Fitch, W. Tecumseh 2006. The biology and evolution of music: a comparative perspective, *Cognition* 100: 173–215.

Mithen, Steven 2005. *The Singing Neanderthals. The Origins of Music, Language, Mind, and Body*. London: Weidenfeld and Nicolson.

Mithen, Steven 2012. Musicality and language, in Maggie Tallerman and Kathleen R. Gibson (eds) *The Oxford Handbook of Language Evolution*. Oxford: Oxford University Press, pp. 296–298.

Patel, Aniruddh D. 2008. *Music, Language, and the Brain*. Oxford: Oxford University Press.

Language Pathology

Ball, Martin J., Michael R. Perkins, Nicole Müller and Sara Howard (eds) 2008. *The Handbook of Clinical Linguistics*. Malden, MA: Wiley-Blackwell.

Danon-Boileau, Laurent and James Grieve 2005. *Children Without Language. From Dysphasia to Autism*. Oxford: Oxford University Press.

Danon-Boileau, Laurent and Kevin Windle 2007. *The Silent Child. Exploring the World of Children Who Do Not Speak*. Oxford: Oxford University Press.

Fromkin, Victoria A. 1973. *Speech Errors as Linguistic Evidence*. The Hague: Mouton.

Ingram, John C. L. 2007. *Neurolinguistics. An Introduction to Spoken Language Processing and its Disorders*. Cambridge: Cambridge University Press.

Stavrakaki, Stavroula (ed.) 2015. *Specific Language Impairment. Current Trends in Research*. Amsterdam: John Benjamins.

Language, Society and Culture

Carroll, John B. 1956. *Language, Thought, and Reality. Selected Writings of Benjamin Lee Whorf*. Cambridge, Mass.: MIT Press.

d'Ettorre, Patrizia and David P. Hughes 2008. *Sociobiology of Communication. An Interdisciplinary Perspective*. Oxford: Oxford University Press.

Enfield, N. J. 2017. *How we Talk. The Inner Workings of Conversation*. New York: Basic Books.

Enfield, N. J., Paul Kockelman and Jack Sidnell (eds) 2014. *The Cambridge Handbook of Linguistic Anthropology*. Cambridge: Cambridge University Press.

Everett, Daniel L. 2008. *Don't Sleep, there are Snakes. Life and Language in the Amazonian Jungle*. New York: Pantheon Books.

Nettle, Daniel 1999. *Linguistic Diversity*. Oxford: Oxford University Press.

Richerson, Peter J. and Morten H. Christiansen (eds) 2013. *Cultural Evolution: Society, Technology, Language and Religion*. Cambridge, MA: MIT Press.

History of Languages; Historical Linguistics and Language Change

Anthony, David W. 2007. *The Horse, the Wheel, and Language: How Bronze-Age Riders from the Eurasian Steppes Shaped the Modern World*. Princeton, NJ: Princeton University Press.

Burridge, Kate and Alexander Bergs 2017. *Understanding Language Change*. London: Routledge.

Bybee, Joan L. 2010. *Language, Usage, and Cognition*. Cambridge: Cambridge University Press.

Bybee, Joan L. 2015. *Language Change*. Cambridge: Cambridge University Press.

Campbell, Lyle 2021. *Historical Linguistics*. Fourth edition. Edinburgh: Edinburgh University Press.

Clackson, James 2007. *Indo-European Linguistics. An Introduction*. Cambridge: Cambridge University Press.

Croft, William 2000. *Explaining Language Change. An Evolutionary Approach*. London: Longman.

Hickey, Raymond (ed.) 2003. *Motives for Language Change*. Cambridge: Cambridge University Press.

Hickey, Raymond (ed.) 2020. *The Handbook of Language Contact*. Second edition. Malden, MA: Wiley-Blackwell.

Deutscher, Guy 2011. *Through the Language Glass: Why The World Looks Different In Other Languages*. London: Arrow Books.

Fortson, Benjamin W., IV 2010. *Indo-European Language and Culture. An Introduction*. Second edition. Malden, MA: Wiley.

Janson, Tore 2012. *The History of Languages: An Introduction*. Oxford: Oxford University Press.

Lass, Roger 1997. *Historical Linguistics and Language Change*. Cambridge: Cambridge University Press.

Pereltsvaig, Asya and Martin W. Lewis 2015. *The Indo-European Controversy: Facts and Fallacies in Historical Linguistics*. Cambridge: Cambridge University Press.

Robinson, Andrew 2007. *The Story of Writing: Alphabets, Hieroglyphs and Pictograms*. Second edition. London: Thames and Hudson.

The World Atlas of Language Structures 2013. *WALS (World Atlas of Linguistic Structures)*. Leipzig: Max Planck Institute for Evolutionary Anthropology.

Language Universals and Language Typology

Aikhenvald, Alexandra Y. and R. M. W. Dixon (eds) 2017. *The Cambridge Handbook of Linguistic Typology*. Cambridge: Cambridge University Press.

Campbell, Lyle and William J. Poser 2008. *Language Classification. History and Method*. Cambridge: Cambridge University Press.

Croft, William 2002. *Typology and Universals*. Second edition. Cambridge: Cambridge University Press.

Hickey, Raymond (ed.) 2017. *The Cambridge Handbook of Areal Linguistics*. Cambridge: Cambridge University Press.

Maddieson, Ian 1984. *Patterns of Sounds*. Cambridge: Cambridge University Press.

Spoken Language

Auer, Peter, Elizabeth Couper-Kuhlen and Friederike Müller 1999. *Language in Time: The Rhythm and Tempo of Spoken Interaction*. Oxford: Oxford University Press.

Blevins, Juliette 2004. *Evolutionary Phonology*. Cambridge: Cambridge University Press.

Goldrick, Matthew, Victor S. Ferreira and Michele Miozzo 2014. *The Oxford Handbook of Language Production*. Oxford: Oxford University Press.

Hickey, Raymond 2023. *Sounds of English Worldwide*. Malden, MA: Wiley-Blackwell.

Hinton, Leanne, Johanna Nichols and John Ohala (eds) 1994. *Sound Symbolism*. Cambridge: Cambridge University Press.

Hyman, Larry M. and Frans Plank (eds) 2018. *Phonological Typology*. Berlin: de Gruyter Mouton.

Okrent, Arika 2009. *In the Land of Invented Languages: Esperanto Rock Stars, Klingon Poets, Loglan Lovers, and the Mad Dreamers Who Tried to Build a Perfect Language*. New York: Spiegel & Grau.

Zsiga, Elizabeth C. 2013. *The Sounds of Language. An Introduction to Phonetics and Phonology*. Malden, MA: Wiley-Blackwell.

Grammar

Audring, Jenny and Francesca Masini (eds) 2018. *The Oxford Handbook of Morphological Theory*. Oxford: Oxford University Press.

Baerman, Matthew, Dunstan Brown and Greville G. Corbett, 2017. *Morphological Complexity*. Cambridge: Cambridge University Press.

Baker, Mark C. 2001. *The Atoms of Language. The Mind's Hidden Rules of Grammar*. New York: Basic Books.

Culicover, Peter and Ray Jackendoff 2005. *Simpler Syntax*. Oxford: Oxford University Press.

Gelderen, Elly van 2017. *Syntax. An Introduction to Minimalism*. Amsterdam: John Benjamins.

Heine, Bernd and Heiko Narrog (eds) 2011. *The Oxford Handbook of Grammaticalisation*. Oxford: Oxford University Press.

Hopper, Paul and Elizabeth Closs Traugott 2003 [1993]. *Grammaticalization*. Second edition. Cambridge: Cambridge University Press.

McGilvray, James 2017. *The Cambridge Companion to Chomsky*. Second edition. Cambridge: Cambridge University Press.

Werning, Markus, Wolfram Hinzen and Edouard Machery 2012. *The Oxford Handbook of Compositionality*. Oxford: Oxford University Press.

Vocabulary and Meaning

Aitchison, Jean 2012. *Words in the Mind. An introduction to the mental lexicon*. Fourth edition. Oxford: Blackwell.

Harley, Heidi 2006. *English Words: A Linguistic Introduction*. Oxford: Blackwell.

Hay, Jennifer 2003. *The Causes and Consequences of Word Structure*. New York: Routledge.

Lakoff, George and Mark Johnson 2003 [1980]. *Metaphors We Live By*. Chicago, IL: University of Chicago Press.

Plag, Ingo 2018. *Word-Formation in English*. Second edition. Cambridge: Cambridge University Press.

Štekauer, Pavol and Rochelle Lieber (eds) 2005. *Handbook of Word-Formation*. Dordrecht: Springer.

Wray, Alison 2002. *Formulaic Language and the Lexicon*. Cambridge: Cambridge University Press.

Index